Frontières en mouvement (Frontem) : Which Models of Cross-Border Cooperation for the EU ?

Border Studies – Borders and European Integration

Vol. 6

Birte Wassenberg (ed.)

Frontières en mouvement (Frontem): Which Models of Cross-Border Cooperation for the EU?

A comparative analysis from a Euro-Atlantic perspective

Peter Lang

Bruxelles - Berlin - Chennai - Lausanne - New York - Oxford

La présente publication a été rendue possible grâce au soutien du Réseau Jean Monnet « Frontières en mouvement (Frontem) » (2019-2023), financée par la Commission européenne (Réf. : 611115-EPP-1-2019-1-FR-EPPJMO-NETWORK).

With the support of the Erasmus+ Programme of the European Union

Jean Monnet Network FRONTEM

Cover illustration: Map of Frontem border regions, University of Strasbourg, 2020.

This publication has been peer reviewed.

ISBN 978-2-87574-926-0
ISSN 2736-2450
ePDF 978-2-87574-927-7
ePub 978-2-87574-928-4
DOI 10.3726/b21864
D/2024/5678/11

Bibliographic information published by "Die Deutsche Nationalbibliothek". "Die Deutsche National Bibliothek" lists this publication in the "Deutsche Nationalbibliografie"; detailed bibliographic data is available on the Internet at <http://dnb.de>.

Acknowledgement

Birte Wassenberg

The Jean Monnet Network *Frontières en mouvement: quels modèles pour l'UE* (Frontem)?[1] was supported by the European Union (EU)'s Erasmus+ program for the period between 2019 and 2023.

I would like to take the opportunity to thank all partners of the network for their input, their team spirit, their availability and investment during these four years, which was not an evidence at a time, when the Covid-19 crisis rendered international cooperation extremely difficult. Travel was hindered by new waves of the pandemic and the Frontem events often had to be rescheduled or organised with the harsh constraints of sanitary security measures. Therefore, I would like to express my gratitude to: Steen Bo Fransen, Martin Klatt and Katarzyna Stoklosa from the University of Southern Denmark, to Anne Thevenet from the Euro-Institut in Kehl, to Joachim Beck from the Hochschule Kehl, to Anthony Soares, from the Centre for Cross Border Studies in Armagh, Mary C. Murphy from University College Cork and Katy Hayward form the Queen's University of Belfast, to Nicolae Paŭn from the Bogdan Vodă University Cluj-Napoca, to Gyula Ocskay from the Central European Service for Cross-Border Initiatives (CESCI) in Budapest, to Jean Peyrony and Jean Rubio from the Mission opérationnelle transfrontalière (MOT), to Fabienne Leloup from the Catholic University of Louvain, to Bernard Reitel from the University of Artois and to Emmanuel Brunet-Jailly and Oliver Schmidtke from the University of Victoria in Canada.

At the University of Strasbourg, I would also like to thank the president of the University, Michel Deneken, who supports our Jean Monnet activities – Jean Monnet Chairs, the Franco-German Jean Monnet Center

[1] See European Commission, Erasmus + Program, Jean Monnet Activites: 611115-EPP-1-2019-1-FR-EPPJMO-NETWORK.

of excellence and Frontem – with great enthusiasm and European conviction. Special thanks go also to three colleagues and friends, without whom the this collective work would not have been possible: Frédérique Berrod, professor of Law at Science Po Strasbourg, who holds a mirror Chair Jean Monnet to my own chair on borders and with whom I discuss, coordinate and manage all Jean Monnet activities in Strasbourg; Joachim Beck, professor of Administrative Sciences and rector of the Hochschule in Kehl, who is co-director of the Franco-German Jean Monnet Center of excellence and with whom I share a long and fruitful collaboration of research on borders and cross-border cooperation; and Anne Thevenet, deputy director of the Euro-Institut in Kehl, who helps us to foster the link between researchers and practioners and who has coordinated two important Frontem activities, the *Toolkit on Models of Border Management and Perception in the EU*[2] and the the *Virtual Frontem Platform*[3] realized by Atelier Limo, which presents the 5 European border regions of Frontem in an interactive manner. Finally, I owe my gratitude to my trainee Simon Vanlichtervelde, former Master student in International Relations at Sciences Po Strasbourg, who has realised the editorial work of this publication: he collected the texts, reviewed the content, communicated with the authors during the peer-reviewing process and prepared the manuscript for the editor. This publication is therefore definitely an outcome of a successful teamwork of the Frontem network and I am very grateful for this fruitful cooperation!

[2] Toolkit retrieved from: <https://centre-jean-monnet.unistra.fr/2023/09/05/toolkit-modeles-de-gestion-de-la-frontiere-et-de-la-perception-sur-lunion-europeenne/>. Accessed 23 December 2023.

[3] Platform retrieved from: <https://atelierlimo.pageflow.io/frontem#311367>. Accessed 23 December 2023.

Table of Contents

Introduction

Birte Wassenberg

The Jean Monnet Network *Frontières en mouvement: quels modèles pour l'UE* (Frontem)? was supported by the European Union (EU)'s Erasmus+ program for the period between 2019 and 2023.

Linking two disciplinary fields that have until now been little connected – Border Studies and European Studies – the network exchanged knowledge and practices on five different models of EU border management and perception: the border between France and Germany, the border between France and Belgium, the border between Ireland and Northern Ireland, the border between Romania and Hungary and the border between Denmark and Germany. A comparative analysis was also conducted with regard to the Canada/US border in order to assess whether the European "models" can be transposed to other regions in the world.

The key question addressed by the network was how to assess the role of the border in the process of European integration when faced with processes of re-bordering and the re-questioning of the model of a "Europe without borders".[1] The Frontem network aimed therefore at offering a critical reading of a borderless Europe. It started from the observation that the EU has developed a unilateral approach to borders, which essentially retains their economic dimension as a barrier, without sufficiently taking into account other aspects, symbolic and political, in particular. The hypothesis put forward was that political borders have never disappeared and that there is an ambivalence of borders in the EU both as places of contact and exchange and means of protection and delimitation. Thanks to the comparative approach to the management

[1] B. Wassenberg, "Europa ohne Grenzen. Mythus oder Realität? Eine historische Analyse", *In* B. Wassenberg (ed.), *Frontières, acteurs et représentations d'Europe (Fare). Mélanges en honneur de Sylvain Schirmann*, Peter Lang, Brussels, 2022, pp. 415–439.

and perception of borders in European cross-border regions, the network wished to develop a more differentiated and multidimensional approach to the border. It therefore took a new look at the role of the border in European integration, considering that there is not a single model of the perception and management of borders in the EU, but that they depend on the specific context of each border area concerned.

The partners of Frontem[2] carried out a cross-analysis of border management and perception in their border areas by organising each a scientific seminar with researchers and a focus group with local stakeholders. This publication not only collects the different contributions to the research seminars, but also takes up the elements of the focus groups. All results of the Frontem network are available on the website of the Franco-German Jean Monnet Center of excellence of the University of Strasbourg.[3]

The publication is structured in six parts. After a general introduction to *Frontières en mouvement (Frontem)* by Birte Wassenberg, parts 2 to 6 take on a territorial approach by analysing five models of cross-border cooperation in different EU border regions.

The first part on "Franco-German Cross-Border Cooperation: A Model for European Integration?" examines the two Franco-German borderlands: the Franco-German-Swiss Upper Rhine Region and the Greater Region (Saar-Lor-Lux-Rhineland-Palatinate-Wallonia). Reine Wakote starts by analysing cross-border cooperation in the Greater Region from a legal point of view. Her article on "The Greater Region Model – Legal Aspects" shows that the European Cross-Border Mechanism (ECBM) – if adopted – could help to create in the Greater Region a "local cross-border law", which would better empower local and regional authorities to participate in actions and projects beyond national borders, in accordance with the principles of international law. Nora Crossey and Florian Weber then examine "Borderlands of Governance in the Saarland-Moselle

[2] The University of Southern Denmark in Sonderborg, the Euro-Institute in Kehl, the Hochschule Kehl, the Babes-Bolyai Cluj University, the Centre for Cross-Border Studies in Armagh in Northern Ireland, the Catholic University of Louvain, the University of Artois (Arras), the Mission opérationnelle transfrontalière (MOT) in Paris, the Central European Service for Cross-border Initiatives (CESCI) in Budapest and the University of Victoria in British Columbia, Canada.

[3] Centre Jean Monnet: <https://centre-jean-monnet.unistra.fr/>. Accessed 30 December 2023.

Border Region – the Roles of Municipalities within Current Dynamics of Cross-Border Cooperation". They observe a gradual "territorialisation" of intercommunal cooperation and identify cross-border structures such as the Eurodistrict SaarMoselle as an "ideal type" of territorial governance. The effects of Covid-19 on cross-border cooperation in the Greater Region are finally evaluated by Julia Dittel and Florian Weber, in their contribution on "A Living Lab of European Integration? Cross-Border Developments in the Greater Region during the Covid-19 Pandemic". They illustrate the extent to which the systematic border controls and the partial border closures caused a standstill in the border region. However, they also reveal that the pandemic contributed to a revival of and a stronger desire for cross-border cooperation and a deepened cross-border dialogue. The perception of the open border as an advantage was a central learning experience in favour of border regions like the Greater Region to be more resilient to crises.

The effects of the Covid-19 pandemic on cross-border cooperation in the Upper Rhine Region are analysed by Philippe Hamman. In his article on "Cross-border Urban Governance in the Upper Rhine Region in the Face of the Covid-19 Pandemic", he underlines that the residential and labour cross-border spaces require a differentiated approach to public policies in a multi-level governance approach. According to him, the health crisis has exacerbated the agglomerating effects of borders as filters of communication and it was not only a test for cross-border urban governance, but also revealed practical interdependences at the European and inter-local levels, whose existence calls for the development of shared, coordinated territorial policies. The legal perspective of cross-border cooperation in the Upper Rhine Region is dealt with by Michael Frey and Sarah Herrmann, in their article on "Legal Elements of a Cross-Border Friendly Legislation in the Upper Rhine". They claim that the efficiency and speed of resolving cross-border problems with a legal background could be significantly improved by the establishment of a legally binding *ex ante* impact analysis focusing on cross-border effects. Also, legal cross-border cooperation could be improved if existing problem-solving procedures within the Franco-Swiss Government Commission and the Committee of cross-border cooperation provided for in the Aachen Treaty were made more binding and quicker by adapting the rules laid down in the ECBM. Finally, Catherine Ledig and Frédérique Berrod give an interesting insight into cross-border cooperation in the Information Technology (IT) Sector in the Upper Rhine Region. In their article on "Feedback on more than 20 Years of Cross-Border Cooperation Experiences in the Upper Rhine in the Information Technology Sector (from 1994 to 2018)", they show that numerous projects were undertaken and produced operational results, but that the IT sector is not "an easy one", as one could believe that IT are not developing within national or local borders, as it is a virtual world. Cross-border cooperation in this sector therefore particularly requires a common sense of necessity on both sides of the border.

The second part of this publication on "Franco-Belgian Cross-Border Cooperation: A Schengen Model of an Open Border?" analyses the Franco-Belgian border with regard to its model function. Fabienne Leloup takes a historical perspective by assessing Franco-Belgian cross-border cooperation with regard to "Memories of the Franco-Belgian Border: Battles, Heritage and Cooperation". Whereas she remarks that cross-border cooperation creates incentives for collective regional or local projects based on collective memories – such as the world wars-, there still seems to remain fragmented memories that end at the border. At a time of resurgent nationalism, she emphasises the importance of examining how collective memories and heritage can shape notions of "us" and "them" at the border. Confirming the Frontem network's assumption of specificity of borderlands, she illustrates how the Franco-Belgian border reveals itself as a singular place, a site of histories that combines diversity and continuity, rewriting and reinterpretation. Cyprien Richer and Patrick Palmier then follow on by testing the hypothesis of the Franco-Belgian border as an "open" space. Their contribution on "Public Transport in EUrometropolis Lille-Kortrijk-Tournai. Is the Border Effect still Visible?" focuses on cross-border transport facilities and shows that, contrarily to expectations, the cross-border metropolis Lille is the only one that is not structured by a public transport network. According to them, the borderland is an "in-between" zone that is not covered by the main urban transport lines. Thus, the border effects are not only visible, but the idea of the Franco-Belgian border as an integrated "open" space is highly questionable. Pauline Pupier, in her article on "Water Governance in the European Natural Park Plaines Scarpe Scheldt. Legal Obstacles and Political Will in Cross-Border Cooperation" is more optimistic: She evaluates the European Natural Park ENPPSS as a successful cross-border cooperation. Although it has no competence in water policy or river management, the ENPPSS wishes to contribute to building up cross-border water governance. It can therefore take on an active role as a mutual trust builder, facilitator of governance and leader of cross-border projects. Finally, Bernard Reitel compares cross-border cooperation between France and Germany and France and Belgium. His article on "The Franco-Belgian and Franco-German Borders. A Brief Comparative Analysis", underlines that the two cross-border areas are characterised by intense interactions, but also by a possible heightening of tensions linked to the persistence of socio-economic differences and the reactivation of more or less marked forms of nationalism. Observing increased restrictions on border crossings for reasons of territorial security by the States and a risk of social polarisation, Bernard Reitel therefore doubts that the models of integrated or "open" border spaces can still be firmly asserted for both borderlands.

The model of an invisible border between Ireland and Northern Ireland is analysed in the third part of this publication on "Cross-Border Cooperation on the Isle of Ireland: A (still) Invisible Border?". Milena Komavora and Katy Hayward start by examining the potential effects of Brexit on the invisible border. Their contribution on "Resisting Re-bordering. Cross-Border Mobilities in an Irish Border Region in Anticipation of Brexit" first emphasizes how the Good Friday Agreement and cross-border cooperation have helped to transform a visible border as a security barrier into an invisible "soft space" of integration. However, they then illustrate how Brexit has provoked deep uncertainty over the future of cross-border mobilities and has revived the social and political tensions centering on the border's legitimacy, thus making the border visible again. This analysis is validated by Anthony Soares in his contribution on "The Visible Invisible Border: Reflections on the Frontem Focus Group Discussions on Mobility across the Ireland-Northern Ireland Border". He underlines that Brexit has raised again the question of managing and monitoring the land border on the island of Ireland. According to him, the Frontem focus groups reveal that this raises the "ghosts of the past" and the memories of a border amid a conflict that it was hoped had been left behind. On land, Brexit has made "the invisible visible", at least psychologically. By trying to keep the land border invisible, this has also made visible the maritime border, which is being fiercely contested by the unionists. Focusing on "Practices of Cross-Border Mobility. The Standing Conference on Teacher Education North and South SCo TENS as a Learning Cross-Border Organization", Kathy Hall then presents an initiative which has helped the teacher education community on the island of Ireland to move on from the political turmoil and violence to the relative peace after the Belfast /Good Friday Agreement in 1998. Without insisting on the "invisible" border, she maintains that the opening-up of conversation among "the geographically close" is generating awareness of both difference and commonality and helps to foster trust, familiarity and mutual under-standing. Finally, Jennifer Todd deals with the question of how a constitutional change to a united Ireland could impact the border area. In her article on "Changing Perspectives on the Irish Border, from the Southern Border Counties. Negotiating New Regional Frames and Constitutional Debate", she first emphasizes that the border has changed from a liminal area between states to an open area without barriers and a regional unit with its own practices, identity, problems and expertise. She then claims that the cross-border experience expressed by the participants of the Frontem focus groups could feed into the constitutional debate in order to assess how different institutional models would tackle cross-border dysfunctions, for example within the health service sector.

The fourth part of the publication deals with "Danish-German Cross-Border Cooperation: A Model for Minority Management?". Steen Bo Frandsen delivers a first insight on the Danish-German model with his article on "The Danish-German Border Region. A Critical Introduction". He reveals that the Danish-German border does no longer function as the clear demarcation as "foreseen" by the nation-state. Indeed, the Schengen border has changed border region reality and the inhabitants of the border region easily cross the border for shopping, trips, cultural events or for meeting friends and relatives. However, according to him, this does not mean that there is a revival of the historic "in between" and that the border is still present, even if the border region has moved quite far from the nation-state border of the past. However, in an account of "Celebrating the Border: The Danish-German Border region and the 100-year Anniversary of the 'Genforening'", Gaia Cardin realises that the border is no longer perceived only as a barrier, but also as a bridge between Denmark and Germany and that the narrative used during the celebration made use of historical events to create a sense of continuity and common belonging – such as the plebiscite in February 1920 or King Christian X's ride on a white horse over the old border in June 1920. The model character of the cross-border region can be underlined here, as the use of these historical events or episodes serves the construction of a regional identity.

Martin Klatt then analyses the model function of the Danish-German border region for minority management. His contribution on "The Danish-German-Frisian Region – A Best Practice Example?" is optimistic. He claims that, from a Danish and German minority perspective, the *status quo* is definitely improved practice, if compared to historical periods of conflict and alienation. But according to him there is still room for improvement, for, if political acknowledgement and social inclusion are for example best practice with regard to Danish, German and Frisian minorities, there are clear deficiencies with regard to the Sinti and Roma. Also, the cultural diversity of the region could be made more visible, especially in a cross-border context. Gunvor Vestergaard, in "Minorities and Cross-Border Meetings from the Perspective of Graenseforeningen (the Danish Border Association)", then emphasizes, that cross-border activities in the Danish-German area are important for the minorities living in the border region. From his point of view, the minority youth's narrative about living bilingual and using their cross-cultural possibilities is also useful for the nation-states and for European cooperation in general. Katarzyna Stoklosa tackles a more specific subject concerning minorities in the borderland: In her article on "Migration Movements of the Poles. Comparative Study of the German-Danish Border Region", she claims that Danish and German minorities do not behave in the same way in the border region as Polish minorities. While for the former, both material and non-material values are exchanged, for the latter, it is mainly a development of their new place of living in terms of material aspects which is important. Also, the Covid-19 restrictions seemed for Polish migrants in the Danish-German border region to represent a limitation not because of the closed German-Danish border but because it was not possible for them to travel to Poland. According to Katarzyna Stoklosa, this fact shows that the German-Danish border region does not have a great importance to Polish migrants living there. Finally, in their article on "Achieving (Im)mobility. Resilience at the German-Danish Border during the Covid-19 Pandemic", Dorte J. Andersen and Ingo Winklershow measure the effect of the Covid-19 pandemic on the border region's population. They show that the cross-border commuters' experiences and stories reveal how and why the tightened border regime matters for them and how they at once live with, attempt to resist, and also openly resent the stress they experienced. They reveal different forms of resilience: some clearly aim at depoliticising border controls, others articulate mobilities as a universal right of European citizens, thus constituting "open borders" less as a way of life for the commuters than as a political project. Their findings confirm the complexity and specificity not only of different border regions, but also within one single border region: in the Danish-German borderland, different versions of resilience indicate that there are different ways of living as a borderlander and that they depend on perceptions of the border and related consequences of closures.

The fifth part of the publication focuses on the "Hungarian-Romanian Cross-Border Cooperation: A Model for Central Europe?". First of all, János Sallai analyseds"Border Control on the Romanian-Hungarian Border in the Socialist Era". His contribution reminds of the dividing function of almost hermetically closed borders between Socialist States during the Cold War. He refers mainly to the legislation on the Hungarian-Romanian border regime of 1964, which provided for demarcation of the State border, maintenance of the boundary line and joint preservation and control of boundary marks. Whereas in the 1980s, the Hungarian-Romanian border relations were considered "friendly", which meant that the two countries would return illegal border crossers to each other as per a secret agreement, by the end of the 1980s, the deteriorated relationships between Romania and Hungary created issues on the border: Hungarian travellers and tourists were for example denied entry at the Romanian-Hungarian border by the Romanian guards without offering any explanation. An overview on the current borderland and cross-border cooperation is then presented by Nicolae Păun and Enya-Andrea Tămaș, in their contribution on "Changing Borders, Shifting Paradigms. The Romanian-Hungarian Border". They reveal that, over the course of the last decades, the perception, the form and the functions of the Romanian-Hungarian border have shifted. Thus, the border region has been for a long time separating the cross-border region, but with the end of socialism, the rigidity of the border eased and since then it has become the site of extensive cross-border cooperation. The model character of this cross-border cooperation seems confirmed, as one can identify a transformation of the border from its barrier function to a connecting bridge. Their contribution also underlines the importance for Romania's accession to the Schengen Area, which could inarguably serve the further "spiritualisation" of the border, while also contributing to the physical realisation of a "Europe without borders". István Polgár and Mircea Brie confirm this analysis in their article on "General Trends and Evolutions from the Past Century Regarding the Perception of the Romanian-Hungarian Border in the Context of the European Territorial

Cooperation". Focusing on the Bihor – Hajdú-Bihar Euroregion, they illustrate the success of territorial cooperation by means of joint programs oriented towards innovation and institutional cooperation. They believe that the Hungarian-Romanian border, since the fall of communism, has changed its status of a hard, close and exclusive border to that of a soft, open and inclusive one. Finally, Melinda Benczi deals with the perception of the Hungarian-Romanian border. Her article on "Outlining the Main Perception Strand on the Hungarian-Romanian Border" is rather pessimistic with regard to the model character of the borderland. She affirms that citizens living in this border sections are not hostile to each other, but that the cooperative spirit is sometimes replaced by competitiveness. According to her, the Hungarian-Romanian border itself often still appears as a barrier and not as a bridge, which could be improved through the sensitive narration of history, the increase of the permeability and the prevalence of "friendly" politics.

The sixth and final part of the publication allows for a Euro-Atlantic comparison of cross-border cooperation in the light of crisis. Jean Peyrony's first assesses two years of border management in Europe after the Covid-19 pandemic and its impact on border regions and their inhabitants. His contribution "Europe: Towards Cross-Border Regions and Multi-Level Governance of Borders" was based on an analysis of the impact of border-related measures taken by Member States in the fight against Covid-19, carried out by Mission Opérationnelle Transfrontalière (MOT) for the European Commission in 2022. Jean Peyrony underlines that the crisis has revealed the nature of cross-border interdependencies, of which public authorities had no clear view and that borders' real nature and daily functioning have proven to be largely misunderstood by public authorities and opinion. He believes that cross-border living areas can better integrate in the future if EU Member States agree to exercise their sovereignty together, at the local level of cross-border regions and with the support of a bottom-up multi-level governance. In their article on "(Mis)Perceptions of the Canada-US borderland", Emmanual Brunet-Jailly and Tatsiana Shaban then show the differences of perception and border management in US-Canada borderlands with regard to Europe, emphasising mainly the shift towards a more securitized border after the terrorist attacks in 2001 and the Covid-19 pandemic. According to their analysis, this creates difficulties for cross-border relations and trade. They underline that, when crossing the border today between the US and Canada, it is notable that 20 years of policy changes have contributed progressively to citizens' perceptions of uncertainty and insecurity. This confirms the assumption, that in Europe, and especially within the EU, cross-border cooperation and resilience in crisis might represent indeed a model function for border perception and management in other areas of the world.

Part 1

Franco-German Cross-Border Cooperation: A Model for European Integration?

The Greater Region Model: Legal Aspects

Reine Wakote

The Greater Region is located on the European backbone and is one of the main border regions of the European Union (EU). Its territory is spread over four EU states and consists of: the *Länder* of Saarland and Rhineland-Palatinate in Germany; the former region of Lorraine (now part of the Grand Est (Greater East) region); the Grand Duchy of Luxembourg, as well as Wallonia, the Wallonia-Brussels Federation and the German-speaking Community in Belgium.

The Greater Region was one of the major scenes of armed conflict at the end of the 19th and 20th centuries. The entities that make up the region are also linked by an economic past centred mainly on the steel industry and, to a lesser extent, on textiles. It covers almost 65,000 km^2 and has a population of almost 11.6 million (representing 2.5 % of the population of the EU). Its Gross Domestic Product (GDP) of 390 billion Euro in 2017 makes up about 2.5 % of the EU's GDP.[1] The Greater Region is home to "11.6 million inhabitants as of 1 January 2015, or 2.3 % of the total population of the EU 28. Rhineland-Palatinate is by far the most populous regional component, followed by Wallonia [...] The former Lorraine region comes third [...] The smallest components [are] Saarland and Luxembourg [...]"[2]. The Greater Region would thus

1 Greater Region, "Chiffres clés", 2017. Retrieved from: <http://www.granderegion.net/En-bref/Chiffres-cles>. Accessed 1 December 2022.

2 "Rhineland-Palatinate is by far the most populous regional component with four million inhabitants, followed by Wallonia (including 76,000 people for the German-speaking Community of Belgium), where almost 3.6 million people live. Third place goes to the former Lorraine region, which had some 2.34 million inhabitants at the beginning of 2013. The smallest components [...] namely Saarland and Luxembourg, have one million and just over half a million inhabitants respectively", see Grande Région, "Chiffres clés", *op. cit.*

be comparable to a medium-sized EU Member State,[3] making its analysis particularly interesting for cross-border cooperation. Especially since it has "the largest number of cross-border workers in the EU. In 2019, almost 250,000 (two hundred and fifty thousand) people crossed a border in the Greater Region every day to commute to their workplace in a neighbouring region, including 197,000 (one hundred and ninety-seven thousand) to Luxembourg alone".[4]

These elements make the Greater Region a major laboratory for the European Union's cohesion policy. Its actors also support a "cross-border forward-looking vision [...] to 2040 [placing the Greater Region in] a competitive position at European and international level [and] reconciling economic development and territorial development".[5] In view of this picture, a question arises: Should the Greater Region be considered as a model for cross-border cooperation, for example as "what serves or should serve as an object of imitation in cross-border cooperation"?[6]

To answer this question, it is first necessary to note the extent to which cross-border cooperation depends on the distribution of jurisdictions between the home state and sub-state entities. This question can be answered in three points: by analysing the distribution of competences within the Greater Region, focusing on the situation in France; then by analysing the impact of this distribution of competences on the current

---

3 F. Allemand, "Le développement économique en région transfrontalière: un cadre institutionnel favorable ? L'exemple de la Grande-Région", *In* R. Wakote (ed.), *La coopération transfrontalière : vers un nouvel élan ? – L'exemple du Grand Est*, Mare et Martin, Paris, 2024 (to be published).

4 Grande Région, "Chriffres clés", *op. cit.*

5 Grande Région, "Une vision prospective transfrontalière pour la Grande Région", VIe réunion ministérielle sectorielle de l'Aménagement du territoire de la Grande Région, 2020. Retrieved from: <http://www.granderegion.net/Actualites/2020/VIe-reunion-ministerielle-sectorielle-de-l-Amenagement-du-territoire-de-la-Grande-Region-Une-vision-prospective-transfrontaliere-pour-la-Grande-R>. Accessed 1 December 2022, cited by F. Allemand, "Le développement économique en region transfrontalière", *op. cit.* The strategic vision and the accompanying operational document make up the two parts of the Greater Region territorial development Plan *(Schéma de Développement Territorial de la Grande Région, SDTGR)*, validated in session by the intermediate summit of the Greater Region on 16 June 2020, on <https://amenagement-territoire.public.lu/fr/grande-region-affaires-transfrontalieres/SDT-GR.html>. Accessed 18 April 2023.

6 See dictionary *Le petit Robert.*

Greater Region model and finally, by identifying the prospects for the evolution of this model.

Cross-Border Cooperation Depends on the Power-Sharing System between the State and the Sub-State Entities

Cross-border cooperation implies cooperation across borders and requires that the principles of international law be respected. According to international law, state sovereignty is vested in the state, which is solely responsible for acting across national borders; sub-state entities do not have such competence. Under international law, a sub-state entity needs the state's authorisation and green light to intervene beyond the borders of the state it belongs to. In France, this empowerment results from the devolution law *(loi de décentralisation)* which was passed on the 6th of February 1992 and has been supplemented several times since.[7] It was completed, among others, by the law (n° 2005-95) of 9 February 2005 on the international cooperation of local authorities and water agencies in the fields of water supply and sanitation (known as the *Oudin-Santini* law), which allows municipalities, Public Establishments of International Cooperation (*Etablissements publics de cooperation internationale EPCI),* and mixed unions to allocate up to 1 % of their budget to cooperation and solidarity actions in the field of water, a possibility that was subsequently extended to the fields of electricity, gas and waste management. This was followed by the law (no. 2007-147) of 2nd February 2007 on the external action of territorial authorities and their groupings (known as the *Thiollière* law). This last law set the general rules for "the external action of territorial authorities and their groupings", removing – most importantly – the notion of "local interest": several court decisions had in fact, in previous years, annulled the decisions of French authorities that had undertaken international solidarity actions on the grounds that the local population in France had no interest in them[8]. These texts were adopted following the adoption of the 1980 Madrid Framework

[7] Orientation law ("loi d'orientation" – n° 92-125) of 6 February 1992 on the territorial administration of the Republic.

[8] For an illustration of "local interest", see the following court decisions: TA Poitiers, 18 November 2004, *M. Charbonneau*, AJDA, 2005, p. 486; CAA Bordeaux, 30 October 2007, *Département des Deux-Sèvres*, AJDA, 2008, p. 198; CAA Versailles,

Convention adopted by the Council of Europe on 21th May 1980, which came into force on 22th December 1981 and is the subject of several protocols

In spite of this empowerment, sub-state authorities cannot encroach upon the state's jurisdiction: the impossibility for them to implement competences falling under state sovereignty. But beyond these considerations in relation to international law, it should essentially be stressed that the national law of the State to which the State belongs regulates, in principle, the competences of the State itself on the one hand, and those that may be exercised by its sub-state entities on the other. The national law of each State thus completes the rules of international law.[9]

In the case of France, Article 34 of the 1958 Constitution states that "the law determines the fundamental principles [...] of the free administration of territorial authorities, their powers and their resources". This provision lays down, within the national framework, the conditions for the distribution of powers between the State and sub-State entities, for example the territorial authorities. It requires the intervention of the legislator for any modification or adaptation.

The law on the New territorial organisation of the Republic (known as the *NOTRe* law), promulgated on the 7th of August 2015, is the basic reference for determining the jurisdiction of each territorial authority (*commune*, *départment*, *région* and *métropoles*) and, in so doing, allows to identify the power-sharing boundaries between territorial authorities and the State. Each local authority can thus only implement the competences that have been legally been devolved to them.

At present, article 2 of the law NOTRe[10] assigns competence for cross-border cooperation to the region; in the case of the Greater Region, this competence therefore falls to the Grand Est region: "The region shall draw up a regional plan for economic development, innovation and internationalisation [...] The plan may contain a cross-border section drawn up

---

31 May 2007, *Commune de Stains*, AJDA, 2007, p. 1549; CAA Marseille, 7 December 1999, *Commune d'Istres*, no 98 MA00236.

[9] For Germany, see in particular, M. Frey, "L'articulation des compétences infra-étatiques en matière de coopération transfrontalière en Allemagne", *In* R. Wakote, *La coopération transfrontalière*, *op.cit.*

[10] Article 2 of the law (n° 2015-991) of 7th August 2015 on the new territorial organisation of the Republic (*loi NOTRe*).

in consultation with the local authorities of neighbouring States [...]".[11] But at the same time, the law also allows the métropoles (defined by the NOTRe law as grouping of municipalities according to inhabitants) to intervene in cross-border cooperation, if they do not adhere to what has been defined by the region.[12] The Metropolis of Metz has thus defined a metropolitan plan including a cross-border section.

Secondly, the law of the 2nd of August 2019 on the competences of the European Collectivity of Alsace (*Collectivité européenne d'Alsace* CEA) grants it competence in cross-border cooperation.[13]

> "[..] the European Collectivity of Alsace is responsible for organising on its territory, as lead partner, the arrangements for joint action by territorial authorities and their public establishments in the field of cross-border cooperation. In this capacity, the European Collectivity of Alsace shall draw up an Alsatian cross-border cooperation plan. In particular, it shall involve the state, the Grand Est region, the Eurométropole of Strasbourg and the other territorial authorities concerned, as well as their groupings (...). This plan includes an operational section on structuring projects. It also includes a section relating to cross-border travel, in particular road, river and rail links for which the European Collectivity of Alsace is involved in the preparation of cross-border infrastructure projects, as well as a section relating to cross-border cooperation in the field of health, drawn up in accordance with the regional health project".[14]

---

11 See Art. L. 4251-13 Code général des collectivités territoriales (CGCT).

12 See Art. L. 4251-15 CGCT: "The guidelines of the regional economic development, innovation and internationalisation plan applicable to the territory of a metropolis mentioned in Chapter VII of Title I of Book II of Part Five, the metropolis of Aix-Marseille-Provence or the metropolis of Lyon are drawn up and adopted jointly by the council of the metropolis concerned and the regional council. In the absence of an agreement, the metropolis shall draw up a strategic guidelines document that takes into account the regional plan. This document takes the place, for the metropolis, of guidelines within the meaning of the second paragraph of Article L. 4251-13 CGCT [...]".

13 Article 2 of the law (n° 2019-816) of 2nd August 2019 on the competences of the European Collectivity of Alsace.

14 See Art. L. 3431-1. Art. L. 3431-2 CGCT then specifies that "the Alsace cross-border cooperation plan shall be defined in accordance with the cross-border section of the regional plan for economic development, innovation and internationalisation. The cross-border cooperation plan of the Eurométropole of Strasbourg (...) is defined in a manner consistent with the Alsace cross-border cooperation plan [...]".

Since the adoption of this law, the department of Moselle has also claimed to have competence over cross-border cooperation and has requested the government to draft a law to be submitted to parliament in that way.[15] Pending the government's response and the outcome of the legislative procedure, Moselle has declared itself a "Eurodepartment".[16]

In addition, the law *NOTRe* of 2015 endowed the municipality (*communes*) with a general competence to intervene in all fields, provided that it is justified by local interest. The municipality can thus intervene in the field of cross-border cooperation.

Finally, the department is responsible for health matters. This implies that it can intervene in the cross-border framework in this field, even though the Covid crisis has highlighted the central role of the municipalities and communes[17].

The adoption of the law of 21 February 2022 (known as the 3DS law)[18] could reinvigorate the question of the articulation of competences in cross-border cooperation. For the first time since the ATR law adopted in 1992, a law includes a title or chapter expressly devoted to cross-border cooperation: Title VII of the 3DS law devotes a Chapter III to cross-border cooperation, containing articles 182–190. This can only reinforce the central place that cross-border cooperation now occupies in the domestic legal system. In addition, the 3 DS law extends the areas likely to be covered by cross-border cooperation: cross-border health cooperation (article 182)[19] is integrated into cross-border cooperation;

---

15 A. Villeneuve, "La Moselle veut devenir un « eurodépartement »", *Le Monde*, 2019, Retrieved from: <https://www.lemonde.fr/politique/article/2019/05/10/la-moselle-veut-devenir-un-eurodepartement_5460366_823448.html>. Accessed 1 December 2022. With this new status, Moselle hopes to be granted competence in the fields of education, mobility, employment and financial support for local commerce, in order to meet the challenges of cross-border cooperation with Germany and Luxembourg.

16 P. Braun, "La Moselle se proclame Eurodépartement", *Les Échos*, 2019, Retrieved from: <https://www.lesechos.fr/pme-regions/grand-est/la-moselle-se-proclame-euro departement-1025917>. Accessed 1 December 2022.

17 V. Sempastous, "Le rôle clé du département dans l'organisation territoriale: les leçons de la crise sanitaire", *In* Droit et Gestion des collectivités territoriales (DGCT), *L'administration locale face à la crise sanitaire*, Le Moniteur, Grale, Paris, 2021, p. 93.

18 Law (no. 2022-217) of 21 February 2022 on Differentiation, Decentralisation, Deconcentration and Various Measures to Simplify Local Public Action (known as the 3DS law).

19 Codified in Article L1434-3 of the Public Health Code.

foreign local authorities and representatives of EGTCs will be able to participate, without voting rights, in the deliberations of departmental commercial development commissions on commercial locations (article 184); establishment of a legal framework for cross-border apprenticeships (article 186)[20] and to facilitate the organisation of cross-border sports competitions or trade fairs[21].

Thus, in France, the various territorial authorities located on the territory of the Greater Region can intervene in cross-border cooperation. This distribution has significant consequences for the Greater Region model.

Scope of the Distribution of Competences for the Greater Region Model

With regard to the situation in France, the first consequence is, of course, a large number of local actors who can intervene in cross-border cooperation. Furthermore, the government, guarantor of the national interest, also remains competent.

This situation contrasts with Luxembourg's, where cross-border cooperation is handled by the government of the Grand Duchy, making it the main interlocutor.

These different levels of competence have an impact both on governance and on the structures set up within the Greater Region. Governance is based on the coexistence of governance at regional level and at local level.

The regional governance of the Greater Region is ensured by the Summit of the Greater Region's executives. For Germany and Belgium, these are the minister-presidents, for France the president of the regional Council of the Greater East and the prefect of region, and for Luxembourg the minister for the Greater Region and the prime minister. The Presidency of the Summit revolves between each of the partner regions in two years mandates. The other sub-state entities (and therefore

---

20 Codified in Article L6235-2 of the French Labour Code.

21 For an evolution of French legislation up to the 3DS law, see R. Wakote "Coopération transfrontalière: quelles consolidations des perspectives inaugurées par la loi ATR?", *In* F. Lerique (ed), *Les fondements de l'administration territorialisée – Les éclairages apportés par la loi d'administration territoriale de la République*, Grale, Paris, 2023, p. 253.

in France, the communes, departments and metropolises) are therefore excluded. Since January 2021, the Greater East region has held the presidency.[22]

The Interregional Parliamentary Council (*Conseil parlementaire interrégional CPI*),[23] created in 1986, brings together the members of the parliaments of the regions that make up the Greater Region; for Luxembourg, it is directly handled by the national parliament. The CPI has only a consultative role on matters relating to cross-border cooperation. The Economic and Social Committee of the Greater Region (*Comité économique et social* de la *Grande Région CESGR*), the second consultative body, is made up of experts and economic and social partners (employers' representatives, trade union representatives and civil or administrative society representatives).

Local governance involves actors according to the projects that they carry out. First of all, *EuRegio SarLorLux+*, created in 1995 as a profit-making association under Luxembourg law, provides a forum for exchange between the municipalities of the Greater Region; it is also supposed to represent municipal interests in the Greater Region. Secondly, the *Sillon lorrain* organises a network of cities and conurbations of Metz, Nancy, Épinal and Thionville, in order to promote innovative cross-border projects within the Greater Region. In addition, the *QuattroPole network* of cross-border cities *(réseau QuattroPole de villes transfrontalières)*, created in 2000, includes Luxembourg and the cities of Metz, Saarbrücken and Trier. Its aim is to promote the development of cross-border cooperation in these cities.

On the other hand, the *SaarMoselle Eurodistrict*, created in 2010 in the form of a European Grouping of Territorial Cooperation (EGTC), brings together, in Germany, the Saarbrücken urban community *(Regionalverband Saarbrücken)*, and in France, seven conurbations *(agglomérations)* and communities of municipalities *(communauté de communes)*. Finally, the Alzette-Belval project, set up under French law as

---

22 Editor's note: since the 31st of January 2023, the presidency is held by the minister-president of the *Land* Rheinland-Pfalz, Malu Dreyer.

23 The Interregional Parliamentary Council (IPC) was created on 16th February 1986 by the presidents of the parliamentary assemblies of the regions that make up the Greater Region: Lorraine, Luxembourg, Rhineland-Palatinate, Saarland and Wallonia and, since 2007, the French Community of Belgium and the German Community of Belgium.

a Public Administrative Establishment (*Etablissement public administratif EPA*), intends to reinvest derelict territories that used to host French steel mills.

In view of this panorama, it is appropriate to ask whether this multiplicity of actors actually contributes to the Greater Region's cross-border project. Such a multiplicity of actors leads to the conclusion that the development sought in the framework of cross-border cooperation is actually imbalanced throughout the territory of the Greater Region[24]. There are several reasons for this.

First, the multiplicity of actors implies different levels of intervention. In Germany, the *Länder* may have control over the definition of the broad lines of cross-border cooperation, since they are allowed to define the free administration of the entities located within their territorial perimeter (notably the municipalities). In France, although the region has competence for cross-border cooperation, the state remains a key player. Indeed, the Regional Development, Innovation and Internationalisation Plan *(Schéma Régional de Développement Économique d'Innovation et d'Internationalisation SRDEII)*, which the region must draw up, possibly with a cross-border component, must be approved by the prefect, the state representative, before being applied. Moreover, in France, one may wonder about the competition between the region and the possibility for the municipalities, metropolises and the European Collectivity of Alsace to intervene in cross-border cooperation. Each of them can thus carry out the projects that they consider as priorities in this field. As regards Luxembourg, "the Grand Duchy plays a special role in the Greater Region, not only because of its central position and economic importance, but also because it is the only state among the members of the interregional cooperation. As a nation-state, Luxembourg has greater

[24] M. Quévit, "La Grande Région et la problématique des rapports 'Etats-nation et region' dans l'Union Européenne", *Revue Internationale de Politique Comparée*, 2005, vol. 12, n°2, pp. 207–221 ; F. Clément, "La représentation politique dans la Grande Région SaarLorLux: institutions, compétences et coopérations", *In* C. Wille (ed.) *Lebenswirklichkeiten und politische Konstruktionen in Grenzregionen – Das Beispiel der Großregion SaarLorLux: Wirtschaft – Politik – Alltag – Kultur*, Transcript Verlag, Bielefeld, 2015, pp. 61–82; A. Villeneuve, "La Moselle partagée entre les vices et les vertus de l'emploi frontalier", *Le Monde*, 4 December 2022, p. 22; B. Jérôme, "Pour les soignants, l'attraction du Luxembourg", *Le Monde*, 25 May 2019, p. 8.

political autonomy than its partners, who are subject to other levels of government"[25].

Secondly, the various component entities of the Greater Region do not envisage cross-border cooperation in the same perspective. Indeed, Luxembourg would be its prime beneficiary. As a /major job provider, it attracts inhabitants of the other territories of the Greater Region: its territory benefits from this cross-border flow. However, its investments in transport and even in training, particularly in the medical field, remain below the benefits it receives. Are there not rivalries within the Greater Region? Do the various members of the Greater Region actually participate in cross-border cooperation? Do they all participate in the same cross-border cooperation? What about the smaller cities that are not in the immediate vicinity of the border?

Thirdly, the Greater Region does not seem to be endowed with a clearly defined perimeter of cross-border cooperation. It does not make up a "cross-border metropolis", but rather a "polycentric cross-border metropolis". Nevertheless, the cities distant from the border do not feel the impact of the border's situation.

It seems that these imperfections of cross-border cooperation in the Greater Region could be corrected as showed by the recent declarations that have been recently agreed upon.

Prospects for the Evolution of the Greater Region Model

Recent developments supported the idea of a favourable evolution within the Greater Region. First of all, on the 11th of December 2018, France and the Benelux countries signed a Declaration of Intent "in favour of intensified cooperation between the Benelux countries and the French Republic, in particular the border regions".[26] This bill was followed by the signature, in 2019, of the Treaty of Aachen, a new Franco-German cooperation and integration treaty. In Chapter 4, articles 13–17

---

[25] C. Pereira Carneiro Filho, "La Grande Région, région transfrontalière européenne", *Confins*, 2012, n° 6, p. 18. Retrieved from: <https://journals.openedition.org/confins/7908>. Accessed 18 April 2023.

[26] "Declaration of intent between the Government of the French Republic and the Governments of the Benelux Union on the development of closer cooperation", Retrieved from: <http://www.benelux.int/download_file/view/7201/6447/>. Accessed 1 December 2022..

are devoted to strengthening cross-border cooperation.[27] Article 13-1 states: "The two States recognise the importance of cross-border cooperation between the French Republic and the Federal Republic of Germany in strengthening links between citizens and businesses on both sides of the border, in particular the essential role of local authorities and other local actors in this respect. They intend to facilitate the removal of obstacles in border territories in order to implement cross-border projects and to facilitate the daily life of the inhabitants of these territories". Article 13-2 provides that:

> "To this end, in compliance with the respective constitutional rules of the two States and within the limits of European Union law, the two States shall provide the local authorities of the border territories and cross-border entities such as Eurodistricts with appropriate competences, dedicated resources and accelerated procedures for overcoming obstacles to the realisation of cross-border projects, in particular in the economic, social, environmental, health, energy and transport fields. If no other means are available to overcome these obstacles, appropriate legal and administrative arrangements, including derogations, may also be granted. In this case, it is up to the two States to adopt the appropriate legislation [...]"

Article 14 of the treaty establishes a Cross-border Cooperation Committee responsible for defining a common strategy for the selection of priority projects, monitoring the difficulties encountered in the border territories and making proposals to remedy them. It took shape on 22 January 2020 (the same day the treaty came into force). Operational since the end of 2020 with the installation of a joint secretariat in the border region, the Committee has gone into action to strengthen cross-border coordination in the face of the health crisis.

Finally, the European Commission adopted the European Cross-Border Mechanism (ECBM)[28] in 2018. This mechanism aimed to

---

27 Treaty between the French Republic and the Federal Republic of Germany on Franco-German Cooperation and Integration, 22 January 2019. Retrieved from: <https://www.diplomatie.gouv.fr/fr/dossiers-pays/allemagne/relations-bilaterales/traite-d-aix-la-chapelle-sur-la-cooperation-et-l-integration-franco-allemandes/>. Accessed 1 December 2022.

28 Proposal for a Regulation of the European Parliament and of the Council establishing a mechanism to remove legal and administrative barriers in a cross-border context, 29 May 2018; E. Evrard, "Comment le droit européen déplace les frontières? La proposition de règlement visant à lever les obstacles juridiques et administratifs à la cooperation transfrontalière", *Annales de géographie*, 2020, vol. 4, n°774, pp. 46–66..

remove legal and administrative obstacles in a cross-border context and thus promote the harmonious development of the entire European Union's territory. Through this tool, the Commission aimed to offer a legal tool to cross-border regions, so that they can find solutions to the difficulties they encounter. The ECBM could have allowed a State to apply the regulations of a neighbouring State on its territory, thus removing legal or administrative obstacles to the realisation of projects involving cross-border cooperation. However, on the 13th of July 2021, the Slovenian Minister for Development and Cohesion Policy (Zvonko Černač), announced the cancellation of this new regulation[29].

These texts could have formed the basis for a genuine legal development of cross-border cooperation in the Greater Region. First of all, by envisaging the adoption of a common legal framework, in particular that of the neighbouring state, they made it possible to remove the legal wavering surrounding the implementation of cross-border cooperation projects. Thus, these texts allowed the European commission to outline a "local cross-border law".

Such a "local cross-border law" would have been based primarily on national law, in particular because of the mechanisms empowering local and regional authorities to participate in actions and projects beyond national borders in accordance with the principles of international law.

This "local cross-border law" would have prompted, particularly in France, a clarification of the division of competences between on the one hand the state and the territorial authorities and on the other hand between the territorial authorities. The state would have given up intervening through the prefect, its representative, and left it to the judge to monitor the compatibility of the norms of "local cross-border law" with national law and international law.

Finally, public services or services issued to the target populations by the local cross-border executives would have been the first area where

---

29 A. Sander, "Abandon par les États membres du Mécanisme transfrontalier européen – Un véritable coup de frein à la coopération transfrontalière", Press Communication, 15 July 2021. Retrieved from <https://annesander.eu/communique-de-presse-abandon-par-les-etats-membres-du-mecanisme-transfrontalier-europeen-un-veritable-coup-de-frein-a-la-cooperation-transfrontaliere-15-07-2021/>. Accessed 1 December 2022.

this "local cross-border law" would have had to be deployed as a priority. These services contribute to the implementation of cross-border cooperation, seed of a cross-border identity in the Greater Region. It is only by achieving this goal that the Greater Region can embody a transposable model for cross-border cooperation.

Bibliography

F. Allemand, "Le développement économique en région transfrontalière: un cadre institutionnel favorable ? L'exemple de la Grande-Région", *In* R. Wakote (ed.), *La coopération transfrontalière : vers un nouvel élan ? – L'exemple du Grand Est*, Mare et Martin, Paris, 2024 (to be published).

E. Evrard, "Comment le droit européen déplace les frontières? La proposition de règlement visant à lever les obstacles juridiques et administratifs à la cooperation transfrontalière", *Annales de géographie*, 2020, vol. 4, n°774, pp. 46–66.

M. Frey, "L'articulation des compétences infra-étatiques en matière de coopération transfrontalière en Allemagne", *In* R. Wakote, *La coopération transfrontalière, op.cit.*

Law (n° 2015-991) of 7th August 2015 on the new territorial organisation of the Republic (*loi NOTRe*).

C. Pereira Carneiro Filho, "La Grande Région, région transfrontalière européenne", *Confins*, 2012, n° 6, p. 18.

A. Sander, "Abandon par les États membres du Mécanisme transfrontalier européen – Un véritable coup de frein à la coopération transfrontalière", Press Communication, 15 July 2021.

V. Sempastous, "Le rôle clé du département dans l'organisation territoriale: les leçons de la crise sanitaire", *In* Droit et Gestion des collectivités territoriales (DGCT), *L'administration locale face à la crise sanitaire*, Le Moniteur, Grale, Paris, 2021, p. 93.

R. Wakote, "Coopération transfrontalière: quelles consolidations des perspectives inaugurées par la loi ATR?", *In* F. Lerique (ed.), *Les fondements de l'administration territorialisée – Les éclairages apportés par la loi d'administration territoriale de la République*, Grale, Paris, 2023, p. 253.

M. Quévit, "La Grande Région et la problématique des rapports 'Etats-nation et region' dans l'Union Européenne", *Revue Internationale de*

Politique Comparée, 2005, vol. 12, n°2, pp. 207–221 ; F. Clément, "La représentation politique dans la Grande Région SaarLorLux: institutions, compétences et coopérations", *In* C. Wille (ed.), *Lebenswirklichkeiten und politische Konstruktionen in Grenzregionen – Das Beispiel der Großregion SaarLorLux: Wirtschaft – Politik – Alltag – Kultur*, Transcript Verlag, Bielefeld, 2015, pp. 61–82.

"Borderlands of Governance" in the Saarland-Moselle Border Region: The Roles of Municipalities within Current Dynamics of Cross-Border Cooperation[1]

Nora Crossey & Florian Weber

Cross-Border Cooperation as a Multi-Level Endeavour

The development of European post-war cross-border cooperation can be understood as an interplay between parallel "top-down" impulses from the national and European level and "bottom-up" initiatives by (border-)regional and municipal authorities.[2] The "top-down" impulses can be seen in the form of treaties and agreements providing the political and structural framework for cross-border cooperation, and the gradual expansion of legal and financial opportunities for subnational actors. Regarding "bottom-up" initiatives, the case of the Greater Region in Western Europe can be considered a "prototype" of the gradual development and institutionalisation of interregional cross-border cooperation within the European Community/European Union[3]. In recent years, a

---

1 The empirical results of our paper are based on a research project focusing on the France Strategy of the Saarland, which was funded by the Saarland State Chancellery in 2019–2021. We sincerely thank for the support.

2 J. Beck, "Cross-border cooperation in Europe as an object in transdisciplinary research. An introduction." *In* J. Beck (ed.), *Transdisciplinary Discourses on Cross-Border Cooperation in Europe*, Peter Lang, Brussels, 2019, pp. 13–29; B. Reitel, P. Pupier & B. Wassenberg, "European policies and border cities" *In* E. Medeiros (ed.), *Border Cities and Territorial Development*, Routledge, New York, 2021, pp. 59–75.

3 W. H. Lorig, S. Regolot & S. Henn (eds.). *Die Großregion SaarLorLux. Anspruch, Wirklichkeiten, Perspektiven*, Springer VS, Wiesbaden, 2016; C. Wille (ed.). *Lebenswirklichkeiten und politische Konstruktionen in Grenzregionen. Das Beispiel*

new "trend" of cross-border cooperation can be observed in the context of this cross-border region, namely the announcement of (complementary) regional strategies by the French and German border regions – the Saarland and the *département* Moselle. Building upon existing traditions of close coordination,[4] these strategies posit cross-border cooperation and integration as a constitutive element of border-regional development and survival, and emphasise the relevance of border regions as cross-border actors in their own right.[5]

The first of these border-regional strategies – the "France Strategy" of Saarland – formulates a broad range of both internal and external objectives. They include in particular the promotion of (functional) multilingualism, the intensification of "horizontal" cross-border cooperation in various policy fields, and the further institutionalisation of "vertical" networks through e.g. joint representation of border-regional interests at the national and European level.[6] The France Strategy – which serves as the point of departure for our article – was conceived explicitly as a "participatory governance project" rather than a top-down government programme, making the active inclusion of stakeholders beyond the regional

---

der Großregion SaarLorLux: Wirtschaft-Politik-Alltag-Kultur, Transcript Verlag, Bielefeld, 2015.

4 D. Hüser, "Die Saar-Politik und die deutsch-französische Geschichte an der Grenze" *In* H. J. Lüsebrink, C. Polzin-Haumann & C. Vatter (eds.), *Alles Frankreich oder was – Die saarländische Frankreichstrategie im europäischen Kontext / La France à toutes les sauces? – La stratégie France de la Sarre dans le contexte européen*, Transcript Verlag, Bielefeld, 2017, pp. 39–67; H. Uterwedde, "Die saarländische Frankreichstrategie im Kontext der aktuellen deutsch-französischen Beziehungen" *In* H. J. Lüsebrink, C. Polzin-Haumann & C. Vatter (eds.), *Alles Frankreich oder was, op.cit.*, pp. 27–39.

5 F. Weber, "Grenzüberschreitende Governance in der Großregion SaarLorLux", *Forum Wohnen und Stadtentwicklung*, vol. 4, 2020, pp. 183–186.

6 Landesregierung des Saarlandes, "Eckpunkte einer Frankreichstrategie für das Saarland", 2014. Retrieved from: <https://www.saarland.de/stk/EN/service/publikationen/Frankreichstrategie.html>. Accessed 21 June 2021.

N. Crossey & F. Weber, "Zur Konstitution multipler *Borderlands* im Zuge der Franreichstrategie des Saarlandes" *In* F. Weber, C. Wille, B. Caesar & J. Hollstegge (eds.) *Geographien der Grenzen. Raüme – Ordnunen – Verflechtungen*, Springer VS, Wiesbaden, pp. 145–166; K. Terrollion & C. Klos, "Die Frankreichstrategie des Saarlandes. Ein deutsches Bundesland auf dem Weg zu einer gelebten europäischen Mehrsprachigkeit", *Notes du Cerfa*, n°168, IFRI, March 2022. Retrieved from: <https://www.ifri.org/de/publications/notes-de-lifri/notes-cerfa/die-frankreichstrategie-saarlandes-ein-deutsches-bundesland>. Accessed 31 March 2023.

authorities crucial to its long-term implementation[7]. Furthermore, it spurred the formulation of a complementary "*Stratégie Allemagne*"[8] by the *département* Moselle,[9] reflecting and underlining a certain "complementarity of the border area".[10]

The contributions of regional and municipal authorities to French-German cross-border governance were explicitly highlighted in the Aachen Treaty of 2019,[11] which pledged to address remaining challenges through adapted legal and administrative provisions for border regions. They are to be explored through the newly established Committee for Cross-Border Cooperation (*Ausschuss für Grenzüberschreitende Zusammenarbeit* (AGZ) – *Comité de Coopération transfrontalière* (CCT)) composed of authorities from all administrative levels, cross-border organisations and parliamentarian representatives.[12] Most recently, the experiences of the Covid-19 pandemic, in particular the discrepancy

---

[7] A. Funk & M. Niedermeyer, "Die Frankreich-Strategie des Saarlandes. Governanc eund territoriale Kohäsion als neue Dimensionen regionaler Entwicklungskonzepte" *In* W. H. Lorig, S. Regolot & S. Henn (eds.), *Die Großregion SaarLorLux. Anspruch, Wirklichkeiten, Perspektiven*, Springer VS, Wiesbaden, 2016, pp. 169–183.

[8] Due to the 2015 territorial reform in which the former *région* Lorraine was integrated into the new *région* Grand Est, the Strategy was never formally adopted. Nonetheless, its central objectives continue to resource and inform policy objectives at the level of the *région* or *département* (see Conseil regional du Grand Est, 2018).

[9] Conseil Régional de Lorraine, "'Stratégie Allemagne de la Lorraine' Rapport final. Synthèse et propositions juin 2015", 2015. Retrieved from: <https://metz.fr/pages/conseil_municipal/seances/cm151029/doc/5_d1445930773387.pdf>. Accessed on 31 March 2023.

[10] A. Funk & M. Niedermeyer, "Die Frankreich-Strategie des Saarlandes", *op.cit.*, p. 172. For a further discussion of the France Strategy as an impulse for renewed coordination, see N. Crossey & F, Weber, "Die Frankreichstrategie des Saarlandes als multisektoraler Impuls für die grenzüberschreitende Zusammenarbeit" *In* K. Mangels & D. Briegel (eds.), *Strategien und Konzepte grenzüberschreitender Raumentwicklung ung Raumplanung*, UniGR-Center for Border Studies, Kaiserlautern, 2022, pp. 9–23.

[11] Bundesrepublik Deutschland, République Française, "Vertrag zwischen der Bundesrepublik Deutschland und der Französischen Republik über die deutsch-französische Zusammenarbeit und Integration (short: Treaty of Aachen)", 2019. See Art. 13 and 14 in particular.

[12] Bundesrepublik Deutschland, Französische Republik, Gemeinsame Absichtserklärung zwischen der Regierung der Bundesrepublik Deutschland und der Regierung der Französische Republik über die grenzüberschreitende Zusammenarbeit in den deutsch-französischen Grenzregionen. Retrieved from: <https://www.agz-cct.eu/agz-cct-de/dokumente?openAccordionId=item-2459684-1-panel>. Accessed 6 December 2021.

between national pandemic management on the one hand, and border-regional particularities and interwovenness on the other[13] emphasised the relevance of a more pronounced "multiscalarity"[14] of French-German cross-border governance.

It is within this context of a current "momentum" of French-German cross-border cooperation,[15] reflected first in the France Strategy and substantiated further by the provisions of the Aachen Treaty and the experiences of the Covid-19 pandemic, that we seek to explore current trajectories of a cross-border multilevel governance from a "bottom-up perspective". Here, we draw particular attention to the contributions of French and German municipalities, whose longstanding cross-border experiences and whose dual role as "mouthpieces" for local concerns and "multipliers" of regional policies make them indispensable contributors to (border-)regional strategies. Our theoretical-conceptual approach stems on the one hand from a conceptualisation of cross-border governance in terms of a "multilevel governance"[16] which distinguishes between more stable, broad-scope territorial governance, and more flexible task-specific functional governance arrangements. In practice, along European border regions, we can often find hybrid rather than ideal types,[17] culminating in "a diverse set of arrangements, a panoply of systems of coordination

---

13 F. Weber, R. Theis & K. Terrolion, *Grenzfahrungen | Expériences transfrontalières. COVID-19 und die deutsch-französischen Beziehungen | Les relations frano-allemandes à l'heure de la COVID-19*, Springer VS, Wiesbaden, 2021; F. Weber, "Cross-border cooperation in the border region of Germany, France, Luxembourg in times of COVID-19", *European Societies*, vol. 24, n°3, 2022, pp. 354–381.

14 A. Marchetti, "Durchbruch für die dezentrale deutsch-französiche Zusammerabeit? Perspektiven nach dem Vertrag con Aachen," *Visions franco-allemandes*, n°30, IFRI, Retrieved from: www.ifri.org/sites/default/files/atoms/files/vfa_30_a._marchetti_zusammenarbeit_aachen_juin_2020.pdf. Accessed 25 February 2021.

15 *Ibid.*

16 L. Hooghe & G. Marks, "Unraveling the central state, but how? Types of multi-level governance", *American Political Science Review*, vol. 97, n°2, 2003, pp. 233–243; T. Chilla, E. Evrard & C. Schulz, "On the territoriality of cross-border cooperation: 'Institutional Mapping' in a multi-level context", *European Planning Studies*, vol. 20, n°6, 2012, pp. 961–980; A. Noferini, M. Berzi, F. Cammonita & A. Durà, "Cross-border cooperation in the EU: Euroregions amid multilevel governance and re-territorialization", *European Planning Studies*, vol. 28, n°1, 2020, pp. 35–56; S. Piattoni, "Multi-level governance: a historical and conceptual analysis", *Journal of European Integration*, vol. 31, n°2, 2009, pp. 163–180.

17 T. Chilla et al. "On the territoriality of cross-border cooperation", *op.cit.;* C. Fricke, "Spatial Governance across border revisited: Organizational forms and spatial

and negotiation, among formally independent but functionally interdependent entities that stand in complex relations to one another and that, through coordination and negotiation, keep redefining the interrelations".[18] On the other, we draw from a constructivist approach to borders as contingent and contextual (re- and de-) bordering processes.[19] From this approach follows a particular interest in "borderlands of governance" as the more or less institutionalised spaces and networks between territorial authorities within which idea(l)s about cross-border spatialities are exchanged, negotiated, and eventually perpetuated within horizontal and vertical policy networks. In our empirical approach, we were therefore particularly interested in the ideational and functional motivations behind cross-border activities, in the perceived challenges to the development of cross-border spatialities, and in perspectives on (the potential of) the Strategy to serve as an impulse for cross-border cooperation within the Saarland-Moselle region.

To explore these aspects, we adopted a combination of qualitative and quantitative methods, namely through 27 semi-guided expert interviews (from January 2020 to March 2021) with mayors from the Saarland (systematized as D1–D12) and the *département* Moselle (F1–F9), as well as additional interviews with Luxembourgian mayors (L1–L2) and administrative experts from the *département* Moselle (DM1), the *région* Grand-Est (RG1) and the *Landkreise* of the Saarland (LK1–LK3).[20] These insights were further complemented through online surveys of Saarland (n=23) and Moselle mayors (n=27). In the following, we discuss perspectives on cross-border spatialities, the experience of national pandemic management in the border regions, and persisting challenges to intercommunal cross-border cooperation, closing with an outlook on the potentials of (border-)regional strategies and other developments to address said challenges and on the gradual "territorialisation" of intercommunal cross-border cooperation.

---

planning in metropolitan cross-border regions", *European Planning Studies*, vol. 23, n°5, 2015, pp. 849–870.

18 S. Piattoni, "Multi-level governance", *op.cit.*, p. 172.

19 C. Wille, "Vom processual shift zum complexity shift: Aktuelle analytische trends der Grenzforschung", *In* D. Gerst, M. Klessmann & H. Krämer (eds.), *Grenzforschung. Handbuch für Wissenschaft und Praxis*, Nomos, Baden-Baden, 2021, pp. 106–121.

20 Two of these provided feedback by e-mail.

Intercommunal Cooperation in the Saarland-Moselle Region

Historically, border regions have experienced economic and social marginalisation due to their location at the fringes of the nation state, and to their recurring exposure to conflict and violence.[21] In the case of the Saarland-Moselle border region, this particular historical background of conflict, animosities and (hybrid) regional identification *vis-à-vis* shifting national affiliation, and more recently, of European integration processes provides a substantial source of motivation for intercommunal cross-border cooperation. For most interview partners (see e.g. F1, F2, F5, F8, D2, D4, D6, D10), considering the background of the territories as a former site of French-German enmity, as "tossed around by history, once by this side, once by the other" (F2), makes cross-border ties even more significant and important, not only as a testimonial to French-German reconciliation, but also as an active bottom-up "construction of Europe" (F2) through a "Europe of small ties" (F1) between municipal actors and border citizens, but also through improving the "on-the-ground" functional cooperation and exchange between the border spaces – "because for me, that's what Europe is all about, we need to implement it and translate it into concrete actions" (F5).

Throughout the past decades, processes of European integration and of overarching shifts towards from top-down government towards (regional) governance have additionally brought to the fore the potential of border regions as sites of economic flows and resources, which is similarly reflected in the tenets of the France Strategy presented in 2014. The interview partners widely share this perspective on the "border setting" as an "added value", noting here in particular the variety of shopping opportunities, sightseeing activities, cultural and culinary offers (e.g. D3, D6, D8, D10, F1, F6). There are however certain asymmetries in why and how people cross the border – German interview partners associate trips across the border with a certain "*savoir vivre*" which makes "life as a whole somewhat richer [...]. As soon as I cross the border, this feeling of something special sets in, even if it's just the bars, the shopping opportunities, the street signs, and [...] I have that right on my doorstep" (D6, see also D7, D10, D11, D12). French border citizens on the other hand

21 B. Caesar & K. Pallagst (eds.), *Cross-border territorial development – Challenges and opportunities*, UniGR-Centre for Border Studies, Laiserlautern, 2018.

"cross the border [...] to do their shopping on the other side" (F8) and to access other services because "for us [...], the big city is Saarbrücken, it's not Metz [which] is 60 kilometres away" (F9), pointing towards a notable asymmetry in cross-border flows and in how people make use of and perceive cross-border spatialities. This is further reflected in the (everyday) relevance the French and German mayors see for their municipalities in the proximity to the border (see Figure 1), wherein a large majority of French mayors see the border proximity as "very relevant" (44,4 %) or "relevant" (40,7 %), while these values are significantly lower for German mayors (21,7 % and 26,1 % respectively).

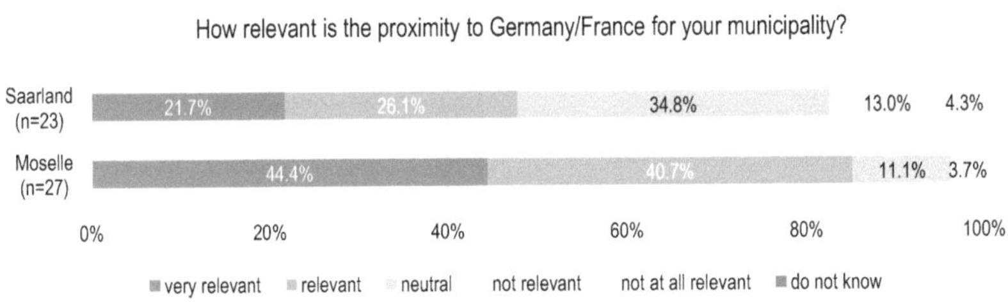

Figure 1. Relevance of the proximity to France or Germany for the municipalities of the border regions.[22]

How can these quite clear differences be interpreted? While German mayors perceive the "border setting" as a "huge opportunity" (D7) and a "stroke of luck" (D10), as "something special for the flair in the city" (D9), for the Moselle municipalities, the differential benefit lies in the role of the Saarland as a place of training, education, and employment, for shopping and other services,[23] and health care. In particular with regard to the latter aspect, French mayors see a certain necessity of improving and expanding existing cooperation projects,[24] notably the

22 Source: own survey and representation 2020.

23 IBA-OIE, "Grenzgängermobilität. Die Arbeitsmarktsituation in der Großregion", 2021. Retrieved from: <https://www.iba-oie.eu/fileadmin/user_upload/IBA-OIE/Publikationen/IBA-Berichte/12._Bericht_2020_21/2021_IBA_OIE_Grenzgaenger.pdf>. Accessed 9 June 2022.

24 Eurodistrict SaarMoselle, "Gesundheit: der Eurodistrict SaarMoselle will einen CORRIDOR MOSAR", 2021. Retrieved from: <https://www.saarmoselle.org/UserFiles/File/presse.de/cp-de-2021/pm-20211118-korridor-mosar-gesundheit-saarmoselle.pdf>. Accessed 2 June 2022. Saarländischer Rundfunk, "Aktueller bericht", 14 September 2022

exchange between the hospitals in Forbach and Völklingen for cardiologic emergencies "which already saves lives which is a significant progress [but] we need to go further" (F5). Within the quantitative survey, 88, 5 % of French mayors vis-à-vis 36,4 % of German mayors see further cooperation within the health care sector as a relevant objective of French-German cooperation.

These cross-border "realities", i.e. the "everydayness" of cross-border flows, as well as a certain ignorance of it at the level of national policymakers were highlighted throughout the first waves of the Covid-19 pandemic. The pandemic saw the (unilateral) reintroduction of border controls on the 16th of March 2020 and closures at the French-German border on the 19th of March 2020, following the classification of the entire *région* Grand Est as a "risk area" by the German Robert Koch Institute.[25] For many interview partners, the reappearance of the visible "border apparatus", the presence of armed German soldiers at the checkpoints, and in particular the lack of communication and coordination was perceived as "anachronistic" (D7), a "painful experience" (RG1, F8, F9), as "shameful" (D9) and a "shock" (D4, F2, F4, L1, DM1) – "the sanitary reasons we understand very well, but not the brutal way in which it was done" (F9). In light of the (heavily-criticised) lack of communication, both between Berlin and Paris, as well as with the regional, municipal and cross-border level – "from one day to the next, the border was closed, no one here was consulted, the Eurodistrict was not consulted, the municipalities weren't consulted, the *départements* weren't consulted" (F2). In other terms, mayors of the border regions, in particular of border municipalities or of municipalities with active town twinnings, reacted swiftly by organising mask deliveries and later patient transfers, joint protests at the border, and through activating vertical networks to the regional and national level (see e.g. D4, D5, D7, D11, F1, F2, F9, L2).[26]

---

25 Robert Koch-Institut, "Täglicher Lagebericht des RKI zur Coronavirus-Krankheit-2019 (COVID-19), Aktualisierter Stand für Deutschland", 11 March 2020. Retrieved from: <https://www.rki.de/DE/Content/InfAZ/N/Neuartiges_Coronavirus/Situationsberichte/2020-03-11-de.html>. Accessed 13 January 2022

26 M. Clivot, "Solidaritätserklärung der saarländischen Bürgermeisterinnen un Bürgermeister nach Frankreich", 2 April 2020. Retrieved from: <https://ec.europa.eu/futurium/en/system/files/ged/gemeinsam_gegen_corona-ensemble_contre_le_covid-19.pdf >. Accessed 20 May 2020; A. Jung & C. Arend, "Gemeinsam gegen Corona / Ensemble contre le Coronavirus. Eine deutsch-französische Initiative

These eventually culminated e.g. in the creation of a "border corridor" within which residents were exempt from certain testing regulations,[27] and the establishment of a French-German test centre "*Goldene Bremm*" in Spring 2021 (coordinated by the Eurodistrict SaarMoselle).[28]

While many interview partners (D9, D12, F4, F7, F9) were concerned about the resurfacing of animosities and hostilities, considering in particular the French-German history – "I got nasty emails from over there, [in which they] insulted us – I don't want to go back to 1940 now, but that's about where we are at the moment" (D12*), others on the other hand argued that "one swallow doesn't make a summer" (F1). These pointed instead to the many initiatives and demonstrations of solidarity, and the at-times "reinvigorating" effect of the pandemic.[29] In one case, a Luxembourgian mayor noted that "if the crisis had not come, I think that [cooperation between the municipalities] would have slowly but surely petered out. Sometimes there would still have been friendly meetings but we wouldn't have discussed functional issues. But now with this crisis, it's picking up again" (L1). For many interview partners, the experience emphatically underlined the relevance of municipal cross-border ties which are "years ahead of those between Paris and Berlin" (D2). Against this backdrop, the necessity of further integrating municipal cross-border activities and experiences into (border-)regional strategies becomes even more evident. The question now remains, which particular challenges remain and how these could be addressed in the context of regional strategies (Figure 2).

---

für Europa / Une initiative franco-allemande pour l'Europe", 2020. Retrieved from : <https://www.bundestag.de/resource/blob/690486/71bf2e61f17326109cccd2ea208c3840/kw15_schaeuble-ferrand_dt_frz_versammlung_erklaerung-data.pdf>. Accessed 17 August 2020.

27 "Saarland weicht von strenger Testpflicht bei Einreise ab. Lockerung im Grenzverkehr", *Saarbrücker Zeitung*, 13 January 2021..

28 H. Maillasson & M. Görtz, "Wie das Saarland mögliche Grenzschließungen verhindern will", *Saarbrücker Zeitung*, 20 February 2021..

29 N. Crossey, "COVID-19 als mögliche Chance. Gemeinden, grenzüberschreitende Beziehungen und die Frankreichstrategie des Saarlandes" *In* F. Weber, R. Theis & K. Terrolion (eds.), *Grenzerfahrungen*, *op.cit.*, pp. 401–415.

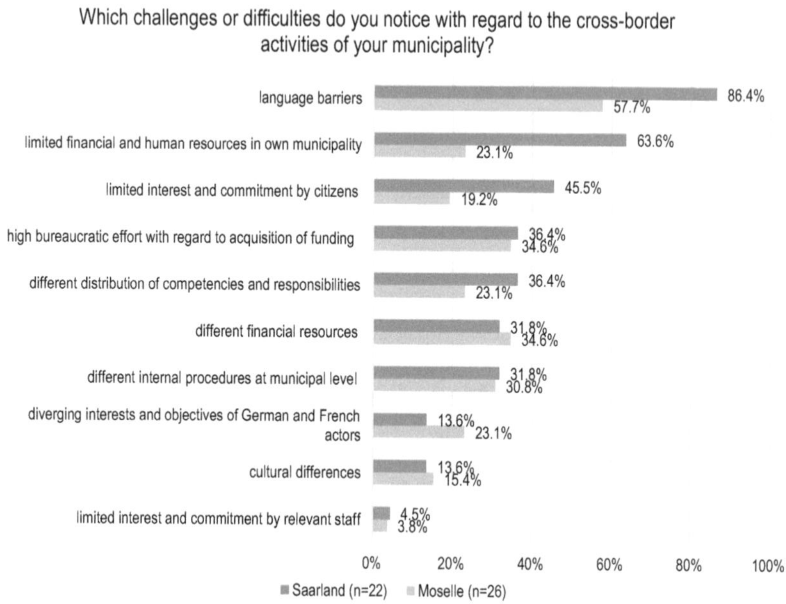

Figure 2. Challenges of intercommunal cooperation.[30]

Here, too, we can see both similarities and divergences of French and German mayors. For instance, language barriers continue to be the most important "hurdle" to cross-border exchange, although less so from a French perspective due to the more pronounced German and dialect competencies within the *département* Moselle – "if I look back over twenty years, I have never taken part in a single meeting where we had to resort to English [...] because our schools and families are still capable of speaking German" (F8). These competencies are, however, declining, which may affect in the long term on-the-ground cooperation – "we already notice the difference [and] that creates an additional barrier to cooperation" (D2). For many interview partners, this effectively underlines the necessity of border-regional strategies, such as the France Strategy, which promote (functional) multilingualism. These strategies should not only aim at facilitating cooperation on the ground, but also to further economic development "because it's a great economic advantage. The regions which are currently the most successful are the ones that are

30 Source: own survey and representation 2020.

perfectly bilingual. [...] There's a growing awareness there – with the France Strategy, the Germans are saying 'more languages, more opportunities', and I share that view" (F2). It is in this vein also that 53,8 % of French vis-à-vis 63,6 % of the German mayors participating in the quantitative survey see the expansion of bilingual education offers as "very relevant".

Further challenges are seen e.g. in the limited financial and human resources within the (German) municipal administrations in particular (63,6 % among German mayors vis-à-vis 23,1 % among French mayors), which further exacerbate the bureaucratic effort of fund acquisition (36,4 % and 34,6 %) as "no one in the municipalities has time to deal intensively [with this jungle of funds]" (D6). For this reason, several interview partners (e.g. D1, D3, D6, D8) call for the establishment of central "contact points" within the ministries to provide advice and support throughout the acquisition process, for the creation of (un-bureaucratic) funding opportunities for projects related to the France Strategy, noting as a potential "best practice-model" the newly-established Franco-German Citizens' Fund. Further proposals for the long-term implementation of the France Strategy include increasing the visibility of cross-border relations and of the France Strategy within the public sphere and within the "vertical networks" (D2, D6, D9), promoting the exchange of municipal actors and administrative staff (D7, D9), to enforce certain objectives as obligatory rather than voluntary (D2). This last point refers to emphasising more strongly the function of the regional level as both encouraging and demanding municipal contribution, seeing as the municipal level is the one that "actually has to implement the France Strategy" (D2, see also D4, D7, D9, F2).

Other challenges related to the persistence of territorially-bound regulations and procedures in a closely interwoven cross-border region go "beyond" the cooperation on the ground, and touch upon the role of cross-border structures within the French-German multilevel governance. A French mayor refers to these as "walls of glass that you don't necessarily see, like fiscal or administrative differences [...] which may be important throughout the implementation of a project" (F2). Here, intercommunal cross-border cooperation has in many cases benefitted from a remarkably high personal motivation of mayors and municipal staff to cooperate, as well as from at times close personal ties between border mayors and municipal staff. One exemplary case illustrating the relevance of these on-the-ground relations as well as the need for appropriate regulations can be found in the planning and participation process surrounding the construction of wind turbines on German ground but in immediate proximity to French municipalities, which was protested against by both French and German citizens due (in part) to an insufficient consideration of the effects on French residents (see F9, D9). Acknowledging that "it's not right that in German law [...] when planning something directly at the border, we pretend as if there's nothing on the other side" (D9), French municipal representatives who had close personal ties to German municipal groups were subsequently included in the German participation process. This was "of no legal value" (F9) but nonetheless mitigated tensions as it effectively shifted the understanding of the conflict from a conflict of interests between French and German border residents to a conflict between the "interwovenness" on the ground, and the lack of appropriate regulations and procedures at the European level: "Why hasn't the EU harmonised this yet?" (D9) "There is no European legislation for issues like these, [...] which is a bit regrettable. [...] There should be a harmonisation of norms and regulations which can be applied to the cross-border [context]" (F9). This example – one of several – underlines the lack of a "fit" between national regulations and planning procedures cultures on the one hand, and "border realities" on the other.[31] Here and in other cases, interview partners express their hope that the AGZ/CCT may work to "define and identify these many 'irritations' [and to] remove these constraints" (F2, see also D1, D4, LK1, DM1). At the same time, French interview partners in particular (see F2, F5, F6, DM1) call for a further devolution of competencies to the level of the *département* and/or the cross-border structures to facilitate the development of a genuine, integrated cross-border space.

> "I hope that this Eurodistrict [SaarMoselle] will be able to evolve into a real metropolis in the decades to come, with specific competences in economic development, tourism [...] and mobility, so that there will be a real coherence between our policies on both sides of the border – because we live in a

---

31 B. Caesar & E. Evrard, "Planungskulturelle Vielfalt in Grenzräumen – Theoretische und methodische Ansätze zur grenzüberschreitenden Raumplanung", *In* F. Weber, C. Wille, B. Caesar & J. Hollstegge (eds.), *Geographien der Grenzen. Räume – Ordnungen – Verflechtungen. (Räume – Grenzen – Hybriditäten),* Springer VS, Wiesbaden, 2020, pp. 95–116; K. Pallagst & A. Hartz, "Raumplanung in Grenzregionen: Gratwanderung zwischen neuen Leitbildern und alten Planungstraditionen?" *In* K. Pallagst & A. Hartz (eds.), *Border Futures – Zukunft Grenze – Avenir Frontière,* Verlag der ARL, Hannover, 2018, pp. 70–87.

shared '*bassin de vie*' which is a daily reality but which, unfortunately, is not recognized by our institutions". (F5)

It is against the background of such and other experiences that municipal actors quite emphatically support the further strengthening of a border-regional representation at the national and European level. Within the quantitative survey, 80,8 % of the French and 63,6 % of the German mayors see further (joint) representation as "very relevant", for instance through the existing joint offices in Paris, Berlin and Brussel, or within the context of the AGZ/CCT.

Conclusion and Outlook

Which insights with regard to current trajectories of intercommunal cooperation and to opportunities provided by (border-)regional strategies can we draw from this admittedly cursory oversight? For one, we can take note that among border municipalities and even those further away with more active town twinnings we can in most cases see a degree of high personal motivation to contribute to French-German relations "on the ground", and beyond that, to a bottom-up construction of a "small-scale Europe", which would significantly benefit from financial and/or personnel support by the regional level. In particular in closer proximity to the "urban hubs" of the border regions, municipal authorities are tackling increasingly complex and functional cooperation projects, pertaining to issues of mobility, health care, tourism or spatial development. Here, the emergence of a "genuine" cross-border space – and with that, French-German public services such as joint day-care facilities, integrated mobility networks, and so on – is somewhat impeded by territorially-bound administrative or legal differences. These differences may potentially be addressed through adapted legal and administrative provisions including derogations as proposed by the Aachen Treaty, which are to be formulated by the AGZ/CCT. Within this context, we can see a certain "momentum" of a simultaneous "bottom-up" processes within which municipal and regional authorities and cross-border structures municipalities, intercommunal cross-border structure adopt a certain "hinge function" (D9) to communicate persisting challenges and to raise awareness of border-specific challenges within national arenas of policy-making. On the other hand, the creation of the AGZ/CCT through the Aachen Treaty "broke up," to some extent, the classical hierarchical structures of communication in favour of a decentralised, multilevel structures of cross-border governance.[32]

Contrasting our empirical insights with our conceptual interest in cross-border multilevel governance, we can over time – and in line with the expanding financial and legal opportunities – observe a gradual "territorialisation" of intercommunal cooperation. . In many border regions, legal instruments, – such as European Economic Interest Groupings (EEIG) introduced in 1985, European Groupings of Territorial Cooperation (EGTC) introduced 2006, accompanied by European and bi- or trilateral treaties such as the Karlsruhe Agreement of 1997 – and financial support – especially, as of 1990, through the Interreg programmes – supported and incentivised the progression of intercommunal cross-border cooperation from singular, often comparably simple and non-committal projects to venture into a broad range of policy-fields.[33]

Cross-border structures such as the Eurodistrict SaarMoselle increasingly correspond to the "ideal type" of territorial governance. However, while there is a significant political motivation among municipal authorities to expand and intensify cooperation[34], they have no mandate or competencies of their own, but are limited by territoriality as the "organising principle of political and social life".[35] As Brunet-Jailly noted, "studying borderland regions is about studying bottom-up processes where local and regional […] political clout, politics and cross-border governance are tightly woven into complex cultural, economic and political structures".[36] Our insights further emphasise how municipal "borderlands of governance" – as an assemblage of networks within which idea(l)s about cross-border spatialities, about affiliation beyond borderlines are

32 A. Marchetti, "Durchbruch für die dezentrale deutsch-französische Zummenarbeit?", *op.cit.*

33 B. Reitel, B. Wassenberg & J. Peyrony. "The INTERREG experience in bridging European territories. A 30-year summary" *In* E. Medeiros (eds.), *European Territorial Cooperation. Theoretical and Empirical Approaches to the Process and Impacts of Cross-Border and Transnational Cooperation in Europe*, Springer International Publishing, Cham, 2018, pp. 7–23.

34 T. Chilla et al., "On the territoriality of cross-border cooperation", *op.cit.*

35 J. Anderson & L. O'Dowd, "Borders, border regions and territoriality: Contradictory meanings, changing significance", *Regional Studies*, vol. 33, n°7, 1999, p. 593.

36 E. Brunet-Jailly, "Power, politics and governance of borderlands", *In* P. Gillis, H. Koff, C. Maganda & C. Schulz (eds.), *Theorizing borders through analyses of power relationship*, P.I.E. Peter Lang, Brussels, 2013, pp. 29–47.

shared and adapted – have both a relevant "horizontal" and "vertical" dimension. On the one hand, notions of cross-border communalities significantly inform and substantiate on-the-ground cooperation and (border-)regional strategies, and are highly relevant to facilitate cross-border development, to navigate conflicts, and so on. On the other hand, these more or less formalised networks are inextricably linked to and in communication with similar (or diverging) notions among higher political-administrative levels. This is evident for instance throughout the first waves of pandemic management. Here, the new "multiscalarity" of French-German cooperation structures may provide new opportunities to represent these cross-border spaces as relevant spaces of political action, and to strengthen their competencies. The establishment of border-specific procedures and regulations might be a step towards the emergence of genuine cross-border spatialities which complement, rather than contradict classical state structures, which are "no longer the only imaginable basis for creating and defining primary political communities and institutions".[37]

Bibliography

J. Anderson & L. O'Dowd, "Borders, border regions and territoriality: Contradictory meanings, changing significance", *Regional Studies*, vol. 33, n°7, 1999, pp. 593–604.

J. Beck, "Cross-border cooperation in Europe as an object in transdisciplinary research. An introduction", *In* J. Beck (ed.), *Transdisciplinary Discourses on Cross-Border Cooperation in Europe*, Peter Lang, Brussels, 2019, pp. 13–29.

J. Blatter, "From 'spaces of place' to 'spaces of flows'? Territorial and functional governance in cross-border regions in Europe and North-America", *International Journal of Urban and Regional Research*, vol. 28, n°3, 2004, pp. 530–548.

[37] J. Blatter, "From 'spaces of place' to 'spaces of flows'? Territorial and functional governance in cross-border regions in Eruope and North-America", *International Journal of Urban and Regional Research*, vol. 28, n°3, 2004, p. 545; S. Piattoni, "Exploring European Union marco-regional strategies through the lens of multilevel governance", *In* S. Gänzle & K. Kern (eds.), *A 'Macro-Regional' Europe in the Making. Theoretical Approaches and Empirical Evidences*, Palgrave Macmillan, Basingstoke, 2016, pp. 75–97.

E. Brunet-Jailly, "Power, politics and governance of borderlands", *In* P. Gillis, H. Koff, C. Maganda & C. Schulz (eds.), *Theorizing Borders through Analyses of Power Relationship*, P.I.E. Peter Lang, Brussels, 2013, pp. 29–47.

B. Caesar & E. Evrard, "Planungskulturelle Vielfalt in Grenzräumen – Theoretische und methodische Ansätze zur grenzüberschreitenden Raumplanung", *In* F. Weber, C. Wille, B. Caesar & J. Hollstegge (eds.), *Geographien der Grenzen. Räume – Ordnungen – Verflechtungen. (Räume – Grenzen – Hybriditäten),* Springer VS, Wiesbaden, 2020, pp. 95–116.

B. Caesar & K. Pallagst (eds.), *Cross-border Territorial Development – Challenges and opportunities*, UniGR-Centre for Border Studies, Laiserlautern, 2018.

N. Crossey, "COVID-19 als mögliche Chance. Gemeinden, grenzüberschreitende Beziehungen und die Frankreichstrategie des Saarlandes" *In* F. Weber, R. Theis & K. Terrolion (eds.), *Grenzerfahrungen*, *op.cit.*, pp. 401–415.

N. Crossey & F. Weber, "Zur Konstitution multipler *Borderlands* im Zuge der Franreichstrategie des Saarlandes" *In* F. Weber, C. Wille, B. Caesar & J. Hollstegge (eds.), *Geographien der Grenzen. Raüme – Ordnunen – Verflechtungen*, Springer VS, Wiesbaden, pp. 145–166.

C. Fricke, "Spatial Governance across border revisited: Organizational forms and spatial planning in metropolitan cross-border regions", *European Planning Studies*, vol. 23, n°5, 2015, pp. 849–870.

A. Funk & M. Niedermeyer, "Die Frankreich-Strategie des Saarlandes. Governanc eund territoriale Kohäsion als neue Dimensionen regionaler Entwicklungskonzepte" *In* W. H. Lorig, S. Regolot & S. Henn (eds.), *Die Großregion SaarLorLux. Anspruch, Wirklichkeiten, Perspektiven*, Springer VS, Wiesbaden, 2016, pp. 169–183.

A. Funk & M. Niedermeyer, "Die Frankreich-Strategie des Saarlandes", *op. cit.*, p. 172. For a further discussion of the France Strategy as an impulse for renewed coordination, see N. Crossey & F, Weber, "Die Frankreichstrategie des Saarlandes als multisektoraler Impuls für die grenzüberschreitende Zusammenarbeit" *In* K. Mangels & D. Briegel (eds.), *Strategien und Konzepte grenzüberschreitender Raumentwicklung ung Raumplanung*, UniGR-Center for Border Studies, Kaiserlautern, 2022, pp. 9–23.

L. Hooghe & G. Marks, "Unraveling the central state, but how? Types of multi-level governance", *American Political Science Review*, vol. 97, n°2, 2003, pp. 233–243; T. Chilla, E. Evrard & C. Schulz, "On the territoriality of cross-border cooperation: 'Institutional Mapping' in a multi-level context", *European Planning Studies*, vol. 20, n°6, 2012, pp. 961–980.

D. Hüser, "Die Saar-Politik und die deutsch-französische Geschichte an der Grenze", *In* H. J. Lüsebrink, C. Polzin-Haumann & C. Vatter (eds.), *Alles Frankreich oder was – Die saarländische Frankreichstrategie im europäischen Kontext / La France à toutes les sauces? – La stratégie France de la Sarre dans le contexte européen*, Transcript Verlag, Bielefeld, 2017, pp. 39–67

W. H. Lorig, S. Regolot & S. Henn (eds.). *Die Großregion SaarLorLux. Anspruch, Wirklichkeiten, Perspektiven*, Springer VS, Wiesbaden, 2016; C. Wille (ed.), *Lebenswirklichkeiten und politische Konstruktionen in Grenzregionen. Das Beispiel der Großregion SaarLorLux: Wirtschaft-Politik-Alltag-Kultur*, Transcript Verlag, Bielefeld, 2015.

A. Marchetti, "Durchbruch für die dezentrale deutsch-französiche Zusammerabeit? Perspektiven nach dem Vertrag con Aachen," *Visions franco-allemandes*, n°30, IFRI.

A. Noferini, M. Berzi, F. Cammonita & A. Durà, "Cross-border cooperation in the EU: Euroregions amid multilevel governance and re-territorialization", *European Planning Studies*, vol. 28, n°1, 2020, pp. 35–56; S. Piattoni, "Multi-level governance: A historical and conceptual analysis", *Journal of European Integration*, vol. 31, n°2, 2009, pp. 163–180.

K. Pallagst & A. Hartz, "Raumplanung in Grenzregionen: Gratwanderung zwischen neuen Leitbildern und alten Planungstraditionen?" *In* K. Pallagst, A. Hartz (eds.), *Border Futures – Zukunft Grenze – Avenir Frontière*, Verlag der ARL, Hannover, 2018, pp. 70–87.

S. Piattoni, "Exploring European Union marco-regional strategies through the lens of multilevel governance", *In* S. Gänzle & K. Kern (eds.), *A 'Macro-Regional' Europe in the Making. Theoretical Approaches and Empirical Evidences*, Palgrave Macmillan, Basingstoke, 2016, pp. 75–97.

B. Reitel, P. Pupier & B. Wassenberg, "European policies and border cities", *In* E. Medeiros (ed.), *Border Cities and Territorial Development*, Routledge, New York, 2021, pp. 59–75.

B. Reitel, B. Wassenberg & J. Peyrony, "The INTERREG experience in bridging European territories. A 30-year summary", *In* E. Medeiros (ed.), *European Territorial Cooperation. Theoretical and Empirical Approaches to the Process and Impacts of Cross-Border and Transnational Cooperation in Europe*, Springer, Dordrecht, 2018, pp. 7–23.

K. Terrollion & C. Klos, "Die Frankreichstrategie des Saarlandes. Ein deutsches Bundesland auf dem Weg zu einer gelebten europäischen Mehrsprachigkeit", *Notes du Cerfa*, n°168, IFRI.

H. Uterwedde, "Die saarländische Frankreichstrategie im Kontext der aktuellen deutsch-französischen Beziehungen", *In* H. J. Lüsebrink, C. Polzin-Haumann, & C. Vatter (eds.), *Alles Frankreich oder was, op.cit.*, pp. 27–39.

F. Weber, "Grenzüberschreitende Governance in der Großregion SaarLorLux", *Forum Wohnen und Stadtentwicklung*, vol. 4, 2020, pp. 183–186.

F. Weber, R. Theis & K. Terrolion, *Grenzfahrungen | Expériences transfrontalières. COVID-19 und die deutsch-französischen Beziehungen | Les relations frano-allemandes à l'heure de la COVID-19*, Springer VS, Wiesbaden, 2021; F. Weber, "Cross-border cooperation in the border region of Germany, France, Luxembourg in times of COVID-19", *European Societies*, vol. 24, n°3, 2022, pp. 354–381.

C. Wille, "Vom processual shift zum complexity shift: Aktuelle analytische trends der Grenzforschung", *In* D. Gerst, M. Klessmann & H. Krämer (eds.), *Grenzforschung. Handbuch für Wissenschaft und Praxis*, Nomos, Baden-Baden, 2021, pp. 106–121.

A Living Lab of European Integration? Cross-Border Developments in the Greater Region during the Covid-19 Pandemic

Julia Dittel & Florian Weber

Introduction: The Covid-19 Pandemic as a Caesura for European Integration in Border Regions?

Since the end of the Second World War, many steps have been taken within Europe towards closer European integration. The development of the European Union (EU) testifies to this. Even though crises, such as the economic crisis, the refugee crisis, or nationalist tendencies caused the emergence of new challenges in the first decades of the 21st century, the open internal borders as an achievement of the Schengen Agreement of 1985, among others, remained a constant.[1] The spread of the SARS-CoV-2 virus in spring 2020 however caused this normality, which was believed to be a given, to falter, as freedom of movement on a global scale was severely restricted in a very short time. In Europe, too, political decision-makers saw the need to act quickly and took extraordinary and far-reaching measures to combat the pandemic. For as the virus spread, so did its (perceived) threat. The internal borders of the Schengen area were suddenly systematically controlled or – in parts – even closed in order to counteract a further spread of the virus. They became a means to protect the population of the respective nation states from the

---

[1] H. Böhm, "Five Roles of Cross-border Cooperation Against Re-bordering", *Journal of Borderlands Studies*, 36(5), 2021, p. 1; J. Jańczak, "Integration De-scaled. Symbolic Manifestations of Cross-border and European Integration in Border Twin Towns", *Journal of Borderlands Studies*, 33(3), 2018, p. 393 ff.; E. Medeiros, M. Guillermo Ramírez, G. Ocskay & J. Peyrony, "Covidfencing effects on cross-border deterritorialism: the case of Europe", *European Planning Studies*, 29(5), 2021, p. 963.

"undesirable" – the Covid-19 virus.[2] Borders were "clearly back in a way that may have seemed unlikely prior to the outbreak"[3] – a "rebordering" or "covidfencing" as Medeiros *et al.* (2021) put it.[4]

The border regulations during the Covid-19 pandemic led to an increased awareness of their political and social significance as well as the power of borders, which today often no longer have their original, demarcating function, especially within Europe. In the EU, they have become more permeable for people, goods, services, and capital. Borders have become detached from maps and – following a constructivist understanding – can be interpreted as social constructs that do not exist "simply" and "fixedly", but are subject to constant change.[5] Before the pandemic, open internal EU borders and "the political support of the interlocking, convergence and intertwining capacity of national peripheries across European borders"[6] had been considered almost irrefutable features of the Schengen area. Within the EU, borders and border areas – borderlands – had thus become centres of growing interdependencies themselves.[7] These regions are characterised by a high degree of exchange

---

2 A. Alemanno, "The European Response to COVID-19: From Regulatory Emulation to Regulatory Coordination?", *European Journal of Risk Regulation*, 11(2), 2020, p. 307; B. Wassenberg, B., "'Return of Mental Borders': A Diary of COVID-19 Closures between Kehl, Germany, and Strasbourg, France", *Borders in Globalization Review*, 2(1), 2020, pp. 114–120.

3 S. M. Radil, J. Castan Pinos & T. Ptak, "Borders resurgent: Towards a post-Covid-19 global border regime?", *Space and Polity,* 25(1), 2021, p. 132.

4 E. Medeiros *et al.*, "Covidfencing effects on cross-border deterritorialism: the case of Europe", *op.cit.*

5 P. Ulrich, "Re-Figuration von Grenzen und Ordnungen im sozialen Raum: Konzeptualisierung eines Analysemodells partizipativer Governance in EU-Grenzregionen", *In* F. Weber, C. Wille, B. Caesar & J. Hollstegge (eds.), *Geographien der Grenzen: Räume – Ordnungen – Verflechtungen*, Springer VS, Wiesbaden, 2020, p. 169; F. Weber, C. Wille, B. Caesar & J. Hollstegge, "Entwicklungslinien der Border Studies und Zugänge zu Geographien der Grenzen", *In* F. Weber et al. (eds.), *Geographien der Grenzen, op.cit.*, p. 4 f.

6 P. Ulrich, "Re-Figuration von Grenzen und Ordnungen im sozialen Raum", *op.cit.*, p. 168. In German: "die politische Unterstützung der Verzahnung, Konvergenz und Verflechtungsleistung nationaler Peripherien über europäische Grenzen hinweg".

7 A. Cooper & C. Rumford, "Monumentalising the Border: Bordering Through Connectivity", *Mobilities,* 8(1), 2013, pp. 107–124; T. Kuhn, "Europa ante portas: Border residence, transnational interaction and Euroscepticism in Germany and France", *European Union Politics*, 13(1), 2012, pp. 94–117; C. Rumford, "Introduction: Theorizing Borders", *European Journal of Social Theory*, 9(2), 2006, pp. 155–169.

between the regions involved. In a highly interconnected economic area such as the EU, the partly uncoordinated national decisions on stricter border regulations adopted in spring 2020 led to serious disruptions and ran counter to the achievements of European integration in a central way.[8] The aforementioned border regions, which are characterised by extensive economic and social interconnections and cross-border mobility, were particularly affected by the consequences of unilateral decisions;[9] Their inhabitants could suddenly actively experience nation-state borders as barriers and as "real".

Against this introductory background, we will examine the effects of the rebordering processes with a focus on the so-called Greater Region in the border region between Belgium, Germany, France, and Luxembourg. In comparison with other cross-border regions in Europe,[10] the historically developed cooperation area is characterised

---

8 T. Albers, V. Köbele-Ennaji, J. Ross & V. Wolfart, "Nationale Reflexe im Angesicht einer europäischen Krise: Die Kontrollen an der deutsch-französischen Binnengrenze während der ersten Welle der COVID-19-Epidemie", *In* F. Weber, R. Theis & K. Terrolion (eds.), *Grenzerfahrungen/Expériences transfrontalières: COVID-19 und die deutsch-französischen Beziehungen/Les relations franco-allemandes à l'heure de la COVID 19*, Springer VS, Wiesbaden, 2021, pp. 367–378; F. Weber & C. Wille, "Grenzgeographien der COVID-19-Pandemie", *In* F. Weber et al. (eds.), *Geographien der Grenzen, op.cit.*, p. 204.

9 J. Beck, H.-J. Drewello, M. Frey, J. Peyrony, B. Reitel, A. Thevenet, B. Wassenberg & F. Weber, "Manifesto on the future of cross-border cooperation in Europe", 2021, p. 1. Retrieved from: transfrontier.eu/site/wp-content/uploads/2021/10/ManifestoGB.pdf. Accessed 1 March 2022; E. Medeiros et al., "Covidfencing effects on cross-border deterritorialism: the case of Europe", *op.cit.;* Mission opérationnelle transfrontalière (MOT) & EU, "The effects of COVID-19 induced border closures on cross-border regions: 20 case studies. Covering the period March to June 2020", 2021. Retrieved from: op.europa.eu/en/publication-detail/-/publication/bf14de68-6698-11eb-aeb5-01aa75ed71a1/language-en. Accessed 21 February 2022; MOT & EU, "The effects of COVID-19 induced border closures on cross-border regions: An empirical report covering the period March to June 2020", 2021. Retrieved from: op.europa.eu/en/publication-detail/-/publication/46250564-669a-11eb-aeb5-01aa75ed71a1/language-en. Accessed 21 February 2022.

10 H. Bähr, H. Schenke & D. Filsinger, "Transnationale Netzwerke und Institutionen in der Großregion SaarLorLux", *In* J. Meyer & L. Rampeltshammer (eds.), *Grenzüberschreitendes Arbeiten in der Großregion SaarLorLux*, Saarbrücken: Universaar, Saarbrücken, 2012, p. 17.

by a high degree of interdependence and is considered the strongest cross-border labour market region in the EU.[11] Cross-border mobility was a part of normal life in the Greater Region before the pandemic.[12] The Covid crisis caused a cesura in cross-border cooperation and revealed its weaknesses. At the same time, it underlined the relevance of cross-border cooperation within the EU. This is why in this context we ask to what extent the positioning of the European Commission in 2021,[13] which in the light of the crisis described border regions as living labs of European integration, can be supported. To contextualise, we firstly present a chronology of the events of spring 2020. We then elaborate on the rebordering and debordering processes that took place in the context of the first wave as well as their practical and symbolic implications for cross-border cooperation. Exemplarily selected measures to combat the pandemic taken during later pandemic phases in the Greater Region are then described. With reference to the presented consequences of the global health crisis on cross-border regions, we conclude by discussing the lessons learned from the events as well as the resulting opportunities for cross-border cooperation and networking in favour of a strengthened cohesion within the EU.

---

11 H. P. Dörrenbächer, "Die Entwicklung grenzüberschreitender Berufsausbildung im Spannungsfeld unterschiedlicher Dimensionen von Distanz – das Beispiel der Großregion", *In* F. Weber et al. (eds.), *Geographien der Grenzen, op.cit.*, p. 119.

12 *Ibid.* See also H. Bähr, H. Schenke & D. Filsinger "Transnationale Netzwerke und Institutionen in der Großregion SaarLorLux", *op.cit.;* F. Weber, "Grenzüberschreitende Governance in der Großregion SaarLorLux", *Forum Wohnen und Stadtentwicklung* 4, 2020, pp. 183–186; C. Wille, "Grenzüberschreitende Alltagspraktiken in der Großregion SaarLorLux: eine Bestandsaufnahme", *In* C. Wille (ed.), *Lebenswirklichkeiten und politische Konstruktionen in Grenzregionen: Das Beispiel der Großregion SaarLorLux: Wirtschaft – Politik – Alltag – Kultur*, Bielefeld: transcript Verlag, Bielefeld, 2015, pp. 133–156.

13 European Commission, "Report from the Commission to the European Parliament, the Council, the European Economic and Social Committee and the Committee of the Regions: EU Border Regions: Living labs of European integration", COM(2021) 393 final, 2021. Retrieved from: eur-lex.europa.eu/legal-content/EN/TXT/PDF/?uri=CELEX:52021DC0393&from=DE. Accessed 10 October 2021.

The First Wave of the Covid-19 Pandemic in 2020 as a Shock for Interconnected Borderlands

Covid-19 and Border Regions – An Introductory Overview

In December 2019, medical experts in the Chinese city of Wuhan became aware of an increased number of respiratory illnesses that could be attributed to a novel virus – SARS-CoV-2. On 23 January 2020, the city itself, and one day later the entire Hubei province, were quarantined to prevent a further spread of the virus. By this time, however, the pathogen had already spread.[14] In Europe, those responsible did not initially assume there was a threat – not even after individual cases were detected in Germany, France, and Italy. Infection routes of infected people were still traceable to China, the "country of origin". As a result of this localisation, the virus still seemed quite distant, both spatially and considering people's everyday lives.[15] Only when the number of cases in northern Italy and the French *région* Grand Est began to rise in spring 2020, and the situation in hospitals threatened to get out of control, the risk perception in Germany changed.[16] Due to the rapid spread of the virus, the disease was reclassified within a short time from a regionally localised health crisis to a pandemic.[17] Governments in Europe hastily responded to the accelerating speed at which the virus spread with restrictions on movement and eventually with increased border controls and even the closure of national borders.[18]

---

[14] E. Brunet-Jailly & M. J. Carpenter, "Introduction to the special issue: Borderlands in the era of COVID-19", *Borders in Globalization Review*, 2(1), 2020, pp. 7–11; S. L. Greer, E. J. King, E. M. d. Fonseca & A. Peralta-Santos (eds.), *Coronavirus Politics: The Comparative Politics and Policy of COVID-19*, University of Michigan Press, Ann Arbor, 2021; C. Wille & R. Kanesu (eds.), *Bordering in Pandemic Times: Insights into the COVID-19 Lockdown*, UniGR-Center for Border Studies, Luxemburg, Trier, 2020.

[15] Weber, F. & C. Wille, "Grenzgeographien der COVID-19-Pandemie", *op.cit.*, p. 192.

[16] *Ibid.*, p. 193.

[17] World Health Organization, "WHO Director-General's opening remarks at the media briefing on COVID-19 – 11 March 2020", 2020. Retrieved from: www.who.int/director-general/speeches/detail/who-director-general-s-opening-remarks-at-the-media-briefing-on-covid-19---11-march-2020. Accessed 3 March 2022.

[18] Medeiros et al., "Covidfencing effects on cross-border deterritorialism: The case of Europe", *op.cit.*

In the light of the rapid increase in cases in that area, the French *région* Grand Est, which borders Germany, Luxembourg, Belgium, and Switzerland, was declared a risk area (*Risikogebiet*) by the German Robert Koch-Institute on 11 March 2020.[19] In response, the German and French ministers of internal affairs decided on 15 March that border crossings should be prevented as far as possible in order to slow down the spread of the virus.[20] After Denmark, the Czech Republic, and Poland had already closed their borders with Germany in the days before, the Federal Republic introduced checks on persons at the borders with France and Luxembourg, among others, on 16 March 2020.[21] France carried out checks at the border to the Grand Duchy, while Luxembourg refrained from border controls to its neighbouring states. On the same day, EU Commission President Ursula von der Leyen proposed a 30-day entry ban for the EU for the first time, thus regulating the external borders.[22] Three days later, the Federal Republic of Germany decided to partially close its borders with France and Luxembourg, among others. For private traffic, certain border crossing points were defined at which e.g. cross-border workers could continue to travel, while other border crossings were closed off by partly temporary barriers.[23] On 20 March 2020,

19 Robert Koch-Institut, "Täglicher Lagebericht des RKI zur Coronavirus-Krankheit-2019 (COVID-19): 11.03.2020 – Aktualisierter Stand für Deutschland", 2020, p. 5. Retrieved from: www.rki.de/DE/Content/InfAZ/N/Neuartiges_Coronavirus/Situationsberichte/2020-03-11-de.html. Accessed 13 January 2022.

20 I. Pigeron-Piroth, E. Evrard & R. Belkacem, "Marché du travail transfrontalier: négocier avec les frontières à l'heure de la crise sanitaire COVID-19", *In* C. Wille & R. Kanesu (eds.), *Bordering in Pandemic Times*, *op.cit.*, p. 20.

21 Bundesministerium des Innern (BMI), "Vorübergehende Grenzkontrollen an den Binnengrenzen zu Österreich, der Schweiz, Frankreich, Luxemburg und Dänemark: Pressemitteilung", 2020. Retrieved from: https://www.bundespolizei.de/Web/DE/04Aktuelles/01Meldungen/2020/03/200315_grenzkontrollen_corona.html. Accessed 27 March 2020.

22 European Commission, "COVID-19: Guidelines for border management measures to protect health and ensure the availability of goods and essential services", C(2020) 1753 final, 2020. Retrieved from: ec.europa.eu/home-affairs/sites/homeaffairs/files/what-we-do/policies/european-agenda-migration/20200316_covid-19-guidelines-for-border-management.pdf. Accessed 10 October 2021.

23 BMI, "Grenzübertritt nur noch an bestimmten Grenzübergangsstellen: Pressemitteilung", 2020. Retrieved from: www.bmi.bund.de/SharedDocs/pressemitteilungen/DE/2020/03/festlegung-grenzuebergangsstellen.html. Accessed 27 March 2020;

Belgium, for its part, introduced border controls with Germany, while no checks were carried out by Germany with this neighbouring country.[24]

These exemplary measures already illustrate the inconsistency and poor coordination of the largely unilateral, national decisions on border controls and border closures, which were adopted gradually – in a "domino effect", so to speak.[25] In this way, freedom of movement within the EU was severely restricted within a few days. Suddenly, different border crossings within the Greater Region were limited to valid reasons such as the mobility of workers and the movement of goods and merchandise, while border crossings for other purposes, such as shopping or visiting family members, were no longer permitted. Pupils and students did not benefit from the same regulation as workers – they were mostly prevented from crossing the border.[26] The uncoordinated national decisions on border controls and closures, which in some cases were taken in an almost panicked manner, made it difficult or even impossible to reach agreements in favour of an EU-wide approach. The lack of coordination and arrangement added to the population's uncertainty and confusion and soon became the main aspect of criticism in the media.[27] Three months of increased border controls within the Schengen area resulted in a number of serious economic disruptions and social tensions, demonstrating "how fragile cross-border cooperation still is after several decades".[28] Against this background, it is hardly surprising that especially in border regions,

---

H. Thewes, "Flatterband und Barrikaden: Grenzschließungen in der Pandemie", *In* F. Weber, R. Theis & K. Terrolion (eds.), *Grenzerfahrungen, op.cit.*, pp. 195–200.

24 "Coronavirus: la Belgique ferme ses frontières sauf pour le fret ou un motif essentiel", *Le Soir*, 20 March 2020.

25 For a detailed overview of the regulations on border regulations, see MOT & EU, "The effects of COVID-19 induced border closures on cross-border regions: An empirical report", *op.cit.*; T. Steffen, "Corona-Überblick: Diese Corona-Maßnahmen gelten in den EU-Staaten", *zeit online*, 20 March 2020.

26 MOT& EU, "The effects of COVID-19 induced border closures on cross-border regions: An empirical report", *op.cit.* p. 86.

27 C. Dylla, "Drei Monate 'Grenzerfahrungen': Was bleibt: Betrachtungen aus der Medienperspektive", *In* F. Weber, R. Theis & K. Terrolion (eds.), *Grenzerfahrungen, op.cit.,* pp. 261–283; S. Freitag-Carteron, "Deutschland, Frankreich, COVID-19 – das Virus und die Grenzregion", *In* F. Weber, R. Theis & K. Terrolion (eds.), *Grenzerfahrungen, op.cit.,* pp. 295–308; Weber, F. & C. Wille, "Grenzgeographien der COVID-19-Pandemie", *op.cit.*,p. 195.

28 J. Beck et al., "Manifesto on the future of cross-border cooperation in Europe", *op. cit.*, p. 2.

national decisions on transnational mobility restrictions were frequently questioned and demands for reopening borders became increasingly vocal – as seen in the Greater Region, for example.[29] These wishes were fulfilled with the end of reinforced border controls with Luxembourg on 16 May and with France, among others, on 15 June 2020.[30]

The Practical Economic Consequences of Border Restrictions in the Greater Region

In the Greater Region, economic and social structures are largely based on transnational exchange and cross-border workers are essential for the functioning of the local economy. Here, the effects of the sudden introduction in border controls and partial border closures were clearly perceptible.[31] The Greater Region (see Figure 1) today covers a total of 65,400 km². It consists of the German States of Saarland and Rhineland-Palatinate, the French territory of Lorraine within the *région* Grand Est, the Grand Duchy of Luxembourg, and parts of Belgium, comprises a population of 11.6 million inhabitants and has a Gross Domestic Product (GDP) of 390 billion euros (year 2017), which corresponds to about 2.5 % of the EU's GDP.[32] Almost 250,000 cross-border commuters regularly cross national borders here.[33] The participating regions regularly exchange information on the level of the Greater Region,

---

29 D. Kirch, "Erste Rufe nach Ende der Grenzschließung", 2020. Retrieved from: rp-sz-epaper.s4p-iapps.com/artikel/918279/16334487. Accessed 31 March 2020; H. Maillasson, "Saarland setzt sich für Grenzöffnungen ein", 2020. Retrieved from: www.saarbruecker-zeitung.de/saarland/blickzumnachbarn/deutschland-und-frankreich-halten-an-grenzkontrollen-fest_aid-50461933. Accessed 8 May 2020; R. Theis, "Bien plus qu'un 'plus jamais!': Quelles leçons nous pouvons tirer de la crise – la chance d'une nouvelle dynamique pour la région frontalière franco-allemande!", *In* F. Weber, R. Theis & K. Terrolion (eds.), *Grenzerfahrungen, op.cit.*, pp. 449–468.

30 F. Weber et al. (eds.), *Geographien der Grenzen, op.cit.*

31 I. Pigeron-Piroth, E. Evrard & R. Belkacem, "Marché du travail transfrontalier", *op. cit.*

32 EVTZ Gipfelsekretariat der Großregion, "Die Großregion in Zahlen", 2021. Retrieved from: www.grossregion.net/Die-Grossregion-kompakt/Die-Grossregion-in-Zahlen. Accessed 16 December 2021.

33 Interregionale Arbeitsmarkt Beobachtungsstelle (IBA)-Observatoire interrégional du marché de l'emploi (OIE), "Die Grenzgängerströme der Großregion ", 2022. Retrieved from: https://www.iba-oie.eu/. Accessed 19 November 2022.

and various forms of institutionalisation have emerged in the regional context, including the European Grouping of Territorial Cooperation (EGTC) Eurodistrict SaarMoselle – a link-up between five municipal clusters of the *département* Moselle and the Saarland Regional Network (*Regionalverband*), the Quattropole City Network (joining Saarbruecken, Metz, Luxembourg, and Trier) or the Tonicités City Network (Arlon, Esch-sur-Alzette, Longwy, Luxembourg, Metz, and Thionville), all of which are based on and benefit from the Europe of open borders.

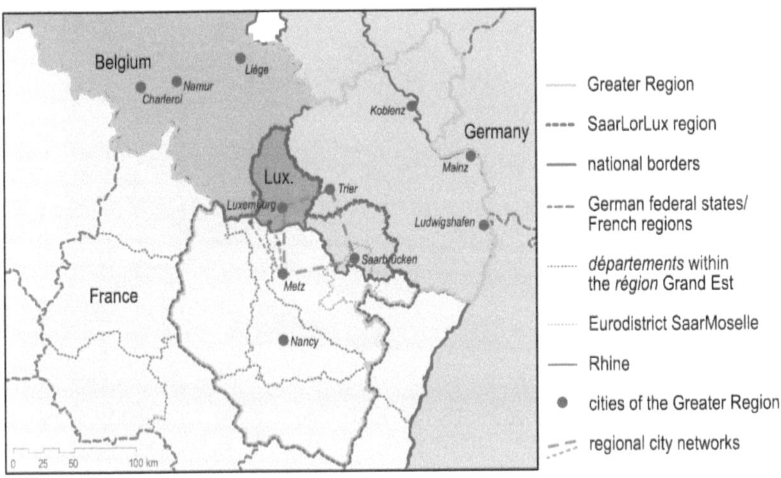

Figure 1: The Greater Region.[34]

Cross-border mobility had previously been a part of everyday life for many workers. Due to the restrictions however, border crossings became more time-consuming: For one thing, detours had to be made in some cases in order to reach an official border crossing point (see Figure 2 for an illustration). For another thing, traffic jams formed due to the controls

[34] Source: Own representation, based on Bundesministerium für Verkehr, Bau und Stadtentwicklung (BMVBS), "Metropolitane Grenzregionen: Abschlussbericht des Modellvorhabens der Raumordnung (MORO). Überregionale Partnerschaften in grenzüberschreitenden Verflechtungsräumen", 2011, p. 27. Retrieved from: www.bbsr.bund.de/BBSR/DE/Veroeffentlichungen/ministerien/BMVBS/Sonderveroeffentlichungen/2011/DL_MetropolitaneGrenzregionen.pdf?__blob=publicationFile&v=2. Accessed 14 December 2021.

and the bundled traffic volume.[35] Local (cross-border) public transport – both regional bus and tram services and international train services – also came to a sudden standstill. Although workers were still allowed to cross the borders between Germany, France, Luxembourg, and Belgium – for which they had to carry a bilingual "commuter certificate" issued by their employer – detours, traffic jams and the interruption of cross-border public transport meant a considerable additional effort for border commuters.[36]

Figure 2: Closed border crossings between Saarland and the *département* Moselle in spring 2020.[37]

---

35 F. Weber & C. Wille, "Grenzgeographien der COVID-19-Pandemie", op.cit., p. 206 f.

36 F. Berrod, B. Wassenberg & M. Chovet, "La frontière franco-allemande au temps du Covid-19: la fin d'un espace commun?", *op.cit.*, p. 41.

37 Source: Photographs by Peter Dörrenbächer, 2020.

In addition to these restrictions of movement, border controls and working from home caused initial uncertainties for employees. Since, formally speaking, the "country of occupation" changes when working from home, questions of insurance and tax law had to be clarified. EU legislation stipulates, for example, that people who work in several countries of the EU can only be covered by one country's social insurance scheme. Prior to the pandemic, working from home had always gone along with the risk of being transferred into the coverage of the social security system of the respective country of residence as soon as more than 25 % of the working time was performed there. In order to guarantee the social security of employees, France and the neighbouring states of Germany, Luxembourg, and Belgium (among others) decided in March 2020 to temporarily suspend this regulation;[38] Moreover, working from home might impact the taxation of cross-border workers. The Organisation for Economic Cooperation and Development (OECD) recommends that taxes be levied only in the country where the work is actually performed.[39] Between Germany and France, the so-called "double taxation agreement" regulates which state has the right of taxation in cross-border labour relations. The bilateral treaty, which has been in place since 2006, stipulates that employees who have the status of "cross-border worker" only pay taxes in their country of residence.[40] People without this status normally had to pay taxes in the country where they work on the given day. In the wake of the Covid-19 pandemic,

---

38 F. Berrod, B. Wassenberg & M. Chovet, "La frontière franco-allemande au temps du Covid-19", *op.cit.*, p. 41 ; Frontaliers Grand Est, "Télétravail et COVID-19 – Evolution des accords", 2022. Retrieved from : frontaliers-grandest.eu/teletravail-et-covid-19-evolution-des-accords/. Accessed 16 January 2022 ; Ministère de l'Economie, des Finances et de la Relance, "Communique de presse: La France s'accorde avec l'Allemagne, la Belgique et la Suisse pour que, dans le cadre de la crise sanitaire liée au Covid-19, le maintien à domicile des travailleurs frontaliers non éligibles aux régimes frontaliers n'entraîne pas de conséquence sur le régime d'imposition qui leur est applicable", 2020. Retrieved from : minefi.hosting.augure.com/Augure_Minefi/r/ContenuEnLigne/Download?id=8BA93857-6354-4B1D-909B-35FA63F14D5E&filename=2165 %20.pdf. Accessed 9 March 2021; I. Pigeron-Piroth, Funk, B. Nienaber, H. P. Dörrenbächer & R. Belkacem, "Der grenzüberschreitende Arbeitsmarkt der Großregion: Der Einfluss der COVID-19-Pandemie", *Informationen zur Raumentwicklung*, 26(2), 2021, p. 81.

39 I. Pigeron-Piroth, E. Evrard & R. Belkacem, "Marché du travail transfrontalier", *op. cit.*, p. 21.

40 Frontaliers Grand Est, "Télétravail et COVID-19 – Evolution des accords", *op.cit.*

the double taxation agreement was extended to those who do not hold the status of "cross-border worker".[41] While this agreement ensured that there were no tax implications between Germany and France, an adapted regulation was needed with Luxembourg: To date, an agreement between France and Luxembourg stipulated that French employees working for a Luxembourg employer could spend a maximum of 29 days per year[42] from home in order to remain taxable only in their country of residence.[43] At the end of March and beginning of April, Luxembourg and its neighbouring countries therefore agreed on a temporary adjustment of these taxation rules to allow working from home.[44]

Although solutions to these emerging problems could be found in order to ensure the best possible continuation of cross-border work, commuters employed in Germany faced several disadvantages: French employees are entitled to receive a short-time allowance. However, while this payment is tax-free for German employees, it is taxable in France.[45] Overall, it was especially the high degree of uncertainty that represented a challenge for many cross-border workers, who were initially unable to assess the practical consequences of border restrictions and remote work. The decisions at short notice left little time to react to the new requirements and to communicate necessary information.

The Covid-19 pandemic thus made existing cross-border interdependencies apparent. As the local economy in the Greater Region is closely interconnected and based on different dependencies, the economic impact

---

41 *Ibid.*

42 A maximum of 24 working days (between Luxembourg and Belgium) respectively 19 working days (between Luxembourg and Germany) could be conducted from home.

43 Ministère de l'Economie, des Finances et de la Relance, "Communique de presse: Les accords amiables concernant les travailleurs frontaliers et transfrontaliers, conclus avec l'Allemagne, la Belgique, le Luxembourg et la Suisse, dans le contexte de la lutte contre la propagation du COVID-19, continueront de s'appliquer jusqu'au 31 décembre 2020", 2020. Retrieved from : minefi.hosting.augure.com/Augure_Minefi/r/ContenuEnLigne/Download?id=FD630F92-59C8-49F2-A831-E538993755E2&filename=102%20-%20CP%20accords%20fiscalit%C3%A9%20travailleurs%20frontaliers%20et%20transfrontaliers%20covid-19.pdf. Accessed 9 March 2021.

44 *Ibid.*

45 Ministère de l'Economie, des Finances et de la Relance, "Communique de presse: La France s'accorde avec l'Allemagne", *op.cit.*, p. 1 ; Pigeron-Piroth et al., "Der grenzüberschreitende Arbeitsmarkt der Großregion", *op.cit.*, p. 82.

of the border restrictions was strong and noticed immediately. Especially right after the uncoordinated introduction of border controls and partial border closures, the question arose as to how supply chains would be affected.[46] Even though it was promptly decided that cross-border shipments of goods shall still be possible, congestions and required detours led to delays. This development was problematic in that it, for example, slowed down the transport of urgently needed medical supplies.[47] Given that the Saarland, as well as Luxembourg, rely on doctors and nurses from Lorraine, the border restrictions created further uncertainty and instability in the medical sector. With around 70 % of the employees in Luxembourg's health sector coming from its neighbouring states,[48] the Grand Duchy appealed to France's responsibility to ensure the border crossing of commuters – especially those working in the medical sector – in order to guarantee the provision of medical care.[49]

While the cross-border movement of workers as well as the movement of goods and merchandise was continuously secured – albeit with certain obstacles and disadvantages – everyday practices, e.g. concerning cross-border shopping came to a standstill during the first wave of the pandemic. This particularly affected the retail structure in the Greater Region, which was specifically aligned toward cross-border commuters. Main reasons for the cross-border flow of customers lie in diverse price and supply structures of products for everyday needs, for example because of the different levels of taxation in the various EU member states.[50] The established cross-border purchasing and consumption practices had been inhibited due to the restriction of free movement of persons. This is a problematic development, as many companies in the Greater Region depend on the purchasing power of cross-border

---

46 A. Sommarribas & B. Nienaber, "Border controls versus European solidarity", *In* G. Mein & J. Pause (eds.), *Self and Society in the Corona Crisis: Perspectives from the Humanities and Social Sciences*, Melusina Press, Esch-sur-Alzette, 2021, pp. 329–339.

47 *Ibid.*, pp. 333–334.

48 C. Wille, "Border(ing)s in Times of COVID-19", *In* C. Wille & R. Kanesu (eds.), *Bordering in Pandemic Times, op.cit.*, p. 12.

49 A. Sommarribas, A. & B. Nienaber, "Border controls versus European solidarity", *op. cit.*, p. 335.

50 See in detail: Wille, C., "Grenzüberschreitende Alltagspraktiken in der Großregion SaarLorLux", *op.cit.*, pp. 139 ff.

consumers.[51] Overall, the uncoordinated, partly spontaneous decisions in the various countries and the already tense political, economic, and social situation posed a central challenge for border region residents. In this context, informing citizens proved essential, because they had to take into account not only the regulations of their own, but also those of the neighbouring country: Relevant information first had to be obtained and verified, then standardised and translated before it could be made available (online) to cross-border workers.[52] This required a special focus on the needs of cross-border workers by identifying and addressing the questions, challenges, and hurdles they saw themselves confronted with. Institutions such as *Frontaliers Grand Est*, the Grand Est Regional Council's information point for cross-border issues, played a central role in this process alongside ministries and regional authorities. During the health crisis, these institutions of cross-border cooperation had the task to communicate and update the constantly changing decisions in a detailed and precise manner, focusing especially on the needs of border region residents.[53]

The Symbolic Significance of the Perceptibly Restrictive Function of Borders in the Greater Region

The social and cultural life in the Greater Region is significantly shaped by cross-border practices. National borders suddenly separated friends, couples, and families. The inhabitants of the borderland were thus affected by restrictions in two ways – the national regulations that applied for any citizen, and the additional restrictions of the cross-border exchange. The interest and concerns of those living in border regions were initially hardly considered in the national political decisions, although *their* realities of life were *particularly* restricted. This resulted in a high

---

51 *Ibid.*, p. 140; See also MOT & EU, "The effects of COVID-19 induced border closures on cross-border regions: An empirical report", *op.cit.*, p. 88.

52 MOT & EU, "The effects of COVID-19 induced border closures on cross-border regions: An empirical report", op.cit., p. 83.

53 *Ibid.*, p. 85.

degree of disappointment and resignation.[54] Inconsistent regulations, some of which were not clearly communicated, contributed to further confusion and frustration. At this point, the symbolic effect of the border regulations becomes apparent, as the border restrictions by the Federal Republic led to a short-term disruption of the good relations between Germany and France.[55] The erection of physical barriers at various border crossings, for example between Saarland and Lorraine, not only meant that borders that had long been invisible became visible, but also perceptible again. This development manifested itself, for example, in the form of resentments by parts of the German population towards French border crossers: In spring 2020, the perception of France as a risk area and thus of the French neighbours as a potential threat led to discriminatory statements and hostility.[56] Headlines such as "Corona enters Saarland from France"[57] promoted this. The broadcasting service *Saarländischer Rundfunk* (SR) ran surveys in November and December 2020 in the Saarland, the *département* Moselle and Luxembourg. In all three areas, respondents felt the intensified border controls had harmed relations, but feelings were not as strong as anticipated. Impairments were perceived as stronger by residents from Luxembourg rather than Lorraine, although the French were reported to have suffered more from resentment than other nationalities.[58] (See Figure 3).

---

54 See in detail: Dylla, C., "Drei Monate 'Grenzerfahrungen': Was bleibt: Betrachtungen aus der Medienperspektive", *op.cit.*; S. Freitag-Carteron, "Deutschland, Frankreich, COVID-19 – das Virus und die Grenzregion", *op.cit.*

55 F. Berrod, B. Wassenberg & M. Chovet, "La frontière franco-allemande au temps du Covid-19", *op.cit.*, p. 41 ff; I. Pigeron-Piroth, E. Evrard & R. Belkacem, "Marché du travail transfrontalier", *op.cit.*, p. 20 f.

56 MOT & EU, "The effects of COVID-19 induced border closures on cross-border regions: An empirical report", *op.cit.*, p. 97 f.

57 H. Maillasson & T. Prommersberger, "Corona kommt von Frankreich ins Saarland: Corona-Fälle an der Grenze", *Saarbücker Zeitung*, 6 March 2020.

58 Saarländischer Rundfunk (SR), "Unterschiedliche Bewertung der Grenzschließungen", 2020. Retrieved from: www.sr.de/sr/home/nachrichten/politik_wirtschaft/saarlandtrend/saarlandtrend_2020/saarlorluxtrend_2020_grenzschliessungen_100.html. Accessed 3 March 2022.

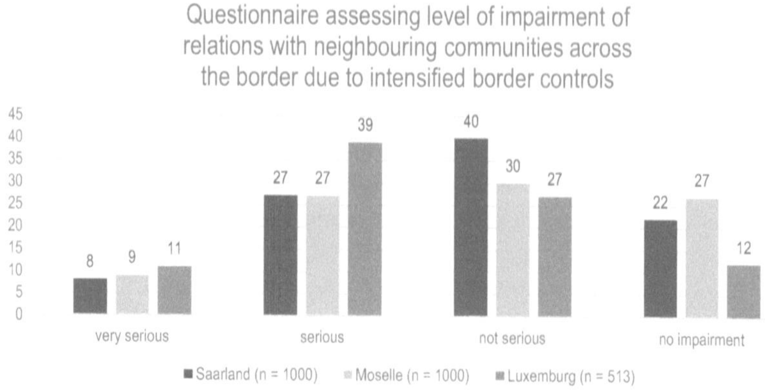

Figure 3: SaarLorLux survey by the Saarland Broadcasting Service *Saarländischer Rundfunk* (SR): "The French-German-Luxembourg borders were closed for a prolonged period in the spring. Was the resulting impairment of relations with neighbouring communities across the border very serious, serious, not serious, or was their no impairment".[59]

The "reflexive retreat towards the national sphere"[60] separated supposedly safe places from supposedly unsafe places and people.[61] By delineating state boundaries, the state defined which space it was responsible for and what lay outside its jurisdiction and responsibility. In this way, drawing borders became a political tool. This action simultaneously raised the question of cohesion and solidarity between the EU states and thus also the state of European integration. The rebordering processes in spring 2020 not only disregarded the economic interdependencies of cross-border relations and their social significance, but they also hindered

59 Source: Own representation, based on SR Survey, 11 December 2020, *In* SR, "Unterschiedliche Bewertung der Grenzschließungen", *op.cit.*

60 N. Crossey, "Corona – neue Herausforderungen und Perspektiven für Grenzraumpolitiken und grenzüberschreitende Governance", *In* C. Wille & R. Kanesu (eds.), *Bordering in Pandemic Times*, *op.cit.*, p. 71.

61 H. Böhm, "Researching cross-border cooperation under the shadow of COVID 19 pandemic: scientific report from e-conferences and blog-reflections produced between 14 March and 21 June 2020", *Pogranicze Polish Borderlands Studies*, 8(2), 2020, pp. 81–86.

cross-border action. Against this background, the admission of French patients from the overburdened hospitals in Grand Est to German hospitals[62] was not only important from a medical perspective, but is also served as a significant symbol of Franco-German solidarity. In addition, various actors in the Greater Region, who were directly affected by the restrictions in everyday cross-border life, criticised the national approaches. They actively demanded that the borders be opened again and emphasised the importance of cross-border cohesion within the Greater Region. In doing so, they clearly expressed their opposition to Euroscepticism and national solo efforts. Beginning of April, Saarland's mayors recorded a statement of solidarity with their French partner communities and posted it on YouTube.[63] Less than three weeks later, the mayors of the border municipalities of the EGTC Eurodistrict SaarMoselle published a joint appeal to Angela Merkel and Emmanuel Macron to enable and ensure a closer coordination and a strengthened cooperation across the Franco-German border.[64] Other solidarity actions and manifestos symbolised the fight for open borders.[65] (Illustrated by Figure 4). The European idea was thus actively brought to life on a local and regional level.

---

62 Deutscher Ärzteverlag, "Deutsche Krankenhäuser nehmen COVID-19-Patienten aus Italien und Frankreich auf: Dienstag, 24. März 2020", 2020. Retrieved from: www.aerzteblatt.de/nachrichten/111286/Deutsche-Krankenhaeuser-nehmen-COVID-19-Patienten-aus-Italien-und-Frankreich-auf. Accessed 20 April 2020.

63 M. Clivot, "Solidaritätserklärung der saarländischen Bürgermeisterinnen und Bürgermeister nach Frankreich: 2 April 2020", 2020. Retrieved from: youtu.be/kgaUYwPbWqA. Accessed 20 May 2020.

64 Eurodistrict SaarMoselle, "Appel des maires allemands et français des communes frontalières de l'Eurodistrict SaarMoselle/Appell der deutschen und französischen Bürgermeister/innen von Grenzgemeinden im Eurodistrict SaarMoselle: 20 April 2020", 2020. Retrieved from: www.youtube.com/watch?v=m0IxSeZI608&ab_channel=StadtSaarbruecken. Accessed 4 January 2022.

65 J. Beck et al., "Manifesto on the future of cross-border cooperation in Europe", *op. cit.*; C. Wille & F. Weber, "Analyzing border geographies in times of COVID-19", *In* G. Mein & J. Pause (eds.), *Self and Society in the Corona Crisis, op.cit.*, pp. 361–385.

Figure 4: Banner of the initiative "keep Schengen alive" in Schengen, Luxembourg, May 2020.[66]

Adapted Measures in the Course of Later Pandemic Phases as Lessons Learned from the First Wave

As shown, the increased border controls and closures introduced in spring 2020 came as a far-reaching shock for the closely intertwined Greater Region, but at the same time resulted in counteractive actions. Due to the fundamental economic and social malfunctions, the importance of open borders for a smooth everyday life within border regions came to the fore of socio-political consciousness at local, regional, national, and European levels. For example, the (state) governments of Baden-Württemberg, Rhineland-Palatinate, Saarland, and the *région* Grand Est drew lessons from the experience of the first wave of the Covid-19 pandemic by signing a "mutual assistance pact for enhanced cross-border cooperation and coordination" on 27 November 2020.[67]

66 Photograph by Florian Weber, 2020.

67 Baden-Württemberg, Rheinland-Pfalz, Saarland & Grand Est, "Beistandspakt für eine verstärkte grenzüberschreitende Kooperation und Koordination / Pacte d'assistance mutuelle pour une coopération et coordination transfrontalières renforcées", 2020. Retrieved from: www.saarland.de/SharedDocs/Downloads/DE/stk/2020-11-27-beistandspakt.pdf?__blob=publicationFile&v=1. Accessed 3 March 2021.

In it, they commit to taking measures in times of crisis that "take into account the human, economic, and social interdependencies in border areas"[68] As a direct response to the developments of spring 2020, the executives advocated in particular for increased cooperation between health systems and the mutual provision of medical resources, such as bed capacities.[69] The partners also emphasised that even outside times of crisis, stronger cohesion and closer cooperation in the border region should be pushed forward.[70] In January 2021, the EU leaders also committed themselves to avoiding border closures and general travel bans within the EU in the future. However, border-related measures to contain the virus, such as compulsory testing, travel restrictions or random border controls for non-essential travel within the EU – for example for the purpose of tourist stays – were still possible.[71] Border restrictions thus remained an "important option in the political decisions to control the pandemic and in the German public"[72] even in later pandemic phases.

By the time of the "second wave" in autumn 2020, however, not only had there been a significant change in awareness regarding cross-border linkages. There was also a more routine approach to the challenges posed by the Covid-19 pandemic.[73] Two crucial developments should be considered when contextualising the events of autumn and winter 2020/2021 in the Greater Region: Firstly, new uncertainties and challenges arose due to emerging virus variants, which brought the option of travel restrictions back to the table. Although these measures initially affected only visitors entering the EU from Great Britain, South Africa and Brazil, the variants soon also impacted the EU's internal borders.[74] Secondly, Germany and France pursued different strategies in dealing with the

[68] *Ibid.*, p. 1. In German: "die menschliche[n], wirtschaftliche[n] und soziale[n] Verflechtungen in den Grenzgebieten berücksichtigen".

[69] *Ibid.*, p. 2.

[70] *Ibid.*

[71] C. Kauber, "Die COVID-19-Krise und die deutschen Grenzen: Erfahrungen aus der ersten und zweiten Welle", *Informationen zur Raumentwicklung*, 26(2), 2021, p. 92.

[72] *Ibid.* In German: "wichtige Option in den politischen Entscheidungen der Pandemiebekämpfung und in der deutschen Öffentlichkeit".

[73] F. Weber, R. Theis & K. Terrolion, "Neue Herausforderungen, alte Grenzen: Wie die COVID-19-Pandemie das deutsch-französische Verhältnis aufwirbelt", *In* F. Weber, R. Theis & K. Terrolion (eds.), *Grenzerfahrungen*, *op.cit.*, pp. 3–18.

[74] *Ibid.*, p. 16 f.

pandemic: The Federal Republic of Germany sought to reduce the number of infections, while the French government chose to allow higher infection rates to avoid stricter measures. These national policies resulted in contrasting developments and a disparity in infection rates on the two sides of the border.[75]

Across the board, the pandemic phases in the Greater Region proceeded without the reintroduction of border closures. Systematic controls at border crossings gave way to randomised checks by the federal police. The Saarland and Rhineland-Palatinate, among others, established a clause within their regulations that allowed cross-border commuters to enter the neighbouring country – in case it was declared a risk area – for a maximum period of 24 hours without the obligation of testing or quarantine.[76] The "small border traffic", i.e. commuter movements for short-term stays in the neighbouring state, could thus be maintained. Regulations on taxation and social security, as introduced in the course of the first wave, were extended.[77] From the second wave onward, European, national, and regional decision-makers fulfilled their promise to take greater account of border regions. Nevertheless, infection rates not running in parallels and the eventual classification of French regions as risk/virus variant areas continued to pose challenges, as illustrated by the following exemplary regulations in autumn/winter 2020 and spring 2021.

Due to rising infection numbers, France introduced a curfew in November 2020. From then on, only shops for daily needs remained open – Germans were still allowed to enter the country for shopping with a special permit form. Since this French form had to be carried at all times, the Covid information page of the Saarland state government provided a German guide on how the form was to be filled. Cross-border mobility for holidays or to visit friends and family members was not permitted, and cross-border workers henceforth needed a certificate

---

75 F. Weber et al. (eds.), *Geographien der Grenzen, op.cit.,* pp. 13 ff; T. Wieder, "France/Allemagne: deux cultures politiques face à la COVID-19", *In* F. Weber, R. Theis & K. Terrolion (eds.), *Grenzerfahrungen, op.cit.*, pp. 285–294.

76 C. Kauber,"Die COVID-19-Krise und die deutschen Grenzen: Erfahrungen aus der ersten und zweiten Welle", *op.cit.,* p. 92.

77 Frontaliers Grand Est, "Télétravail et COVID-19 – Evolution des accords", *op.cit.*

from their employer.[78] When the *département* Moselle was declared a virus variant area (*Virusvariantengebiet)* in March 2021, the Federal Republic introduced a testing requirement to enter Germany – also for cross-border commuters. In order to prevent far-reaching disruptions to cross-border traffic, the Saarland government set up the Franco-German test centre *Goldene Bremm* in Saarbruecken, directly adjacent to the border, with the participation of other actors and under the Eurodistrict SaarMoselle within a few days.[79] (see Figure 5). There was no systematic border control, but the requirement to present a negative test certificate no older than 48 hours led to renewed criticism.[80]

Figure 5: The Franco-German Covid 19 test centre at the Goldene Bremm border crossing.[81]

In contrast to implemented measures in spring 2020, dynamic adjustments to the measures were continually made in response to infection rates in 2021 – for example, by introducing the obligation for Covid-testing for cross-border commuters. Also during the second and third wave, the nation state remained the most important frame of reference for

78 H. Maillasson, "Nach Lothringen nur mit Bescheinigung", *Saarbücker Zeitung,* 2 November 2020.

79 For more details see F. Weber et al. (eds.), *Geographien der Grenzen, op.cit.,* p. 15.

80 H. Maillasson, "Warum es keine systematischen Kontrollen an der Grenze gibt", *Saarbrücker Zeitung,* 17 March 2021.

81 Photograph by Eva Nossem, 2021.

dealing with the pandemic, so that the local level continued to depend on national decisions. Yet, both the type of restrictions and the *modus operandi* had changed compared to the decisions of spring 2020: the concerns of border regions were given more weight in the regulations than had been the case during the first wave. Cross-border workers remained exempt from most border restrictions. A stronger vertical and horizontal communication and coordination, i.e. on one hand between national and regional decision-makers, and on the other, particularly on the local level, in cross-border exchanges, could also be observed.[82] These adjustments can be framed as the results of central learning experiences from the tense situation of the first wave.[83] The European idea of exchange and mutual support was thus promoted in the cross-border regional context.

Conclusion and Outlook: The Covid-19 Pandemic as an Impulse Generator in Favour of Resilient Cross-Border Linkages?

As of spring 2020, border barriers, fences, and warning tapes have made the national borders between European nation states visible and tangible once again. The uncoordinated measures of the EU states in order to contain the spread of the Covid-19 virus put cross-border cooperation and European integration to a hard test, especially because the dynamics of border regions were initially given too little consideration in national decisions. The political reactions "disregarded the cross-border nature of European borderlands".[84] The Covid-19 pandemic and the associated rebordering processes thus revealed the fragility of borderlands resulting from a tension between a general dependence on national decisions on the one hand and close cross-border networking on the other: First of all, it is important to note that health policy and thus also infection control fall within the remit of EU member state.[85] Thus, it is not surprising that

---

82 C. Kauber,"Die COVID-19-Krise und die deutschen Grenzen: Erfahrungen aus der ersten und zweiten Welle", *op.cit.*, p. 93.

83 *Ibid.*, p. 92; F. Weber et al. (eds.), *Geographien der Grenzen, op.cit.*

84 H. Böhm, "Five Roles of Cross-border Cooperation Against Re-bordering", *op. cit.*, p. 15.

85 F. Weber, "Das Coronavirus und die Erosion von Gewissheiten", *In* C. Wille & R. Kanesu (eds.), *Bordering in Pandemic Times, op.cit.*, p. 35.

nation states remained the main frame of reference during the Covid-19 pandemic. At the same time, the functioning of border regions is largely based on the existence of open internal EU borders, so that the border controls and partial border closures implemented in the course of the first wave of the pandemic caused far-reaching disruptions in the everyday lives of border region residents.

The experiences of the first wave revealed that renationalisation, unilateralism, and a lack of intergovernmental agreements can significantly disrupt the functioning of border regions. This is by no means an issue of secondary importance, as today about 40 % of EU citizens live in border regions and almost two million people regularly cross national borders.[86] The example of the Greater Region shows the extent to which the systematic border controls and the partial border closures, as well as the uncertainties that accompanied them, brought life there to a standstill. The pandemic acted as a caesura for the interconnected space. At the same time, it manifested the inevitable need for coordination and action in favour of resilient transnational cooperation within the EU.[87] Based on the regulations implemented during later pandemic phases, which were characterised by increased cross-border communication and coordination, it became apparent that lessons had been learned from the shock of this first wave. The adaptations of measures in the course of later pandemic phases indicate that initial lessons have been learned from the experiences. They can be understood as a sign of joint efforts to align or at least coordinate national policies – whenever possible – to take greater account of cross-border interdependencies. Against this background, border regions stand for the idea of living labs of European integration, which thus also function as indicators. The "European project" is in fact "deeply rooted in border regions where the everyday notion of a Europe without borders is forged".[88] The pandemic contributed to a revival of

---

86 European Committee of the Regions, "Report on the Public Consultations on the Future of Cross-Border Cooperation", 2021, p. 3. Retrieved from: portal.cor.europa.eu/egtc/about/Documents/4525_COTER_Report_on_the_Consultations-Future_of_CBC.pdf. Accessed 11 October 2021.

87 J. Kajta & E. Opiłowska, "The Impact of Covid-19 on Structure and Agency in a Borderland: The Case of Two Twin Towns in Central Europe", *Journal of Borderlands Studies*, 37(4), 2021, pp. 699–721.; E. Opiłowska, "The Covid-19 crisis: The end of a borderless Europe?", *European Societies*, 23(sup1), 2021, pp. 589–600.

88 Medeiros et al., "Covidfencing effects on cross-border deterritorialism: the case of Europe", *op.cit.*, p. 979.

and a stronger desire for cross-border cooperation and a deepened cross-border dialogue.[89] The perception that "the border is a handicap when it creates obstacles, but an advantage when it is open"[90] can be framed as a central learning experience in favour of border regions that are more resilient to crises – and this also applies to the Greater Region. In this way, the Covid-19 pandemic can unfold a positive side: By not only reaffirming existing processes of integration but also stimulating new forms of cross-border understanding and cooperation, the pandemic can be understood as a source of impetus and, prospectively, ideally as an opportunity.

Nonetheless, the challenges to be faced are still manifold and require action from the European to the local level.[91] Accordingly, there is a lot to be done and considered – and it remains to be seen which of these requirements can and will be realised. In the long term, it will be important to focus more on the characteristics of border regions and the specific needs of their inhabitants. Therefore, border regions should rather be regarded as interconnected areas and less as individual spaces separated from each other by a national border. The EU has the task of providing information, ensuring coordination and promoting dialogue and partnership between the individual member states and thus especially within border regions by pursuing a cohesion policy.[92] In the long run, it is also important to promote projects that coordinate cross-border crisis management, develop and update emergency scenarios regularly, and take border regions into account in their decision-making.[93] In addition, there is a need for strengthened horizontal *and* vertical communication and coordination. Against this backdrop, structures of cross-border multilevel

89 For example: C. Kauber,"Die COVID-19-Krise und die deutschen Grenzen: Erfahrungen aus der ersten und zweiten Welle", *op.cit.*

90 J. Peyrony, J., "Grenzschließungen als Folge von COVID-19: Empirischer Bericht für den Zeitraum März bis Juni 2020", *Informationen zur Raumentwicklung*, 26(2), 2021, p. 106. In German: "die Grenze ein Handicap ist, wenn sie Hindernisse schafft, aber ein Vorteil, wenn sie offen ist".

91 J. Beck et al., "Manifesto on the future of cross-border cooperation in Europe", *op. cit.*, pp. 3 ff.

92 European Commission, "COVID-19: Guidelines for border management measures to protect health and ensure the availability of goods and essential services", *op.cit.*, pp. 143 ff; H. Handler, *Krisengeprüftes Europa: Wie wir die Solidarität in der EU stärken können*, Springer Gabler, Wiesbaden, 2021, p. 234 f.

93 J. Peyrony, "Grenzschließungen als Folge von COVID-19: Empirischer Bericht für den Zeitraum März bis Juni 2020", *op.cit.*, pp. 102 ff.

governance, that incorporate different political and administrative levels, including the EU, and that will place greater emphasis on political decisions at the supra- and subnational level, should be promoted in order to contribute to an increased resilience of cross-border interdependencies.[94] The extent to which level mismatches can have an inhibiting effect must also be considered. In the Greater Region, for example, a nation state (Luxembourg), two federal states with their own governments (Rhineland-Palatinate and Saarland), a region with several *départements* with relatively little room for independent actions (Lorraine/Moselle) and parts of Belgium with specific possibilities for action come together, which makes coordination processes highly complex. From a scientific perspective, the Covid-19 pandemic raises the question of the longer-term effects on cooperation and cohesion in border regions. It is also necessary to address the factors that determine resilience and to what extent the interconnected areas can function as strong living labs of European integration. Further, in-depth research is needed to address these questions.

Bibliography

T. Albers, V. Köbele-Ennaji, J. Ross & V. Wolfart, "Nationale Reflexe im Angesicht einer europäischen Krise: Die Kontrollen an der deutsch-französischen Binnengrenze während der ersten Welle der COVID-19-Epidemie", *In* F. Weber, R. Theis & K. Terrolion (eds.), *Grenzerfahrungen/Expériences transfrontalières: COVID-19 und die deutsch-französischen Beziehungen/Les relations franco-allemandes à l'heure de la COVID-19*, Springer VS, Wiesbaden, 2021, pp. 367–378.

A. Alemanno, "The European response to COVID-19: From regulatory emulation to regulatory coordination?", *European Journal of Risk Regulation*, 11(2), 2020, pp. 307–316.

H. Bähr, H. Schenke & D. Filsinger, "Transnationale Netzwerke und Institutionen in der Großregion SaarLorLux", *In* J. Meyer & L.

[94] On this, for example : N. Crossey & F. Weber, "Handlungsempfehlungen zur weiteren Gestaltung der grenzüberschreitenden Kooperation im deutsch-französischen Verflechtungsraum/Recommandations d'action pour les orientations futures de la coopération transfrontalière dans le bassin de vie franco-allemand", *Policy Paper*, Vol. 4, University of Saarland, 2021; Weber, F., "Das Coronavirus und die Erosion von Gewissheiten", *op.cit.*, p. 36.

Rampeltshammer (eds.), *Grenzüberschreitendes Arbeiten in der Großregion SaarLorLux*, Saarbrücken: Universaar, Saarbrücken, 2012, pp. 17–78.

F. Berrod, B. Wassenberg & M. Chovet, "La frontière franco-allemande au temps du Covid-19: la fin d'un espace commun?", *The Conversation,* 16 April 2020.

H. Böhm, "Researching cross-border cooperation under the shadow of COVID 19 pandemic: scientific report from e-conferences and blog-reflections produced between 14 March and 21 June 2020", *Pogranicze Polish Borderlands Studies*, 8(2), 2020, pp. 81–86.

H. Böhm, "Five roles of cross-border cooperation against re-bordering", *Journal of Borderlands Studies*, 36(5), 2021, pp. 1–20.

E. Brunet-Jailly & M. J. Carpenter, "Introduction to the special issue: Borderlands in the era of COVID-19", *Borders in Globalization Review*, 2(1), 2020, pp. 7–11.

A. Cooper & C. Rumford, "Monumentalising the border: Bordering through connectivity", *Mobilities,* 8(1), 2013, pp. 107–124.

N. Crossey, "Corona – neue Herausforderungen und Perspektiven für Grenzraumpolitiken und grenzüberschreitende Governance", *In* C. Wille & R. Kanesu (eds.), *Bordering in Pandemic Times: Insights into the COVID-19 Lockdown,* UniGR-Center for Border Studies, Luxemburg, Trier, 2020, pp. 69–72.

N. Crossey & F. Weber, "Handlungsempfehlungen zur weiteren Gestaltung der grenzüberschreitenden Kooperation im deutsch-französischen Verflechtungsraum/Recommandations d'action pour les orientations futures de la coopération transfrontalière dans le bassin de vie franco-allemand", *Policy Paper,* vol. 4, University of Saarland, 2021.

C. Dylla, "Drei Monate 'Grenzerfahrungen': Was bleibt: Betrachtungen aus der Medienperspektive", *In* F. Weber, R. Theis & K. Terrolion (eds.), *Grenzerfahrungen/Expériences transfrontalières: COVID-19 und die deutsch-französischen Beziehungen/Les relations franco-allemandes à l'heure de la COVID-19*, Springer VS, Wiesbaden, 2021, pp. 261–283.

H. P. Dörrenbächer, "Die Entwicklung grenzüberschreitender Berufsausbildung im Spannungsfeld unterschiedlicher Dimensionen von Distanz – das Beispiel der Großregion", *In* F. Weber et al. (eds.), *Geographien der Grenzen: Räume – Ordnungen – Verflechtungen*, Springer VS, Wiesbaden, 2020, pp. 117–142.

S. Freitag-Carteron, "Deutschland, Frankreich, COVID-19 – das Virus und die Grenzregion", *In* F. Weber, R. Theis & K. Terrolion (eds.), *Grenzerfahrungen/Expériences transfrontalières: COVID-19 und die deutsch-französischen Beziehungen/Les relations franco-allemandes à l'heure de la COVID-19*, Springer VS, Wiesbaden, 2021, pp. 295–308.

S. L. Greer, E. J. King, E. M. d. Fonseca & A. Peralta-Santos (eds.), *Coronavirus Politics: The Comparative Politics and Policy of COVID-19*, University of Michigan Press, Ann Arbor, 2021.

H. Handler, *Krisengeprüftes Europa: Wie wir die Solidarität in der EU stärken können*, Springer Gabler, Wiesbaden, 2021.

J. Jańczak, "Integration de-scaled. Symbolic manifestations of cross-border and European integration in Border Twin Towns", *Journal of Borderlands Studies*, 33(3), 2018, pp. 393–413.

J. Kajta & E. Opiłowska, "The impact of Covid-19 on structure and agency in a borderland: The case of two twin towns in central europe", *Journal of Borderlands Studies*, 37(4), 2021, pp. 699–721.

C. Kauber,"Die COVID-19–Krise und die deutschen Grenzen: Erfahrungen aus der ersten und zweiten Welle", *Informationen zur Raumentwicklung*, 26(2), 2021, pp. 86–93.

T. Kuhn, "Europa ante portas: Border residence, transnational interaction and Euroscepticism in Germany and France", *European Union Politics*, 13(1), 2012, pp. 94–117.

E. Medeiros, M. Guillermo Ramírez, G. Ocskay & J. Peyrony, "Covidfencing effects on cross-border deterritorialism: the case of Europe", *European Planning Studies*, 29(5), 2021, pp. 962–982.

MOT & EU, "The effects of COVID-19 induced border closures on cross-border regions: 20 case studies. Covering the period March to June 2020", 2021.

MOT & EU, "The effects of COVID-19 induced border closures on cross-border regions: An empirical report covering the period March to June 2020", 2021.

E. Opiłowska, "The Covid-19 crisis: The end of a borderless Europe?", *European Societies*, 23(sup1), 2021, pp. 589–600.

J. Peyrony, "Grenzschließungen als Folge von COVID-19: Empirischer Bericht für den Zeitraum März bis Juni 2020", *Informationen zur Raumentwicklung*, 26(2), 2021, pp. 94–107.

I. Pigeron-Piroth, E. Evrard & R. Belkacem, "Marché du travail transfrontalier: négocier avec les frontières à l'heure de la crise sanitaire COVID-19", *In* C. Wille & R. Kanesu (eds.), *Bordering in Pandemic Times: Insights into the COVID-19 Lockdown,* UniGR-Center for Border Studies, Luxemburg, Trier, 2020, pp. 17–23.

I. Pigeron-Piroth, Funk, B. Nienaber, H. P. Dörrenbächer & R. Belkacem, "Der grenzüberschreitende Arbeitsmarkt der Großregion: Der Einfluss der COVID-19–Pandemie", *Informationen zur Raumentwicklung*, 26(2), 2021, pp. 74–85.

S. M. Radil, J. Castan Pinos & T. Ptak, "Borders resurgent: Towards a post-Covid-19 global border regime?", *Space and Polity,* 25(1), 2021, pp. 132–140.

C. Rumford, "Introduction: Theorizing borders", *European Journal of Social Theory*, 9(2), 2006, pp. 155–169.

A. Sommarribas & B. Nienaber, "Border controls versus European solidarity", *In* G. Mein & J. Pause (eds.), *Self and Society in the Corona Crisis: Perspectives from the Humanities and Social Sciences*, Melusina Press, Esch-sur-Alzette, 2021, pp. 329–339.

R. Theis, "Bien plus qu'un 'plus jamais!': Quelles leçons nous pouvons tirer de la crise – la chance d'une nouvelle dynamique pour la région frontalière franco-allemande!", *In* F. Weber, R. Theis & K. Terrolion (eds.), *Grenzerfahrungen/Expériences transfrontalières: COVID-19 und die deutsch-französischen Beziehungen/Les relations franco-allemandes à l'heure de la COVID-19*, Springer VS, Wiesbaden, 2021, pp. 449–468.

H. Thewes, "Flatterband und Barrikaden: Grenzschließungen in der Pandemie", *In* F. Weber, R. Theis & K. Terrolion (eds.), *Grenzerfahrungen/Expériences transfrontalières: COVID-19 und die deutsch-französischen Beziehungen/Les relations franco-allemandes à l'heure de la COVID-19*, Springer VS, Wiesbaden, 2021, pp. 195–200.

P. Ulrich, "Re-Figuration von Grenzen und Ordnungen im sozialen Raum: Konzeptualisierung eines Analysemodells partizipativer Governance in EU-Grenzregionen", *In* F. Weber et al. (eds.), *Geographien der Grenzen: Räume – Ordnungen – Verflechtungen*, Springer VS, Wiesbaden, 2020, pp. 167–190.

B. Wassenberg, " 'Return of mental borders': A diary of COVID-19 closures between Kehl, Germany, and Strasbourg, France", *Borders in Globalization Review*, 2(1), 2020, pp. 114–120.

F. Weber, "Grenzüberschreitende Governance in der Großregion SaarLorLux", *Forum Wohnen und Stadtentwicklung*, 4, 2020, pp. 183–186.

F. Weber, R. Theis & K. Terrolion, "Neue Herausforderungen, alte Grenzen: Wie die COVID-19-Pandemie das deutsch-französische Verhältnis aufwirbelt", *In* F. Weber, R. Theis & K. Terrolion (eds.), *Grenzerfahrungen/Expériences transfrontalières: COVID-19 und die deutsch-französischen Beziehungen/Les relations franco-allemandes à l'heure de la COVID-19*, Springer VS, Wiesbaden, 2021, pp. 3–18.

F. Weber & C. Wille, "Grenzgeographien der COVID-19-Pandemie", *In* F. Weber et al. (eds.), *Geographien der Grenzen: Räume – Ordnungen – Verflechtungen*, Springer VS, Wiesbaden, 2020, pp. 191–223.

F. Weber, C. Wille, B. Caesar & J. Hollstegge, "Entwicklungslinien der Border Studies und Zugänge zu Geographien der Grenzen", *In* F. Weber et al. (eds.), *Geographien der Grenzen: Räume – Ordnungen – Verflechtungen*, Springer VS, Wiesbaden, 2020, pp. 3–22.

T. Wieder, "France/Allemagne: deux cultures politiques face à la COVID-19", *In* F. Weber, R. Theis & K. Terrolion (eds.), *Grenzerfahrungen/Expériences transfrontalières: COVID-19 und die deutsch-französischen Beziehungen/Les relations franco-allemandes à l'heure de la COVID-19*, Springer VS, Wiesbaden, 2021, pp. 285–294.

C. Wille, "Grenzüberschreitende Alltagspraktiken in der Großregion SaarLorLux: eine Bestandsaufnahme", *In* C. Wille (ed.), *Lebenswirklichkeiten und politische Konstruktionen in Grenzregionen: Das Beispiel der Großregion SaarLorLux: Wirtschaft – Politik – Alltag – Kultur*, transcript Verlag, Bielefeld, 2015, pp. 133–156.

C. Wille, "Border(ing)s in times of COVID-19", *In* C. Wille & R. Kanesu (eds.), *Bordering in Pandemic Times: Insights into the COVID-19 Lockdown*, UniGR-Center for Border Studies, Luxemburg, Trier, 2020, pp. 11–16.

C. Wille & R. Kanesu (eds.), *Bordering in Pandemic Times: Insights into the COVID-19 Lockdown*, UniGR-Center for Border Studies, Luxemburg, Trier, 2020.

C. Wille & F. Weber, "Analyzing border geographies in times of COVID-19", *In* G. Mein & J. Pause (eds.), *Self and Society in the Corona Crisis: Perspectives from the Humanities and Social Sciences*, Melusina Press, Esch-sur-Alzette, 2021, pp. 361–385.

Cross-Border Urban Governance in the Upper Rhine Region in the Face of the Covid-19 Pandemic

Philippe Hamman

Cross-national relationships between European cities involve multiple repertoires of action.[1] The trinational Upper Rhine region, which is the focus of this chapter, puts this in sharp relief, with its areas of urban continuity across national borders that include Strasbourg (France) – Kehl (Germany), and Saint-Louis (France) – Weil am Rhein (Germany) – Basel (Switzerland).[2] A number of cross-border projects have been implemented there: public transport networks,[3] cultural events and exchanges

---

1 R. Dormois, *Les politiques urbaines. Histoire et enjeux contemporains*, Presses universitaires de Rennes, Rennes, 2015, pp. 235–259.

Ph. Hamman & S. Ruß-Sattar, "Les répertoires d'action transfrontaliers des communes françaises et des communes allemandes: une mise en parallèle, de 1950 à nos jours", *Revue des sciences sociales*, n° 60, 2018, pp. 14–25.

2 M. Blanc & Ph. Hamman, "La construction politique métropolitaine dans le Rhin Supérieur: Coopérations intercommunales et transfrontalières. L'exemple de l'agglomération de Strasbourg" *In* C. Maurer & A. Starck-Adler (eds.), *L'Espace rhénan, pôle de savoirs/Rheinischer Raum, Wissensraum*, Presses universitaires de Strasbourg, Strasbourg, 2013, pp. 223–244.

Ph. Hamman, "Qu'est-ce qu'une Euro-région? Identifications et frontières au défi de l'institution dans l'espace du Rhin supérieur" *In* A. Koukoutsaki-Monnier (ed.), *Identités (trans)frontalières dans l'espace du Rhin supérieur*, Éditions universitaires de Lorraine, Nancy, 2014, pp. 53–76.

See also the website of the Mission opérationnelle transfrontalière (MOT). Retrieved from: <http://www.espaces-transfrontaliers.org/ressources/themes/agglomerations/agglomerations-3/>. Accessed 2 January 2024.

3 See the extension of Strasbourg's tramway to the German city of Kehl since 2017: Ph. Hamman, "Frontières fluviales ou espaces-frontières? Regards

between citizens (through sport clubs, schools, grassroots organisations, etc.) drawing on sometimes decades-old twinning programs, repurposed within a greater movement of bottom-up Europeanisation.[4] Such initiatives are particularly concerned with producing the signifiers of a "memory of peace",[5] especially in territories that have been historically contested and have swung back and forth between French and German sovereignty, as has been the case of Alsace-Moselle, which was annexed by Germany between 1871 and 1918[6] and effectively again between 1940 and 1945.[7]

As a result, these border areas are conurbations of the in-between and have their own quality; they are not merely dividing lines between two territories or States on a map.[8] In sociology, the concept of "border" relates to several interconnected dimensions:

- Borders are analysed as a dialectic between circulations and boundaries, regarding concrete human, social, economic, political and/or cultural issues.[9]

---

sociologiques", *Les cahiers du GEPE (Groupe d'Étude sur le Plurilinguisme Européen)*, n° 11, 2019. Retrieved from: <https://ouvroir.fr/cpe/>. Accessed 2 January 2024.

4 Ph. Hamman, "Les jumelages de communes, miroir de la construction européenne 'par le bas' ", *Revue des Sciences Sociales*, n° 30, 2003, pp. 92–98.

Ph. Hamman & S. Ruß-Sattar, "Les répertoires d'action transfrontaliers…", *op. cit.*

5 See the testimonies of actors collected and analysed by Ph. Hamman & B. Wassenberg, *Mémoire d'Europe – Mémoire de paix. Témoignages de la région frontalière d'Alsace*, Franz Steiner Verlag, Stuttgart, 2021, pp. 255–479.

6 Ph. Hamman, "Figures de l'État dans l''entre-deux': Le *Reichsland* d'Alsace-Lorraine, 'laboratoire' d'une politique de germanisation entre France et Allemagne (1871–1918)" *In* A. Chatriot & D. Gosewinkel (eds.), *Les figures de l'État en Allemagne et en France. 1870–1945 / Figurationen des Staates, Deutschland und Frankreich, 1870 bis 1945*, Oldenbourg, München, 2006, pp. 141–158.

7 Ph. Hamman, "Les jumelages de communes Moselle-Charente: Territoires et identités frontalières France-Allemagne à l'heure de l'Europe", *Bulletin de l'Institut de la Décentralisation*, n° 87, 2001, pp. 15–32.

8 Ph. Hamman, "Penser la différence à l'aune des espaces-frontières: une proposition d'analyse sociologique" *In* L. Denooz & S. Thiéblemont-Dollet (eds.), *Le moi et l'autre*, Presses universitaires de Nancy, 2011, pp. 29–39.

9 C. Rumford (ed.), *Citizens and Borderwork in Contemporary Europe*, Routledge, London and New York, 2014, pp. 1–12.

- Borders are sites of conflicts, socio-economic inequalities as well as hybridisation processes, i.e. social transactions at different levels:[10] in addition to national borders, there are regional or local borders, which are all the result of social inclusion and exclusion dynamics.
- Borders are both material (spatial boundaries, economic issues, conflicts of interest) and symbolic (social inclusion within a group, or exclusion; conflicts of values, interculturalism).
- Lastly, "border" is a discursive term that may prompt various political, popular and academic perceptions which overlap but never coincide; these divergences are precisely part of the border experience. "Border effects" are related to all these dimensions.[11]

It is important to account for the social thickness of border spaces and the cooperations forged in them to consider what has been described as the "model" or "laboratory" of the Upper Rhine region in Europe.[12] Local authorities are nodal actors of these cooperations, as are the workers who cross the border daily between their place of residence and their place of work, physically speaking as well as in terms of labour, tax laws and health coverage, between two systems that never fully coincide, despite the existence of European coordination regulations.[13]

---

10 Ph. Hamman, "From 'Multilevel Governance' to 'Social Transactions' in the European Context", *Swiss Journal of Sociology*, vol. 31, n° 3, 2005, pp. 523–545.

11 Ph. Hamman, "Borders and Cross-Border Cooperation in Europe from a Sociological Perspective" *In* J. Beck (ed.), *Transdisciplinary Discourses on Cross-Border Cooperation in Europe*, Peter Lang, Bern, 2019, pp. 123–145.

12 Ph. Hamman & B. Wassenberg, *Mémoire d'Europe – Mémoire de paix*, *op. cit.*, pp. 164–166.

13 These include Regulations (EEC) No 1408/1971 and No 883/2004 on the coordination of social security systems.

See Ph. Hamman, *Sociologie des espaces-frontières. Les relations transfrontalières autour des frontières françaises de l'Est*, Presses universitaires de Strasbourg, Strasbourg, 2013, pp. 123–189.

Ph. Hamman, "Cross-border worker" *In* B. Wassenberg & B. Reitel (eds.), *Critical Dictionary on Cross Border Cooperation in Europe*, PIE-Peter Lang, Brussels, 2020, pp. 268–271.

Ph. Hamman, "Les travailleurs transfrontaliers, entre pratiques et représentations autour des frontières françaises de l'Est" *In* N. Dziub (ed.), *Le transfrontalier: pratiques et représentations*, Éditions et Presses universitaires de Reims, Reims, 2020, pp. 163–177.

In terms of governance, cross-border cooperations have multi-level implications, both (inter)local and international.[14] Horizontal dynamics develop between different actors within an urban area, or in the context of competition between local authorities. This applies to economic development, which sees centralities and peripheries being reshaped. Examples include the increase in the flows of cross-border workers from Southern Alsace to Basel in the 1990s and its economic impact, leading to a saturation of the public transport network and a rise of housing prices in the French part of the cross-border urban area.[15] Vertical processes are also at work, for instance when (big) cities commit to European and international relations, as in the case of Strasbourg's efforts to become the "European Green Capital".[16]

This multi-scalar territorialisation has been particularly apparent since 2020, as the impact of the Covid-19 pandemic on transnational relations makes itself felt. This chapter analyses the challenges and initiatives related to the handling of the public health crisis in cross-border urban areas of the Upper Rhine. It draws on a corpus of national and regional print and online press articles collected since 2020 and on early studies on the topic. These residential and labour areas, which are subject to everyday uses, require a differentiated approach to public policies in a multi-level governance approach. This will be shown in three steps: having discussed the re-emergence of borders in cross-border urban areas, I will focus first on cooperation on public health matters and then on socio-economic interactions.

---

14 Ph. Hamman, "Gouvernance", *Publictionnaire*, 28 February 2020. Retrieved from: <http://publictionnaire.huma-num.fr/notice/gouvernance/>. Accessed 27 April 2023 ; Ph. Hamman, *Gouvernance et développement durable*, De Boeck Supérieur, Louvain-la-Neuve, 2019, pp. 114–126.

Ph. Hamman (ed.), *Sustainability Governance and Hierarchy*, Routledge, London and New York, 2019, pp. 3–59.

15 D. Chassignet, "L'intégration transfrontalière et ses conséquences spatiales dans le Sud-Alsace", *Revue Géographique de l'Est*, vol. 36, n° 2, 1996, pp. 113–131.

16 Strasbourg made it to the final round but ultimately lost its bid. See O. Claudon, "Strasbourg veut le titre de capitale verte européenne", *Dernières Nouvelles d'Alsace*, 5 May 2019.

The Resurgence of Borders in Cross-Border Urban Areas as a Result of the Pandemic

Currently, in Europe, borders are not only materialisations of a boundary; they are also structures of constraints and opportunities[17] owed to differences in the standards, practices and representations that shape them. Organised cross-border urban cooperation is a fairly recent phenomenon. It took until the Karlsruhe Accords of 23 January 1996 to see a strictly cross-border legal basis enshrined, as signatory states allowed local authorities to establish, within a given perimeter, ad hoc structures named Local Groupings for Cross-border Cooperation, which they did in the Upper Rhine area. A decade later, Regulation (EEC) n°1082/2006 introduced the European Grouping of Territorial Cooperation (EGCT), an EU tool for the entire Union, applicable not only in cross-border areas.[18]

In the context of an ongoing metropolisation process in Strasbourg and Basel, there is a noticeable gap between an advanced functional integration exemplified by the circulation of cross-border workers and a lagging political integration regarding inter-regional institutions, which have no significant policy-making powers.[19] The governance of cross-border urban areas is characterised by asymmetries between stakeholders, leading to constant territorial transactions that have played out during the pandemic between policy-makers and representatives of the local populations.

The Covid-19 crisis began locally when, on 11 March 2020, the Robert-Koch Institute, Germany's federal health reporting body, designated the French Grand Est region as "at risk" for the virus, like China and Italy. The federal Minister of the Interior introduced strict restrictions:[20]

---

[17] D. Duez & D. Simonneau, "Repenser la notion de frontière aujourd'hui. Du droit à la sociologie", *Droit et société*, vol. 98, n° 1, 2018, pp. 37–52.

[18] See www.espaces-transfrontaliers.org. Accessed 2 January 2024.

[19] B. Dupeyron, *L'Europe au défi de ses régions transfrontalières. Expériences rhénane et pyrénéenne*, Peter Lang, Bern, 2008, pp. 190–241.

H. Nonn, *L'Alsace actuelle. Développement régional et métropolisation depuis les années 1950*, Presses universitaires de Strasbourg, Strasbourg, 2015, pp. 55–114.

[20] For a legal analysis: L. Jaillet, "'Fermeture' de la frontière franco-allemande et Covid-19", *Revue générale du droit on line*, n° 51886, 2020. Retrieved from: www.revuegeneraledudroit.eu/?p=51886. Accessed 2 January 2024.

entry on German territory from France would be authorised only for diplomats and cross-border workers in possession of an ad hoc certificate. The Minister argued that in the absence of a joint European decision, acting at the national level was necessary: "One shouldn't be surprised that countries are closing their borders when they can't see a solution to protect their own population". His French counterpart struck a reassuring tone in response, claiming that "These checks do not mean that our common border is closing".[21] There was in effect little coordination, no more than between the German *Bund* and *Länder* (federal and state level), where, as of mid-March 2020, measures ranged from a lockdown in Bavaria to no restrictions in North Rhine-Westphalia.[22]

Practical repercussions in cross-border urban areas came swiftly. The German city of Kehl, across the Rhine from Strasbourg, immediately recommended its France-based employees and Kehl residents working in France or with children in school in Strasbourg to "stay home". Cooperation between French customs and German police at the border was suspended, and joint controls in TGV/ICE high-speed trains were halted in border stations. German companies asked their German employees to refrain from crossing into France and their French employees not to come into work in Germany (as did the famous Europa Park theme park).[23]

Public health measures thus had social consequences at the cross-border level. Yet, they came from German institutions and did not involve consultations or exchanges. On the very day the Robert-Koch Institute issued its statement, the regional health agency for Grand-Est pointed out that five of the region's nine departments were "practically

---

21 S. Roland, "L'Allemagne verrouille à son tour ses frontières", *Libération*, 15 March 2020. Retrieved from: <https://www.liberation.fr/planete/2020/03/15/l-allemagne-verrouille-a-son-tour-ses-frontieres_1781776/>. Accessed 2 January 2024.

22 F. Berrod, B. Wassenberg & M. Chovet, "La frontière franco-allemande au temps du Covid-19: la fin d'un espace commun?", *The Conversation*, 16 April 2020. Retrieved from: <https://theconversation.com/la-frontiere-franco-allemande-au-temps-du-covid-19-la-fin-dun-espace-commun-136467>. Accessed 2 January 2024.

23 S. Mazzucotelli, "Coronavirus: le Grand Est déclaré comme zone à risque par un institut allemand", *Le Républicain Lorrain*, 11 March 2020.

R. Bloch, "En Alsace, les travailleurs transfrontaliers font les frais du coronavirus", *Les Échos*, 12 March 2020.

F. Recker, "Coronavirus: à la frontière franco-allemande, le retour des 'vieux démons'", *Public Sénat*, 21 April 2020.

not affected" by Covid-19,[24] evidencing that the decision had been made strictly on a German institutional level. Likewise, Swiss authorities decided to channel cross-border traffic starting on the 17th of March, by closing a dozen border crossings with Alsace, including those located in areas of urban continuity, such as the one on Kohlenstrasse between Basel and Huningue. The statement they issued cited an openly national motive: "This measure is intended to protect the Swiss population and to maintain the capabilities of the Swiss healthcare system".[25]

The Franco-German border was also reinstated with the Strasbourg-Kehl Eurodistrict. Starting on the 16th of March, upon entry in Kehl, the German police controlled each individual – an example is provided here by this slice of life reported by a local journalist: "Thursday April 23rd, 10.51 a.m., a German regional train bound for Offenburg leaves the Strasbourg train station. Three men are on board. At the second stop, the journey ends for them. When controlled by the German police, the three passengers are found not to have a certificate of housing or work in Germany. They are thus ordered to go back to France on foot."[26] In a powerful symbol, at the end of the Beatus Rhenanus Bridge over the Rhine, fences prevented access to pedestrians, and the cross-border tramway line no longer connected the two cities.[27]

In response to what were perceived as blanket measures, some local French and German political representatives mobilised to defend cross-border urban continuity on the ground against "top-down" dictates. The 30th of April call by 15 Alsatian and Mosellan members of the government majority to the French and German Ministers of the Interior reads as an attempt to regain control in the face of the imposed return of the national border, citing the risk of disruption of the cross-border living area, the conflict-laden legacy of Franco-German history and the "European ideal": "The closure of the border between our two friendly countries has reopened wounds of Franco-German history that took over fifty years for our ancestors to heal. […] Only preventive measures

---

24 *Ibid.*

25 J.-C. Meyer, "Bâle: une dizaine de petits postes frontières fermés", *L'Alsace*, 17 March 2020.

26 K. Guillaume, "Grillages, contrôles et incompréhensions: le retour de la frontière franco-allemande", *Rue89Strasbourg*, 25 April 2020.

27 *Ibid.* See also : F. Berrod, B. Wassenberg & M. Chovet, "La frontière franco-allemande au temps du Covid-19", *op. cit.*

and European mutual aid, such as taking in French patients in German, Luxembourgian or Swiss hospitals, can truly save lives".[28]

Such stances reflected two levels of tension. First, there was a tension between policymakers, particularly between the national and local levels. During the Franco-German Parliamentary Assembly's May 28th session, the two ministers of the Interior vowed to give more consideration to the specificity of cross-border urban areas: "There is a significant impetus for progress on these questions. It will allow us to be more reactive and to act on the finer level of micro-regions, which will no longer require closing borders".[29] They acknowledged that a "consensus is needed on public health matters", in response to an Alsatian member of parliament's impassioned appeal: "What became of Franco-German friendship, what happened to Europe, when on March 16th, the border was closed off by gates and police officers? [...] All of a sudden, the memory of a painful border along the Rhine [...] resurfaced along with the memory of a dark time we thought had ended forever."[30]

The return of the everyday experience of the border also brought back perceptions of the Other, informed by jealousy and resentment in tight-knit areas. It bears reminding that during the 1990s, many German nationals moved across the border to small towns in Alsace. Enticed by lower housing and land prices on the French side, they often settled in fairly isolated housing developments without necessarily knowing much French, which contributed to reawakening some tensions, associated with the perception of Germans as having a higher purchasing power.[31]

As Anne Rasmussen put it, "behind the question of quarantine, the history of states' uses of their borders lurked";[32] at the same time, the situation could also be seen as "the end of a taboo" for the European Union: "Europe was built on the utopia of the disappearance of borders;

---

28 S. Mazzucotelli, "Réouverture des frontières avec la Sarre et l'Allemagne pour le 11 mai?", *Le Républicain Lorrain*, 30 April 2020.

29 S. Mazzucotelli, "Frontière franco-allemande: la levée des restrictions confirmée pour le 15 juin", *Le Républicain Lorrain*, 29 May 2020.

30 F. Florin, "Coronavirus: réouverture complète et sans restriction de la frontière franco-allemande le 15 juin", *France 3 Grand Est*, 31 May 2020.

31 Ph. Hamman, "Les travailleurs transfrontaliers, entre pratiques et représentations...", *op. cit.*

32 A. Rasmussen, "Coronavirus: Derrière l'enjeu de la quarantaine se profile toute l'histoire de l'usage que les États ont fait de leurs frontières", *Le Monde*, 6 March 2020.

it has worked to gradually devaluate them. [...] However, the European Union is now faced with the task of redefining the function of its borders."[33] Historically, borders have been entrusted with three intersecting functions: protection (initially, protecting a territory from the enemy, like the walls on the edges of cities in the Middle Ages), control, and its corollary, passage. Borders play a filtering role in the exchanges across it, and differentiate between models, systems of norms and technical schemes; individuals may experience this as a constraint, but they can also take advantage of it, having for instance the opportunity to be better paid.[34] There is a dialectic at work between spatial proximity and social distance, reminiscent of Norbert Elias and John Scotson's analysis in *The Established and the Outsiders.*[35]

Elected representatives on both sides of the Rhine responded to mounting tensions by launching a Franco-German manifesto entitled "Borders are closing but not our hearts". The president of the Alsace-Lorraine cross-border workers association in Haguenau reported that: "Some people have told us that they are not allowed to have lunch in the same canteen as the German workers and some employers have already started to lay people off". The Grand Est regional councillor in charge of transport also pointed out that "Some [cross-border workers] have reported spending up to two hours at the border crossing and being subjected to finicky checks pertaining to their tires or the validity of their vehicle safety inspection – nothing to do with the coronavirus". On the German side, the mayor of Birkenhördt emphasised the local context: "These finicky

---

33 F. Berrod & P. Bruyas, "Union européenne: la frontière comme antidote à l'épidémie?", *The Conversation*, 29 March 2020. Retrieved from: <https://theconversation.com/union-europeenne-la-frontiere-comme-antidote-a-lepidemie-134844>. Accessed 2 January 2024.

34 Ph. Hamman, "Frontières: regards sociologiques", *Textes et Documents pour la Classe*, n° 1127, 2020, pp. 68–71.

G. Hamez, F. Morel-Doridat, K. Oudina, M. Le Calvez, M. Boquet, N. Dorkel, N. Greiner & S. de Pindray d'Ambelle, "La frontière 'nationale' brouillée par le Covid-19", *The Conversation*, 1 April 2020. Retrieved from: <https://theconversation.com/la-frontiere-nationale-brouillee-par-le-covid-19-134443>. Accessed 2 January 2024.

35 N. Elias & J. Scotson, *The Established and the Outsiders*, SAGE, New York, 1994, pp. 1–36. In their analysis of the English community of Winston Parva in the late 1950s/early 1960s, they pointed out a "racism without racists" and an "exclusion without an economic divide", which can only be understood in relation to territorialised processes of construction of "village" social cohesion.

checks at the borders are intolerable. They are often performed by agents who come from other regions and have no knowledge of local history".[36] The press published multiple stories about these tensions. For instance, the mayor of Gersheim (Germany), near the French border, told the T-online news website that "some people have been stopped in the street and received abuse, there are French people who are afraid of coming here now", citing a slur overheard in a supermarket: "Go back to your Coronaland!"[37]

In addition to economic relations, social relations are also at stake, then, as was demonstrated by the reopening of the Franco-German border on the 15th of June 2020. Mayors and collectives staged symbolic marches, and ceremonies were held on the Three Countries Bridge which connects the cities of Huningue (France) and Weil-am-Rhein (Germany), and on the Mimram footbridge between Strasbourg and Kehl. The elected officials gathered on the 15th of June at the former border crossing between Schweigen (Germany) and Wissembourg (France) drew one conclusion from the experience: "The biggest issue in this situation has been lack of communication. The impact on people's everyday lives was not considered at all", said the mayor of Birkenhördt. "We'll have to see how these borders will disappear in people's minds now. It will take some time", an Alsatian regional councillor added.[38]

This is a particularly relevant question given that the pandemic has periodically resurged, sparking the reinstatement of restrictions at the borders: for instance, in late August 2020, when the Swiss Federal Office for Public Health said that adding France to its list of at-risk countries

---

[36] R. Willhelm, "Coronavirus: tensions à la frontière franco-allemande, un manifeste contre la haine recueille 200 signatures", *France 3 Grand Est*, 19 April 2020.

F. Sergeur, "Refoulée à la frontière entre la France et l'Allemagne, elle passe la nuit dans sa voiture", *Capital*, 3 May 2020.

[37] C. Richard-Le-Bris, "Coronavirus: l'Allemagne condamne des actes anti-français à sa frontière", *RTL*, 11 April 2020.

P. Thibaut, "Coronavirus: des Français insultés dans la Sarre, région frontalière allemande", *RFI*, 12 April 2020.

F. Recker, "Coronavirus: à la frontière franco-allemande", *op. cit.*

[38] "Réouverture de la frontière: une cérémonie sur la passerelle des Trois Pays à Huningue", *L'Alsace*, 15 June 2020.

F. Grandon, "Coronavirus: les frontières rouvrent très lentement entre la France et l'Allemagne", *France 3 Grand Est*, 16 May 2020.

was "a possibility", or when France introduced a second national lockdown starting on the 29th of October 2020 and Germany began its own partial lockdown on the 2nd of November. The implementation of these measures elicited divergent interpretations. On the afternoon of the 30th of October, the prefect for the Bas-Rhin department of Alsace told the press that Alsatians could no longer go shopping in Germany and only cross-border workers were allowed to cross; later in the day, the German authorities announced that no new restrictions to circulation in Baden-Wurttemberg, Rhineland-Palatinate and Saarland would be imposed for anyone coming from France for under 24 hours, while recommending to avoid tourist or personal convenience trips.[39]

Things became even more complex after the 15th of December 2020 for two sets of reasons. First, the lifting of the lockdown in France (replaced by a curfew) coincided with the decision to impose a new lockdown at the German federal level on the 16th of December. Second, the *Länder* again introduced different rules, as had been the case in the spring of 2020. While a resident of Alsace was allowed to cross the border for under 24 hours without showing a negative PCR test result, rules pertaining to their domicile varied: "Baden-Wurttemberg demands proof of principal or secondary residence in Alsace, Rhineland-Palatinate requires no particular place of residence, and Saarland demands that you reside in the 'border region' ", as the regional press put it.[40]

In the case of Baden-Wurttemberg, an "essential reason" was required – the rule was summed up as follows by the Minister-President: "Einkaufen ja! Shopping Tour nein!" (meaning roughly "Grocery shopping yes, window shopping no") – with police checks announced.[41] While intended as general, these measures were again decided unilaterally: they targeted both Alsatians and Baden-Wurttemberg residents who might be tempted to cross the border for Christmas shopping; the latter were

---

39 "Finalement, les Alsaciens pourront faire des courses de première nécessité en Allemagne", *Dernières Nouvelles d'Alsace*, 31 October 2020.

F. Haby, "Déplacements transfrontaliers: ce qui reste autorisé", *Dernières Nouvelles d'Alsace*, 6 November 2020.

40 "Courses en Allemagne: on fait le point", *Dernières Nouvelles d'Alsace*, 15 December 2020.

41 *Ibid.*

asked to isolate for ten days upon return, in order to discourage them.[42] The "Umbrella movement", which brings together a number of associations and Europhile personalities (*Unir l'Europe*, the Young European Federalists, etc.) with a shared commitment to Franco-German reactions, issued a statement in response, deploring that "the government of the *Land* of Baden-Wurttemberg does not explain in what way its decision is absolutely necessary in light of the public health situation" and attacking "protectionist measures from another era […] antithetical to the spirit of the European Union", disregarding the reality of cross-border *vivre-ensemble* (literally living together, meaning community well-being): "These restrictive measures, imposed from the top on one side of the border or the other, are meaningless to those who live here. […] We need to have a different state of mind in 2021, make everyone in the Rhine region understand that we have to restore our community life!"[43] Cross-border urban cooperation has intertwined public health, socio-economic and socio-cultural implications.

Public Health Cooperation, an Under-Invested Field in Cross-Border Urban Areas?

The testimony of a physician residing in Strasbourg and working in the hospitals of Rastatt and Baden-Baden, which appeared in the regional media in May 2020, is particularly interesting in this regard. He expressed reservations about the border closing, confirming the difference in the two countries' health situation: "Closing borders reassures people and we know how to do it. […] Is it really useful though? […] The Germans got scared. But we did see the difference, my wife works at the civil hospital in Strasbourg, and she's been doing management. It's been less flagrant for us in Germany."[44]

This quote alludes to three levels of cross-border public health cooperation. The first is the EU framework. The EU has no direct competence

42 "Le Bade-Wurtemberg veut empêcher le shopping transfrontalier", *Dernières Nouvelles d'Alsace*, 22 December 2020.

P. Von Ditfurth, "Nun also doch: Baden-Württemberg schränkt den Grenzverkehr ein", *Badische Zeitung*, 22 December 2020.

43 C. Joubert, "Un collectif alsaco-badois: 'les frontières doivent rester ouvertes sans restriction'", *Dernières Nouvelles d'Alsace*, 27 December 2020.

44 F. Grandon, "Coronavirus: les frontières rouvrent très lentement", *op. cit.*

on public health matters; it only has jurisdiction to harmonise national rules, especially concerning drug safety, and a "supporting competence" to propose forms of coordination.[45]

The second level of cooperation is that of national healthcare systems, in terms of capacities and organisation. Regarding the latter, the difference is very sharp: Germany had 25,000 intensive care beds available when the crisis began, and 40,000 later, whereas France only had 5,000 initially and 15,000 at best later, which had effects on saturation. The same applies to testing: up to 500,000 tests were performed weekly in Germany in March and April 2020, compared to only 35,000 to 80,000 in France. This national contrast poses a first challenge for joint management in cross-border areas.[46] There is also an organisational one. The case of France's regional health agencies (ARS) is telling. According to a report submitted to the Minister of Health in 2008, they were precisely established to ensure "a unified, responsible steering of the territorial healthcare system", which was previously divided between state services, the national social security body and the regional agencies for hospitalisation. Yet, elected representatives criticised their functioning during the crisis, beginning with the Grand Est officials who spoke on the 30th of April to the Senate's delegation for local authorities. The president of the Haut-Rhin departmental council used the term "tsunami" in her discussion of early crisis management, and argued that the ARS should be placed under the prefect's authority to operate more effectively.[47]

This was yet another iteration of the debate on the need for a "leader" within the French landscape of plural, shared territorial competences, correlated with the relationships between national authorities and elected officials on the ground. The latter have seen their role reassessed since the beginning of the pandemic.[48] Yet, the question keeps recurring. In

---

45 A. Frachon, "La crise du coronavirus doit amener l'Europe à changer de logiciel économique", *Le Monde*, 19 March 2020.
F. Berrod & P. Bruyas, "Union Européenne", *op. cit.*

46 P. Maurice, "Coronavirus: la solidarité franco-allemande à l'épreuve", *Éditoriaux de l'Ifri*, April 2020, pp. 1–2. Retrieved from: <https://www.ifri.org/fr/publications/editoriaux-de-lifri/coronavirus-solidarite-franco-allemande-lepreuve>. Accessed 2 January 2024.

47 See <https://www.banquedesterritoires.fr/covid-19-face-aux-collectivites-il-faut-un-chef-de-file-unique-et-cest-le-prefet>. Accessed 2 January 2024.

48 "First, to address national shortcomings in terms of health protection, but also to provide emergency responses in each of their competence areas. Amid this major crisis,

response to a resurgence in viral circulation in late August 2020, when the Bas-Rhin prefect made mask wearing mandatory in the department's thirteen towns with a population over 13,000, a decision officially made "in close cooperation with public health authorities and after exchanging with elected representatives", several of the mayors concerned publicly complained about not having been consulted.[49]

The third level of cooperation is local. Gaps in the availability of medical specialists and hospital equipment such as MRI scanners have been abundantly noted before. As of late 2019, the European Consumer Centre in Kehl reported that patients were asked to wait an average of a month for an emergency MRI scan in France, compared to a few days in Germany. While the Directive 2011/24/EU has facilitated access to cross-border healthcare, allowing a French patient to have consultations with a German doctor reimbursed by French social security, for some procedures, such as MRIs, prior approval by the health insurance fund remains mandatory in France.[50]

Hence these words by the head of the Pamina Eurodistrict (Palatinat-Mittler Oberrhein-Nord Alsace): "The only positive aspect about the border closure is that it has shown us everything that isn't working; for instance, the mutualisation of healthcare is non-existent. We would have needed to be aware of intensive care capacities on both sides of the Rhine. If we improve on this, if we pool our resources and we constantly stay in touch, we'll no longer close borders". The arrangements made to transfer Alsatian patients with serious forms of Covid-19 was cited as practical evidence of this: "For our part, locally, with [the mayor of Karlsruhe], we have already managed to arrange for the transfer of 4 Covid patients from Strasbourg to Karlsruhe, thanks to our relationships and without waiting for national decisions."[51] Considering the saturation of Alsatian hospitals, the Minister of Health for Baden-Wurttemberg asked "all hospitals in the *Land* to take in patients from the cross-border region between France and

all involved, both at State and local level, have realised the importance of working in close collaboration", as *Le Monde* columnist Patrick Roger put it. See P. Roger, "Coronavirus: entre l'État et les collectivités locales, une coopération pas toujours tranquille", *Le Monde*, 23 April 2020.

49 "La préfète impose le port du masque dans les principales communes du Bas-Rhin", *Dernières Nouvelles d'Alsace*, 28 August 2020.

50 See <https://www.cec-zev.eu/de/startseite/>. Accessed 2 January 2024.

51 F Grandon, "Coronavirus: les frontières rouvrent très lentement", *op. cit.*

Germany that are requiring respiratory assistance as far as existing capacities allow them".[52] The president of the departmental council for Haut-Rhin publicly applauded this decision: "I was very touched to receive a very quick and positive response from the neighbouring Swiss cantons and that the Minister-President called hospitals in Baden-Wurttemberg to take in Alsatian patients in intensive care. […] We are a true cross-border territory, in good times and in times of crisis alike."[53] She went on to call for further cross-border urban cooperation in the healthcare field: "We should set up a healthcare cooperation scheme that will, in times of crisis and in normal times, allow our hospitals to support each other, to share their practices and their knowledge."[54] Most local elected representatives in the Grand Est region have concurred with this view, like the mayor of Riedseltz (Bas-Rhin): "The positive thing is, there is a potential for harmonisation here".[55]

Depending on the territory considered, the crisis has highlighted "economic interdependences (regarding the supply of goods, the sudden disruption of cross-border or seasonal work) or functional ones (the healthcare system […] is heavily dependent on French cross-border workers)", according to a report by the Transfrontier Operational Mission (MOT).[56] Urban revitalisation and the cross-border healthcare economy are interrelated, including on the level of cross-border workers' flows and of their own social protection schemes.

---

52 "Coronavirus: l'Allemagne est prête à accueillir des malades français", *LCI*, 21 March 2020.

53 P. Hugues, "Coronavirus: les hôpitaux allemands à la rescousse de l'Alsace", *Le Point*, 29 March 2020.

P. Maurice, "Coronavirus…", *op. cit.*, pp. 3–4.

While the transfer of some 130 patients to Germany, sometimes in military planes from the Bundeswehr, which showed that local initiatives were supported at the national level, made headlines, around 200 patients were also transferred from Alsace and Lorraine to Luxembourg and Switzerland: MOT, *La crise du covid-19 aux frontières: retours d'expérience du réseau de la MOT*, report, June 2020, p. 15. Retrieved from: <http://www.espaces-transfrontaliers.org/la-mot/publications/la-crise-du-covid-19-aux-frontieres-retours-dexperience-du-reseau-de-la-mot/>. Accessed 2 January 2024.

54 "Les hôpitaux allemands accueillent des patients français", *La Dépêche du Midi*, 26 March 2020.

55 F. Recker, "Coronavirus: à la frontière franco-allemande", *op. cit.*

56 MOT, *La crise du covid-19 aux frontières, op. cit.*, p. 25.

Socio-Economic Interdependences in Cross-Border Urban Areas

While the phenomenon of cross-border work epitomises the principle of the free movement of persons in Europe, it is concentrated in small areas. There are over 40 % of transnational commuters in the population in employment around Wissembourg, at the Baden-Wurttemberg border, and around Saint-Louis, near the industrial hub of Basel.[57] In such proportions, cross-border work has a key influence in structuring urban development, spurring investments, economic and demographic growth, cross-border consumption, lower unemployment rates, etc.; but also in destructuring it by causing housing and land prices to rise for non-cross-border residents, traffic congestion, and other nuisances.[58]

The Covid-19 crisis has impacted these socio-economic interactions in multiple ways. First, the very status of cross-border worker is based on cross-border commuting, which posed challenges when German bosses forced employees to stay home: "Some have ended up with fifteen unpaid leave days on their pay slips; others were forced to go on vacation leave", the president of the Association of Alsace-Lorraine cross-border workers in Haguenau said, knowing that Germany's partial unemployment scheme has experienced "harmonisation issues" with France in fiscal matters.[59] Issues also arose with remote working, which was largely encouraged during the lockdowns. A French cross-border worker contributes to the German social security plan if they are employed in that country and if they spend under 25 % of their time working remotely from France. The French Minister of Labour has had to invoke "force majeure" to justify their continued affiliation to the German regime beyond that threshold, in an initial agreement of the 19th of March 2020.[60] The Franco-German

[57] INSEE Grand Est, May 2022. See <https://www.insee.fr/fr/statistiques/6444588>. Accessed 2 January 2024.

[58] Ph. Hamman, *Sociologie des espaces frontières, op. cit.*

Ph. Hamman, "Cross-border worker", *op. cit.*

Ph. Hamman, "Les travailleurs transfrontaliers, entre pratiques et représentations…", *op. cit.*

[59] F. Recker, "Coronavirus: à la frontière franco-allemande", *op. cit.*

[60] F. Berrod, B. Wassenberg & M. Chovet, "La frontière franco-allemande au temps du Covid-19", *op. cit.*

MOT, *La crise du covid-19 aux frontières, op. cit.*, pp. 14–15.

and Franco-Swiss agreements to consider remote working from one's country of residence as cross-border work were subsequently extended.[61]

Additionally, cross-border urban areas are characterised by crisscrossing flows, and by the importance of mobility. Although they are fewer than those who do the opposite, there are also border residents that have kept their job in their country of origin (Germany, Luxembourg or Switzerland) and made a home in France, taking advantage of often lower land and housing prices, or of the opportunity to live in a big city like Strasbourg. When unilaterally decided border checks were implemented, this diversity of situations was not fully considered. For instance, when restrictive measures for entry into Germany were taken in the spring of 2020, while French citizens were turned away, cross-border workers asked to present their special authorisation or *Passierschein*, and their vehicles were sometimes searched, German citizens residing in Alsace were able to cross simply by showing their German passport. Cross-border workers were also not allowed to make purchases in German stores during that same period.[62] This resulted in complex individual situations: "For me, a person who lives in France, it is a bit tricky. I learned on Thursday that I would have to go back to work on Friday, and I didn't pack a lunch. And I realised that I'm not allowed to buy anything or to go into a store in Kehl. Even though I'm working in a store that receives customers all day."[63] Prominent local and regional elected representatives denounced this as discrimination: "I am ashamed of Baden-Wurttemberg. […] This has been the worst thing about this period, this discrimination against the French", said the mayor of Karlsruhe, much like the president of the Haut-Rhin department, who cried: "This is just unimaginable and it's discrimination!"[64]

Ultimately, in the urban areas of north-eastern France, the vicissitudes resulting from the public health crisis of spring 2020 have paradoxically

---

61 See <https://frontaliers-grandest.eu/accueil/salaries/france-allemagne/coronavirus-covid-19/teletravail-covid19/>. Retrieved from: <https://www.netpme.fr/actualite/frontaliers-en-teletravail-maintien-des-accords-amiables-jusquen-septembre-2021/>. Accessed 2 January 2024.

62 J.-P. Tranvouez, L. Klethi & B. Bour, "Coronavirus: la frontière allemande rouverte à Petite-Rosselle", *France 3 Grand Est*, 1 April 2020.

63 F. Grandon, "Coronavirus: à Kehl, ville d'Allemagne voisine de Strasbourg, les habitants s'équipent de masques avant le 27 avril", *France 3 Grand Est*, 22 April 2020.

64 Grandon F., "Coronavirus: les frontières rouvrent très lentement", *op. cit.*

exacerbated the agglomerating effects of the borders (despite their closure), especially considering that "the border has a powerful filtering effect that is owed more to the challenges of communication between the actors rather than of the importance of control by States."[65] On a broader level yet, the crisis highlights the question of the meaning and scope of cooperation in cross-border regions: should those be designed to boost economic development and/or to become areas of shared solidarity? In that sense, borders are a "reflection of overall transformations in governance",[66] both at the cross-border level and more largely in a multi-level perspective[67] encompassing local, national and European authorities. In many ways, although they have been heralded as "laboratories" of European construction (in terms of free movement, territorial cohesion, economic interdependences, etc.), these spaces remain noticeably undervalued at the European policy level (beyond the dedicated funds and schemes) and in national official political stances (beyond bilateral diplomacy and symbolic cases of decentralised cooperation, inherited from twin cities programs).

These regulatory instabilities show what living in such an in-between state means on a daily basis and evidence the multiplicity of cooperations at work in these urban areas; retail stores are a particularly telling example. Indeed, while death stats and images from hospitals effectively differed, socio-economic interactions between the two sides of the border were thrown into stark relief. Kehl, for instance, was emptied out of the Alsatian customers whose presence usually makes stores and businesses thrive. A tree nursery owner at the town market put it very clearly on the 4th of April 2020: "We make a living from the Alsatians, it's very tough for us right now. […] The future is bleak, I'm not sure we'll make it and avoid going bankrupt."[68] Territorial transactions are always partial and provisional; it remains to be seen now to what extent the effects of the coronavirus crisis will be characterised by transactions in continuity or

---

65 B. Reitel, T. Perrin & P. Pupier, "La métropolisation transfrontalière" *In* F. Moullé (ed.), *Frontières*, Presses universitaires de Bordeaux, Bordeaux, 2017, pp. 215–216.

66 F. Leloup & S. Pradella, "Gouvernance, gouvernance multiniveaux et gouvernance transfrontalière", *In* F. Moullé (ed.), *Frontières*, *op. cit.*, p. 146.

67 F.-M. Poupeau, *Analyser la gouvernance multi-niveaux*, Presses universitaires de Grenoble, Grenoble, 2017, pp. 9–15.

68 F. Grandon, "Coronavirus: Kehl en Allemagne, la voisine de Strasbourg, n'est pas confinée mais souffre de l'absence des Français", *France 3 Grand Est*, 4 April 2020.

rupture in national and local policy within urban areas along the French borders.[69]

The case of cross-border workers illustrate practical compromises that are made, unmade and remade. During the lockdowns of November 2020 in France and Germany, cross-border workers were still allowed to travel across the border upon presentation of a certificate by their employer.[70] Likewise, in early February 2021, as German federal authorities were considering classifying France as a "high-incidence zone", the *Land* of Baden-Wurttemberg quickly issued a statement to say that it was looking into the option of bearing the cost of antigen tests for cross-border workers.[71] 300,000 tests were effectively made available to employers in late March, as the obligation of showing proof of a negative test to cross the border entered into force (until the 23rd of May[72]). Alsatian cross-border workers were also entitled to a partial exemption to the German rule on the tests' validity period – they were asked to do two tests a week as opposed to one every 48 hours. "Thanks to our tireless talks with the German authorities, including the presidents of the *Länder* on the border and my German counterpart, we have been able to mitigate the impact of this classification, putting the priority on preserving the everyday and work lives of French cross-border workers. We will keep up this cooperation process in the hope that these restrictions will stay for as short a period of time as possible", the French Assistant Minister for European Affairs declared.[73]

These small arrangements have also been informed by practical constraints in areas with high number of cross-border workers. "Between having to go to a pharmacy to get tested every two or three days and having a test provided by your company at your place of work, it's not the same thing in my view. People were helpless. There's been a lot of stress, and it could've been relieved by informing them", an Alsatian

69 For an overarching perspective, see MOT, *La crise du covid-19 aux frontières, op. cit.*

70 Likewise, Alsatian students in Franco-German secondary schools, in Freiburg, for instance, were asked to show a certificate from their school: F. Haby, "Déplacements transfrontaliers", *op. cit.*

71 "Si le passage de la frontière se durcissait, le Bade-Wurtemberg prêt à payer les tests antigéniques des frontaliers", *Dernières Nouvelles d'Alsace*, 5 February 2021.

72 "Covid-19: l'Allemagne n'oblige plus les frontaliers à se faire tester", *Le Républicain Lorrain*, 23 May 2021.

73 J. Steinhauser, "Frontière: votre test svp !", *Dernières Nouvelles d'Alsace*, 26 March 2021.

cross-border worker points out. Pharmacies located near the German border were overextended, as in Wissembourg: "It's complete mayhem. So many cross-border workers coming for a test. Most of them complain. It requires a lot of organisation"; "We couldn't manage appointments on the phone any longer, we had to set up an online appointment system. If all cross-border workers have to get tested continually and there's no openings left, I don't know what people are going to do"; "We're overwhelmed by requests".[74] Similar bottlenecks were observed on the German side, particularly in Kehl, near Strasbourg, leading the president of the Freiburg district to officially ask French cross-border workers to favour testing at the workplace or during their commute, or in the offices of German physicians.[75]

It is worth noting that on the 1st of August 2021, when the German federal government reinstated restrictions at the borders as the risk of a fourth wave of Covid-19 loomed with the emergence of the Delta variant, cross-border workers were exempted from the mandatory test rule; so was anyone crossing the border for under 24 hours. During its busy summer season, the highly popular theme park in Rust, near Strasbourg, made a point to announce that "the new federal regulation for entry on the German territory provides for no restrictions for Europa-Park visitors".[76]

This exemplifies the complexity inherent in the transposition of measures at different levels, when national and cross-border repertoires are embedded in a supra-national pandemic context. Yet, this is also a confirmation of the reality of the border as a specific space, characterised by socio-economic interdependences embodied by daily flows of commuters and shoppers. We know that globalisation processes do not make borders disappear: they produce new perimeters of action and new dividing lines that are more specialised that in the Westphalian state (as in the example of health controls), and that are closely connected to "the need to gain an awareness of the territories that we inhabit".[77] At the inter-regional

---

74 "Le grand flou des tests Covid pour les frontaliers", *Dernières Nouvelles d'Alsace*, 29 March 2021.

75 "Les travailleurs frontaliers appelés à se faire tester sur leur lieu de travail", *Dernières Nouvelles d'Alsace*, 3 April 2021.

76 "Un pass sanitaire… sauf pour les frontaliers et les séjours cours", *Dernières Nouvelles d'Alsace*, 31 July 2021.

S. Wiessler, "Covid-19: test négatif, certificat de vaccination… l'Allemagne durcit l'entrée sur son territoire", *Le Républicain Lorrain*, 31 July 2021.

77 M. Foucher, "Le retour des frontières", *Études*, n° 9, 2017, pp. 7–18.

scale of the Upper Rhine region, this phenomenon directly interacts with dynamics of production of borderlands, their outlines and their representations, which themselves may contest established national subdivisions. These frictions are increasingly intertwined and of a social and political nature.

Ultimately, the Covid-19 crisis is not simply a test for cross-border urban governance; it also reveals practical interdependences at the European and inter-local levels, whose existence calls for the development of shared, coordinated territorial policies.

Bibliography

F. Berrod & P. Bruyas, "Union européenne: la frontière comme antidote à l'épidémie?", *The Conversation*, 29 March 2020.

F. Berrod, B. Wassenberg & M. Chovet, "La frontière franco-allemande au temps du Covid-19: la fin d'un espace commun?", *The Conversation*, 16 April 2020.

M. Blanc & Ph. Hamman, "La construction politique métropolitaine dans le Rhin Supérieur: Coopérations intercommunales et transfrontalières. L'exemple de l'agglomération de Strasbourg" *In* C. Maurer & A. Starck-Adler (eds.), *L'Espace rhénan, pôle de savoirs/Rheinischer Raum, Wissensraum*, Presses universitaires de Strasbourg, Strasbourg, 2013, pp. 223–244.

D. Chassignet, "L'intégration transfrontalière et ses conséquences spatiales dans le Sud-Alsace", *Revue Géographique de l'Est*, vol. 36, n° 2, 1996, pp. 113–131.

R. Dormois, *Les politiques urbaines. Histoire et enjeux contemporains*, Presses universitaires de Rennes, Rennes, 2015, pp. 235–259.

D. Duez & D. Simonneau, "Repenser la notion de frontière aujourd'hui. Du droit à la sociologie", *Droit et société*, vol. 98, n° 1, 2018, pp. 37–52.

B. Dupeyron, *L'Europe au défi de ses régions transfrontalières. Expériences rhénane et pyrénéenne*, Peter Lang, Bern, 2008, pp. 190–241.

N. Elias & J. Scotson, *The Established and the Outsiders*, SAGE, New York, 1994, pp. 1–36.

M. Foucher, "Le retour des frontières", *Études*, n° 9, 2017, pp. 7–18.

G. Hamez, F. Morel-Doridat, K. Oudina, M. Le Calvez, M. Boquet, N. Dorkel, N. Greiner & S. de Pindray d'Ambelle, "La frontière 'nationale' brouillée par le Covid-19", *The Conversation*, 1 April 2020.

Ph. Hamman, "Les jumelages de communes Moselle-Charente: Territoires et identités frontalières France-Allemagne à l'heure de l'Europe", *Bulletin de l'Institut de la Décentralisation*, n° 87, 2001, pp. 15–32.

Ph. Hamman, "Les jumelages de communes, miroir de la construction européenne 'par le bas' ", *Revue des Sciences Sociales*, n° 30, 2003, pp. 92–98.

Ph. Hamman, "From 'Multilevel Governance' to 'Social Transactions' in the European Context", *Swiss Journal of Sociology*, vol. 31, n° 3, 2005, pp. 523–545.

Ph. Hamman, "Figures de l'État dans l''entre-deux': Le *Reichsland* d'Alsace-Lorraine, 'laboratoire' d'une politique de germanisation entre France et Allemagne (1871–1918)" *In* A. Chatriot & D. Gosewinkel (eds.), *Les figures de l'État en Allemagne et en France. 1870–1945 / Figurationen des Staates, Deutschland und Frankreich, 1870 bis 1945*, Oldenbourg, München, 2006, pp. 141–158.

Ph. Hamman, "Penser la différence à l'aune des espaces-frontières: une proposition d'analyse sociologique" *In* L. Denooz & S. Thiéblemont-Dollet (eds.), *Le moi et l'autre*, Presses universitaires de Nancy, 2011, pp. 29–39.

Ph. Hamman, *Sociologie des espaces-frontières. Les relations transfrontalières autour des frontières françaises de l'Est*, Presses universitaires de Strasbourg, Strasbourg, 2013, pp. 123–189.

Ph. Hamman, "Qu'est-ce qu'une Euro-région? Identifications et frontières au défi de l'institution dans l'espace du Rhin supérieur", *In* A. Koukoutsaki-Monnier (ed.), *Identités (trans)frontalières dans l'espace du Rhin supérieur*, Éditions universitaires de Lorraine, Nancy, 2014, pp. 53–76.

Ph. Hamman & S. Ruß-Sattar, "Les répertoires d'action transfrontaliers des communes françaises et des communes allemandes: une mise en parallèle, de 1950 à nos jours", *Revue des sciences sociales*, n° 60, 2018, pp. 14–25.

Ph. Hamman, "Borders and cross-border cooperation in Europe from a sociological perspective", *In* J. Beck (ed.), *Transdisciplinary Discourses on Cross-Border Cooperation in Europe*, Peter Lang, Bern, 2019, pp. 123–145.

Ph. Hamman, *Gouvernance et développement durable*, De Boeck Supérieur, Louvain-la-Neuve, 2019, pp. 114–126.

Ph. Hamman (ed.), *Sustainability Governance and Hierarchy*, Routledge, London and New York, 2019, pp. 3–59.

Ph. Hamman, "Cross-border worker" *In* B. Wassenberg & B. Reitel (eds.), *Critical Dictionary on Cross Border Cooperation in Europe*, PIE-Peter Lang, Brussels, 2020, pp. 268–271.

Ph. Hamman, "Les travailleurs transfrontaliers, entre pratiques et représentations autour des frontières françaises de l'Est", *In* N. Dziub (ed.), *Le transfrontalier: pratiques et représentations*, Éditions et Presses universitaires de Reims, Reims, 2020, pp. 163–177.

Ph. Hamman, "Frontières: regards sociologiques", *Textes et Documents pour la Classe*, n° 1127, 2020, pp. 68–71.

Ph. Hamman & B. Wassenberg, *Mémoire d'Europe – Mémoire de paix. Témoignages de la région frontalière d'Alsace*, Franz Steiner Verlag, Stuttgart, 2021, pp. 255–479.

Ph. Hamman, "Frontières fluviales ou espaces-frontières? Regards sociologiques", *Les cahiers du GEPE (Groupe d'Étude sur le Plurilinguisme Européen)*, n° 11, 2019.

L. Jaillet, "'Fermeture' de la frontière franco-allemande et Covid-19", *Revue générale du droit on line*, n° 51886, 2020.

F. Leloup & S. Pradella, "Gouvernance, gouvernance multiniveaux et gouvernance transfrontalière", *In* F. Moullé (ed.), *Frontières*, Presses universitaires de Bordeaux, Bordeaux, 2017, pp. 143–164.

P. Maurice, "Coronavirus: la solidarité franco-allemande à l'épreuve", *Éditoriaux de l'Ifri*, April 2020, pp. 1–2.

MOT, *La crise du covid-19 aux frontières: retours d'expérience du réseau de la MOT*, report, June 2020.

H. Nonn, *L'Alsace actuelle. Développement régional et métropolisation depuis les années 1950*, Presses universitaires de Strasbourg, Strasbourg, 2015, pp. 55–114.

F.-M. Poupeau, *Analyser la gouvernance multi-niveaux*, Presses universitaires de Grenoble, Grenoble, 2017, pp. 9–15.

B. Reitel, T. Perrin & P. Pupier, "La métropolisation transfrontalière", *In* F. Moullé (ed.), *Frontières*, Presses universitaires de Bordeaux, Bordeaux, 2017, pp. 215–216.

C. Rumford (ed.), *Citizens and Borderwork in Contemporary Europe*, Routledge, London and New York, 2014, pp. 1–12.

Legal Elements of a Cross-Border Friendly Legislation in the Upper Rhine

Michael Frey & Sarah Herrmann

Cross-border citizens living in border regions often face legal and administrative obstacles in their cross-border everyday reality. Using the example of the Franco-German border region on the Upper Rhine, the present text categorises the actual problems and shows the current legal channels for solving the problem.

On this basis, also in the light of the Aachen border treaty concluded between Germany and France, legal options are presented as to how such problems can be avoided *ex-ante* and solved more efficiently *ex-post* by means of cross border friendly legislation.

The text builds on a presentation by the author at the Jean Monnet-Centre of Excellence Research Workshop *Frontières en mouvement* (Frontem)[1] on the subject.

Description of the Actual State of Play

Types of Cross-Border Legal Problems in the Upper Rhine

If legal problems with cross-border implications arise in a cross-border region or better in a cross-border region such as the Upper Rhine, they usually and first arise with citizens who are active across the border – or, in order to formulate it under European law, nationals of European

1 Centre d'Excellence Jean Monnet franco-allemand, Jean Monnet Network "Frontiers in motion". Retreived from : <http://centre-jean-monnet.unistra.fr/the-project>. Accessed 24 April 2023.

Union (EU) Member States or third countries (such as Switzerland) who make use of their EU fundamental freedoms (such as the freedom of movement of goods, services, establishment or capital or, last but not least, the free movement of workers) or their fundamental rights (such as a family visit to the neighbouring country).

The coronavirus crisis has once again shown this impressively.

In total, three types of problems, provided they are of a legal origin, can be divided into three broad groups: First of all, and from a practical point of view, these are problems with the application of the law, which are characterised by the fact that the legislators had anticipated a legally unobjectionable, i.e. lawful solution to the problem in the legislative process and that only the bodies responsible for the application of the law, such as public authorities, do not apply it correctly, with the result that the action taken against the citizen is unlawful. These problems occur more frequently in cross-border practice compared to purely national law application because of the more complex legal situation, but only in individual cases or in a larger number of similar individual cases.

Secondly, problems often arise in the legislative process. The legislator or legislators have already failed to identify or incorrectly resolve a legal problem in cross-border situations in the context of legislation. The authorities responsible for the application of the law – in the absence of competence to overturn the law – apply correctly the law assigned to them. Nevertheless, this leads to the unlawfulness of the actions of the public authorities *vis-à-vis* the citizen.

There is, of course, a third group: problems which, from the point of view of cross-border citizens, are perceived as problems, but which do not lead to illegality as a result of incorrect legislation or misapplication of the law. This includes, first of all, problems with more complex and possibly uncoordinated double administrative procedures as a result of the cross-border situation, which make it more burdensome for citizens to find a legal solution, but also 'perceived' problems reported by citizens, which cannot be included in one of the categories already mentioned.

From the point of view of practice, it can be concluded that the "*Oberrhein* model"[2] uses the citizen as a vehicle for identifying

---

2 For example the Upper Rhine Conference model: J. Beck, *Transnationale Kooperationskultur in der grenzüberschreitenden Zusammenarbeit am Oberrhein – Zur Konzeption eines empirischen Untersuchungsansatzes, in* D. Beck, U. Becker-Beck, J. Beck

problems – a process that is characterised by the fact that the EU or the Upper Rhine states first legislate and only then try to identify, analyse and remedy any problems that may arise in the application of the law. That is to say, a subsequent *ex-post* problem solution and not, for example, the unevenly elegant way of avoiding *ex-ante* problems already in the legislative process by means of a legislative impact analysis which also anticipates cross-border issues and by means of appropriate solutions, such as the experimentation or derogation provisions also provided for in Article 13(2) of the Aachen Treaty.

The Long Way to Solve the Problem

From the point of view of the concerned citizen, there are several ways of solving the problem.

Beforehand the classic path: The citizen resolves his question before the competent court. This path is practically always the correct path, at least from a legal point of view, but is lengthy, expensive and cumbersome from the citizen's point of view. Sometimes even to this extent, the citizen may prefer strategies to avoid problems, e.g. by accepting the illegal solution or changing their way of life, when looking at costs and revenues.

At regional level, he is likely to turn to theINFOBEST-network, which has been operating successfully in this field for 30 years, with his problem at the Upper Rhine even sooner or later. The colleagues there will inform and advise him in a comprehensive, life-oriented and independent manner and forward it accordingly to the competent specialised authorities.[3] It is conceivable that in the course of this process our citizens will also turn to one of the four Eurodistricts in order to raise political awareness about the problem. Both institutions – the Infobest network and the Eurodistricts – should also try, as far as possible in terms of time and content, to document and legally process the problem and pass it on in their turn. For example, to the Expert Committee on Cross-Border

& A. Dussap (eds.) *Kultur der grenzüberschreitenden Verwaltungszusammenarbeit – Eine empirische Modellstudie am Beispiel der Oberrhein-Region*, Speyerer Arbeitsheft n°221, Speyer, 2015, p. 16.

3 INFOBEST Kehl/Strasbourg, 25 Jahre INFOBEST Kehl/Strasbourg, 2018. Retrieved from: <https://www.infobest.eu/fileadmin/data/user_upload/25_Jahre_INFOBEST_Kehl-Srasbourg_Jubil%C3%A4umsbroschure.pdf>. Accessed 24 April 2023.

Workers of the Upper Rhine Conference (*Expertenausschuss Grenzgänger der Oberrheinkonferenz – groupe d'experts Travailleurs frontaliers de la Conférence du Rhin Supérieur)*, from there to the German-French-Swiss Government Commission or the new Committee on Cross-Border Cooperation (*Ausschuss für Grenzüberschreitende Zusammenarbeit (AGZ) – Comité de Coopération transfrontalière (CCT*)) established by the Aachen Treaty. Both aim at finding a solution at national level, under the leadership of the Ministries of Foreign Affairs, with the involvement of national ministries. These paths are characterised by the fact that there are neither defined procedures with defined deadlines nor any kind of right to a problem-solving at the end of the procedure – or in short: informal, without notice and unsuccessful.

At European level, there is at least a digitally accessible body "Solvit" with which regional actors are not connected to problems of application of law.[4] For selected legislative problems, the "b-solutions" project launched by the AEBR and the Commission was able to provide independent legal expertise, whose task was to deal with the problem in law and to come up with proposals for solutions.[5]

However, major breakthroughs have not yet been achieved in these paths.

There was also a draft regulation on the so-called European Cross-border Mechanism (ECBM), which, unlike the provisions of the Aachen Treaty and the previous proceedings of the Upper Rhine Conference and the Government Commission, provides for a clearly defined procedure for dealing with problems and binding solutions, but which clearly lacks the political will of the nation states.[6]

---

4 European Commission, "Solvit". Retrieved from: <https://ec.europa.eu/solvit/index_en.htm>. Accessed 24 April 2023.

5 AEBR, "b-solutions". Retrieved from: <https://www.aebr.eu/projects/b-solutions/>. Accessed 24 April 2023.

6 European Commission, Proposal for a regulation of the European Parliament and of the Council on a mechanism to resolve legal and administrative obstacles in a cross-border context, COM/2018/373 final-2018/0198 (COD), 2018. Retrieved from <https://eur-lex.europa.eu/legal-content/EN/TXT/?uri=CELEX:52018PC0373>. Accessed 24 April 2023; Council of the European Union, Proposal for a Regulation of the European Parliament and of the Council on a mechanism to resolve legal and administrative obstacles in a crossborder context, 6009/20, 2020. Retrieved from: <https://data.consilium.europa.eu/doc/document/ST-6009-2020-INIT/en/pdf>. Accessed 24 April 2023.

Whether these pathways meet the criteria of good governance,[7] is at least doubtful.

Legal Elements of a Solution – Towards a Cross-Border Friendly Legislation

On the basis of the problem description, possible legal approaches to a solution, summarised under the term "cross border friendly legislation", should now be gathered and evaluated.

The common feature of all the solutions outlined below is that they are carried out at an early stage (i.e. already in the context of the legislative process and not only afterwards in the application of the law) and systematically in the formal (i.e. in principle taken into account in each legislative procedure) and substantive sence (in terms of content all areas of legislation are comprehensively and not specific to individual cases) and on the basis of a cross-border coordinated toolkit of standards.

Prior Rather Than Later: Identification of Standards with Cross-Border Effects and Explicit Cross-Border Impact Assessment in the Legislative Process through a Focused Regulatory Assessment

Instead of today's procedure described above, in which the cross-border effects of legislation are only identified, documented and transmitted after the legislative process has been completed at the application stage and at regional level, cross-border effects of legislative proposals should already be examined during the legislative process for their cross-border effects as part of a legislative impact assessment.

Already today, at European level, as well as in many Member States and, of course, in Switzerland, new legislation is being examined in the context of the legislative process by means of a regulatory impact assessment[8] on various aspects, such as their financial impact, on the economy

---

[7] R. Dolzer, M. Herdegen & B. Vogel (eds.). *Good governance. Gute Regierungsführung im 21. Jahrhundert*, Freiburg im Breisgau, 2007.

[8] European Parliament, "Regulatory Impact analysis (RIA) – Developments and current practices in the EU member states, on the EU level and in selected third

and on public finances. The same should be done for the cross-border effects of the provision.

This could be justified by the importance of the internal market and thus by the impact on the economy (Article 16 TFEU et seq.) of virtually all provisions with cross-border effects. However, in order to do so, it would first have to be known through the regulatory impact assessment and thus whether a legislative amendment would have cross-border effects.

By means of this procedure, all rules with cross-border effects could first be identified and, subsequently, legislation considering this dimension could identify at an early stage the number of problems encountered in cross-border practice and thus reduce it in anticipation by appropriate legislative measures.

Systematic Approach Instead of Case-Based Solution Procedure

In addition to a cross-border-friendly, technical and policy-specific legislation based on the regulatory impact assessment described above, which focuses on cross-border effects and which systematically considers all legislative proposals with potential cross-border effects already during the legislative phase, but especially those under public law (e.g. in the german law on protection against infection: the *Infektionsschutzgesetz* (IfSG)), cross-border problems in the application of the law could be avoided more efficiently.

Building on this, a general authorisation to adopt regionally different rules to avoid cross-border problems would have to be included in all laws with cross-border implications. The use of that option may be subject to a number of conditions, which may be known, for example, from the provisions of Article 13(2) of the treaty of Aachen: no breach of higher-ranking legislation. This concerns primarily the provisions of national constitutional law, but also those of EU law.

---

countries", 2020. Retrieved from: <http://www.europarl.europa.eu/RegData/etudes/etudes/join/2002/317602/DG-4-JURI_ET(2002)317602_EN.pdf>. Accessed 24 April 2024.

France has already inserted a constitutional experimentation clause in article 37-1 and article 72 of the French Constitution with the so-called *loi 3DS* in it's constitution.[9]

Alternatively, a regional competence to derogate could be envisaged rather than a decentralised integration into the numerous specialised laws with potential cross-border effects, also by means of a centrally formulated and effective manner in all the relevant technical acts.

Excursion: The Legal Framework for Regionally Different Rules for Cross-Border Situations

In practice, the use of such opening or experimentation clauses or, more generally, regionally divergent rules for cross-border situations is often argued that they would create new rather than less legal-administrative boundaries.

The question of whether and, if so, to what extent regionally divergent provisions to prevent or remedy cross-border problems are permissible, depends on the higher-ranking rules of international law, and above all EU and State constitutional law, which lays down the framework for those rules.

International Law

The limits of such opening clauses under international law stem above all from the principle of sovereignty.

The principle of the sovereignty of subjects of international law implies, in principle, that the States involved are of equal rank[10] and, therefore, that there is no hierarchy between the States. The principle of sovereignty falls within external and internal sovereignty, i.e. States are free to cooperate with external partners or not and to what extent.[11] They

---

9 Conseil d'État, "Les expérimentations: comment innover dans la conduite des politiques publiques", les études du conseil d'État, 2019, p. 10. Retrieved from: <https://www.dalloz-actualite.fr/document/etude-du-conseil-d-etat-1>. Accessed 24 April 2024.

10 B. Fassbender & A. Bleckmann, *Article 2(1)*, in B. Simma (ed.), *The Charter of the United Nations, A Commentary*, 2002, as quoted in K. Doehring, *Völkerrecht: ein Lehrbuch*, Heidelberg, 2004, p. 83.

11 T. Schweisfurth, *Völkerrecht*, Vienna, 2006, p. 351.

are also, in principle, free as regards the national legal and administrative form of a policy area (e.g. through the delegation of legislative or administrative powers or the establishment of public authorities).

It follows from the State's right to the non-intervention of other States (the prohibition of intervention under international law) that, in principle, any public intervention by a State or its subdivisions in the territory of another State is contrary to international law, unless an international authorisation agreement so permits.[12]

Therefore, if foreign law is to be applied by means of such an opening clause, this requires, in principle an international treaty.[13] As this applies to both general and technical opening clauses, this will no longer be dealt with separately.

EU Law

Under EU law, such rules would have to be assessed primarily in the light of mandatory, often specialised, EU law.

Furthermore, the provisions would have to comply with the prohibitions of discrimination under EU law under primary and secondary law, such as the fundamental freedoms, but also under Article 18 TFEU. In view of the purpose described above to implement the single internal market, there should therefore be no general concern about a corresponding difference in treatment by placing cross-border situations in a better position than purely internal ones.

'The discrimination of nationals' induced by EU law is not open to criticism under EU law and, as a result of the primacy of the application, is subject only to national constitutional criteria.[14]

---

12 M. Frey, *Konzepte für eine grenzüberschreitende regionale Governance am Beispiel der trinationalen Region am Oberrhein,* VBlBW, Stuttgart, 2009, p. 43.

M. Frey, *Les concepts d'une gouvernance régionale transfrontalière – l'exemple de la région trinationale du Rhin supérieur*, in B. Wassenberg, *Vivre et penser la coopération transfrontalière,* Stuttgart, 2009, p. 331.

13 An example of such a international treaty in the Upper Rhine region is the Badische Bahnhof Basel, H. Schneider, *Grenzüberquerungen des öffentlichen Rechts Deutschland – Schweiz,* in G. Müller (ed.), *Staatsorganisation und Staatsfunktion im Wandel: Festschrift für Kurt Eichenberger zum 60. Geburtstag,* Basel, 1982, pp. 673 ff.

14 C. Riese & P. Noll, *Europarechtliche und verfassungsrechtliche Aspekte der Inländerdiskriminierung,* NVwZ, Frankfurt am Main, 2007, p. 516.

German Constitutional Law

In the core of national constitutional law, two sets of questions are concerned: on the one hand, whether such rules satisfy the general rules of the rule of law resulting from Article 20(3) of the Basic Law *(Grundgesetz)* and Article 80(1) of the Basic Law, in particular the requirements of the rule of law, such as the precision[15] and clarity of the rule[16]; on the other hand, to what extent such provisions are compatible with the equality framework conditions, in particular the requirement of equal rights under Article 3(1) of the Basic Law. The former concerns in particular the general derogations referred to above, i.e. cross-cutting rules, whereas the equality aspects are also relevant in the case of individual special legislation. Against this background, a distinction needs to be made between general and specific derogations.

General Derogations

General derogations (for instance a general exception to prevent and remedy cross-border problems in this case) must first be examined in the light of the principle of primacy deriving from Article 20(3) of the Basic Law. Under that provision, the executive must, in principle, be bound by the legal framework. The primacy of the law would therefore be infringed if the legislature generally authorised the executive to derogate not only on a case-by-case basis but from an indeterminate number of statutory provisions.[17] In that regard, a general power of derogation for the executive would be problematic in this respect and would, moreover, amount to a discretionary power of the executive, which is also in principle prohibited. Furthermore, such general derogations would be problematic in the light of the constitutional principle of precision. This requires the

---

15 BVerfGE 62, 169/183; BVerfGE 80, 103/107f; 149, 293 recital 77; Robbers BK 2128.

16 BVerfGE 99, 216/243; BVerfGE 103, 21/33; BVerfGE 108, 1/20; BVerfGE 114, 1/53.

17 BVerfG, NJW 1959, 235 (236) quoted in S. Missling, H. Lange, S. Michaels & M. Wiese, *Möglichkeiten und Grenzen einer probeweisen Weiterentwicklung spezifischer Regelungen des Energiewirtschaftsrechts im Wege sog. "Regulatorischer Innovationszonen"*, EnWZ, Munich, 2016, p. 203.

legislature to have sufficiently precise legislative provisions to make the action of the executive based on them foreseeable and predictable.[18]

Such a rule must also be assessed in the light of the principle of equal treatment laid down in Article 3(1) of the Basic Law. This requires treating substantially equal things equally and substantially unequal things unequally. Any unequal treatment must be justified by factual reasons which are proportionate to the objective of differentiation and the extent of the unequal treatment.[19]

A factual reason for unequal treatment of cross-border situations with purely national situations could be – induced by EU law – the enforcement of the single market by abolishing disadvantages caused by the crossing of borders. It is true that this factual ground would in general be capable of justifying a difference in treatment between categories of persons as to the extent to which there is in fact a difference in treatment which is no longer justified, it will then have to be examined on a case-by-case basis.

The limit on the possibility of justifying such a difference in treatment is only exceeded in the case of "evident injustices", which can no longer be reconciled with an "approach based on justice"[20] or the "founded general vision of justice of society."[21] On the other hand, this limit should not normally be exceeded in the case of general derogations or opening clauses.

It must therefore be stated that general derogations appear to be problematic, particularly in the light of the requirements of the rule of law regarding the primacy of the law and its precision, but appear to be possible in principle but to be examined on a case-by-case basis.

---

18 BVerfGE 56, 1 (12); BVerfG 108, 52 (75); BVerfGE 110, 33 (53) quoted in S. Missling et al., *Möglichkeiten und Grenzen, op.cit.*, p. 203.

19 So- called "new formula", BVerfG 42, 64 (72).

20 BVerfGE 1, 264 (276) quoted by S. Missling et al., *Möglichkeiten und Grenzen, op.cit.*, p. 203.

21 BVerfG 14, 64 (72) quoted by S. Missling et al., *Möglichkeiten und Grenzen, op.cit.*, p. 203.

Clearly Defined and Binding Problem-Solving Procedure for the Remaining Problems (Which Arise in the Application of the Law)

While the two points mentioned above already intervene *ex-ante* in the context of the legislative procedure, the point set out below is intended to deal with cross-border problems which arise only later in the application phase.

In essence, as explained above, there are similar procedures at the Upper Rhine: This is, first of all, a core task of the German-Franco-Swiss Government Commission, which is suspended at national level, to regulate border neighbour issues,[22] and thus also of the Regional Conference of Upper Rhine,[23] whose task is to prepare the Commission's work.

In addition, Aachen's contract with the Committee of cross-border cooperation (AGZ-CCT) has also created a mechanism to that effect.[24]

Last but not least, the ECBM[25] would also have been a mechanism to address such problems.[26]

While the first two procedures (Government Commission and AGZ-CCT) are neither clearly defined nor binding as regards both the problem-solving procedures and the solutions found in that procedure, the ECBM would have included a binding procedure for the definition, resolution and reshuffling of this situation.[27]

---

22 Bundesanzeiger, Bundesgesetzblatt II, Bekanntmachung der Vereinbarung zwischen den Regierungen der Bundesrepublik Deutschland, der Französischen Republik und der Schweizerischen Eidgenossenschaft über die Bildung einer Kommission zur Prüfung und Lösung von nachbarschaftlichen Fragen, 1976, pp. 194–198.

23 Staatsministerium Baden-Württemberg, Baden-Württemberg in Europa – Bericht an den Landtag von Baden-Württemberg über die Europapolitik der Landesregierung im Jahre, Drucksache 12/1428, 1996, p. 13.

24 Europäisches Zentrum für Föderalismus-Forschung Tübingen, *Jahrbuch des Föderalismus 2020,* Baden-Baden, 2020, pp. 378–382.

25 European Commission, Proposal for a regulation of the European Parliament and of the Council on a mechanism to resolve legal and administrative obstacles in a cross-border context, COM/2018/373 final-2018/0198 (COD), 2018.

26 A. Engl & E. Evrard, "Agenda-setting dynamics in the post-2020 cohesion policy reform: the pathway towards the European cross-border mechanism as possible policy change", *Journal of European Integration* vol. 42, n°7, 2020, pp. 917–935.

27 F. Sielker, "The European Commission's proposal for a cross-border mechanism (ECBM): Potential implications and perspectives", *Journal of Property Planning and Environmental Law*, n°3, 2018..

The result would therefore be to try at bilateral and trilateral level to transfer the procedural elements developed at European level under the ECBM to the German-French-Swiss-Luxembourg-Belgian border region accordingly.

Conclusion

As far as citizens and regional cooperation partners are concerned, the efficiency and speed of resolving cross-border problems with a legal background, which has often been judged unsatisfactory from the point of view of citizens and regional cooperation partners, could not only be significantly improved by means of the instruments set out above, but in many cases already avoided *ex-ante*:

First of all, the establishment of a legally binding impact analysis focusing on cross-border effects, in particular at national level, can already significantly reduce the number of problems that arise later in the application phase. Furthermore, by integrating a regionally limited derogation/opening clause, cross-border systemic failures could be avoided by systematically applying to all laws with cross-border effects identified as part of the regulatory impact assessment and, in the event of subsequent system failures, eliminated. Ultimately, the problem-solving procedure already established today in the tasks of the German-French-Swiss Government Commission for the regulation of border neighbour issues and the AGZ-CCT provided for in the Aachen Treaty would have to be made more binding and quicker by adapting the rules laid down in the ECBM accordingly.

The numerous, sometimes complex, issues cannot be resolved by means of a problem-solving mechanism, which only intervenes at the stage of the application of the law, with a problem-solving mechanism

---

S. Dühr, "Zukunft der europäischen Zusammenarbeit in der Raumentwicklung – Arbeitspaket (AP) 2: Identifizierung möglicher Schwerpunkte der europäischen Raumentwicklung unter der deutschen Ratspräsidentschaft der EU, Leistungsbaustein 2.4: Rechtsinstrument zur grenzüberschreitenden Zusammenarbeit", BBSR BBR, 2018, p. 21. Retrieved from: <https://www.bbsr.bund.de/BBSR/DE/forschung/programme/moro/studien/2016/zusammenarbeit -raumentwicklung/dl-lb2-4-rechtsinstrument-grenz%C3 %BCberschr-zusammarb.pdf?__blob=publicationFile&v =1>. Accessed 24 April 2023.

on a case-by-case basis, with an indefinite and substantively non-binding procedure.

Irrespective of this, the regional cross-border level would need to be enabled to carry out a legal analysis of the problems that still arise within a reasonable time and independently. It could, for example, be possible to provide the committee of experts of the Upper Rhine Conference and the AGZ-CCT with a budget which, as in the case of b-solutions, can be used to draw on the legal expertise of independent experts (b-solutions upper-Rhine/regional). The handling could be based on b-solutions and the cooperation fund of the Upper Rhine Conference.

The peers of regional cross-border cooperation define a generally accepted governance for dealing with cross-border problems, including all agents and the Eurodistricts. In doing so, INFOBEST would need to be strengthened as a network and the role of a living situation based one-stop agency would have to be maintained.

Switzerland or rather their cantons and their institutions should be involved on an equal footing in this process.

In any event, the aim and the measures appear to be legally clear and practicable. The political level, in particular the national level, must assume responsibility. The current governance to solve legal problems with cross-border implications is not satisfactory and also falls short of the political ambition of Member States. A simple further development of existing procedures does not appear to be sufficient here, but rather a systematic and early *ex-ante* solution on the basis of a focused impact analysis, a regional opening clause to be inserted systematically and a binding problem-solving procedure based on the ECBM.

The structures at regional level (INFOBESTs, Upper Rhine Conference, AGZ-CCT) would need to be upgraded for this problem-solving procedure. For example, a b-solutions regional project could enable legal issues to be assessed in a timely manner by independent experts.

Bibliography

J. Beck, *Transnationale Kooperationskultur in der grenzüberschreitenden Zusammenarbeit am Oberrhein – Zur Konzeption eines empirischen Untersuchungsansatzes, in* D. Beck, U. Becker-Beck, J. Beck, & A. Dussap (eds.), *Kultur der grenzüberschreitenden Verwaltungszusammenarbeit – Eine empirische Modellstudie am Beispiel der Oberrhein-Region*, Speyerer Arbeitsheft n°221, Speyer, 2015.

Bundesanzeiger, Bundesgesetzblatt II, Bekanntmachung der Vereinbarung zwischen den Regierungen der Bundesrepublik Deutschland, der Französischen Republik und der Schweizerischen Eidgenossenschaft über die Bildung einer Kommission zur Prüfung und Lösung von nachbarschaftlichen Fragen, 1976.

Conseil d'État, Les études du conseil d'État, *Les expérimentations: comment innover dans la conduite des politiques publiques*, 2019.

R. Dolzer, M. Herdegen & B. Vogel (eds.). *Good governance. Gute Regierungsführung im 21. Jahrhundert*, Freiburg im Breisgau, 2007.

S. Dühr, "Zukunft der europäischen Zusammenarbeit in der Raumentwicklung – Arbeitspaket (AP) 2: Identifizierung möglicher Schwerpunkte der europäischen Raumentwicklung unter der deutschen Ratspräsidentschaft der EU, Leistungsbaustein 2.4: Rechtsinstrument zur grenzüberschreitenden Zusammenarbeit", BBSR BBR, 2018.

A. Engl & E. Evrard, "Agenda-setting dynamics in the post-2020 cohesion policy reform: The pathway towards the European cross-border mechanism as possible policy change", *Journal of European Integration*, vol. 42, n°7, 2020, pp. 917–935.

European Parliament, "Regulatory Impact analysis (RIA) – Developments and current practices in the EU member states, on the EU level and in selected third countries", 2020.

Europäisches Zentrum für Föderalismus-Forschung Tübingen, *Jahrbuch des Föderalismus 2020,* Baden-Baden, 2020, pp. 378–382.

B. Fassbender & A. Bleckmann, *Article 2(1)*, in B. Simma (ed.), *The charter of the United Nations, a commentary*, 2002.

M. Frey, *Konzepte für eine grenzüberschreitende regionale governance am Beispiel der trinationalen Region am Oberrhein,* VBlBW, Stuttgart, 2009.

M. Frey, *Les concepts d'une gouvernance régionale transfrontalière – l'exemple de la région trinationale du Rhin supérieur*, in B. Wassenberg (ed.), *Vivre et penser la coopération transfrontalière,* Stuttgart, 2009, pp. 331–345.

S. Missling, H. Lange, S. Michaels & M. Wiese, *Möglichkeiten und Grenzen einer probeweisen Weiterentwicklung spezifischer Regelungen des Energiewirtschaftsrechts im Wege sog. "Regulatorischer Innovationszonen"*, EnWZ, Munich, 2016.

C. Riese & P. Noll, *Europarechtliche und verfassungsrechtliche Aspekte der Inländerdiskriminierung*, NVwZ, Frankfurt am Main, 2007.

H. Schneider, *Grenzüberquerungen des öffentlichen Rechts Deutschland – Schweiz*, in G. Müller (ed.), *Staatsorganisation und Staatsfunktion im Wandel: Festschrift für Kurt Eichenberger zum 60. Geburtstag*, Basel, 1982, pp. 673ff.

T. Schweisfurth, *Völkerrecht*, Vienna, 2006.

F. Sielker, "The European Commission's proposal for a cross-border mechanism (ECBM): Potential implications and perspectives", *Journal of Property Planning and Environmental Law*, n°3, 2018.

Staatsministerium Baden-Württemberg, Baden-Württemberg in Europa – Bericht an den Landtag von Baden-Württemberg über die Europapolitik der Landesregierung im Jahre, Drucksache 12/1428, 1996.

Feedback on More Than 20 Years of Cross-Border Cooperation Experiences in the Upper Rhine in the Information Technology (IT) Sector (from 1994 to 2018)

Catherine Ledig & Frédérique Berrod

The Information Technology (IT) sector stands a field of successful cross-border cooperation in the Upper Rhine region. Catherine Ledig has launched many projects in this field during the last 20 years; her feedback clearly shows that such cooperation requires confident networks to mobilise such European funds. This article will reconsider these experiences in the light of the persistence of borders and the influence of Interreg to diminish border effects in the Upper Rhine, a model of cross-border relationships.

Catherine Ledig has been the director of the *Association pour le développement des entreprises et des compétences* (ADEC), a former economic development agency dedicated to digital sector for the French *département* of the *Bas-Rhin* until 2018. Its primary goals were to accelerate Information and Communication Technology (ICT) or sustainable development projects' success by its different tasks, more specifically assistance to innovative projects and skill network. This project built in line with the European ICT policy (the development of Ecommerce and the digital internal market) and European Union (EU) funding, mainly through the Interreg programs.

The ADEC has played the role of coordinator of a well-structured network, composed of:

- Administrative bodies (comprising Switzerland in the framework of the Upper Rhine region)

- Chambers of commerce (as the Chambre de Commerce et d'Industrie of the *région* Alsace or the Industrie und Handelskammer Südlicher Oberrhein)
- Universities (Faculty of Law, University of Strasbourg)
- Clusters (Cyber Forum, IT-FORUM Rhein Neckar; Medien Forum Freiburg or Rhenatic)
- Apecialised bodies of cross-border cooperation (Centre européen de la consommation/Zentrum für Europäischen Verbraucherschultz e. V.)

The ADEC also constituted a center of specific tools used to boost cross-border cooperation: Delphi studies, shared knowledge management platforms, cross-border seminars with experts selected according to criteria specific to cross-border cooperation, a deep knowledge of the culture of the other and a mindset oriented cross-border business and IT.

ADEC has participated over 20 years in more than 40 European projects characterised by public-private partnerships combining local and European resources of financing. The most significant implemented cross-border projects at the level of the Upper-Rhine region are:

- Gateway 2002–2005
- Nice Pamina 2005–2008
- IT 2 RHINE 2009–2011
- IT2RHINE Ecommerce 2012–2015
- Entrepreneurship beyond borders 2015–2018.

Catherine Ledig gave feedback on those projects, which can be summarised in three main points. The obstacles to cross-border cooperation appear still persistent in the Upper Rhine region even if the IT sector represents a good field to develop new practical tools for cross-border cooperation actors. EU funding has definitely been necessary to build such networks beyond borders.

The Obstacles to Cross-border Cooperation are Still Persistent in the Upper Rhine Region

The EU institutions often consider that the Upper Rhine region is a model for cross-border cooperation. It nevertheless is a border region with many obstacles to free movement of goods or services because of the existence of different border effects.

The persistent borders are of various natures. This outcome appears particularly accurate concerning Ecommerce in the Upper Rhine, studied in the "IT2Rhine-Ecommerce" project. For example, parcels sent from Strasbourg to Kehl have to pay the tariff of international post shipments, which is a financial burden for E-Shops. The project showed that osts could only be diminished by an agreement between the French undertaking and a German partner to guarantee a fairer price and the development of ecommerce. Even the Internet website must be adapted for Ecommerce as confidence does not have the same visual translation on both sides of the border. As is always the case in border regions, many administrative obstacles exist due to the existence of two different sets of law and regulations on each side of the border. The EU can help to solve these divergent regulations but only since the very last years did it create a solid and systematic policy of harmonisation to give birth to a proper digital single market. Another classical border effect, which consists of different languages, reinforces the legal obstacles and may result in a burden for cross-border cooperation.

Different business cultures and different levels of maturity produce other border effects, particularly when cross-border cooperation occurs on emergent markets. The "IT2Rhine" projects also underlined the need for a stronger implication of elected bodies into cross-border cooperation, which is not often a "natural" national or even regional political priority.

European programs developed by the ADEC clearly showed that such obstacles exist, even in the IT sector. However, those projects also demonstrate that this sector is a real opportunity to "forget" many obstacles, and that it 'i a field for efficient de-bordering processes. It was interesting to test how Internet and digital technologies can transform current German and French societies into post-industrial societies based on information and data. It was the aims of the Gateway project, consisting in testing the understanding and the perception of citizens on each side of the border. This type of disruptive technology can only exist if citizens living at borders understand their functioning and integrate them in their everyday life. As has been the case during the Covid-19 crisis, one of the outcomes of the project was that there is a clear common economic and cultural border region in the Upper Rhine. On this first basis, it was possible to implement other EU projects.

The "IT2Rhine" projects developed by the ADEC were feasible, as based on a long-term cooperation at the border triangle between France, Germany and Switzerland. The field of cooperation, i.e. the IT sector, was particularly rich, because of the specific strength of this economic area in the Upper Rhine region. Since 2009, it was a common goal to move forward across borders to structure the IT sector and boost innovation. There was a clear need for cross-border cooperation, embedded in the development strategy of IT clusters. Cross-border cooperation was a clear need in strengthening innovation in this emerging market. The idea was to create an open and global digital convergence for several usages and activities on the border.

One of the pre-existing conditions to boost such projects was the existing cross-border IT cluster activities, additionally supported by the "IT2Rhine" projects. In January 2009, it represented therefore more than 15 institutions and IT networks in the whole Upper Rhine, which agreed on a joint project presented to the European Commission and accepted in the framework of the Interreg programs. More than 2000 undertakings and start up in the IT sector in the Upper Rhine were represented in the network and it was one of the key points for the success of the project.

The "IT2Rhine" projects was the opportunity to organise regular meetings between stakeholders to exchange good practices and to coordinate the network to deliver. The Interreg programs supported the costs of such an organisation and the extended network provided for the co-financing implied by such programs. It resulted into the creation of the first Transnational ICT region between 2009 and 2011 through the federation of the existing IT clusters and the conception of common sustainable solutions to face the major challenges of this sector. This is a prefiguration of the sovereign digital market created by the EU institutions since 2015 through a comprehensive policy of harmonisation of the European market. The border region can therefore reinforce the competitiveness and the excellence of the IT sector.

The Need to Develop Practical Guidelines for Actors

The projects led by ADEC always implied trinational and multi-disciplinary teams, with a special attention to public/private links. The "IT2Rhine 2020 ecommerce" project followed this approach aiming at facing the challenges of cross-border electronic business traffic in the Upper Rhine region and finding suitable solutions already in 2012, following the EU Commission's Digital Agenda (2010).

The methodology used was based on specific investigations and a practice-oriented approach in the border area on the reasons why electronic business traffic was not developing further. After such an inquiry, the design of the project lead to provide concrete support to develop Ecommerce, practical solutions further elaborated in the "IT2Rhine 2020 Ecommerce" (2012–2014). For example, one of the achievements of the study was to propose the elaboration of a confidence label to reinforce the decision to buy on the website of the neighbour country along the border. The research also encompassed practical information to build bilingual websites designed for Ecommerce in the Upper Rhine or to give proper information to consumers on both sides of the border. After a study of the ecommerce context, the team of experts wrote good practice guides and diffused them to cross-border cooperation actors through seminars and professional training.

This experience clearly shows that providing operational solutions is crucial to guarantee an accelerated knowledge management or intelligence business watch. Deliverables of the project were written and translated into several languages to use as guidelines. This appears a significant condition to allow the EU innovation policy to be closer to other territories. It is important to reflect on good practices to understand very concretely obstacles to free trade at an internal EU border. One of the important achievements of the projects consists in thinking to reproducible solutions. Imply all stakeholders is of utmost importance for dissemination but also for the replication of the solutions on other borders (for example in the Alpine space for the ADEC projects).

All these projects implied vivid research and policy because the ADEC always reintegrated the projects' results into university teaching, particularly at master degree level.[1] This is key because academic programs incorporate the projects' results in the competences of future

---

1 See for example the Master Programm on Ecommerce, at the Law Faculty of Strasbourg. Retrieved from: <https://www.droit.unistra.fr/formation/diplomes-detat/master/mentions-et-parcours-masters/administration-economique-et-sociale/commerce-electronique/>. Accessed 27 June 2023.

actors. Cross-border cooperation is therefore more "normal" than it used to be. It is a chance to systematise relationships beyond borders, as well as important to get rid of border effects. It finally shows that borders are specific spaces of development of IT innovations, following the 2017 Communication of the Commission on border regions.

Such a methodology worked well because the network was based on a reinforced team with a real team spirit. Cross-border cooperation takes time and is linked to people much more than institutions or programs. These elements are necessary but not sufficient if one does not have people willing to cooperate and discover what is different behind borders.

The Opportunity of EU Funding to Build Networks beyond Borders

Cash flows are often the main obstacles to cross-border cooperation. EU funding means to find finances for such a development and to give a strong impetus to find ways to limit border effects. In financial terms, it is difficult in such projects to find the 50 % co-financing by partners. In the projects led by ADEC, this cash flow was "offered" by the association. Experience shows that building a strong financial partnership is important before trying to obtain EU funds. It is also essential to have such a structure to be able to deal with the duration of projects and the complexity of administrative deadlines of the EU and the *région* Alsace combined.

The second opportunity of EU funding is to build bridges on borders to experience the reality of EU integration. This implies a capacity to hear the different partners of the network and to get out of ones' zone of comfort. That is why it is important to build a multi-disciplinary team, as well as to feel and understand the needs of the neighbouring country. ADEC had organised its governance to incorporate within the board of directors, a German partner, Pr. Thomas Fischer, who had a right to vote on decisions. It proved a wise and valuable external advice. It was also far easier to have a shared direction of European projects and of the network necessary for each project.

This question of intercultural management was also primordial within the ADEC for its own recruitment, its administrative and technological architecture. It was important to work as Europeans to be able to manage the whole network. Building links with experts of cross-border cooperation also proved a major advantage. ADEC had for example worked very closely with the EGTC Pamina and its director Patrick Harster, who participated and shared his knowledge of cross-border mechanisms.

Projects initiated in the Upper Rhine region around IT developments also implied a common linguistic capacity. For Franco-German cooperation, the common language was Globish. Every member of the network lost a bit of fluidity in the oral exchanges but it was a nice tool to discuss face-to-face and to build in common concrete solutions. The geographical proximity generated by a team built on the border certainly reinforced such outcome. In any case, German and French were used for deliverables.

After such a rich experience, Catherine Ledig was able to underline three main drawbacks in the participation to EU projects. Cross-border cooperation takes time and stabilise experts in French teams was more difficult than in German ones (in German teams, the majority of experts are still present in 2021 and involved for more than 10 years in their job). Facing the duration of projects and their multiplication therefore represents a true challenge.

EU projects of the ADEC always had an academic outcome. It is a major point to have more experts of cross-border cooperation that is definitely a "small world". This lack of competence is the second main trend detected by Catherine Ledig.

The third and last lesson learnt for more than 20 years of cross-border projects is the perpetual gap between the teams actually working on the ground and the absence of consideration by politicians of the work made in the different projects. This may be due to constant changes in public policies on each side of the border. In any case, cross-border cooperation needs the impetus of the EU to be more effective but a strong Franco-German political commitment in Berlin, Paris, Strasbourg and Kehl is constantly required.

Concluding Remarks

This look in decades of cross-border cooperation in the field of IT showed that numerous projects were undertaken and produced many operational results. The importance of building new competences based on such cooperation is a clear outcome of this experience, at the level of Master degrees.

Nevertheless, cross-border cooperation is not a continuous process of reunification of border regions. Cross-border cooperation needs a strong political will, which is very sensitive to administrative organisation. For example, the administrative reorganisation of regions or communal cooperation structures in France diminished the importance of "départements" and induced new imbalances between administrative scales in France and Germany.

Cross-border cooperation also requires the perception of positive impact on economic activities by French and German undertakings, associations and political authorities. In this respect, the contemporary "air du temps" is not so much in favour of such a cooperation; it is due to a younger generation less aware of the benefits of cross-border living spaces. They lose the sense of such cooperation because it does not appear as necessary as it was in the 1980s. They do not understand which chance they lose when they renounce to cross-border cooperation.

Finally yet importantly, cross-border cooperation requires a common sense of necessity on both sides of the border. The IT sector was not an easy one. It is easy to consider that IT are not developing within national or local borders, as it is a virtual world. Nevertheless, it was not easy to convince the different stakeholders that cross-border cooperation was necessary because the degree of awareness of the digital transition was not the same in France and in Germany.

This look in decades of cross-border cooperation is interesting because it shows that all stakeholders need to constantly redraw the border to avoid too stringent obstacles to economic development. However, they should not wait for quick results. Cross-border cooperation needs long-term perspective to be elaborated; it is the condition to develop a sustainable space of economic and social development for the common benefit of France and Germany in an ever-closer EU.

Part 2

Franco-Belgian Cross-Border Cooperation: A Schengen Model of an Open Border?

Memories of the Franco-Belgian Border: Battles, Heritage and Cooperation

Fabienne Leloup

Although the subtitle of this article – "battles, heritage and cooperation" – refers to the structure of the text below, this article may have alternatively been entitled "The Franco-Belgian border: What a (hi) story", followed by a double exclamation and a question mark. This subtitle would have then reflected the diversity of the historical narratives to which the Franco-Belgian border – like any border – refers, a diversity firmly embedded in collective memory and reinforced, in particular, by how different sites and intangible phenomena come to be granted World Heritage status.

The question that underpins this article can be formulated as follows: does the region covering the Franco-Belgian border produce a common history or separate histories? This question may seem trivial on the face of it. After all, this border is – except during periods of terrorism or pandemic – so easily crossed physically. Moreover, it has been the subject of more than 5,000 partnerships since 2000 within the framework of projects financed by European territorial cohesion policy as detailed below. Beyond this functional cooperation and the creation of Franco-Belgian cross-border institutions, what can be said about social representation of this area's history? Is it distinct to each side of the border or is it somewhat similar?

The scope of the question at hand is wide. The discussion here, however, is limited to three battles and three UNESCO World Heritage properties. Moreover, the position adopted is that of a Belgian researcher who works in economic and political sciences.[1] The data analysed have been

---

[1] I would like to thank Philippe Walkowiak for his comments and advice on this chapter. Any inaccuracies found within the text are mine.

drawn from a number of studies carried out thanks to collaborations instituted by the Franco-German Jean Monnet Excellence Centre in Strasbourg, the Jean Monnet Borders in Motion Network (Frontem), the Franco-Belgian scientific interest group *Institut Frontières et Discontinuités* (IFD) and the Jean Monnet European Capitals of Culture and Cross-Border Urban Cohesion Network (CECCUT). The information is based on tourist brochures, media, and field trips to the areas in addition to scientific sources.

The first section of this article outlines the major phases of the definition and delimitation of the political border between the French Republic and the Kingdom of Belgium. The second section provides some details about the kinds of cooperation that link both sides of the Franco-Belgian border together, including the creation of recent European institutions of cross-border cooperation, thereby highlighting how open and porous this border really is. The third section concentrates on three battles that took place in the Franco-Belgian border region and discusses the meanings these events have acquired in collective memory. The fourth section examines the process of granting heritage status as a means of recognising and valorising memorial traces and it considers three examples that have been included on the UNESCO World Heritage List. Finally, section five links these examples to the question of cross-border cooperation in the area and in more general terms.

Six cases are discussed in this article to trace the collective memories that influence the Franco-Belgian border region. As detailed in the text, these specific examples do not provide an exhaustive analysis of the memories that exist along the border. However, the question of whether such memories serve as the basis for a common history or separate histories is addressed and the results provide fruitful lessons concerning the uses of memory and heritage and their potential connection with (cross-)border exchanges and cooperation.

Defining the Franco-Belgian Border

Without going into minute detail to trace the various occurrences that led to the delimitation of the Franco-Belgian border, it is important to define the demarcation of the Franco-Belgian border and to highlight some of the major steps that explain how the 620-kilometre-long border that separates the French Republic and the Kingdom of Belgium today came into being.[2]

In political terms, the territory that roughly corresponds to present-day Belgium was initially part of the Holy Roman Empire and it was gradually merged into a single entity at the instigation of the Dukes of Burgundy in the 14th and 15th centuries. The House of Habsburg, which counted Charles V – born in the county of Flanders in 1500 – among its members, inherited this entity. Its seventeen provinces were amalgamated in 1549. The seven Northern provinces (the United Provinces) seceded from the Southern Netherlands or Spanish Netherlands in 1581.[3] From the 17th century onwards, the Kingdom of France annexed various territories of the Spanish Netherlands following the Treaty of Aix-la-Chapelle in 1668 and the Treaties of Nijmegen in 1678–9; with the Peace of Utrecht in 1713, the Southern Netherlands passed into the hands of the Habsburgs of Austria.[4]

At the end of the 18th century, France invaded the Southern Netherlands and the Principality of Liege, and they remained in French hands until 1814–15; in the 19th century, following the lead of independence movements in Ireland and Greece, similar movements forced the secession of the Belgian provinces from the Netherlands and the establishment of an independent state in 1830.[5] On 19 April 1839, the Treaty of London was signed between the Netherlands and Belgium, as well as other European powers, granting international recognition of Belgium's independence, delineating the territory of the Belgian state and established the Grand Duchy of Luxembourg.[6] The border between France and Belgium has been largely stable since 1815, although the press did report that a boundary marker had been moved 2.2 metres into French territory in 2021.[7]

---

2 F. Alabrune, "Les frontières de la France", *Pouvoirs*, vol. 2, n°165, 2018, pp. 51–64.

3 M. Malia, "La révolte des Pays-Bas, 1566–1609", *In* Bury, L. (ed.) *Histoire des revolutions*, Tallandier, Paris, 2008, pp. 157–176.

4 X. Mabille, *Histoire politique de la Belgique*, Centre de recherche et d'information socio-politiques, Brussels, 2000.

5 *Ibid..*, p. 87.

6 *Ibid.*

7 M. Bercksmans, J. Gasparutto, C. Vanpée & M. Wegnez, "Belgique: la frontière avec la France repousée à cause d'une border déplacée", *France Info*, 2021. Retrieved

Defining the Franco-Belgian Cross-Border Region

A priori, the Franco-Belgian border cuts across a region that has been considered "cross-border" for decades. The "cross-border" adjective[8] reflects the possibility of crossing, of a return to proximity between two neighbouring regions separated by a border between states, and even the emergence or re-emergence of shared traits.[9] In geographical terms, the Franco-Belgian border passes through plains and forests, rarely corresponding to any natural boundary, with some exceptions such as the Lys between Armentières and Comines-Warneton and the Honnelle at Quiévrain and, even then, it crosses them over. Open at times to facilitate, for example, trade or worker flows, closed or more closely monitored at other moments, this border has reflected the national policies of the two states throughout the 19th, 20th and 21st centuries, as well as broader European policies and European regional cooperation.

In his analysis of mixed marriages between Belgian and French people over several decades, G. Hamez has shown that European openness has not significantly altered relations in the cross-border region. It has, however, spread them further afield within both countries, leading to a change in spatial scale, which has shifted from the regional or local (i.e. the cross-border region) to a binational scale (i.e. inhabitants coming from and residing anywhere in Belgium or France).[10]

In more general terms, since its demarcation, the frontier in the Franco-Belgian cross-border region – as it is the case in many of them – has either been used as the gateway to a protected area, a discontinuity between asymmetric situations, or an almost nonexistent line, depending on various national and the European policies. Even if this article does

from : <https://www.francetvinfo.fr/monde/europe/belgique-la-frontiere-avec-la-france-repoussee-a-cause-d-une-borne-deplacee_4620095.html>. Accessed 23 March 2023.

8 I use the term "transboundary" when discussing World Heritage listings as this is the term used by UNESCO, whereas the term 'cross-border' is used in a more general sense.

9 B. Reitel & F. Leloup, "Borders and cross-border cooperation. A political and geographical point of view", *In* J. Beck (ed.), *Transdisciplinary Discourses on CrossBorder Cooperation in Europe*, Peter Lang, Brussels, 2019, pp. 83–119.

10 G. Hamez, *Du transfrontalier au transnational: Approche géographique. L'exemple de la frontière franco-belge*. Geography thesis. Université Panthéon-Sorbonne, Paris I, 2004.

not propose an exhaustive analysis of flows or projects, it is useful to illustrate these different situations.

During the 19th and the beginning of the 20th centuries and the expansion of the mining basins both in Belgium and France, the border served as a tool either used to attract people coming from Belgium to France – including during periods of strikes – or to limit the arrival of foreign, albeit neighbouring, labour. The free movement of European workers has opened up the border in a more structured way without erasing all the differences in status. Today, thousands of individuals cross the border daily to work. For example, in 2022, 31,657 French residents crossed the border every day to work in Wallonia and 6465 in Flanders, while 1027 Flemish and 7222 Walloon travelled in the opposite direction daily to work.[11] As for companies, a number of these have set up operations close to the border – such as Toyota,which established a factory in Valenciennes, a city "close to the Franco-Belgian border", in 1998[12] – to attract workers from both countries.

Cross-border consumer purchases increased due to price, taxation or availability differentials, all of which being the result of national policies. At the same time, illegal trade, smuggling and other types of criminal activity still exist at the border. As the *musée de la vie frontalière* (museum of border life) in Godewaersvelde (France) illustrates, the border has adapted rather than modified regional practices or customs for long time, and this includes illegal practices.[13]

Since 1990, Europe has established specific instruments known as Interreg programs and especially Interreg A in order to help border regions achieve rates of development that are as high as the regions which are located closer to the political and decisional centres of the Member States. The objective is to push local and regional stakeholders to limit barriers to the free movement of goods, people, capital and services and

---

11 IWEPS, "Travailleurs frontaliers par pays de travail ou de residence", 2023. Retrieved from: <https://www.iweps.be/indicateur-statistique/travailleurs-frontaliers-pays-de-travail-de-residence/>. Accessed 5 March 2023.

12 Toyota, Histoire. Retrieved from: <https://www.toyota.fr/production-valenciennes/qui-sommes-nous/histoire>. Accessed 5 March 2023.

13 Nord Découverte, "Le musée de la vie frontalière", 2023. Retrieved from: <https://nord-decouverte.fr/musee-de-la-vie-frontaliere-godewaersvelde/>. Accessed 10 March 2023.

to promote cross-border projects.[14] The Franco-Belgian cross-border region has been highly active since the first steps of these programs and has implemented more than 5,000 Interreg projects. For instance, even if cross-border cooperation on health exists in other cross-border regions, this type of cooperation has been developed from the beginning of the Interreg programs in the Franco-Belgian area, resulting in a shift of focus from individual flows to inter-hospital agreements and a binational framework agreement.[15] Such cross-border cooperation broadens the realm of the possible but relies on mutual trust, an ability to speak the same language, the learning process and the goodwill among stakeholders.

In order to capitalise on positive cross-border results beyond Interreg funding and to amplify the effects of individual or entrepreneurial initiatives, the need for official cross-border agreements or institutions has increased. In this region, this need has been met by Franco-Belgian agreements (such as the Brussels Convention as a legal form for cross-border cooperation),[16] as well as by soft institutions, such as the organised zones for cross-border access to healthcare (ZOASTs)[17] or cross-border police cooperation. This formalisation of cooperation was also reinforced by the creation of new European groupings. More specifically, the first European Grouping of Territorial Cooperation (EGTC) to have been established was the Eurometropolis Lille–Kortrijk–Tournai, created in 2006. The Franco-Belgian border region is the focus of two other official cross-border organisations: the Greater Region (including the State of Luxembourg, Wallonia, Saarland, Rheinland-Pfalz and Lorraine) and the EGTC West-Vlaanderen/Flanders-Dunkerque-Côte d'Opale (including the province of Western Flanders, the districts of Dunkerque, Calais, Saint-Omer, Boulogne-sur-mer and Montreuil and the cities of Sailly sur la Lys, Fleurbaix, Laventie and Lestrem).[18]

---

14 S. Sassano, "Interreg", *In* B. Wassenberg & B. Reitel (eds.), *Critical Dictionary on Borders, Cross-Border Cooperation and European Integration,* Peter Lang, Brussels, 2020, pp. 542–548.

15 E. Delecosse, F. Leloup & H. Lewalle, *European cross-border cooperation on health: Theory and practice,* European Commission publications, Brussels, 2017.

16 B. Reitel, "French-Belgian Border", *In* B. Wassenberg & B. Reitel (eds.), *Critical Dictionary on Borders, op.cit.*, pp. 488–493.

17 E. Delecosse et al., *European cross-border cooperation on health, op. cit.*

18 E. Delecosse, L. Delhuvenne & F. Leloup, "L'Eurométropole Lille-Kortrijk-Tournai, instrument au service de l'institutionnalisation de la coopération transfrontalière", *Courrier hebdomadaire du CRISP,* vol. 1, n°2526–2527, 2022, pp. 5–90.

All of these examples show that, in addition to frequent daily exchanges, functional cooperation and innovative funded projects, cross-border formalisation has been encouraged and established by national authorities as well as local stakeholders in order to sustain and amplify cross-border dynamics. However, in the context of this porous and peaceful border, which history is reflected by collective memory nowadays? Is this memory still fragmented? Is the recognition of heritage anchored in a common past or one that is distinct and still limited by national boundaries? To address these questions, section three considers the examples of three battles that took place before the Kingdom of Belgium became independent, while section four looks at three UNESCO World Heritage listings, all six of which are located in the Franco-Belgian border region and connected with the people and history of this area.

Battles and Collective Memory in the Franco-Belgian (Cross-)Border Region

The Border and Representations

As a political device, the border is both an instrument of geographical delimitation and territorial discontinuity, the mark of a social construct and a symbol of differentiation.[19] Even if, in the Franco-Belgian context that concerns us, it is barely demarcated, it is nevertheless perceptible, thanks to road signs or even different telephone operators.[20] As the research of Sylvie Considère and Thomas Perrin in 2017 and others has demonstrated, rules, practices, artefacts, as well as broadcasts, information, discourse and feelings shape the experiences and realities of border regions and contribute to the creation of social representations. The study of collective memory as developed by Brady Wagoner and Ignacio Bresco[21] questions common identities that can emerge from these representations; work in this field reveals how narratives conveying collective

[19] F. Leloup, "La frontière construction politique", *In* F. Moulle (ed.), *Frontières*, Presses universitaires de Bordeaux, Bordeaux, 2017, pp. 71–103.

[20] S. Considère & F. Leloup, "Les pratiques au quotidien ou comment les habitants font avec la frontière", *In* S. Considère, S. & T. Perrin (eds.), *Frontières et représentations sociales*, L'Harmattan, Louvain-la-Neuve, 2017, pp. 27–59.

[21] B. Wagonder & I. Bresco, "Conflict and Memory: The Past in the Present", *Peace and Conflict: Journal of Peace Psychology*, vol. 22, n° 1, 2016, pp. 3–4.

memories can give meaning to individual memories, objects, monuments and spaces. The special issue edited by Wagoner and Bresco[22] analyses intergroup conflict dynamics, highlights the role of memory and how groups represent their past to define themselves.

Since its creation, Belgium has never been at war with France and both were allies during the world wars, but battles involving the French took place in the area, leaving their mark on the Franco-Belgian border region. Three of these battles[23] have been chosen in order to analyse the memorial framework that is associated with them. A very brief historical overview of each battle is provided for context, before their place in current collective memory and the traces which are still visible today are analysed.

The three battles that have been selected – the siege of Mons, the Battle of Jemappes and the Battle of the Golden Spurs – were fought on the territory of present-day Belgium, but they took place before Belgium gained its independence.

French Cannonballs

In March and April 1691, the French besieged the medium-sized town of Mons, a city in the former of County of Hainaut which is located 15 km from the present-day Franco-Belgian border on the Belgian side. This battle was known as the siege of Mons. In 1688, just a few years after the treaties of Aix-la-Chapelle and Nijmegen, Louis XIV of France embarked on a series of military campaigns and formed a plan to capture the city.[24] The city was besieged from 15 March to 10 April 1691. The cannons fired 57,673 cannonballs and the mortars hurled 5,750 bombs at the city.[25] On 10 April, the surrendering conditions were agreed upon, and the city was evacuated. Six years later, Louis XIV gave up the territories

---

22 B. Wagonder & I. Bresco, "Conflict and Memory", *op.cit.*

23 The elements presented here have been summarised in F. Leloup, "*La Frontière franco-belge*", *In* B. Wassenberg (ed.), *Frontières, acteurs et répresentations d'Europe,* Borders and European Integration, Peter Lang, vol. 5, 2022, pp. 225–240.

24 Crédit Communal de Belgique, *1691 : Le siège de Mons par Louis XIV,* Crédit Communal, Brussels, 1991.

25 B. Van Mol, "L'état des canons après le siege", *In* Crécit Communal de Belgique, *1691, op.cit.*, p. 6.

he had annexed whose majority returned to the Spanish Netherlands and Charles II of Spain.

Today, it is still possible to see the cannonballs in some of Mons' fortified walls as well as in the city museum.[26] Pointed out by city guides, they are evoked as "vestiges of the war between France and the future Belgium. The cannonball thus became a symbol of Mons" resistance. It serves to illustrate the difference between "them" (the French) and "us" Belgians. Nevertheless, Belgium did not exist at the time of the siege and Louis XIV's wars of conquest were waged against the possessions of the Spanish kingdom, which included Mons. These narratives reinforce the memory of a "Belgian place" defending itself against France. Therefore, they highlight how these objects are now used by Belgians to define themselves as different, how they have become part of collective memory and how Belgium represents its past to define its national identity.

Louis XIV's cannonballs appear as precursors of a distinction between France and Belgium; they insist on the division of the two sides and maintain a secessionist memory of "Belgian" resistance against the invader. The French cannonballs in Mons are, thus, symbols of the past conflict and have been mobilised in contemporary narratives to perpetuate it.

Jemappes and the First French Republic

Not far from Mons, a monument entitled "Le Coq" commemorates the Battle of Jemappes in memory of Franco-Belgian friendship, in connection with the so-called "Battle of Jemappes".[27]

This Battle of Jemappes took place on 6 November 1792 and was part of the wider French Revolutionary Wars and is considered to be the first major victory of the armies of the Republic. Initially, the French objective in this engagement was not to annex new territory but promote the emergence of a belt of republics as stipulated president of the Diplomatic Committee, Jacques-Pierre Brissot, if these new political formations established governments based on the sovereignty of the people; the then French Minister of Foreign Affairs, Charles-François Dumouriez,

---

26 Crédit Communal, *1691*, *op.cit.*, p. 110.

27 Traces de France, "Le monument au coq à Jemappes (Wallonie)", 2019. Retrieved from: <https://tracesdefrance.fr/2019/10/26/le-monument-au-coq-a-jemappes/>. Accessed 5 March 2023

consequently proclaimed the independence of a (Belgian) republic on 8 November 1792; however, internal dissension, a lack of external support and a shift in the policy of the French National Convention transformed this independence into annexation in 1793.[28]

In 1911, a monument, *Le Coq*, was unveiled in Jemappes to celebrate Franco-Belgian friendship to coincide with the *Congrès international des Amitiés françaises* (International Congress of French Friendships); this congress was organised under the auspices of a committee in which Jules Destrée, a major player in the Walloon Movement,[29] was involved. This movement aimed to promote the specific characteristics of Wallonia, a French-speaking region in Belgium that was distinct from Flanders and Brussels. The monument was considered a symbol of the friendly relations between the French and Walloons, their proximity and their shared values. All signs of conflicts or annexation were erased. In 1989, the Walloon movement *Wallonie Région d'Europe* tried to rekindle interest in this event but without success.[30]

Nowadays, the monument is only visited by French tourists seeking traces of this battle which was so symbolic for the armies of the Republic. Few residents of Jemappes, let alone Belgians from elsewhere, know the location nor the meaning of the monument. *Le Coq* thus commemorates neither the premature independence of the Belgian state, nor an alliance with France in support of this independence. Initially a witness to Franco-Belgian *amitié* (friendship) and then a certain type of Francophilia promoted by Walloon regionalism, the monument – although located in Belgium – now essentially belongs to France's memorial heritage. For French people, it perpetuates the memory of the victories of the French revolution and the links with neighbouring friendly areas, such as Wallonia. For Walloons or Belgians more broadly, it holds little to no significance.

---

28 X. Mabille, *Nouvelle histoire politique de la Belgique*, *op. cit.*, pp. 48–9.

29 The Walloon Movement is a political regional movement developed in the 19th century in order to defend the existence of Wallonia the southern part of the country, of a Walloon identity and of the defence of the French.

30 F. Joris & F. Marchesani, "Le monument au coq à Jemappes". 2009. Retrieved from <https://tracesdefrance.fr/2019/10/26/le-monument-au-coq-a-jemappes/>. Accessed 10 March 2023.

From the Battle of the Golden Spurs to the 11 July Day of Celebration

In the west Flemish town of Kortrijk, the Groeninge Monument that was erected in 1902 commemorates the Battle of the Golden Spurs or *Guldensporenslag*, as it is known in Dutch.

From the end of the 12th century onwards, a number of battles were fought between the Counts of Flanders and the King of France. The Battle of the Golden Spurs took place on 11 July 1302 on the plains of the Lys and involved several thousand French combatants and about the same number of soldiers from Flemish towns, as well as Namur, Liege, Zeeland and even England.[31] The Flemish troops routed Philip the Fair's army and legend has it that they captured 500 pairs of golden spurs, from which the battle derives its name. The County of Flanders briefly gained its independence, but France reconquered the territory in 1304 and did not relinquish it until Francis I ceded the county to Charles V's Holy Roman Empire.

At the time of Belgium's independence, symbols of the country's heroic history were sought and the Battle of the Golden Spurs came to be celebrated as a Belgian victory over the occupier given the romantic tendencies of the period.[32] During this period, romanticism, using intuition and emotion, assigned a high value to the achievements of heroes and notions of historical or natural inevitability. At the end of the 19th and beginning of the 20th centuries, the focus on this victory in Flanders homed in on the struggle between Flemish and French combatants, becoming a symbol of Flemish liberation. As the Belgian historian Vincent Dujardin has argued, the battle not only came to represent independence from France, but, by extension, liberation from French-speaking Belgium.[33] Without delving too deeply into the evolution of the federal structure of the Kingdom of Belgium, it is important to mention that, from the 1960s onwards, a series of revisions resulted in a transition away from a unitary and centralised Belgium to a federal state, which was made up of

---

31 C. Istasse, "Histoire, mémoire et identité: Les fêtes nationales, régionales et communautaires en Belgique", *Courrier hebdomadaire du CRISP*, n° 2412–2413, 2019, p. 33.

32 C. Istasse, "Histoire, mémoire et identité", *op. cit.*

33 RTBF, "Le 11 juillet 1302: une victoire belge devenue mythe flamand", Retrieved from <https://www.rtbf.be/article/le-11-juillet-1302-une-victoire-belge-devenue-mythe-flamand-6424733>. Accessed 5 March 2023.

three regions and three communities: the Flemish Region, the Brussels-Capital Region and the Walloon Region, plus the Flemish Community, the French Community and the German-speaking Community. In 1973, the date of the Battle of the Golden Spurs – 11 July – was selected as the Day of the Flemish Community and has been a public holiday since in the Flemish Region.

The Battle of the Golden Spurs is nowadays remembered less as a victory over France than over "francisation". The Groeninge Monument has thus been appropriated as the memorial heritage of Flemish nationalism; it is thrust into the centre-stage during the annually celebrated Day of the Flemish Community festival where it serves as a backdrop for Flemish elected officials.[34] In this third case, the battle is recalled and even celebrated as a victory, not against France as another country, but against France and French-speaking influence more broadly. The memory of this battle is mobilised to assert a common Flemish identity. In other words, the narrative, monument and festival serve to reinforce and perpetuate such a representation.

These three examples illustrate how the memories of these battles have been perpetuated. Their significance has been evolving over time, the narratives have been reshaped, thereby resulting in different meanings and evoked for the purposes of identity construction. In all three cases, France is opposed to "Belgium". After all, France is still cast as the invader in narratives recounted about the cannonballs of Mons (even if Belgium did not exist at the time). For the Battle of Jemappes, the ideology of the French revolution and fleeting Belgian independence sparked the memory of a specific Walloon and French friendship, but it now only features in the present-day French Republican narrative. In the case of the Battle of the Golden Spurs, the memory of the conflict was invoked in the context of a Belgian national identity in the 19th century, but it has since evolved. Nowadays, it is a cornerstone of Flemish identity.

---

34 C. Istasse, "Histoire, mémoire et identité", *op.cit.*

Heritage and Borders in the Franco-Belgian Border Region

World Heritage, Transboundary Properties and the Border Region

As emphasised by several authors, including David Lowenthal, heritage is not history; it celebrates the past, it provides representations of the past in a specific ways, with specific meanings and tailors it to present-day purposes.[35] Physical objects or buildings related to the three battles discussed above can be defined as elements of heritage: they are conserved in order to preserve and commemorate memories of the past and to pass them from one generation to the next. As previously seen, they shape ideas about the past and these ideas may evolve along with a community's identity. Notably, in the case of the Battle of Jemappes, French revolutionary ideology and the meaning of battle faded away for Belgian inhabitants while the status of the opponent has been redefined in contemporary Flemish narratives surrounding the Battle of the Golden Spurs.

In this section, three examples of heritage located in the Franco-Belgian border region are analysed. They have been chosen for two reasons. First, they are concerned with cross-border buildings or practices or, in other words, they are part of the traditions and the history of both Belgium and northern France. Second, they have been recognised by UNESCO as being of outstanding universal value and have, thus, been granted World Heritage status, either as transboundary or as national property, which ensures their world-wide exposure.

Historical or cultural heritage status bolsters memory for material or immaterial monuments, places, or events. This status is seen as a form of inheritance for locals that has been passed from previous generations and is worth preservation for present and future generations. It includes physical objects, places and even practices. The process of granting World Heritage status to a monument or tradition tends to stabilise, anchor and perpetuate the memory it represents; through the process of selection and the *mise en scène* proposed, institutional actors valorise a monument

---

35 OpenLearn, "What is heritage", Retrieved from <https://www.open.edu/openlearn/history-the-arts/history/what-heritage/content-section-2.1>. Accessed 25 March 2023.

or tradition by endowing it with a particular meaning, a chosen history inserted into a coherent narrative. Granting heritage status is always linked to a discourse and a moment based on the past that still resonates in the present and is determined by preferences and political choices.[36] The selection process and meanings assigned to these objects, places or practices are always partial and evolving as they are based on contemporary judgments and cultural values.

A World Heritage listing reinforces the valorisation of a particular object, place or practice and its accompanying narrative thanks to UNESCO's international standing and the worldwide scope of the label. This label designates heritage, cultural or natural, tangible or intangible, that is judged to be of exceptional interest to humanity and which reflects unique specificities rooted in local history. It includes cultural heritage sites, historic cities or cultural landscapes, natural sacred sites or museums, handicrafts or documentary and digital heritage, oral traditions or traditional medicine, among others.[37]

In most cases, heritage listings are used to commemorate material or immaterial components defined within the borders of a single state. UNESCO's current list counts 1157 properties, of which 43 are defined as "transboundary", meaning properties which are located on the territory of concerned "States Parties" sharing adjacent borders. 22 properties are classified as "transnational" which refers to properties that are located in several places throughout several countries. Examples of transnational UNESCO properties include the "Great Spa Towns of Europe and the Architectural Work of Le Courbusier, an Oustanding contribution to the Modern movement".[38] It is important to note that the proposal for the nomination to the World Heritage List is always prepared by "State Parties" and are prepared and submitted jointly by State Parties in case of transboundary or transnational properties. As shown in one of the following case studies, provisions exist to broaden the scope of a

[36] C. Lacour, F. Leloup & L. Moyart, "Culture, patrimoine, savoirs: facteurs dynamiques de développement: Introduction", *Revue d'économie régionale et urbaine*, n°5, 2014, pp. 785–799.

[37] UNESCO, retrieved from <https://whc.unesco.org/en/35/>. Accessed 5 March 2023.

[38] UNESCO, "World Heritage List", Retrieved from <https://whc.unesco.org/en/list/>. Accessed 23 March 2023.

previously listed World Heritage property so that it may be classified as "transboundary".[39]

The three heritage cases discussed here are situated in the Franco-Belgian border region. The analysis of these sites fits into the socio-economic history of the area which is much the same on both sides of the border: the growth of cities in the Middle Ages and the industrial exploitation of coal that shaped the development of the North of France and Belgium in relatively similar ways. The examples selected are the network of 56 belfries in Belgium and France included on the World Heritage List; the separate inclusion of the Nord Pas-de-Calais mining basin on the French side and four Walloon mining sites on the Belgian side; and, finally, the Belgian recognition of carillons. For each of these heritage sites, the narrative that justifies their exceptionality and, therefore, the UNESCO label is outlined and the positioning of this recognition, whether it is cast as cross-border or national, is discussed.

From 32 to 56 Belfries

The first heritage example concerns the belfries of Belgium and France. In 1999, 32 Belgian belfries were included in the UNESCO World Heritage List and in 2005, 23 belfries located in northern France, as well as the Gembloux belfry in Belgium, joined them.[40]

The belfries were built between the 10th and 17th centuries. They represent the attainment of civil liberties in cities, specifically in north-western Europe.[41] These architectural masterpieces, constructed in Romanesque, Gothic, Renaissance or Baroque styles, were erected in the heart of the city as a counterpoint to the keep, which symbolised the aristocracy and the bell-tower that was the symbol of the Church. They represented the power of the commune: they have been recognised as universal heritage because they are unique examples of urban and public architecture. As major features of the medieval town, they still play a significant role in

[39] As detailed in UNESCO, *Operational Guidelines for the Implementation of the World Heritage Convention*, World Heritage Centre, Paris, 31 July 2021, art. 136.

[40] UNESCO, "Belfries of Belgium and France", Retrieved from <https://whc.unesco.org/en/list/943/>. Accessed 23 March 2023.

[41] B. Bove, F. Lachaud, P. Monnet & J.-B. Santamaria, "Villes et construction étatique au Moyen Âge", *Histoire urbaine*, n° 1, special issue, 2021, pp. 7–28.

the development of their towns today, whether they serve as a clock, a carillon, a tourist attraction, or a cultural artefact.[42]

Initially, only 32 Belgian belfries were recognised before another 24 French and Belgian belfries were added to the list. In this way, the case of the belfries has become a form of transboundary recognition. The cross-border nature of this heritage has led to the establishment of an annual Franco-Belgian meeting of those responsible for the restoration and maintenance of the belfries so that they may share examples of best practices in terms of management and renovation, even if they do not meet more than once a year. This transboundary heritage status commemorates the collective memory related to the common development of the individual towns as a shared past and draws a transnational line which ignores the border.

Mining Basin and Mining Sites

Our second example concerns the Nord-Pas de Calais Mining Basin on the one hand and the Major Mining Sites of Wallonia on the other, both of which were recognised as UNESCO World Heritage sites in 2012. Four sites have been selected in Belgium, three of which are located in Hainaut, a Belgian province that borders France. The French mining basin selected covers 120 km squared located in Nord-Pas de Calais, which borders Belgium.[43] They provide evidence of the common coal mining history by highlighting their architectural, technological, landscape and social significance.

Coal mining, a defining feature of the industrial revolution of the 19th and 20th centuries, had a profound effect on the living spaces where it was developed. This was not just in relation to the industrial sites themselves or the mining methods and technologies used, but also about the evolution of urban planning and transformation of the wider landscapes

---

42 C. Lamour, F. Durand, P. Bosredon, F. Leloup, T. Perrin, N. Popa & C. Tursie, *European Capitals of Culture and Cross-border Urban Cohesion: Best Practices Guide and Toolkit for Evaluation*, Luxemburg, 2022. Retrieved from: <http://www.ceccut.eu/en/home/>. Accessed 25 April 2023.

43 UNESCO, "Nord-Pas-de-Calais Mining Basin", Retrieved from <https://whc.unesco.org/fr/list/1360>. Accessed 5 March 2023. UNESCO, "Major Mining Sites of Wallonia", Retrieved from <https://whc.unesco.org/fr/list/1344>. Accessed 5 March 2023.

through the construction of canals and the appearance of slag heaps.[44] The collieries dramatically reshaped urban areas, housing and ways of life in northern France and Belgium, as well as other areas in France and in Europe. They also resulted in high rates of migration, such as Belgian workers travelling to France during the first stages of mining developments in the region. The heritage listing of these former coal mining regions sheds light on the working conditions, as well as the workers' lives in the industry and wider environment.[45]

UNESCO's recognition of the exceptional character of this area's heritage proceeded in a distinct manner on the French and Belgian sides. In both cases, these heritage sites exhibit the deep change in buildings, technology and landscape although the distinctiveness of the Nord-Pas de Calais Mining Basin also resides in its: "associat[ion] with events, with living traditions, with ideas, or with beliefs [and] with artistic works".[46] In 2012, the two sites have been listed separately, despite the close degree of similarity between the buildings, technology, landscape, working class conditions and social struggles highlighted by the heritage listing. Coexisting independently, they are managed by different authorities with distinct objectives on either side of the border. Moreover, they are included in different tours which overlook the collective past of the sites, including their close-knit economic and labour ties.

Carillons

In 2014, UNESCO recognised carillon culture as the best way to safeguard intangible heritage in Belgium. Applied to Belgium as a whole, but not to France, the listing acknowledges the transmission and enhancement of this historical cultural phenomenon that originated in the Spanish Netherlands and dates back to the 16th century.[47]

---

[44] P. Tilly, "Fleuves et Canaux dans la zone franco-belge entre 1814 et 1914: vers une redéfinition des espaces?", *Revue du Nord*, vol. 3, n° 416, 2016, pp. 577–599.

[45] S. Lembre, "Terres de travail, terrains d'histoire", *Revue du Nord*, vol. 2, n° 435, 2020, pp. 277–288.

[46] UNESCO, "The Criteria for Selection", Retrieved from <https://whc.unesco.org/en/criteria/>. Accessed 5 March 2023.

[47] UNESCO, "La sauvegarde de la culture du carillon: preservation, transmission, échange et sensibilisation", Retrieved from <https://ich.unesco.org/fr/BSP/la-sauvegarde-de-la-culture-du-carillon-preservation-transmission-echange-et-sensibilisation-01017>. Accessed 5 March 2023.

Although their chiming gives rhythm to the daily life of communes in northern France, Flanders, Brussels and Wallonia, carillons on the French side of the border have not been granted World Heritage status. UNESCO's listing is restricted to within the Kingdom of Belgium's border, thereby omitting the tradition which is still practised on the other side of the border.

The carillons are often located in belfries, therefore, their recognition could have been connected but, until now, only Belgian carillons have been granted World Heritage status.

Battles, Heritage and European Cooperation

The memorial elements analysed, whether they be battles or UNESCO World Heritage sites, tell us about memory in the Franco-Belgian border region. Although the battles discussed predated the existence of Belgium as a state, they have been mobilised to reinforce French republicanism, as well as Belgian, Walloon or Flemish nationalism. As for the World Heritage listings, while these may point to the shared social, cultural and economic history of northern France and Belgium, two out of the three properties have served to commemorate national or regional pasts within the borders of a single state. Conversely, the Franco-Belgian area has become more and more porous through the development of formal cross-border institutions, largely as a result of European Union policies and European territorial cooperation.[48]

These examples suggest that memories and historic heritage are mainly mobilised in national contexts to sustain national or regional identities. In order to link these examples and the question of cross-border cooperation, it is worth analysing the presence of European cooperation, as based on Interreg programmes, in connection with such cases. Indeed, Interreg programmes have sought to foster horizontal links across borders and sustain cohesion in cross-border regions.

First, we can draw attention to the so-called Interreg project "A 300-year border".[49] This project aimed to develop a series of cultural activities

---

[48] B. Reitel & B. Wassenberg, *Territorial Cooperation in Europe – A Historical Perspective*, European Commission, Brussels, 2015.

[49] Interreg IV France-Wallonie-Vlaanderen, "300ans de frontière". Retrieved from <http://www.interreg4-fwvl.eu/admin/upload/project/pdf/229-fr.pdf>. Accessed 3 March 2023.

over two years related to the tercentenary of the Treaty of Utrecht, cast as a foundational moment of the Franco-Belgian border as it heralded an end to the War of the Spanish Succession.[50] What is interesting is that, at that time, the border was celebrated in terms of good neighbourliness. The Interreg project included the transformation of former border posts into artistic workshops and works of art aimed at re-conciliating French and Flemish sides of the border.

The case of the coal mining and mining landscapes has been addressed in various Interreg projects during the last two programming periods. As part of Interreg IV, the *Agir pour la connaissance, l'évaluation, l'interprétation et la gestion du Patrimoine naturel et culturel du Bassin Minier franco-wallon* (APBM) project, which translates as "Acting for the knowledge, evaluation, interpretation and management of the natural and cultural heritage of the Franco-Walloon mining basin", aimed to manage the cultural and natural heritage of Franco-Walloon mining areas.[51] The *Itinéraire de Culture Industrielle* (ICI) project, which roughly translates as "Industrial culture trail", created an industrial culture trail based on Franco-Belgian mining heritage which were eventually listed among the four Walloon mining sites that were granted World Heritage status by UNESCO in 2012.[52] The *Destination Terrils* project (INTERREG V) focused on 75 slag heaps and aimed to develop tourism and the economy around these coal mining relics while preserving their heritage and ecology.[53]

Regarding the carillons, it should be noted that, since 2020, the West-Vlaanderen/Flandre-Dunkerque Côte d'Opale EGTC has been supporting various Interreg micro-projects in order to revalorise the art of carillon music: "as French and Flemish Flanders are home to the largest number of carillons in the world, as the birthplace of the carillon can be

50 Alabrune, F., "Les frontières de la France", *op. cit.*

51 Interreg IV France-Wallonie-Vlaaderen, "Agir pour la connaissance, l'évaluation, l'interprétation et la gestion du patrimoine naturel et culturel du basin minier franco-wallon". Retrieved from <http://www.interreg4-fwvl.eu/admin/upload/project/pdf/101-fr.pdf>. Accessed 3 March 2023.

52 Interreg IV France-Wallonie-Vlaanderen, "ICI – Itinéraire de culture industrielle". Retrieved from <http://www.interreg4-fwvl.eu/admin/upload/project/pdf/83-fr.pdf>. Accessed 3 March 2023.

53 Destination Terrils. Retrieved from <https://www.destinationterrils.eu/fr/>. Accessed 3 March 2023.

found in this region, [and] since most of the belfries that house them are recognised as UNESCO World Heritage sites".[54]

These examples demonstrate that a shared past is required, at least where the development of Interreg projects is concerned. The reason why these local operators used this common history is not clear. Perhaps local stakeholders are more aware and more interested in commemorating collective memories; or the Interreg programmes only create functional opportunities that are shaped by their potential to attract tourists rather than their ability to resonate with locals. Answering this question clearly requires further research.

A closer look at the Interreg programmes reveals a number of other projects that are linked to the historical question of the border. More specifically, a group of four projects is worthwhile mentioning in terms of collective, possibly cross-border, memory. These were included in the Interreg IV programme (2007–2013) and dealt with the First or Second World War: *Mémoire de la Grande Guerre*[55] (i.e. "Memory of the Great War") and *La Grande Guerre, corps et âmes de paix* (i.e. "The Great War: bodies and souls of peace") were dedicated to commemorating the centenary of the First World War.[56] *Transmussites 14–45* and *Mémoire commune du XXème siècle* (i.e. "Common Memory of the 20th century") were focused on the development of museum infrastructures and joint research in relation to both global conflicts.[57] Belgium and France were allies during these wars and these projects were designed to encourage joint commemoration.

---

54 GECT West-Vlaanderen Flandre-Dunkerque-Côte d'Opale, "Le ciel comme sale de concert – L'histoire des carilloneurs". Retrieved from <https://www.egts-gect.eu/fr/actualites/le-ciel-comme-salle-de-concert-lhistoire-des-carrilloneurs>. Accessed 3 March 2023.

55 Interreg IV France-Wallonie-Vlaanderen, "Mémoire de la grande guerre". Retrieved from <http://www.interreg4-fwvl.eu/admin/upload/project/pdf/16-fr.pdf>. Accessed 3 March 2023.

56 Interreg IV France-Wallonie-Vlaanderen, "La grande guerre". Retrieved from <http://www.interreg4-fwvl.eu/admin/upload/project/pdf/210-fr.pdf>. Accessed 3 March 2023.

57 Interreg IV France-Wallonie-Vlaanderen, "Mémoire commune du XXème siècle". Retrieved from <http://www.interreg4-fwvl.eu/admin/upload/project/pdf/20-fr.pdf>. Accessed 3 March 2023. Interreg IV France-Wallonie-Vlaanderen, "Transmussites 14–45". Retreived from <http://www.interreg4-fwvl.eu/admin/upload/project/pdf/21-fr.pdf>. Accessed 3 March 2023.

In all these cases, cooperation from both sides of the border resulted in joint actions, essentially with touristic objectives. They illustrate the desire of actors to work together sometimes for one-off or short-term events and induce mainly a functional cooperation.

Conclusion

The question at the centre of this article was whether the Franco-Belgian border region produces a common history or separate histories. In other words, do the historic battles that ravaged the region and common heritage commemorate distinct narratives defined within the borders of a single state in spite of the openness of the border and the high degree of exchanges, cross-border cooperation and formal European institutions? Our set of six case studies is limited but provides interesting results where collective memory is concerned, both in relation to how it is shaped by commemoration and dominant narratives, as well as how it is invoked to obtain World Heritage status for various sites, objects and practices.

The three battles and their commemoration have fed nationalist or regionalist identities, they have amplified narratives espousing a rhetoric of "us" against "them", glorifying locals' resistance against the outsiders' power. In all these cases, they certainly have not promoted cross-border perspectives. In the case of the World Heritage properties, the first step to getting on UNESCO's list always focuses on national recognition. Heritage that is submitted for listing is linked to shared histories or culture, such as the belfries and their symbolic link to the rise of civil liberties in French and Belgian cities or the industrialisation of the mining basins in northern France and Belgium. Despite this fact however, their granting of World Heritage status remains overwhelmingly restricted to national boundaries.

Cross-border cooperation, such as that funded by Interreg programs, creates incentives for and supports collective regional or local projects based on collective memories (such as the world wars). The jury is still out, however, on how sustainable these projects and the networks they forge are; and whether they can influence representations and memories beyond their quantifiable outputs, which are evaluated based on elements such as audience, activity and building figures.

While, analysing the way in which representations account for the past or the way heritage listing is carried out is not intended to erase any battle or opposition. Cross-border cooperation in Europe builds bridges and creates opportunities for exchange and working together. In doing so, it encourages the creation of an increasingly integrated functional space but does not address questions of representation or collective histories.

At a time of resurgent nationalism and discounting the illusion of "borderlessness", such partial results tend to require more in-depth reflection on how collective memories and heritage can shape notions of "us" and "them" at the border. Beyond exchanges, flows and other cross-border actions, the border reveals itself as a singular place, a site of histories – in the *plural* – that combines diversity and continuity, rewriting and reinterpretation. Avenues of research are emerging; they focus on these narratives and representations that will provide us with in-depth answers to crossing national borders and linking heritage, (hi) stories and borders.

Bibliography

F. Alabrune, "Les frontières de la France", *Pouvoirs*, vol. 2, n°165, 2018, pp. 51–64.

B. Bove, F. Lachaud, P. Monnet & J.-B. Santamaria, "Villes et construction étatique au Moyen Âge", *Histoire Urbaine*, n° 1, special issue, 2021, pp. 7–28.

S. Considère & F. Leloup, "Les pratiques au quotidien ou comment les habitants font avec la frontière", *In* S. Considère, S. & T. Perrin (eds.), *Frontières et représentations sociales*, L'Harmattan, Louvain-la-Neuve, 2017, pp. 27–59.

Crédit Communal de Belgique, *1691 : Le siège de Mons par Louis XIV,* Crédit Communal, Brussels, 1991.

E. Delecosse, F. Leloup & H. Lewalle, *European Cross-border Cooperation on Health: Theory and Practice,* European Commission publications, Brussels, 2017.

E. Delecosse, L. Delhuvenne & F. Leloup, "L'Eurométropole Lille-Kortrijk-Tournai, instrument au service de l'institutionnalisation de la coopération transfrontalière", *Courrier hebdomadaire du CRISP*, vol. 1, n°2526–2527, 2022, pp. 5–90.

G. Hamez, *Du transfrontalier au transnational: Approche géographique. L'exemple de la frontière franco-belge*. Geography thesis. Université Panthéon-Sorbonne – Paris I, 2004.

C. Istasse, "Histoire, mémoire et identité: Les fêtes nationales, régionales et communautaires en Belgique", *Courrier hebdomadaire du CRISP*, n° 2412–2413, 2019.

C. Lacour, F. Leloup & L. Moyart, "Culture, patrimoine, savoirs: facteurs dynamiques de développement: introduction", *Revue d'économie régionale et urbaine*, n°5, 2014, pp. 785–799.

C. Lamour, F. Durand , P. Bosredon, F. Leloup, T. Perrin, N. Popa & C. Tursie, *European Capitals of Culture and Cross-border Urban Cohesion: Best Practices Guide and Toolkit for Evaluation*, Luxemburg, 2022.

F. Leloup, "La frontière construction politique", *In* F. Moulle (ed.), *Frontières*, Presses universitaires de Bordeaux, Bordeaux, 2017, pp. 71–103.

F. Leloup, "*La Frontière franco-belge*" *In* B. Wassenberg (ed.), *Frontières, acteurs et répresentations d'Europe,* Borders and European Integration, Peter Lang, vol. 5, 2022, pp. 225–240.

S. Lembre, "Terres de travail, terrains d'histoire", *Revue du Nord*, vol. 2, n° 435, 2020, pp. 277–288.

X. Mabille, *Histoire politique de la Belgique*, Centre de recherche et d'information socio-politiques, Brussels, 2000.

M. Malia, "La révolte des Pays-Bas, 1566–1609" *In* Bury, L. (ed.) *Histoire des revolutions*, Tallandier, Paris, 2008, pp. 157–176.

B. Reitel, "French-Belgian Border" *In* B. Wassenberg & B. Reitel (eds.), *Critical Dictionary on Borders, Cross-Border Cooperation and European Integration,* Peter Lang, Brussels, 2020, pp. 488–493.

B. Reitel & F. Leloup, "Borders and cross-border cooperation. A political and geographical point of view", *In* J. Beck (ed.), *Transdisciplinary Discourses on CrossBorder Cooperation in Europe*, Peter Lang, Brussels, 2019, pp. 83–119.

B. Reitel, & B. Wassenberg, *Territorial Cooperation in Europe – A Historical Perspective*, European Commission, Brussels, 2015.

S. Sassano, "Interreg", *In* B. Wassenberg & B. Reitel (eds.), *Critical Dictionary on Borders, Cross-Border Cooperation and European Integration,* Peter Lang, Brussels, 2020, pp. 542–548.

P. Tilly, "Fleuves et Canaux dans la zone franco-belge entre 1814 et 1914: vers une redéfinition des espaces?", *Revue du Nord*, vol. 3, n° 416, 2016, pp. 577–599.

UNESCO, *Operational Guidelines for the Implementation of the World Heritage Convention*, World Heritage Centre, Paris, 31 July 2021.

B. Wagonder & I. Bresco, "Conflict and memory: The past in the present", *Peace and Conflict: Journal of Peace Psychology*, vol. 22, n° 1, 2016, pp. 3–4.

Public Transport in the Eurometropolis Lille-Kortrijk-Tournai: Is the Border Effect Still Visible?

Cyprien Richer & Patrick Palmier

Introduction: Is the Border Effect Still Relevant in Cross-Border Metropolises?

Borders are geographical limits at the heart of territorial appropriation dynamics. Since antiquity, they have been associated with the idea of delimiting space "in the sense of an extremity beyond which the unknown often begins".[1] However, it was not until the birth of modern States that they were defined as limits to sovereignty. Today, this definition has been challenged by the creation of international spaces of cooperation and the free movement of people and goods. The trend towards more porous borders is particularly visible in the European Union thanks to the single market and the Schengen Area. Consequently, borders have come to be considered privileged places of exchanges and encounters,[2] while continuing to mark political, legal and, indeed, cultural differences of varying degrees. This transnational dynamic has regularly been interrogated by research in the field of geography because the apparent "devaluation" of borders[3] is never exempt from social, spatial or political discontinuities on either side of a State's border.

---

1 J. Lévy & M. Lussault, *Dictionnaire de la Géographie*, Belin, Paris, 2013, p. 414.

2 Groupe Frontière, "La frontière, un objet spatial en mutation". Retrieved from: <https://www.espacestemps.net/articles/la-frontière-un-objet-spatial-en-mutation>. Accessed 4 July 2023.

3 M. Foucher, *La République Européenne*, Belin, Paris, 2000.

More specifically, on the scale of urban systems, it is worthwhile considering the role of metropolisation in transcending these territorial limits. The metropolisation process is characterised by a concentration of management activities (economic, political and cultural, among others) and higher tertiary functions at the top of the urban hierarchy. By relying on powerful connections (transport and telecommunications), metropolises assemble and reassemble systems of networks between territories.[4] Indeed, mobility is a major development challenge, and requires the existence of efficient public transportation networks, which is an important factor in the formal integration of cross-border conurbations.[5]

In other words, will metropolisation, which promotes ever greater mobility, fluidity and interdependence, signal the end of borders? The answer is obviously complex and varied: the presence of an international border within metropolises can serve as both a barrier and an opportunity to creating new links and generating new exchanges.[6] The case of cross-border metropolises (Lille, but also Strasbourg, Basel or Geneva) is a singular one[7]: the border, although devalued and easily crossed, continues to exist (political, legal differences, etc.). In the case of the Eurometropolis Lille-Kortrijk-Tournai, are the public transport systems integrated? Or, on the contrary, do discontinuities persist?

The aim of this chapter is to examine whether there is a border effect on mobility between France and Belgium, and, more specifically, between the three entities responsible for local mobility of the Eurometropolis Lille-Kortrijk-Tournai: The *Métropole Européenne de Lille* (MEL), also known as the "European Metropolis Lille" in English, Flanders and Wallonia. Do integration factors in the Eurometropolis (devaluation of borders, absence of natural barriers, integrated economic functioning, urban continuity, etc.) combine with integration through the networks, notably public transport?

---

4 J. C. Gallety (ed.), *Le processus de métropolisation et l'urbain de demain*, collection Essentiel n°15, Lyon, 2013.

5 A. Bayer & B. Reitel, "La gouvernance des transports publics comme mesure de l'intégration transfrontalière en Europe de l'Ouest. L'exemple des trois Eurodistricts de Sarrebruck, Strasbourg et Bâle", *Géo-Regards*, n°4, 2011, pp. 77–93.

6 B. Reitel, C. Sohn & O. Walther, "Analyse compare de l'intégration métropolitaine transfrontalière en Europe", *Territoire en movement*, 2007, pp. 62–75.

7 P. Pupier, *Recompositions scalaires en Europe occidentale: la construction de régions métropolitaines transfrontalières dans l'Eurorégion Transmanche et le Rhin Supérieur*, Geography thesis, Université d'Artois, Arras, 2021.

To achieve our aim, we draw on several kinds of data: mobility surveys to examine differences between the various entities of the Eurometropolis; an analysis of integration relating to information and pricing; and, above all, a comparative measurement of public transport accessibility. Our investigations led us to question the role of the integrated public transport network within the Eurometropolis given cross-border mobility issues.

Eurometropolis Lille-Kortrijk-Tournai: What Kind of Spatial Integration through Transport?

Mobility in Metropolises and Cross-Border Conurbations

Metropolisation is a process that raises new questions about territories. This phenomenon is now crucial to how we understand urban transition.[8] On the one hand, advanced and strategic services, which attract the "creative classes",[9] are concentrated in the heart of metropolises, generating the bulk of these areas' economic growth.[10] On the other hand, metropolitan spaces continuously expand their influence across shifting territories with blurred contours, the limits of which depend on the criteria and indicators used to measure them.[11]

The term "metropolis" refers to an urban form that is characterised by a large-scale urban area. It should not be confused with the administrative status assigned to groupings of municipalities with more than 400,000 inhabitants in France. Its functional space, which generally covers several public action spheres, tends to make metropolises "ungovernable".[12] The intensification of metropolitan mobility[13] requires the

---

[8] F. Bachelet, P. Menerault & D. Paris (eds.), *Action publique et projet métropolitain*, L'Harmattan, Paris, 2006.

[9] R. Florida, *Cities and the Creative Class*, Routledge, London, New York, 2005.

[10] L. Davezies, *La crise qui vient: la nouvelle fracture territoriale*, Le Seuil, Paris, 2012.

[11] D. Pinson & S. Thomann, *La maison en ses territoires. De la ville à la ville diffuse*, L'Harmattan, Paris, 2002 ; A. Bretagnolle, "La naissance du périurbain comme catégorie statistique en France. Une perspective internationale", *L'Espace Géographique*, 2015, pp. 18–37.

[12] B. Jouvre & C. Levèvre (eds.), *Métropoles ingouvernables*, Elsevier, Paris, 2002.

[13] L. Vodoz, B. Pfister & C. Jemelin, *Les territoires de la mobilité. L'air du temps*, Presses polytechniques et universitaires romandes, Lausanne, 2004.

creation of high-performance transportation systems that limit congestion. Controlling the traffic flow is all the more complex when an urban system's area of influence extends beyond national borders, as is the case of cross-border polycentric metropolitan regions.[14]

Nevertheless, the stakes are high as analyses of mobility attest to the specificities of cross-border commuting. In particular, commuting in cross-border areas is characterised by extensive car use and the long distances covered to arrive at work.[15] In the Luxembourg cross-border region, private car is by far the most common mode of transport employed by cross-border commuters.[16] Motorway congestion is a recurring problem and public transport has struggled to expand or is underperforming. The same applies to the Geneva region and the Franco-Swiss Jura Arc, where private cars dominate cross-border travel.[17] Public transport services are poorly adapted to needs[18] and cross-border commuters contribute to the heavy congestion of the local road network.[19] Studying the case of Basel, Yann Dubois' thesis has highlighted three structural weaknesses that can be found in other cross-border regions: "large differences in living

---

14 ESPON, *Metroborder, Cross-border Polycentric Metropolitan Regions,* Final Report, European Union, 2010.

15 C. Enaux & P. Greber, "Beliefs about energy, a factor in daily ecological mobility?", *Journal of Transport Geography*, 2014, pp. 154–162; G. Drevon, O. Klein, L. Gwiazdzinski & P. Gerber, "Frontière et espace de vie: comparaison de deux faisceaux de mobilité quotidienne", *Espace Populations Sociétés*, 2015. Retrieved from : <https://doi.org/10.4000/eps.6045>. Accessed 21 March 2023; P. Gerber, T.Y. Ma, O. Klein, J. Schiebel & S. Carpentier, "Cross-border residential mobility, quality of life and modal shift: a Luxembourg case study", *Transportation Research Part A: Policy and Practice*, 2017, pp. 238–254.

16 É. Auburtin, "Anciennes frontières, nouvelles discontinuités: les impacts du développement du travail frontalier sur les populations et les territoires du Nord lorrain", *Espace Populations Sociétés*, 2005.

17 A. Vigani, "Transports, frontière et développement territorial de la Regio Insubrica", *Géo-Regards*, 2011, pp. 95–112.

18 A. Audikana, V. Kaufmann & M. A. Messer, "Governing the Geneva tram network: Making decisions without making choices", *Journal of Urban Technology*, 2015, pp. 103–124; S. Munafò, V. Kaufmann, D. Christie, S. Vincent-Geslin & E. Ravalet, "Dispositions et usages de l'automobile et des transports publics entre 1994 et 2011: Analyses des cas de Berne, Genève et Lausanne", *Revue d'Économie Régionale & Urbaine*, 2015, pp. 753–782.

19 A. Vigani, "Transports, frontière et développement territorial de la Regio Insubrica", *op. cit.* ; E. Barth, *How international borders affect local public transport*, thesis, ETH, Zurich, 2014.

conditions between cross-border and non-cross-border commuters; significant traffic flows; the functional and social specialisation of a conurbation's territories".[20] Dubois concludes by adding that these weaknesses highlight the need for some form of flow regulation and more hands-on management to handle the impacts of open borders. The difficulty of developing efficient transnational public transport often leads to a shortage of alternatives to private cars. In the end, traffic conditions, transport times and the absence of alternatives leave individuals vulnerable and dependent on their cars, which exacerbates the environmental impact of mobility.

The question of transport regulation, a recurring theme in metropolitan areas,[21] arises from a different perspective when the metropolis physically and functionally straddles two countries. The challenge of managing transport is particularly complex in the absence of shared governance between actors on either side of the border.[22] Yann Dubois and Patrick Rérat (2012)[23] have pointed out that a border separates two distinct institutional systems. Yet, the existence of an international border in a functional area of everyday transport can be an opportunity to create new forms of governance, forge new links and generate

20 Y. Dubois, *Les frontières au quotidien? Modes de vie dans l'agglomération trinationale de Bâle*, thesis, Laboratoire de sociologie urbaine, École Polytechnique Fédérale de Lausanne, Lausanne, 2017, p. 20.

21 C. Gallez & P. Menerault (eds.), *Recomposition intercommunale et enjeux des transports publics en milieu urbain*, PREDIT report on the ADEME-INRETS convention, 2005 ; N. Douay, *La planification urbaine à l'épreuve de la métropolisation: enjeux, acteurs et stratégies à Marseille et Montréal*, urban planning thesis, Université de Montréal & Aix-Marseille Université, Montréal, Marseille, 2007 ; P. Estèbe, *Gouverner la ville mobile. Intercommunalité et démocratie locale*, Presses Universitaires de France Paris, 2008 ; M. Claux, *La ville mobile gouvernée: intercommunalités, politiques de déplacements et trajectoires de développement dans la région urbaine marseillaise*, thesis, Université de Marseille, Marseille, 2014 ; S. Döry & A. Decoville, "Gouvernance and transportation policy networks in the cross-border metropolitan region of Luxembourg: A social network analysis", *European Urban and Regional Studies*, 2007, pp. 69–85.

22 M. Guérois, A. Pavard, A. Bretagnolle & H. Mathian, "Les temps de transport pour délimiter des aires urbaines fonctionnelles? Une investigation critique à partir de trois métropoles Européennes", *Belgeo*, 2016. Retrieved from : <https://doi.org/10.4000/belgeo.17789>. Accessed 23 May 2023.

23 Y. Dubois & P. Rérat, "Vivre la frontière: les pratiques spatiales transfrontalières dans l'Arc jurassien francosuisse", *Belgeo*, 2012. Retrieved from : <http://dx.doi.org/10.4000/belgeo.6249>. Accessed 21 March 2023.

new exchanges.[24] While some inter-communal bodies are now called "European Metropolises" or have been working to establish European Groupings of Territorial Cooperation (EGTC), the different actors are yet to find an optimal form of cross-border governance to manage mobility.[25] Resources dedicated to intervention in this area remain modest and the tools for observing and analysing cross-border realities are still rare. This plethora of actors and action frameworks leads to complex, often multi-level and multi-channel governance, and results in diverging interests.[26] In all cases, cross-border spaces can be understood as privileged sites for observing how transport systems are restructured and how their territorial governance evolves.[27]

Eurometropolis Lille-Kortrijk-Tournai: A Polycentric Cross-Border Conurbation?

Eurometropolis Lille-Kortrijk-Tournai is a European Grouping of Territorial Cooperation (EGTC) composed of the Métropole Européenne de Lille, four Belgian inter-communal bodies (two in Flanders, two in Wallonia) and various other administrative structures on either side of the border: one French *département* and two provinces in Belgium; one French *région* and three federated Belgian entities (the Walloon Region, the French Community and the Flemish Region, which partly merged with its Community in 1980); and the two states. Created in 2007, the

---

[24] B. Reitel, C. Sohn & O. Walther, "Analyse comparée de l'intégration métropolitaine transfrontalière en Europe", *op. cit.*

[25] A. Beyer, *La construction d'un espace européen des transports. Du principe de souveraineté national aux modalités de son dépassement*, HDR Université Paris IV Sorbonne, Paris, 2014.

[26] V. Biot, "Les systèmes de gouvernance des territoires transfrontaliers: la mise en œuvre du règlement européen sur les GECT", *Belgeo*, 2013. Retrieved from: <http://dx.doi.org/10.4000/belgeo.10565>. Accessed 23 May 2023; F. Bachelet, P. Menerault & D. Paris, *Action publique et projet européen, op. cit.* ; J. Nelles & F. Durant, "Political rescaling and metropolitan governance in cross-border regions: Comparing the cross-border metropolitan areas of Lille and Luxembourg", *European Urban and Regional Studies*, 2014, pp. 104–122.

[27] B. Jouve, *Coopération transfrontalière et services urbains en réseaux,* thesis, LATTS-Université Paris-Val de Marne, Parusn 1992 ; A. Beyer, *La construction d'un espace européen des transports, op. cit.* ; B. Reitel, *Villes et frontières: un jeu de construction de territoires*, HDR, Université de Nanterre, Paris, 2010.

group encompasses 14 partners from four institutional levels covering 157 communes.

Historically, this particular EGTC is the result of several successive forms of cross-border cooperation in the Eurometropolis. The *Conférence Permanente Intercommunale Transfrontalière* (COPIT), established in 1991, was the first joint tool to be developed. An association under French law since 2001, it was a steering and consultation body operating in the same area as the future EGTC (covering five inter-communal structures, four Belgian, one French). After an experimental phase aimed at implementing a cross-border development policy, the framework for cross-border cooperation in the Franco-Belgian metropolis evolved further with the Brussels Agreement's entry into force in 2005. On 1 June 2006, the members of the COPIT created a Local Cross-border Cooperation Grouping (LCCG) called "Lille Eurométropole franco-belge" to succeed the COPIT association. At the same time, a Franco-Belgian parliamentary working group developed the constitution of a "Eurodistrict", which led to the establishment of an EGTC in 2007.

In the meantime, the presidents of 23 French and Belgian public bodies signed a declaration of intent for the Lille Metropolitan Area in 2005. Associating the "southern" part of the former mining basin with the cross-border area around Lille, this "informal transnational cooperation body was later dissolved after a poor assessment in 2019 and it remains an example of an institutional structure that was not supported by a genuine cross-border political project supporting its development".[28]

The city of Lille is the principal hub of this polycentric urban system of over 2 million inhabitants. The Lille metropolis encompasses overlapping geographical features that can be understood from different scales, each bearing meaning in relation to the concept of "metropolis".[29] These scales range from the city of Lille, which is "symbolised in its metropolitan dimension by Euralille", to its urban region, which includes, "in addition to the Belgian sector of the cross-border metropolis, the coastline of the Opal Coast and the eastern part of the Nord department, constituting, within an hour or so of the metropolis, a functional regional system with a very high urban density (over 300 inhabitants/km^2)".[30]

---

[28] S. Barles & M. Dumont, *Métabolisme et métropole. La métropole lilloise entre mondialisation et interterritorialité*, 2014, Autrement, Paris, p. 98.

[29] D. Paris, "Lille, de la métropole à la région urbaine", *Mappemonde*, 66, 2002, pp. 1–7.

[30] *Ibid.*, p. 1.

Since a long time, the Nord and Pas-de-Calais *départments* have been identified as a study area corresponding to the Lille metropolitan area.[31] These two departments count approximately 4 million inhabitants also includes Belgian cross-border territory around Tournai, Mouscron and Kortrijk, a sector at the limit of Brussels' and Gent's influence.

Residential sprawl and the development of large industrial and commercial zones have led to the gradual formation of an almost continuous urban belt from the French towns of Arras and Cambrai in the south to the Belgian towns of Roeselare and Waregem in the north. Internal relations within this cluster have become increasingly intense and spatially expansive; not only do they involve employment and real estate markets, but they also encompass training, commerce, services and leisure activities, among others. These internal relations are complex, multi-polarised and vary considerably depending on the themes and territories concerned. None of them can be defined strictly along the lines of a centre-periphery relationship, even if they undeniably resemble the ever-expanding phenomenon of metropolisation.

Sabine Barles and Marc Dumont (2021)[32] have described several autonomous territorial layers that make up the present-day Lille metropolitan system: The first layer is "historically made up of the main dense urban hub of the Lille system (20 % of inhabitants) combined with a double linear conurbation representing 40 % of inhabitants"[33]; it covers Lille and Roubaix-Tourcoing on the northeast side and Villeneuve-d'Ascq in the east. The second layer which is added to this first dense structure comprises "the cross-border and European metropolis, a territory of experiences and projects implemented institutionally within the framework of the Eurometropolis".[34] The third layer refers to another scale, that of the "urban area", made up of "networks of towns linked together in a broad system of functional interdependence: the southern arc – comprising the mining basin, Arras, Béthune, Lens, Douai and Valenciennes – the Cambrai and Maubeuge regions and the Opal Coast, which is repositioned as part of a vast extended coastline that includes Dunkirk in the framework of this new region' ".[35]

---

31 F. Damette (ed.), *La région du Nord-Pas-de-Calais, Villes et Systèmes urbains*, Report for the Agence de Développement et d'Urbanisme de Lille-Métropole, Lille,1997.

32 S. Barles & M. Dumont, *Métabolisme et métropole*, *op. cit.*

33 *Ibid.*, p. 24.

34 *Ibid.*, p. 25.

35 *Ibid.*

These three layers can be contextualised by considering the transport networks that help shape their practical and political realities. As far as the MEL urban area is concerned, this can be seen from the development of metro networks, first between Lille and Villeneuve d'Ascq, then between Lille and Roubaix and, later, Tourcoing. These two light rail lines "are perhaps first and foremost vehicles for unifying the metropolitan area".[36] At the level of the urban region, it has been the Regional Express Transport (known as *Transport Express Régional* or TER in French) that has played a structuring role. The anticipated regionalisation of Nord-Pas-de-Calais at the end of the 1990s revitalised the train network and transformed it into a metropolitan resource. The innovation of TER-GVs (TGVs used for high-speed regional services) was driven by regional planning considerations with the political objective of ensuring that various conurbations of Nord-Pas-de-Calais were within an hour's travel from the Lille metropolis.[37] Even if Nord-Pas-de-Calais' merger with the *région* Picardie has altered regional strategies, recent projects for a Metropolitan Express Service (MES) or a Metropolitan Regional Express Network (known as *Réseau Express Régional* or RER in French) highlight rail transportation's enduring importance in this expanded urban system.

Unlike the central urban hub and the urban region, however, the cross-border and European metropolis does not appear to be supported by any overarching public transportation infrastructure. Despite the urban continuity between Tourcoing and Mouscron (60,000 inhabitants), the metro line ends just a few metres short of the border. Moreover, the MEL decided to proceed with the construction of high-frequency tram and bus lines around the Lille hub and the northeast side of the Eurometropolis (Roubaix-Tourcoing) as recently as 2019. The final routes were approved in 2022 and they will not cross the border.

In this first section, we highlighted the particular problems related to mobility in cross-border metropolises. The polycentric, multi-scale nature of the Eurometropolis makes it difficult to apprehend mobility

---

36 P. Menerault & A. Barré, "De la description des circulations à la territorialisation des réseaux: 35 ans de géographie des transports dans 'Hommes et Terres du Nord'", *Territoire en mouvement*, 2006, pp. 45–53.

37 S. Delmer, V. Facchinetti-Mannone, P. Menerault & C. Richer, "L'appropriation locale de la grande vitesse ferroviaire en Europe: Trajectoires croisées: Anvers, Lille, Metz-Nancy, Saragosse", *Revue d'histoire des chemins de fer*, 2012, pp. 319–348.

issues at the EGTC level. To examine the situation further, it is necessary to look more closely at mobility practices and conditions in the territory.

Mobility in the Eurometropolis: Between Contrasts and Discontinuities

Contrasting Mobility Profiles

In France, local mobility surveys comply with a national standard certified by Cerema.[38] Cerema certified mobility surveys (EMC2) provide a snapshot of travel practices in a territory every decade or so. For cross-border metropolises, their challenge is to find a common research method with partners on both sides of the border in order to carry out an assessment on a relevant scale. In 2006, the MEL established a partnership with Wallonia to carry out a survey, but not with Flanders. In Wallonia, the survey was conducted using the Walloon Mobel methodology, a self-administered mail survey with subsequent follow-ups. In Belgian Flanders, the MEL expanded its methodology to survey border communes. In 2016, a partnership was established with Flanders and Wallonia to carry out a survey using the Cerema method by telephone (in French and Dutch) on the perimeter of the Lille-Kortrijk-Tournai Eurometropolis.

An initial finding of this survey was that mobility (number of trips per person per day) was lower in the Walloon (3.65) and Flemish (3.18) regions than in the MEL (4.07). The higher proportion of public transport in the MEL (12 %) than in the Belgian regions (between 5 and 6 %) could be explained by the difference in transport supply between the territories (See Figure 1). The MEL represents a dense urban hub covered by metro and tram lines, whereas the territories surveyed in Flanders and Wallonia are comprised of medium-sized towns and rural areas with no major public transport networks apart from the train. The survey also found that the proportion of residents who got around by cycling was significantly higher in Flanders than in the MEL and Wallonia. Some striking results emerged that were

[38] The Centre for Studies and Expertise on Risks, the Environment, Mobility and Urban Planning, or Cerema, is a French public establishment of an administrative nature operating under the joint authority of the French Ministry of Ecological and Solidarity Transition and the Ministry for Territorial Cohesion.

related to, for example, differences in bike use; it was 10 times higher in Menen than in Halluin, even though there is a continuous built-up area at the border between these two communes. Conversely, people in the MEL were twice as likely to walk than in Flanders or Wallonia. In Wallonia, given the low bike use and the lack of structured public transport services, private cars dominated local mobility with a modal share of 73 % (driver and passenger), as opposed to 63 % in Flanders and 57 % in the MEL.

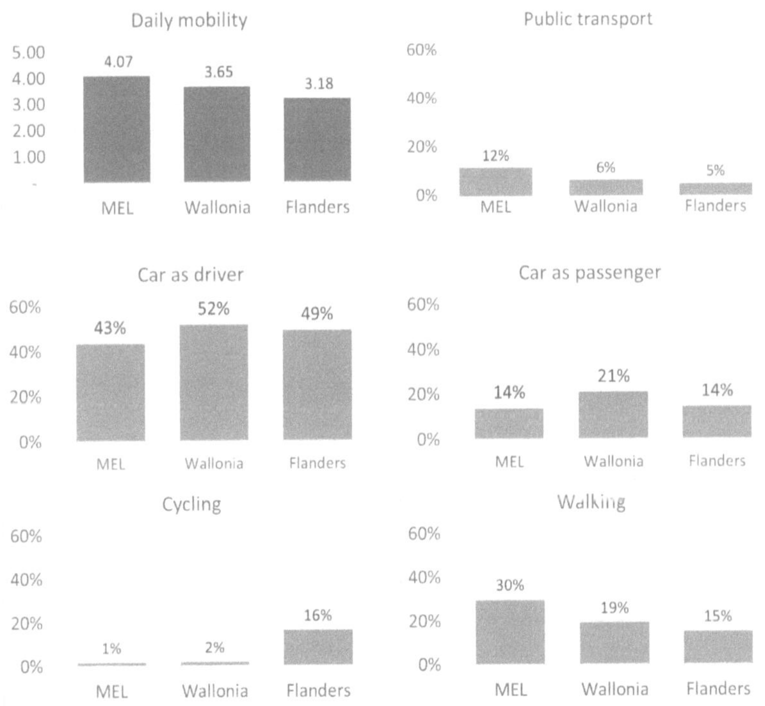

Figure 1. Comparison of daily mobility in the three Eurometropolis territories.[39]

[39] Source : EMC² Métropole Européenne de Lille 2016, Graphs drawn up by Richer and Palmier- 2023.

In the MEL, a *Diagnostic Énergie Émissions des Mobilités* (DEEM) was carried out using the data from 2016 Cerema Certified Mobility Survey. The method applied provided detailed and precise analyses all the way down to each leg of each trip. Emission factors were added at a disaggregated level for each leg of every trip. The aim was to establish a quantitative assessment of the environmental impacts of travel on the scale of an urban region.[40] In 2016, most of the greenhouse gas (GHG) emissions linked to mobility were generated by passenger transport by car (nearly 70 %) compared with 30 % for freight, although the latter was increasing more rapidly. Overall, transport-related GHG emissions rose by 7.2 % between 2006 and 2016. For travellers, the breakdown of GHG emissions by type of trip showed that 72 % of the environmental impact was linked to the density of mobility within the MEL (72 %). However, exchange trips (between the MEL and another territory) and transit trips (between two territories passing through the MEL) accounted for 28 % of GHG emissions, even though they only accounted for 11 % of all trips. Consequently, the impact of external journeys was particularly important for environmental balance, as these were longer and more dependent on travel by car.

A Persistent Border Effect on Flows

The dynamics of mobility flows attest to a border effect. For MEL residents, cross-border traffic is lower in volume than those with other territories (south of the metropolis). This traffic also covers shorter distances between the MEL's border areas and, especially, Wallonia (Comines-Warneton, Mouscron, Tournai); there is very little traffic between the MEL and Flanders, which emphasises the language divide between these areas. The flows of Walloon residents reflect those of MEL residents. There are local exchanges within the border sectors (Comines-Warneton-Mouscron and Tournai), but a similar border effect is perceptible for traffic between Flanders and Wallonia. On the other hand, commuting between Brussels, Flanders and Wallonia is significant. Residents of Flanders commute almost exclusively within the Flemish region and thus rarely commute to French-speaking areas, whether they be in Belgium

---

[40] B. Grosjean, "Les ambiguïtés de l'Eurométropole Lille-Kortrijk-Tournai, à travers ses représentations régionales", *Belgeo*, 2019. Retrieved from : <https://doi.org/10.4000/belgeo.33151>. Accessed 23 May 2023.

or in France. In the Eurometropolis, there are, therefore, two types of boundaries that cross this EGTC; "one is political, state-based, the other is linguistic, cultural, and the two do not overlap."[41]

Maps of private car travel also show that flows are much reduced near the international border, as well as between Flanders and Wallonia. Conversely, flows are very heavy between both Flanders and Wallonia and the Brussels-Capital Region. Concerning heavy goods vehicle traffic, this trend is less pronounced, but flows are significant between the Netherlands, Flanders, France axis along the A1 freeway, which corresponds to the route of long-distance goods flows from North Sea ports (Antwerp, Rotterdam) towards France and the Mediterranean regions.

Gradually Replacing Rail with Road

The cross-border rail network has gradually contracted due to the closure of several lines. Up until the 1930s, there were six Franco-Belgian rail links. The local cross-border links between Armentières and Comines (Belgium), Tourcoing-Roubaix-Wattrelos and Herseaux, Lille and Comines (Belgium), and Tourcoing and Menen (Belgium) were closed to passenger traffic in 1937, 1939, 1955 and 1971 respectively. Today, there are still two international rail links between Lille and Kortrijk and Lille and Tournai that passengers can take to cross the border. In the second half of the 20th century, road and motorway links underwent rapid development, particularly from the 1970s onwards with the construction of the A22 (E17) to Kortrijk and the A27 (E42) to Tournai. During the 1980s and 1990s, other roads to Comines, Menen and Dottignies were improved. Some segments, however, took a long time to complete, such as the cross-border stretch of the M700, which was opened in 2005 between Wattrelos and Estaimpuis, 13 years after the preceding segment was opened.

The development of the car has been accompanied by the production of cross-border maps so motorists can find their way in other countries. Companies, such as Michelin, have long been concerned with cross-border issues. Today, thanks to the development of satellite imagery and new technologies, road information no longer stops at the border. The border is now relatively imperceptible to motorists; border posts have

41 *Ibid.*, p. 2.

disappeared and road signs are fairly similar from one country to another. These different elements demonstrate the gradual integration of cross-border practices, especially where road links are concerned.

In the end, transportation choices and mobility practices illustrate that road transport plays a key role in the running of the Eurometropolis. These dynamics call into question public policies related to the consequences and limits of the private car's hegemony in cross-border mobility. In the following section, the analysis continues with public transport accessibility considered as an "indicator of the degree of formal integration of cross-border conurbations".[42]

The Difficulty of Cross-Border Integration through Public Transport Networks

Overlapping Public Transport Authorities

In a multipolar Eurometropolis with a population of 2.2 million, one might expect public transport to play a structuring role. However, public transportation in this area has struggled to contribute to closer-knit spatial integration and offer a credible alternative to car transport. There are several factors that have limited the ability of public transportation to play a leading role here. First, institutional overlap has held back greater intermodal organisation.

In Belgium, the regions are responsible for mobility and manage their public transport networks through different operators (the TEC for Wallonia, De Lijn for Flanders and the STIB for Brussels-Capital) (See Figure 2). The rail network is managed by the SNCB at a federal level. In France, responsibility for mobility is shared between two levels of organising authority: one at a local, generally inter-communal scale (for Lille, the Métropole Européenne de Lille) for bus, tram and metro services; the other at a regional level for intercity trains and coaches (*région* Hauts-de-France). The Eurometropolis' public transport offer therefore requires effective coordination between five major network operators (the SNCF, the SNCB, Ilévia, the TEC and De Lijn). However, public transport

42 A. Beyer & B. Reitel, "Les systèmes de transports publics comme mesure de l'intégration transfrontalière", *op. cit.*

network plans are limited to the respective territories of each operator without including the neighbouring areas.

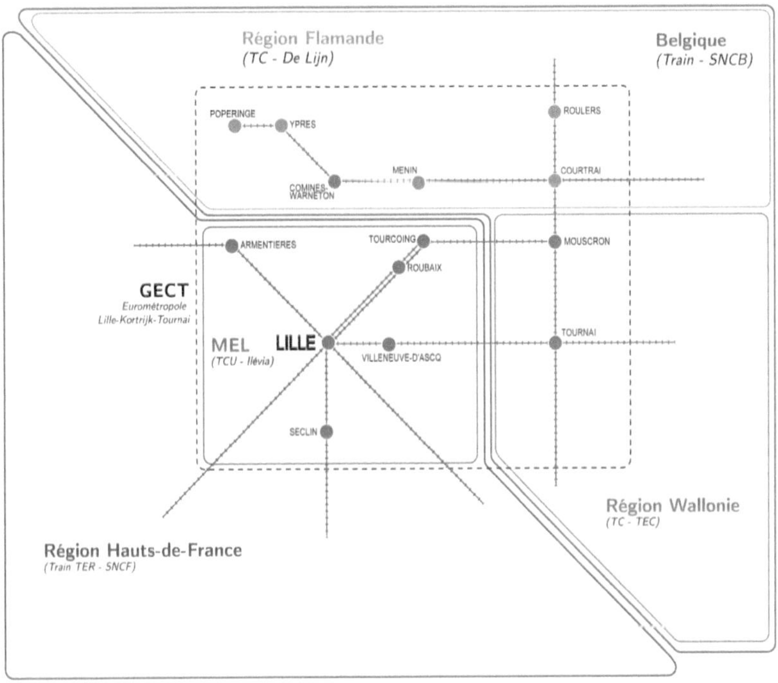

Figure 2. Authorities responsible for transport in the Eurometropolis.[43]

Crossing the Border by Public Transport: Rare and Subject to Conditions

Only one bus line and two train lines can really be considered cross-border lines within the Eurometropolis. Where buses are concerned, the most fully developed example is the Mouscron-Wattrelos-Roubaix (MWR) cross-border line. The Conférence Permanente Intercommunale Transfrontalière (COPIT) – which benefited from the first European Interreg programmes (I and II) – introduced this cross-border bus service between Mouscron and Wattrelos in 1992 and it was extended to

[43] Various sources, Graphs drawn up by Richer and Palmier, 2023.

Roubaix in 1995.[44] The MWR line was a "genuine new cross-border line"[45] according to Frédérique Durand. It was created specifically to meet a cross-border need and was not simply a redevelopment of an existing line for a start. Moreover, it has been co-managed and co-financed by two operators (Ilévia and the TEC).[46] This line appears in the MEL network plan, but in the form of a schematic representation in Belgium. Furthermore, this line is subject to a specific, more expensive fare system and does not fall within the scope of social pricing or free travel for 18–25-year-olds in the MEL (See Figure 3).

	MEL	**MWR Line (no connections)**	**MWR Line (with connections)**
Single Ticket	1.80€	2.25€	2.45€
Pack of 10 tickets	15.40€	18.20€	20.20€
Travel Pass	54.00€ / month	13€ / week	Unavailable
Travel Pass – 18 years	free	13€ / week	Unavailable

Figure 3. MWR line pricing compared to the Ilévia network (MEL).[47]

Other bus lines cross the border, but they are simply extensions of a particular operator's line. For example, as of 2022, the Flemish Region has been trialling the extension of a bus line to Armentières station (Line 72 Ieper-Le Bizet) with the aim of improving Flemings' access to the French rail network. However, the line does not currently appear on the MEL map, and travellers have to purchase tickets in Flanders (i.e. a De Lijn ticket) or with the De Lijn smartphone application. The service runs a return trip every 2 hours, which is adjusted according to the arrival and departure of trains at Armentières station.

It is also worth noting that several of the MEL operator's main lines (express buses called "liane") have terminuses close to the border. However, the continuity of urbanisation and the presence of stations on

44 F. Durand, "Émergence d'une gouvernance métropolitaine transfrontalière au sein de l'Eurométropole Lille-Kortrijk-Tournai. Analyse des relations entre les organisations dans le domaine du transport public transfrontalier", *Annales de Géographie*, 2013, pp. 290–311.

45 *Ibid.*, p. 294.

46 *Ibid.*

47 Source: Ilévia 2023, Richer and Palmier, 2023.

the other side of the border could encourage extensions. This is the case for the L90 line (terminus at Comines, 1 km from Comines (B) station in Belgium), the L4 and L91 lines (terminus at Halluin, 2 km from Menen station) and the L8 link (terminus at Tourcoing, 4.5 km from Mouscron station).

Regarding train travel, the links between Lille and Kortrijk and between Lille and Tournai are included in the respective maps of the TER Hauts-de-France and the SNCB. However, not all intermediate stations are indicated. The lack of a truly cross-border map makes it difficult to travel from one intermediate station to another as they only appear on the map of the country to which they belong.

In Hauts-de-France, the rail network is powered by a 25kW alternating current, whereas in Belgium the network is powered by a 3000V direct current. This means that trains servicing the cross-border lines require dual-mode locomotives, as is the case for SNCB trains, or diesel locomotives, which have a far less favourable carbon footprint. Today, all cross-border services are operated by SNCB trains with dual-mode locomotives.

As with the MWR bus line, cross-border train fares are complex and specific. By design, the fare policies of Hauts-de-France and Belgium differ, so setting prices for the cross-border line requires agreement between all concerned parties: the Hauts-de-France Region has reached an arrangement with the SNCB so that the fares offered for cross-border links are identical on both sides of the border. However, as the table (Figure 4) below shows, the price of cross-border tickets can still vary depending on the purchasing platform used: For the Lille-Tournai line, the prices on the regional TER, SNCB and TrainLine sites are identical as advertised. However, the SNCF website, SNCF Connect, sells the same single ticket for €8.20 instead of €4.60, which is 78 % more expensive. Discrepancies also exist in the way tickets can be purchased. For instance, it is not possible to purchase a ticket from the regional TER website for travel between Baisieux in France and Froyennes in Belgium, two intermediary stations on the Lille-Tournai line. Moreover, SNCF Connect charges €6.10 for a ticket between these stops instead of €3.70.

On the Lille-Kortrijk line, the differences are even more pronounced. A Lille-Kortrijk ticket cannot be purchased on the regional TER website, whereas it is available for Lille-Tournai. While the SNCB and Trainline sites charge the same fare of €5.30 for a second-class ticket, SNCF

Connect charges €8.90. On the Tourcoing-Mouscron cross-border route, SNCF Connect's fare of €10.60 is more expensive than the Lille-Kortrijk fare, although services covering the latter line travel a greater distance.

	SNCF-Connect	TER Hauts-de-France	SNCB	Trainline
Lille-Tournai	8,20€	4,60€	4,60€	4,60€
Baisieux-Froyennes	6,10€	Unavailable	3,70€	3,70€
Lille-Kortrijk	8,90€	Unavailable	5,30€	5,30€
Tourcoing-Mouscron	10,60€	Unavailable	3,70€	3,70€

Figure 4. Comparative pricing of cross-border rail lines.[48]

The principle of systematically repeating routes and timetables at regular intervals, which was adopted by the *région* Hauts-de-France in 2020, has enabled rail services to be restructured along similar lines to those in Belgium. Although this interval system has not been perfected, it has had a significant impact on cross-border train travel. Previously, on the Lille-Kortrijk route, some trains spent 20 minutes waiting at Mouscron station to align themselves with the Belgian timetable. There are approximately 78 trains per day travelling between Tournai and Mouscron and 66 travelling between Tournai and Kortrijk. Conversely, there are only 32 and 35 trains covering the cross-border lines between Lille and Kortrijk and Lille and Tournai respectively.

Ultimately, research on mobility as a service (MaaS) allows us to assess the integration of transport in the Eurometropolis. MaaS consists of grouping all mobility services available in an area on a single platform, which may be a smartphone application or a website. The first stage of MaaS concerns the integration of information, the second refers to the integration pricing and the third involves the integration of mobility services and governance.[49] In the case of the Eurometropolis, the situation is currently at the minimal, first stage: information sharing. Trip planners still only partially cover the range of transport options on offer in

---

48 Source: Various operator websites March 2023, Richer and Palmier, 2023.

49 J. Sochor, H. Arby, I. C. Karlsson & S. Sarasini, "A topological approach to mobility as a Service: A proposed tool for understanding requirements and effects, and for aiding the integration of societal goals", *Research in Transportation Business and Management*, 2018, pp. 3–14.

neighbouring regions, and the lack of a shared platform means that transport networks cannot be identified beyond institutional borders.

Measuring Accessibility Via Public Transport in the Eurometropolis

To contribute to discussions surrounding the degree of cross-border public transport networks' integration, we propose an original method to measure territorial accessibility. Accessibility is "essentially a spatial concept which aims to assess the effort required to travel through space to reach a place where a resource is located".[50] This spatial potential will be measured here by examining the timetable structure of public transport services.

We approach accessibility through an analysis of timetable data. This data can provide us with a more detailed picture of mobility potential and is more closely linked to usage constraints than other indicators, such as shortest trip time. To carry out this assessment, we have identified and analysed various timetable databases. In France, public transport timetables are made available by Hauts-de-France Mobilités, which provides them to the Hauts-de-France Regional Directorate for the Environment, Planning and Housing (DREAL), which, in turn, adds them to a database. A considerable amount of timetable data is also available at the national access point.[51] Equivalent timetables are also available in Belgium (from De Lijn, the TEC, the SNCB and the STIB), either by direct download or by signing a user agreement. Currently, these data can be used to calculate cross-border multimodal routes or cross-border accessibility maps by using, for example, the "networks" plugin of the QGIS Geographic Information System, which was developed for this purpose by Cerema with an open-source license. This plugin can be used to create multimodal networks which combine services that are both road-based (walking, cycling, car, etc.) and based on timetables and calendars (public transport, train, plane, ferry, etc.). This tool can also be used to analyse multimodal services and produce multimodal accessibility maps in the

50 A. L'Hostis & A. Conesa, "Définir l'accessibilité intermodale", *In* A. Banos & T. Thevenin (eds.), *Systèmes de transport urbains. Caractérisation de l'offre et estimation de la demande*, Hermès-Lavoisier, Cachan, 2010, p. 24.

51 Retrieved from: <https://transport.data.gouv.fr/>. Accessed 4 July 2023.

form of isochrones or catchment areas.[52] The isochrone maps are generated according to the "Limewise" algorithm.[53]

A cross-border multimodal network for the Eurometropolis was thus compiled based on data from the following operators: Ilévia (bus, metro, trams in the MEL), the SNCF (the TER in the *région* Hauts-de-France), the SNCB (trains in Belgium), the TEC (public transport in the Walloon Region) and De Lijn (public transport in the Flemish Region). We have not included private coach or high-speed rail services in our analysis as they have no impact on trips limited to within the Eurometropolis.

As there are many potential elements to assess, we have chosen a model that allows us to focus on the hourly accessibility of the Eurometropolis' three main cities: Lille, Kortrijk and Tournai. Consequently, the map (Figure 5) is made up of three superimposed measurements that correspond to public transport arrivals before 9 a.m. in these three cities. The zone depicted covers the areas where travel time does not exceed one hour to reach one of the three destinations before 9 a.m.

Optimal routes have been calculated using generalised time primarily to set weightings for walking time (50 % penalty) and transfer time (10-minute penalty) (See Figure 5). Generalised time is the same as a time period in which each of its components (time on board a vehicle, waiting time, walking time, transfer time) is associated with a weighting that makes it possible to simulate actual user behaviour. The parameters chosen were defined based on exploratory research aimed at reconstructing, through modelling, trips recorded on the Lille Metropole public transport network.[54] These measurements are in line with other studies that have attempted to estimate users' perceptions of connections.[55] Walking time is penalised by 50 % in generalised time. According to this model,

---

52 P. Palmier, "Musljw. Logiciel de calcul d'accessibilité multimodale SETRA", Cerema, Lyon, 2011. Retrieved from: <http://www.bv.transports.gouv.qc.ca/mono/1173266.pdf>. Accessed 23 May 2023.

53 P. Palmier, "LIMEWISE: A deterministic line-based interpolation methodology for a realistic multimodal accessibility representation", *Geomatics Workbooks*, vol. 12, FOSS4G Europe, Corno, 2015, pp. 1–12.

54 P. Palmier, *Étude sur les flux en transports collectifs urbains sur Lille Métropole*, Report for LMCU, CETE Nord-Picardie, Paris, 2009, p. 51.

55 F. Dobruskes, M. Hubert, F. Laporte & C. Veiders, "Réorganisation d'un reseau de transport collectif, ruptures de charge et mobilités éprouvantes: l'expérience bruxelloise", *Articulo*, 2011. Retrieved from : <https://doi.org/10.4000/articulo.1844>. Accessed 23 May 2023.

a user will prefer a 20-minute trip on public transport to a 15-minute journey by foot.

$t_{generalised} = C_{tc}.t_{tc}+c_{waiting}.t_{waiting}+c_{walking}.t_{walking}+c_{connection}.t_{connection}$

$t_{generalised}$: Generalised time

t_{tc} : Time on public transport

$t_{waiting}$: Waiting time

$t_{walking}$: Walking time

$t_{connection}$: Sum of minimum connecting time: Minimum connection time: 2 minutes

c_{tc} : Weighting of time on public transport: 1

$c_{waiting}$: Weighting of waiting time: 1

$c_{walking}$: Weighting of walking time : 1.5

$c_{connection}$: Weighting of minimum connection time: 5

Figure 5. Description of the generalised time calculation.[56]

Furthermore, "available" walking time must also be determined. The maximum allotted walking time was set at 30 minutes for an entire trip, which is equivalent to walking 2 km (no U-turns, walking speed set at 4 km/h). This maximum limit is supported by the results of mobility surveys, which generally show that the modal share of walking for a trip remains significant up to 2 km, but negligible beyond this distance. Therefore, the cumulative walking time to reach a public transport stop from home, to make any connections and to reach the final destination (Lille, Kortrijk or Tournai stations) must not exceed 30 minutes. If a trip exceeds this threshold, travel by public transport is deemed unfeasible.

Influence of Institutional Borders on Public Transport Accessibility

Our results demonstrate that accessibility is particularly consistent within the respective institutional boundaries of the Eurometropolis. Beyond the three main cities, few areas display much accessibility potential for trips to Lille, Kortrijk and Tournai. In other words, outside of these cities, it is difficult to benefit from favourable accessibility conditions to

---

[56] Source : Richer and Palmier, 2023.

cross a border, whether it be between France and Belgium or Flanders and Wallonia.

There are two notable exceptions: the cross-border areas around Mouscron and Baisieux stations. Both of these areas provide comprehensive access to Lille, Kortrijk and Tournai. Access to the Belgian rail network and also provides regular connections to Kortrijk.

Under the conditions tested, we found that Kortrijk was accessible within an hour from Lille and around the railway axis at Tourcoing and Roubaix, close to the border. It should also be noted that the towns of Roubaix and Tourcoing do not fall within the Tournai accessibility area. These results show that public transport services are primarily directed towards Lille.

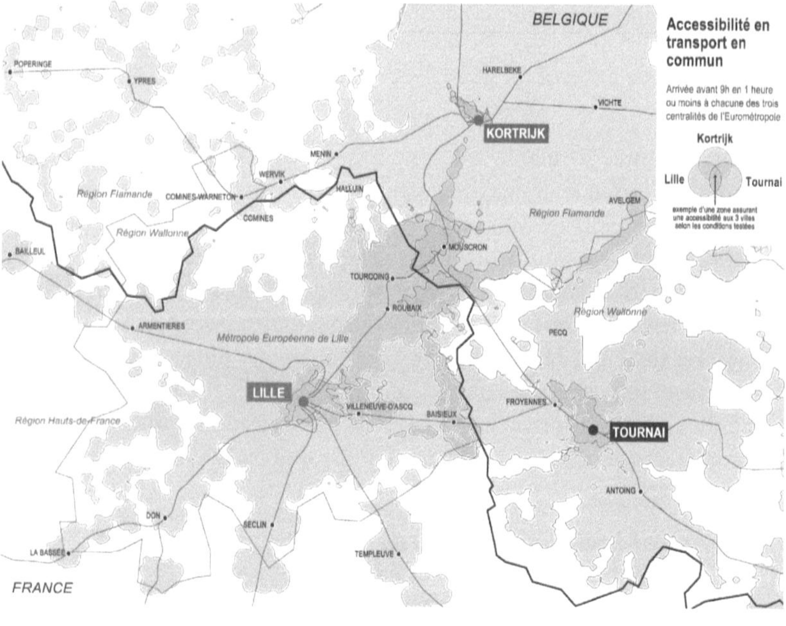

Figure 6. Measuring accessibility by public transport in the Eurometropolis: arrival before 9am in 1 hour or less to the three main cities, Lille, Kortrijk or Tournai.[57]

[57] Source: Base de données régionale de l'offre de transport multimodale – millésime 2022 – DREAL Hauts-de-France, Richer and Palmier, 2023.

Lille's accessibility area covers the entire MEL. Some areas to the west (in the Weppes) and north of Lille, towards Comines, are more unevenly covered. The Lille-Comines railway line was discontinued in 2019 and buses operating along this corridor are less convenient in terms of travel time. As a result, the French town of Comines is neither integrated into the Lille area (an averages bus trip takes 68 minutes), nor the Kortrijk area as it takes 12 minutes to walk from Comines station in France to Comines station in Belgium and catch a connecting train on the Belgian rail network. To the south and west, Lille's accessibility extends beyond the MEL along the rail network.

The bus network around Kortrijk provides a relatively large and homogenous accessibility area. The Flemish city is also accessible from the main hubs to the west around Ypres and Poperinge, primarily by train. The network is more fragmented around Tournai in Wallonia, which is due to the lower density of urban centres located close to the city.

In short, there is a clear border effect between France and Belgium, but also between Flanders and Wallonia. The situation is straightforward; where public transport is concerned, the institutional framework imposes itself on the functional territory of the Eurometropolis. According to Jean-Marc Offner and Pierre Zembri, various politico-administrative levels and the operators' areas of responsibility prioritise intra-territorial routes (particularly between the centre and the periphery) and neglect those that cross borders,[58] as is the case in the Eurometropolis.

Conclusion and Prospects

This chapter has examined the territorial dynamics at work in the Eurometropolis Lille-Kortrijk-Tournai. Studying mobility with a particular focus on measures of accessibility confirms the persistence of a border effect. Possibilities for public transport travel within this Eurometropolis are hindered by the administrative boundaries of the organising authorities. There are numerous aspects that indicate that there is a real cleavage at the border, namely in terms of information provision, ticketing differences and network coverage. Improvements to public transport services

---

[58] J. M. Offner & P. Zembri, "Les réseaux, entre territoires fonctionnels et territoires institutionnels", *NETCOM: Réseaux, communication et territoires*, 1994, pp. 480–485.

in the three territories that make up the Eurometropolis have not helped improve cross-border links.

In fact, over the long term, it is no longer public transport that structures Eurometropolitan integration, but road connections, which have been expanded, become more accessible and are free of charge. The discontinuation of certain rail lines has not been adequately offset by cross-border bus routes. Finally, no new cross-border route has been established since the "MWR" line (1992) which is now over 30 years old. These discontinuities ensure that travel by car has a considerable advantage over public transport for those who wish to cross national or regional borders within the Eurometropolis.

Of the three layers that make up the current Lille metropolitan system,[59] the cross-border metropolis is the only one that is not structured by a public transport network. It occupies an "in-between" zone that is neither covered by the main urban transport lines (bus, metro, tramway), nor is it fully incorporated within the "Metropolitan RER" that is oriented towards the southern rim of the former mining basin.

Actors in the Eurometropolis are attempting to establish frameworks for cooperation that will provide effective cross-border governance in relation to mobility. The Eurometropolis Agency recently organised a meeting on the theme of mobility and public transport (February 2023). This issue is an important one in the European grouping, which has encouraged dialogue between Walloon, Flemish and French actors, as well as the main transport operators. However, although the assessment of the situation is widely accepted, the measures undertaken to improve it remain modest or fragmented; this state of affairs is hindering the creation of a genuine cross-border network. In the end, regional and local authorities have been unable to reach an agreement over the removal of obstacles or guarantee cross-border services.[60] As a result, the ambiguity of the Eurometropolis, as identified by Bénédicte Grosjean (2019),[61] remains unresolved. On the one hand, there is a desire to be fully involved in the process of metropolisation by establishing a "multicultural regional entity for the purposes of visibility and attractiveness".[62] On the other

---

59 S. Barles & M. Dumont, *Métabolisme et métropole*, *op. cit.*

60 A. Beyer & B. Reitel, "Les systèmes de transports publics comme mesure de l'intégration transfrontalière", *op. cit.*

61 B. Grosjean, "Les ambiguïtés de l'Eurométropole Lille-Kortrijk-Tournai", *op. cit.*

62 *Ibid.*, p. 15.

hand, the Eurometropolis remains the "sum of its territories", a "mosaic of juxtaposing fragments" characterised by its inability to implement an efficient cross-border public transport network.

Bibliography

É. Auburtin, "Anciennes frontières, nouvelles discontinuités: les impacts du développement du travail frontalier sur les populations et les territoires du Nord lorrain", *Espace Populations Sociétés*, 2005.

A. Audikana, V. Kaufmann & M. A. Messer, "Governing the Geneva tram network: Making decisions without making choices", *Journal of Urban Technology*, 2015, pp. 103–124.

F. Bachelet, P. Menerault & D. Paris (eds.), *Action publique et projet métropolitain*, L'Harmattan, Paris, 2006.

S. Barles & M. Dumont, *Métabolisme et métropole. La métropole lilloise entre mondialisation et interterritorialité*, 2014, Autrement, Paris.

E. Barth, *How international borders affect local public transport*, thesis, ETH, Zurich, 2014.

A. Beyer, *La construction d'un espace européen des transports. Du principe de souveraineté national aux modalités de son dépassement*, HDR Université Paris IV Sorbonne, Paris, 2014.

A. Bayer & B. Reitel, " La gouvernance des transports publics comme mesure de l'intégration transfrontalière en Europe de l'Ouest. L'exemple des trois Eurodistricts de Sarrebruck, Strasbourg et Bâle ", *Géo-Regards*, n°4, 2011, pp. 77–93.

V. Biot, "Les systèmes de gouvernance des territoires transfrontaliers: la mise en œuvre du règlement européen sur les GECT", *Belgeo*, 2013.

A. Bretagnolle, "La naissance du périurbain comme catégorie statistique en France. Une perspective internationale", *L'Espace Géographique*, 2015, pp. 18–37.

F. Damette (ed.). *La région du Nord-Pas-de-Calais, Villes et Systèmes urbains*, Report for the Agence de Développement et d'Urbanisme de Lille-Métropole, Lille,1997.

L. Davezies, *La crise qui vient: la nouvelle fracture territoriale*, Le Seuil, Paris, 2012.

S. Delmer, V. Facchinetti-Mannone, P. Menerault & C. Richer, "L'appropriation locale de la grande vitesse ferroviaire en Europe: Trajectoires croisées: Anvers, Lille, Metz-Nancy, Saragosse", *Revue d'histoire des chemins de fer*, 2012, pp. 319–248.

F. Dobruskes, M. Hubert, F. Laporte & C. Veiders, "Réorganisation d'un reseau de transport collectif, ruptures de charge et mobilités éprouvantes: l'expérience bruxelloise", *Articulo*, 2011.

S. Döry & A. Decoville, "Gouvernance and transportation policy networks in the cross-border metropolitan region of Luxembourg: A social network analysis", *European Urban and Regional Studies*, 2007, pp. 69–85.

N. Douay, *La planification urbaine à l'épreuve de la métropolisation: enjeux, acteurs et stratégies à Marseille et Montréal*, urban planning thesis, Université de Montréal & Aix-Marseille Université, Montréal, Marseille, 2007.

G. Drevon, O. Klein, L. Gwiazdzinski & P. Gerber, "Frontière et espace de vie: comparaison de deux faisceaux de mobilité quotidienne", *Espace Populations Sociétés*, 2015.

Y. Dubois & P. Rérat, "Vivre la frontière: les pratiques spatiales transfrontalières dans l'Arc jurassien francosuisse", *Belgeo*, 2012.

Y. Dubois, *Les frontières au quotidien? Modes de vie dans l'agglomération trinationale de Bâle,* thesis, Laboratoire de sociologie urbaine, École Polytechnique Fédérale de Lausanne, Lausanne, 2017.

F. Durand, "Émergence d'une gouvernance métropolitaine transfrontalière au sein de l'Eurométropole Lille-Kortrijk-Tournai. Analyse des relations entre les organisations dans le domaine du transport public transfrontalier", *Annales de Géographie*, 2013, pp. 290–311.

C. Enaux & P. Greber, "Beliefs about energy, a factor in daily ecological mobility?", *Journal of Transport Geography*, 2014, pp. 154–162.

ESPON, *Metroborder, Cross-border Polycentric Metropolitan Regions,* Final Report, European Union, 2010.

P. Estèbe, *Gouverner la ville mobile. Intercommunalité et démocratie locale*, Presses Universitaires de France Paris, 2008 ; M. Claux, *La ville mobile gouvernée: intercommunalités, politiques de déplacements et trajectoires de développement dans la région urbaine marseillaise*, thesis, Université de Marseille, Marseille, 2014.

R. Florida, *Cities and the Creative Class*, Routledge, London, New York, 2005.

M. Foucher, *La République Européenne*, Belin, Paris, 2000.

J. C. Gallety (ed.). *Le processus de métropolisation et l'urbain de demain*, collection Essentiel n°15, Lyon, 2013.

C. Gallez & P. Menerault (eds.). *Recomposition intercommunale et enjeux des transports publics en milieu urbain*, PREDIT report on the ADEME-INRETS convention, 2005.

P. Gerber, T. Y. Ma, O. Klein, J. Schiebel & S. Carpentier, "Cross-border residential mobility, quality of life and modal shift: A Luxembourg case study", *Transportation Research Part A: Policy and Practice*, 2017, pp. 238–254.

B. Grosjean, "Les ambiguïtés de l'Eurométropole Lille-Kortrijk-Tournai, à travers ses représentations régionales", *Belgeo*, 2019.

M. Guérois, A. Pavard, A. Bretagnolle & H. Mathian, "Les temps de transport pour délimiter des aires urbaines fonctionnelles? Une investigation critique à partir de trois métropoles Européennes", *Belgeo*, 2016.

B. Jouve, *Coopération transfrontalière et services urbains en réseaux,* thesis, LATTS-Université Paris-Val de Marne, Parusn 1992.

B. Jouvre & C. Levèvre (eds.), *Métropoles ingouvernables*, Elsevier, Paris, 2002.

A. L'Hostis & A. Conesa, "Définir l'accessibilité intermodale", *In* A. Banos & T. Thevenin (eds.), *Systèmes de transport urbains. Caractérisation de l'offre et estimation de la demande*, Hermès-Lavoisier, Cachan, 2010, pp. 24.

J. Lévy & M. Lussault, *Dictionnaire de la Géographie*, Belin, Paris, 2013.

P. Menerault & A. Barré, "De la description des circulations à la territorialisation des réseaux: 35 ans de géographie des transports dans 'Hommes et Terres du Nord' ", *Territoire en mouvement*, 2006, pp. 45–53.

S. Munafò, V. Kaufmann, D. Christie, S. Vincent-Geslin & E. Ravalet, "Dispositions et usages de l'automobile et des transports publics entre 1994 et 2011: Analyses des cas de Berne, Genève et Lausanne", *Revue d'Économie Régionale & Urbaine*, 2015, pp. 753–782.

J. Nelles & F. Durant, "Political rescaling and metropolitan governance in cross-border regions: Comparing the cross-border metropolitan areas of Lille and Luxembourg", *European Urban and Regional Studies*, 2014, pp. 104–122.

J. M. Offner & P. Zembri, "Les réseaux, entre territoires fonctionnels et territoires institutionnels", *NETCOM: Réseaux, communication et territoires*, 1994, pp. 480–485.

P. Palmier, *Étude sur les flux en transports collectifs urbains sur Lille Métropole*, Report for LMCU, CETE Nord-Picardie, Paris, 2009.

P. Palmier, "Musljw. Logiciel de calcul d'accessibilité multimodale SETRA", Cerema, Lyon, 2011.

P. Palmier, "LIMEWISE: A deterministic line-based interpolation methodology for a realistic multimodal accessibility representation", *Geomatics Workbooks*, vol. 12, FOSS4G Europe, Corno, 2015, pp. 1–12.

D. Paris, "Lille, de la métropole à la région urbaine", *Mappemonde*, 66, 2002, pp. 1–7.

D. Pinson & S. Thomann, *La maison en ses territoires. De la ville à la ville diffuse*, L'Harmattan, Paris, 2002.

P. Pupier, *Recompositions scalaires en Europe occidentale: la construction de régions métropolitaines transfrontalières dans l'Eurorégion Transmanche et le Rhin Supérieur*, Geography thesis, Université d'Artois, Arras, 2021.

B. Reitel, *Villes et frontières: un jeu de construction de territoires*, HDR, Université de Nanterre, Paris, 2010.

B. Reitel, C. Sohn & O. Walther, "Analyse compare de l'intégration métropolitaine transfrontalière en Europe", *Territoire en movement*, 2007, pp. 62–75.

J. Sochor, H. Arby, I. C. Karlsson & S. Sarasini, "A topological approach to Mobility as a Service: A proposed tool for understanding requirements and effects, and for aiding the integration of societal goals", *Research in Transportation Business and Management*, 2018, pp. 3–14.

A. Vigani, "Transports, frontière et développement territorial de la Regio Insubrica", *Géo-Regards*, 2011, pp. 95–112.

L. Vodoz, B. Pfister & C. Jemelin, *Les territoires de la mobilité. L'air du temps*, Presses polytechniques et universitaires romandes, Lausanne, 2004.

Water Governance in the European Natural Park Plaines Scarpe Scheldt. Legal Obstacles and Political Will in Cross-Border Cooperation

Pauline Pupier

Between Lille in France and Mons in Belgium lies the European Natural Park Plaines Scarpe Scheldt (ENPPSS). It is made up of villages, natural spaces and wetlands fed by a dense network of rivers. In terms of water management, international agreements provide common references. European law and the most important document in that matter, the Water Framework Directive, in particular harmonise legislation. However, they do not erase the specificities of Belgian and French rules. The ENPPSS is also crossed by an international border, internal to the European Union (EU), which ecosystems ignore.

In law, even more than in political science or geography, the border is the limit of the territory governed by national sovereignty. The Belgian and French legal systems are juxtaposed at the border.[1] In the north of the Park lies the Natural Park Plains of the Scheldt. Belgian water law applies there, and more specifically Walloon water law because water management is a regional competence in this federal state. The Regional Natural Park Scarpe-Scheldt isin the south of the Park, where French law defines the rules that apply. There is therefore a legal asymmetry[2] within the cross-border region covered by the ENPPSS. What are the difficulties

---

[1] D. Nordman, "Note critique. Géographie ou histoire d'une frontière: la frontière franco-belge", *Annales. Histoire, Sciences Sociales* 32(3), 1977, pp. 433–444.

[2] E. Evrard & T. Chilla, T., "Asymétries institutionnelles et territoriales. Quelles conséquences pour la gouvernance transfrontalière?", *In* J. Beck & B. Wassenberg (eds.), *Vivre et penser la coopération transfrontalière. (Volume 6): Vers une cohésion territoriale?*, Franz Steiner Verlag, Stuttgart, 2014, pp. 63–78.

associated with this legal asymmetry? What legal, regulatory and administrative obstacles hinder cross-border cooperation in the management of watercourses?

The ENPPSS wishes to bring together the territorial stakeholders concerned and initiate cross-border water governance. To this end, it has requested expertise under the European Commission's initiative "b-solutions". The aim is to identify the legal obstacles to cross-border cooperation and to consider possible solutions. The analytical documentation work and the final expert report are used in this chapter. A complementary methodology inspired by political geography and the social sciences is also used, since fieldwork, ten interviews and email exchanges were conducted with representatives of key players in the area.

In order to set out the context of watercourses in the ENPPSS, the challenges related to watercourses are illustrated by the analysis of three case studies and the legal framework of their happening. A major pollution along the Scheldt, the failures of a joint flood prevention project and cross-border overflows reveal the legal obstacles and possibilities offered by international, European and national law. The inconsistencies, misunderstandings and tensions identified in water-related cross-border cooperation stem mainly from the asymmetry of competences of waterway managers between France and Belgium. Therefore, the ENPPSS can play an active role as a mutual trust builder, a facilitator of governance and a leader of cross-border projects. The analysis shows that the various cooperation initiatives on watercourses are hampered by legal asymmetries. Consequently, the stakeholders concerned would support a new form of water governance under the lead of the Park.

Water Governance: A Long-Term Project of the European Natural Park Plaines Scarpe Scheldt

A Mixed Territory Composed of Natural, Rural and Urban Areas

The Park Plaines Scarpe Scheldt is a European Nature Park where water defines ecosystems. It covers 950 km^2 between France and Belgium (See Figure 1. The Belgian part of the Park belongs to the Province of Hainaut in Wallonia and the French part to the *Nord Département* in the Hauts-de-France Region. The maps in Figures 1 and 2 highlight that the ENPPSS is structured by 44 km of border and by two major rivers, the Scheldt and its tributary the Scarpe[3], that flow from south to north, from France through Belgium to the North Sea:

- The Scheldt, or Escaut in French, an international river of 350 km, rises in France at an altitude of only 95 m. As illustrated by the map of its International River Basin District in illustration 3, the Scheldt flows through Belgium from south to north. It ends in the North Sea at the delta of the Rhine, Meuse and Scheldt rivers in the Netherlands. Partly canalised, the Scheldt is an important European waterway with more than 250 dams and locks. In some places, it has a natural course.[4]
- The Scarpe is one of its left bank tributaries. It is about 100 km long and joins the Scheldt at Mortagne-du-Nord, a French municipality bordering Belgium. It is also partly canalised.[5]

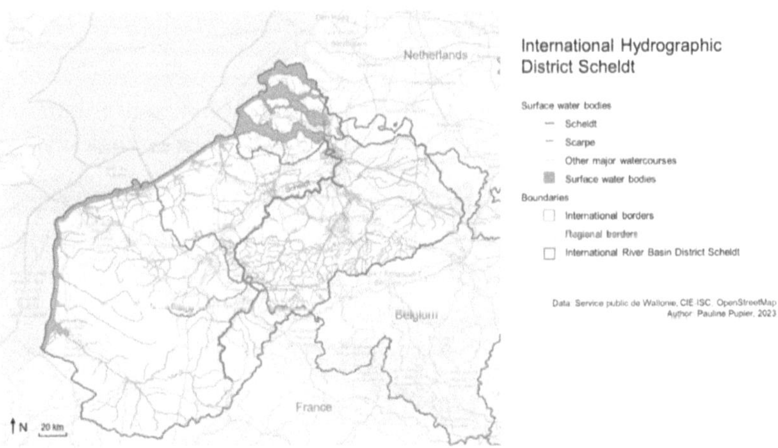

Figure 1: Map of the International Scheldt River Basin District.[6]

---

3 EGTC ENPPSS, data retrieved from; <https://plaines-scarpe-escaut.eu/?TerritoirE>. Accessed 2 January 2024.

4 International Scheldt Commission, data retrieved from: <https://www.isc-cie.org/fr/a-propos/le-district-hydrographique-international-de-lescaut/>. Accessed 2 January 2024.

5 French National Administration Service for Water Data and Reference Systems (SANDRE), data retrieved from: <https://www.sandre.eaufrance.fr/>. Accessed 2 January 2024.

6 Source: P. Pupier, 2023.

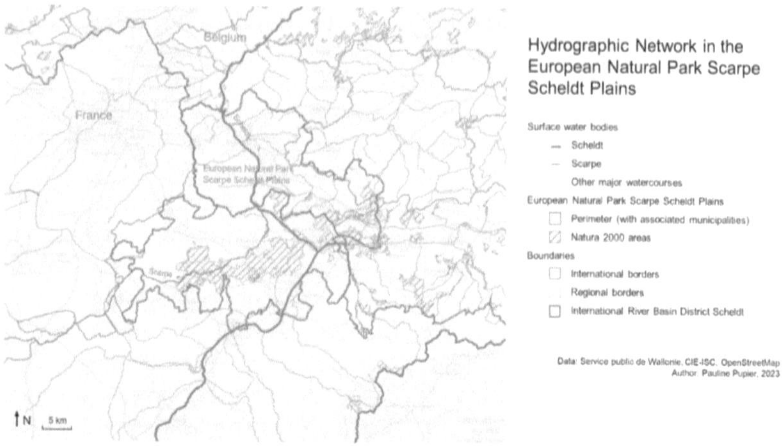

Figure 2: Map of the hydrographic network in the European Natural Park Plaines Scarpe Scheldt.[7]

The ENPPSS is an area of plains and lowlands. Numerous tributaries and small rivers generate a diversity of wetlands of great ecological value (marshes and peat bogs, natural ponds, reservoirs and artificial lakes, wet meadows, alluvial forests and wet forests). The Park has a remarkable ecosystem with a high biodiversity. Almost a third of the Park's area is classified as a European Natura 2000 area and two sites carry the international Ramsar Convention label for the protection of wetlands.[8]

The landscape of ENPPSS is heterogeneous with wooded areas, rural areas and urbanised areas. Unlike other nature parks, the ENPPSS is characterised by a high population density (307 inhabitants/km^2). 292,000 people live in the 127 communes and villages of the Park. The

7 Source: P. Pupier, 2023.

8 Natura 2000 is a European label based on a uniform regulatory framework for protection throughout the EU. Retrieved from: <https://ec.europa.eu/environment/nature/natura2000/index_en.htm>. Accessed 2 January 2024. The 1971 Ramsar Convention is an international treaty for the protection and wise use of wetlands. Retrieved from: <https://www.ramsar.org/about/the-convention-on-wetlands-and-its-mission>. Accessed 2 January 2024. Nature Parks are governed by national legislative frameworks. The name European Nature Park does not correspond to a legal definition.

medium-sized cities of Valenciennes, Douai and Tournai are located on the immediate periphery of the Park.

Water is therefore omnipresent in the ENPPSS, both above and below ground. In today's climate, it is abundant in all seasons. The landscape of wide, flat plains in the ENPPSS is particularly prone to flooding. The alluvial plain of the Scarpe meets a bottleneck in the area around Saint-Amand-les-Eaux in France. Further on, the Scheldt plain narrows again towards Antoing in Belgium. In autumn, winter and spring, flooding episodes are frequent and dangerous. They have been historically documented.[9]

As indicated by the maps, the water issues are eminently cross-border: the Scheldt catchment area extends beyond the borders. Belgium is downstream of the Scarpe and of the French Scheldt. Some small rivers, on the other hand, have their source in Belgium and flow towards France before joining the Scheldt. The border line follows the river bed in several places. The Elnon borders on the west of the Park for about 10 km. And the Hogneau (also called *L'Honneau* or *L'Aunelle*) borders on the east of the Park for 6 km. In addition, there are small border stream: the Calonne and the Vergne Noire and a number of brooks especially in the marshes of La Canarderie (FR) and Harchies (BE). According to a French representative, the border watercourses are the most preserved and the most meandering because any intervention requires cross-border cooperation.

Institutionalised and Ambitious Cross-Border Cooperation

The ENPPSS is the result of 40 years of Franco-Belgian cooperation. On 7 October 1983, the Walloon and Nord-Pas-de-Calais Regions concluded a memorandum of understanding to promote the creation of a "cross-border nature park" for the Scarpe and Scheldt plains. All levels of local government were involved (the French *Départements*, the Belgian Provinces and municipal authorities on both sides). The "Hainaut Cross-Border Natural Park" was created in 1996. After they jointly wrote the 2010–2022 Charter of the Hainaut Cross-border Park, the two nature parks signed a Partnership Agreement on 8 January 2010.

9 T. Ghils, *Mémoire des inondations sur le bassin versant de la Scarpe aval. Étude historique pour le Parc naturel régional Scarpe Escaut et le SAGE Scarpe Aval*, Tournai, 2015.

On 6 September 2021, the institutional cooperation evolved into a European Grouping of Territorial Cooperation (EGTC). It took the French name of *Parc Naturel Européen Plaines Scarpe Escaut* (ENPPSS), which official denomination in English is "European Natural Park Plaines Scarpe Scheldt". The ENPPSS has only two members: the two French and Belgian counterpart parks.

First, the Regional Natural Park Scarpe-Scheldt (*Parc Naturel Régional Scarpe-Escaut* (PNRSE)) in France. The executive body of the Regional Natural Park Scarpe-Scheldt is exclusively composed of elected officials from the Hauts-de-France Region, the *Département* du Nord, and the municipalities and inter-municipalities in its territory. Created on 13 September 1968, it is the oldest French Regional Natural Park. It is implementing its fourth charter. It is entitled "Charter 2010–2022. Regional Natural Park Scarpe-Escaut. Hainaut Cross-Border Park. Another life is invented here".[10] The charter presents "the multi-partner management of the territory classified as a Regional Natural Park" and "the deployment of its cross-border ambition as a Hainaut Cross-Border Natural Park".

Second, the Natural Park Plains of the Scheldt (*Parc naturel des Plaines de l'Escaut* (PNPE)) in Belgium. The Management Commission of the Natural Park Plains of the Scheldt (asbl) gathers about twenty public and private members. The territorial authorities are the seven municipalities of the Belgian Park as well as the Province of Hainaut and the intermunicipal agency IDETA. Other structures and NGOs active in the area chair with representatives from the tourist, cultural, agricultural and ecological sectors, and in particular the Scheldt-Lys River Contract. The Belgian Park was created in 1996 as a counterpart to the neighbouring French Park and at the same time as the cross-border Park of Hainaut. The missions of the Natural Parks include "the implementation of European programmes and European territorial cooperation" as well as "cross-border collaboration with similar areas in neighbouring regions or countries".

10 "Charte 2010–2022 Parc naturel régional Scarpe-Escaut et Parc transfrontalier du Hainaut", Parc Naturel Régional Scarpe Escaut 2009. Approved by the Nord-Pas-de-Calais Regional Council on 1 February 2010 and validated by Ministerial Decree n°2010–1021, Saint-Amand-les-Eaux, 30 August 2010.

Although the two parks should be active in protecting and preserving the ecosystem, they do not have direct responsibility for this key mission of managing water and watercourses.

An Ambition of Cross-Border Water Governance

According to its statutes,[11] the Franco-Belgian Park ENPPSS pursues the objective of "facilitating, promoting and leading cross-border cooperation between its members on the territory". It acts with a "disinterested aim".

> "As such, the Association conducts projects within the scope of its members' missions as defined in the Regional Natural Park Scarpe-Escaut charter, renewed by virtue of ministerial decree no. 2010-1021 of 30 August 2010, and the management plan of the Natural Park Plains of the Scheldt, adopted by the Walloon Minister of Agriculture, Nature and Rurality on 1 April 2015. These projects aim to consolidate the cross-border identity of the territory concerned. The grouping shall define an action plan and ensure its implementation."

The ENPPSS carries out missions of general interest for cross-border cooperation in the field of the missions of the French and Belgian Parks, that is, to enhance natural and cultural heritage, promote economic development, spatial planning and quality of life and inform and educate residents and visitors. However, it does not exercise its own competences in biodiversity nor water management.

The French Charter of the Regional Natural Park Scarpe-Escaut and the Belgian Management Plan of the Natural Park Plains of the Scheldt have already been written in a coordinated manner. The "Action Plan 2022" of the ENPPSS does present a common roadmap. Action No. 4 on natural heritage concerns "water governance" and aims to "position the EGTC on this issue", i.e. "continue the lobbying work with the various institutions and maintain exchanges and dialogue with the technical partners".[12] The details of this action are given in Figure 3.

---

11 "Statuts du Groupement Européen de Coopération Territoriale Parc Naturel Européen Plaines Scarpe Escaut", Parc Naturel Européen Plaines Scarpe Escaut 2021, Notarial deed, Tournai, 3 June 2021.

12 "Plan d'actions 2022", Parc Naturel Européen Plaines Scarpe Escaut, Péruwelz, 2021.

Figure 3: Action plan 2022 of the ENPPSS: Action n°4 of the natural heritage on "water governance".[13]

Because it experienced several difficulties and backlashes, the ENPPSS solicitated an expertise in the *b-solutions* initiative. Initiated by the European Commission, *b-solutions* aims to identify the legal and administrative obstacles to cross-border cooperation and to suggest solutions. Two initial interviews with the management and staff of the ENPPSS helped to narrow down the understanding of the major issue and thus identify the main obstacle to the border. A major objective of the ENPPSS in its request for expertise is to analyse the obstacles to good water governance on its territory. In order to do this, the ENPPSS wishes in a first step to identify the obstacles to cross-border cooperation in the joint management of non-navigable watercourses.

[13] Source: Parc Naturel Européen Plaines Scarpe Escaut 2021.

The landscape, the ecosystem and the biodiversity of ENPPSS are based on water. But water resources are used for many different purposes. The aquifer and the rivers are essential for drinking, industry and agriculture. The Scarpe and especially the Scheldt are major waterways for European transport. These different uses of water are represented by different actors and generate conflicts of use that are likely to increase with climate change.

Water Management Issues in the Park

Three case studies allow us to depict the difficulties encountered in cross-border cooperation and to evaluate the border obstacle in the management of non-navigable watercourses. They were chosen in accordance with the interviews of local actors and with level of law affecting the cross-border cooperation. First, the Scheldt pollution refers to international agreements. Second, the Interreg project "ElnonTransfrontalier" suffered from differences between French and Belgian regulation. Third, the overflow of the Jard Canal was due to a lack of cross-border maintenance coordination.

These recent cases are not representative of a rather advanced cross-border cooperation but they reveal that the difficulty of cooperation primarily lies in the heterogeneity of the actors and in the asymmetries of their legislation. Even if stakeholders concerned are well identified,[14] poor cultural and administrative inter-knowledge and the absence of guided coordination are direct consequences.

Pollution from the Tereos Factory

An International Ecological Disaster Due to an Industrial Accident

During the night of 9 to 10 April 2020, the dam of a settling basin at the Tereos sugar factory in Escaudoeuvres (FR) broke (see map in Figure 4). A very large volume of beet decantation water (around 90,000 m^3) spilled onto the neighbouring municipalities.[15] Located on the banks

---

14 M. Commaux, *La gouvernance environnementale de l'eau sur l'Escaut en contexte transfrontalier*, Master's thesis, University of Artois, Arras, 2022.

15 C. Mir & P. Noualhaguet, *Rupture d'une digue d'un bassin de lagunage de la sucrerie Tereos le 9 avril 2020 à Thun-Saint-Martin (59) et pollution de l'Escaut*, Report for

of the Scheldt, the basin is 25 km upstream from the ENPPSS entrance in Valenciennes and 50 km upstream from the Franco-Belgian border. The wave of polluted water spread to the Scheldt river and flowed downstream. Observations and reactions were hampered by the containment measures in force in France at the time due to an epidemic peak of Covid-19. But the French Office for Biodiversity (OFB) noted a catastrophic impact on the environment and in particular on fish populations as early as 12 April. The asphyxiation of a very large part of the aquatic fauna and flora was observed as far as the northern limit of the ENPPSS at Antoing (see maps in Figure 5). Tereos collected 8.4 tonnes of dead fish in France. The OFB measured a mortality representing a 50 % decrease in the number of species and a 90 % decrease in numbers. It is estimated that in total, between 70 and 100 tonnes of fish died on the French and Walloon portion, resulting in one of the biggest industrial and ecological disasters in Europe.

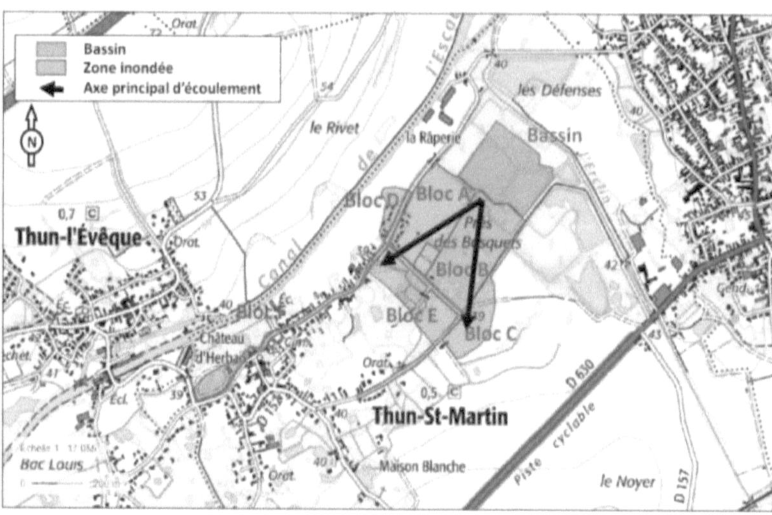

Figure 4: Map of local flooded areas causing the Scheldt pollution of 9 April 2020.[16]

the Ministry of Ecological Transition, General Council for the Environment and Sustainable Development, Paris, 2020.

16 Source : C. Mir & P. Noualhaguet, P., *Rupture d'une digue d'un bassin de lagunage de la sucrerie Tereos le 9 avril 2020 à Thun-Saint-Martin (59) et pollution de l'Escaut, op. cit.*

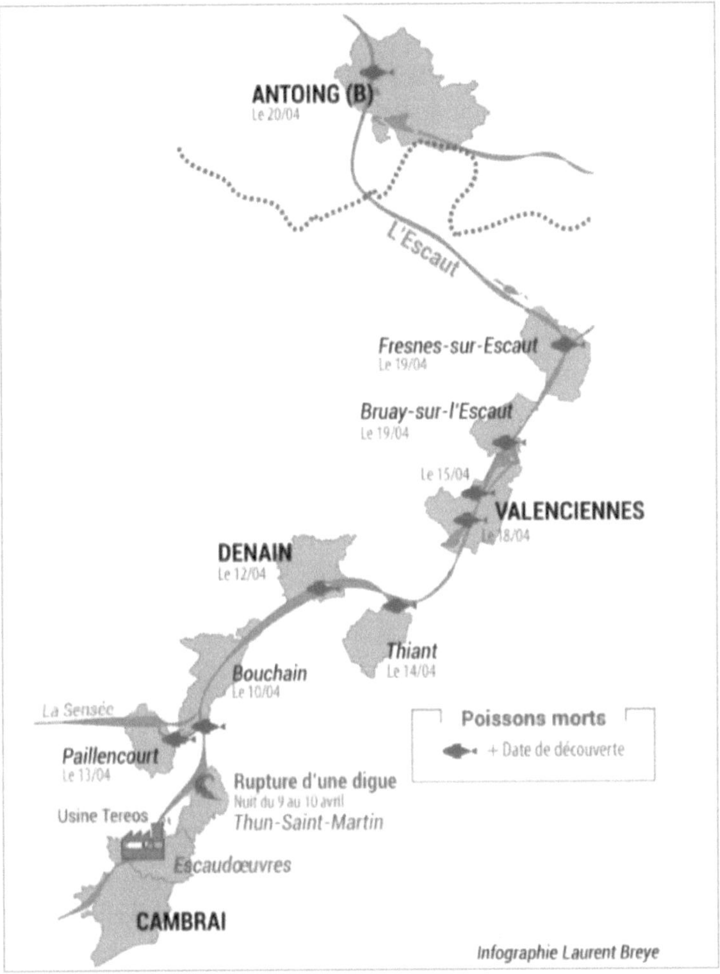

Figure 5: Dates and locations of fish kills of the "Large-scale pollution".[17]

For various reasons (minimisation of the accident by Tereos, underestimation of the dangers by the French authorities, impossibility of observations and poor coordination between administrations due to the confinement of the Covid...), the French authorities did not inform

---

17 Source: *Voix du Nord*.

the downstream states of the pollution as required by the International Scheldt Agreement. On 20 April, the Walloon and Flemish authorities activated the Scheme of Warning and Alert for the Scheldt of the International Scheldt Commission (ISC). Thanks to a continuous monitoring system, they had detected the movement of a water front with low oxygen content. They implemented emergency rescue measures (oxygenation, capture and sheltering of fish).

Indeed, the international river basin of the Scheldt is covered by the Agreement on the protection of the river Scheldt signed on 26 April 1994 in Charleville-Mézière. Signed by France, the Netherlands and the three Belgian federated regions (Vlaanderen, Wallonie and Brussels-Capital), it intends to ensure the quality of the Scheldt's waters by preventing pollution and improving the overall state of the river's waters. The text does not go in the direction of integrated management and reaffirms the principle of territorial sovereignty:

> "In the spirit of the Helsinki Convention on the Protection and Use of Transboundary Watercourses and International Lakes, the Contracting Parties shall cooperate in a neighborly spirit, keeping in mind their common interests as well as each other's special interests, in order to preserve and improve the quality of the Scheldt".[18]

The specificity of this agreement is that it sets up a cooperation between the States and Regions bordering the Scheldt: the International Scheldt Commission (ISC) is responsible for organising the exchange of information, harmonising and centralising measurements, producing a joint management action programme and setting up warning and alert networks. Since the 2000s, it has also ensured the mutual and multilateral coordination of Member States' implementation of their obligations under the Water Framework Directive.

The alert mechanisms of the International Scheldt Commission were not activated by France during the pollution by the Tereos factory in 2020. The ISC reacted afterwards at the request of the Belgian authorities. This questions the reactivity and relevance of this body to act on the scale of the ENPPSS.

---

18 Agreement on the protection of the (River) Scheldt, 26 April 1994, Art. 2.1.

Administrative Proceedings in France Despite International Law

As the deconcentrated representation of the State in the local authorities, the *Préfet* led a process of consultation, reaction and feedback bringing together numerous French and Belgian stakeholders.

Following the initial malfunctions and after the first prefectoral emergency orders, the Prefecture brought together a group of experts responsible for analysing the impact. It is made up of the relevant French government departments, associations, local stakeholders (French and Belgian Natural Parks among others) and Belgian partners.

The Prefecture has also set up a Steering Committee whose aim is to carry out the restoration of the Scheldt over at least seven years. It should also provide feedback to improve the management of the alert between State services but also with the Belgian authorities and local authorities. It brings together local elected officials, Walloon and Flemish partners, including the Regions and the Province of Hainaut, TEREOS and all the stakeholders concerned (environmental protection associations, regional Natural Parks, SMAPI, etc.).

Finally, the Prefecture initiated the "environmental liability" procedure against Tereos, relying for the first time on Directive 2004/35/EC of 21 April 2004. According to the French Environmental Code (L.162-11), the Prefect assesses, approves and prescribes the appropriate remedial measures. By an order dated 31 August 2021, the *Préfet* du Nord ordered the company Tereos to restore 10 ha of land located near the Scheldt river, in order to recreate areas favourable to reproduction. The company will first have to restore two initial sites, including the Rodignies dead branch, which borders Belgium.

The approach taken by the Prefecture has been welcomed by various actors. However, this ad-hoc procedure reveals the weaknesses and limits of regular cross-border cooperation in water governance. Even if coordination is now happening between France and Belgium, the emergency measures have not worked beyond the border and the measurement and restoration actions are confined to the French territory.

A French Trial for Belgian Damages

In parallel with the administrative procedure, legal proceedings were initiated for "ecological damage". Since 2020, numerous criminal complaints have been filed in France and Belgium with 21 civil parties: associations and public authorities, including the ENPPSS through the two French and Belgian parks, and also the *Vlaamse Milieumaatschappij* (Flemish Environment Agency) and the Walloon Region.

The trial had a cross-border dimension, as the Belgian prosecutor's office relinquished jurisdiction to the French courts. The Lille Criminal Court thus took into consideration the compensation claims of the Belgian civil parties. These could be particularly high, since the ecological restoration measures ordered by the *Préfet* did not concern Belgian territory.

In the judgment of 12 January 2023, the Tereos factory was found guilty of five criminal offences that led to the pollution of the Scheldt. The court in Lille ordered it to pay a fine of 500,000 euros and more than 9 million euros in damages to the Walloon Region. This extremely high sum will have to finance the measures to repair the damage caused to the Scheldt ecosystems planned by Wallonia. The company then appealed.

Alternatively to this French trial, international conventions offered other proceeding mechanisms. A founding text of international law on the management of cross-border rivers[19] is the New York Convention on the Law of the Non-Navigational Uses of International Watercourses adopted by the United Nations on 21 May 1997.

It follows on from and incorporates provisions of the 1992 Convention on the Protection and Use of Transboundary Watercourses and International Lakes. Signed in Helsinki, the latter was drawn up by the United Nations Economic Commission for Europe. It applies to the entire EU. The Helsinki Convention is the first binding document to obliged States-Parties to apply a precautionary principle and a "polluter pays" principle.

Going a step further, the New York Convention enacts a community of interest in the resource and its environment. It establishes a limit to the sovereignty of a State over the water in its territory. States-Parties also undertake not to cause significant damage: they have the "the right to

---

19 F. Lasserre & Y. Vega Cárdenas, "L'entrée en vigueur de la Convention de New York sur l'utilisation des cours d'eau internationaux: Quel impact sur la gouvernance des bassins internationaux?", *Revue québécoise de droit international*, 29(1), 2018, pp. 85–106.

utilize the watercourse and the duty to cooperate in the protection and development".[20] In the event of damage or disputes, the Convention provides for emergency, dispute resolution and compensation mechanisms.

> "Where significant harm nevertheless is caused to another watercourse State, the States whose use causes such harm shall [...] take all appropriate measures, [...] in consultation with the affected State, to eliminate or mitigate such harm and, where appropriate, to discuss the question of compensation".[21]

France acceded to the New York Convention on 24 February 2011. Belgium, on the other hand, has never ratified the text and is therefore not a State Party. According to several authors[22], Belgium's vote in the General Assembly was counted as an abstention, although it subsequently declared that it was in favour of the Convention. However, no federal government has ratified since. It is possible that governmental instability in Belgium and the fact that the Regions are competent for water policy have kept the New York Convention off the political agenda.

As it stands and without ratification by Belgium, the legal possibilities opened up by the New York Convention cannot therefore be mobilised on the Franco-Belgian territory of the ENPPSS nor on the resolution of the Tereos pollution. The Helsinki Convention, on the other hand, can be used as a framework to justify and organise bilateral cooperation on a river or sub-basin. However, the literature underlines the extent to which these international texts suffer from institutional weaknesses[23] in relation to the will of States. Consequently this asymmetry does not

---

20 UN Convention on the Law of the Non-navigational Uses of International Watercourses, 21.5.1997, Art. 5.

21 *Ibid.*, Art. 7.2.

22 M. Kamto, "L'entrée en vigueur de la Convention sur le droit relatif à l'utilisation des cours d'eau internationaux à des fins autres que la navigation", *Revue juridique de l'environnement*, 42, 2017, pp. 13–36 ; A. Schwabach, A., "The United Nations Convention on the Law of Non navigable Uses of International Watercourses. Customary International Law and the Interest of Developing Upper Riparians", *Texas International Law Journal*, 1998, pp. 257–258.

23 A. Wolf, K. Stahl & M. Macomber, "Conflict and Cooperation Within International River Basins: The Importance of Institutional Capacity", *Water Resources Update* 31, 2003, p. 125:2. Retrieved from: <https://www.researchgate.net/publication/245767420_Conflict_and_Cooperation_Within_International_River_Basins_The_Importance_of_Institutional_Capacity>. Accessed 2 January 2024.

prevent cross-border cooperation, but it does deprive it of an international legal basis.

Interreg Project "ElnonTransfrontalier"

An Exemplary Project to Prevent Flooding

The Interreg "ElnonTransfrontalier" project is an operational example of cross-border cooperation on watercourse management in the ENPPSS. Launched within the framework of the Interreg V France-Wallonie-Vlaanderen programme, it aims at an "integrated management of surface waters of the cross-border Elnon basin". A budget of 2.8 million euros, 50 % of which is financed by the European Regional Development Fund (ERDF), was planned for the period 2018–2022. SMAPI, a French inter-municipal watercourse manager and the Belgian Province of Hainaut were the two main project partners.

The objective was to prevent flooding in the cross-border Elnon catchment area. Over the past 25 years, the river has experienced eight floods, affecting houses to the west of the ENPPSS in the villages of La Glanerie and Howardries (BE) and Mouchin, Rumegies and Lecelles (FR). French and Belgian watercourses managers wrote a cross-border management plan for the Elnon, together with the National Park. The Interreg project intended to implement the action plan. First, the project had the educational purpose to improve the local population's awareness of the flood risk. Second, it intended to develop technical exchanges between the project partners to better coordinate their actions. Third, the main action also included the development of four flood expansion zones (three ZECin France and one ZIT[24] in Belgium, as illustrated in Figure 6). Well-constructed and very relevant, the concrete project plan can be described as exemplary for cross-border cooperation.

---

[24] ZEEC means *Zone d'Expansion de Crue*, or Flood Expansion Area in France; whereas ZIT means *Zone d'Immersion temporaire*, or Temporary Immersion Area in Belgian. Both are very similar landscaping work.

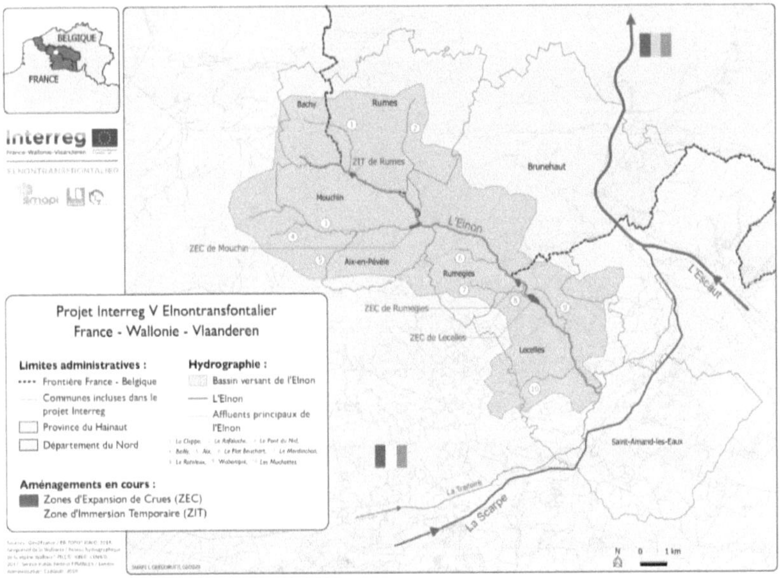

Figure 6: Map of the Interreg project "ElnonTransfrontalier".[25]

Despite the Covid-19 health crisis, which delayed the project's progress and in particular its communication, the awareness and prevention actions for the general public and schools were particularly successful, as reported in the local press. On the technical side, numerous meetings took place between the technical departments. Planning and environmental studies were carried out and land negotiations made it possible to acquire the land required for the construction site. The Belgian ZIT was inaugurated on 7 December 2022 during the closing event of the Interreg project (Figure 7).

25 Source: *Hainaut Ingénieurie Technique* (HIT), Province of Hainaut, 2022.

Figure 7: The Glanerie ZIT in the municipality of Rumes (BE).[26]

The Differences between the French and Belgian River Management Systems

Exchanges between the French and Belgian stakeholders highlighted the great differences in the organisation of river management systems. France relies on a territorial distribution[27] by hydrographic basin, whereas Belgium distributes the competences in a categorical logic between administrative levels according to the size of the watercourse. The French legal and regulatory framework rely on long binding administrative procedures with high environmental standards. The Belgian one is more operational and flexible with the participation of NGOs and the civil society.

For a navigable waterway, such as the Scheldt, the French national services are directly responsible. The Prefecture du Nord and its

26 Source: HIT, Province of Hainaut 2022.

27 N. Jacquin, *SAGE en France: quelles démarches territoriales de gestion de l'eau dans les autres pays européens? Etude de cas de 6 pays et 5 régions européennes*, International Office for Water, Limoges, 2014.

administrations are therefore a major interlocutor. Another State administration, French Waterways (*Voies Navigables de France*, VNF), is responsible for navigation on the large-gauge network of the Scheldt and for river tourism on the Scarpe.

A competence of "Management of Aquatic Environments and Flood Prevention" (*Gestion des Milieux Aquatiques et Prévention des Inondations* GEMAPI) has recently been defined, which is the responsibility of the inter-municipalities (metropolises, urban communities, agglomeration communities, communities of communes). It was introduced in the Environment Code by the latest decentralisation laws. In the French territory of the ENPPSS, two inter-municipalities exercise the GEMAPI competence: Valenciennes *Métropole* and the *Communauté d'Agglomération* Porte du Hainaut which has delegated its competence to the *Syndicat Milieux Aquatiques et Prévention des Inondations Scarpe-aval et Bas-Escaut* (SMAPI). In application of the water law n°92-3 of 03 January 1992, water management is organised by territorial planning documents which are enforceable against administrations and third parties.

In Belgium, with the various State reforms, the water sector has been transferred from the competence of the federal State to the exclusive competence of the Regions.[28] In the Belgian territory of the ENPPSS, the Walloon Region is therefore responsible for water policy as defined by the Water Code of the Walloon Environment Code.

The management and financing of watercourse maintenance is divided between levels of government according to a categorical logic that recalls the principle of subsidiarity. This does not concern the catchment area but the minor bed of the watercourse. Thus the responsibility for managing a single river is divided between several authorities upstream and downstream. It emerged from the interviews that this specificity is little understood by the French stakeholders. Watercourses are divided and classified (Article D.35 of the Water Code) into navigable or non-navigable categories. Non-navigable rivers are categorised according to the size of their catchment area in accordance with the decree of 4 October 2018: Class 1 watercourses are the responsibility of the Region; Class 2, of the province; Class 3, of the municipality.

---

[28] D. Aubin & F. Varone, "La gestion de l'eau en Belgique. Analyse historique des régimes institutionnels (1804–2001)", *Courrier Hebdomadaire Du CRISP*, n° 1731–1732(26), 2001, p. 5.

In the ENPPSS area, the Wateringues riverbank owners' associations[29] are responsible for and bear the costs of class 3 watercourses and unclassified ditches.

The "ElnonTransfrontalier" project achieved quite a progress in the mutual understanding of legal competences and procedures on both side of the border, even if they remain different.

The Contradiction between the European Project Logic and the French Administrative Instability

However, none of the three French ZECs could be completed within the timeframe of the Interreg project. As a result, the project only consumed 70 % of its budget, with a decrease in co-financing and an additional cost of 120,000 Euro for the French partner.

The Covid-19 pandemic is not the only reason. In addition, the partners encountered difficulties specific to cross-border cooperation in the asymmetry of legal and regulatory procedures. First, the French GEMAPI mission was revised during the project implementation period. The inter-municipalities received the competence when these texts came into force on 1 January 2018, which became mandatory and exclusive on 1 January 2020. The French partners, and SMAPI in the first instance, have therefore had to deal with legal changes to their status and authorisation procedures during the Interreg project.

Second, the requirements in terms of fauna-flora impact studies and ecological compensation measures are very high and time-consuming. Finally, he prefectoral decree authorising the development of the three French ZECs and specifying the numerous compensatory measures for the environment was only published on 1er March 2023, a semester after project closure. It is interesting to note that it provides for the project to be accompanied by long-term Franco-Belgian cooperation:

> "Measure A5: The beneficiary of this authorisation updates, together with the Belgian authorities, the transboundary and multi-annual management plan for the Elnon catchment area: by promoting a comprehensive approach to flood management through the maintenance and ecological restoration of

---

29 Technically, wateringues are ditches designed to drain wetlands. A wateringue also refers to a public authority. Within its boundaries, it is responsible for the creation and maintenance of the water system to promote agriculture and hygiene and to prevent flooding.

watercourses; by studying the natural overflow areas of the watercourses, in partnership with the owners and farmers of the riverbank; raising awareness among agricultural stakeholders to improve the management of soil erosion and agricultural runoff issues; In order to complement the action of the ZECs currently being developed."

In comparison, the Belgian approach of the technical services of the Province of Hainaut, implemented by its agency *Hainaut Ingénieurie Technique* (HIT), appeared very rapid. It did not include the same authorisation requirements with fauna-flora impact studies or compensatory measures. The lightness of the Belgian procedure with regard to environmental studies and the protection of ecosystems was also pointed out in a letter from the Scheldt-Lys River Contract to the Province of Hainaut.

These major differences between French and Belgian regulations have created strong tensions between the project partners. They are surprised that the Water Framework Directive, the presence of Natura 2000 areas and the scope of the ENPPSS do not lead to greater harmonisation. The Interreg project highlighted the different administrative cultures, the regulatory asymmetry and the bureaucracy of European projects. In results both sides became tense and accused each other of incompetence: the French judged the Belgians to be messy, lax and lazy, while the Belgians judged the French to be bureaucratic, cowardly and obtuse. For some of the people involved, "ElnonTransfrontalier" was ultimately a bad experience of cross-border cooperation that created interpersonal conflicts and exacerbated national stereotypes and even nationalistic resentments.

Canal du Jard and Vergne

Franco-Belgian Interdependence Where the Downstream Flooded the Upstream

Around the border village of Wiers (BE) in the municipality of Péruwelz, the watercourses Verne de Basècles, Verne de Bury and Vergne are also known in French as Courant Blanc and Courant Noir. Indeed, they flow towards France to the southwest and then join the Canal du Jard (see map in Figure 8). The Jard is a French canal that flows into the Scheldt at Flines-lès-Mortagne (FR). It is maintained by the inter-municipalities on their territory. Due to the French administrative divisions, two inter-municipalities are responsible. Valenciennes *Métropole* and its GEMAPI department are responsible for almost the entire length of the river. Only the last 2.5 km on the territory of the *Communauté d'Agglomération de la Porte du Hainaut* are under the responsibility of the SMAPI.

However, the Canal du Jard has become very congested due to the lack of coordination in maintenance between the two French administrations. In 2022, the water could not flow properly, causing overflows in the Belgian catchment area and also in the French ponds further upstream of the Canal du Jard. These backflows reduced the oxygenation of the watercourses and suffocated the ecosystems, resulting in fish mortality.

On this part of the ENPPSS, the watercourse networks are in a Franco-Belgian interdependence where the downstream has impacted the upstream.

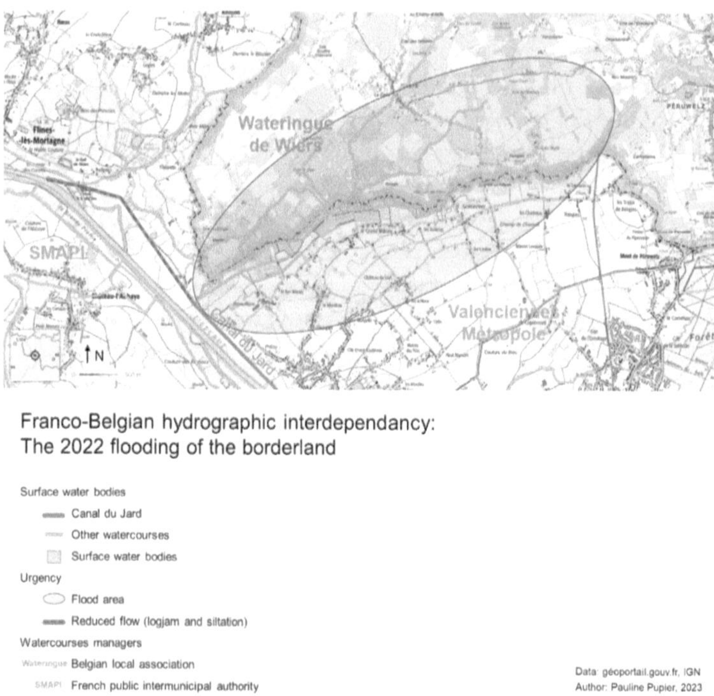

Figure 8: Franco-Belgian hydrographic interdependency: the 2022 flooding of the borderland.[30]

[30] Source: P. Pupier, 2023.

Interpersonal Relationships at Local Scale

The President of the Belgian Association of Riverbank Owners (*Association des propriétaires riverains*) of the Wiers Wateringue contacted the two French services directly by telephone for an emergency intervention. Although they don't know each other very well, these people have already met through institutional cross-border cooperation processes. It is a strength that the local players know how to identify the right contacts on both sides of the border and that they spontaneously contact each other directly. These can be seen as good interpersonal relationships that facilitate informal cross-border cooperation. The French actors concerned were invited to a meeting by the Mayor of Péruwelz. After clarification of competences and costs, it became clear that the GEMAPI service of Valenciennes *Métropole* had cleaned the large section of the Jard on its responsibility area and that the SMAPI was not well aware of the 2.5 km under its jurisdiction. The SMAPI proceeded in November 2022 with the same work on the last small congested portion. They removed logjams, cleared trees and branches and cleaned up the silt, resulting in a reduction in the water level of almost 30 cm.

Despite clear competences on French and Belgian sides, maintenance, anticipation, coordination unfortunately failed at the scale of the local cross-border catchment area. It is the local Belgian associations that have suffered and noted the lack of French maintenance. The individual commitment of a man and interpersonal relationships directly activated the coordination between the two French actors responsible for the Canal du Jard.

Coordination Based on the Water Framework Directive

However, Directive 2000/60/EC of the European Parliament and of the Council of 23 October 2000 establishes a very important common legal level for Franco-Belgian cooperation in the management of non-navigable watercourses. It harmonises the objectives and mechanisms for the management of watercourses in a binding way and gives priority to environmental protection and sustainable water use (Art. 1). Article 2 provides common definitions: The reference scale of the directive is the river basin where water management should be coordinated.

The Water Framework Directive requires the Member States of the European Union to integrate the cross-border dimension of water in their management and to strengthen cooperation between countries (Art. 3). To this end, international river basin districts have been delimited. These districts cover the length of the hydrographic network but also the surface area of the drained territory. The Framework Directive makes explicit the possibility of an international management plan, as for the Scheldt, or a sub-basin management plan.

> "2. In the case of an international river basin district falling entirely within the Community, Member States shall ensure coordination with the aim of producing a single international river basin management plan. Where such an international river basin management plan is not produced, Member States shall produce river basin management plans covering at least those parts of the international river basin district falling within their territory to achieve the objectives of this Directive. [...]
>
> 5. River basin management plans may be supplemented by the production of more detailed programmes and management plans for sub-basin, sector, issue, or water type, to deal with particular aspects of water management. [...]".[31]

Based on these two paragraphs, a management plan for an international sub-basin overlapping with the ENPPSS could be considered. It would give to the ENPPSS a leadership position and an opportunity to coordinate the management of cross-border watercourses.

Overcoming the Obstacle: From Legal Solutions to Political Action

Summary: Rebordering Processes Due to Legal Asymmetries

The main obstacle to the proper management of non-navigable watercourses in the ENPPSS is the asymmetry of the Belgian and French systems in this area (See Figure 9). There are major differences in the spatial division of responsibilities, in the counterpart competent authorities, in

---

[31] Directive 2000/60/EC of the European Parliament and of the Council of 23 October 2000 establishing a framework for Community action in the field of water policy, Art. 13.

the length of procedures and in the balance between development and environmental issues.

Legal level	France	Belgium
International law	Helsinki Convention of 1992 New York Convention of 1997	
EU legislation	Water Framework Directive	
National law:		
- Competence	State competence and inter-municipal competence	Regional competence
- Organisation	Territorial river management system	Categorical river management system
- Project procedure	Hierarchical centralised procedures, long project timeframe	Operational agility of the local provincial agency
- Objective	Strong and binding environmental focus	Utilitarian management of water as a resource

Figure 9: Summary table of asymmetries on water law at the Franco-Belgian border.[32]

This has three consequences that complicate cross-border cooperation: First, the lack of knowledge of the neighbouring system persists among the actors and the foreign legal, regulatory, and political framework generates incomprehension. Second, the exchange and project approaches are hampered by the lack of a mutual knowledge base and risk failing. Third, tensions can build up, reinforcing nationalistic stereotypes and even animosity towards the neighbour.

It is not uncommon in cross-border cooperation in the European Union for partners to be confronted with such asymmetries regardless of the issue at hand. The academic literature analyses international and EU borders as ambivalent socio-spatial processes.[33] These are based on the concepts of "debordering" and "rebordering".[34] "Debordering" is made

32 Source: P. Pupier, 2023.

33 D. Newman, "The lines that continue to separate us: Borders in our 'borderless' world", *Progress in Human Geography*, 30(2), 2006, pp. 143–161.

34 H. Van Houtum & T. Van Naerssen, "Bordering, Ordering, and Othering", *Tijdschrift Voor Economische En Sociale Geografie*, 93(2), 2002, pp. 125–136.

visible by European policies and discourses on the removal of internal EU borders.[35] The "rebordering" which characterises the processes of reinforcing barrier effects and closing borders are less known. However, the two are concomitant and their effects do not cancel each other out. Cross-border cooperation follows the same processes and ambivalences.[36]

Perspectives: ENPPSS as a Facilitator of Water Governance

In this study, the ENPPSS pursues the objective of improving the cross-border management of non-navigable watercourses and, in the long term, of structuring water governance on its Franco-Belgian territory.

Legal and administrative solutions can be envisaged in the application of a single national law or in the search for binding international or European references. However, cross-border cooperation is still an optional policy based on the voluntary commitment of the actors. Thus, the best way to improve cross-border cooperation is to consolidate bilateral knowledge and restore mutual trust through exchange. This policy of small steps can be based on projects of initially modest ambition and on a policy of influence with willing partners. It can only be successful in the long term.

According to the analysis, the difficulty of cooperation in the management of non-navigable watercourses lies in the asymmetry of legal systems, the heterogeneity of the competent actors, their poor cultural and administrative inter-knowledge and the absence of guided coordination. The ENPPSS can act on these last two points by positioning itself as a facilitator of multi-level governance[37] of water on its perimeter. As it has no competences in the field of water, the ENPPSS can position itself as a neutral actor driven by the general interest. Its recent change in status to an EGTC does not provide any decisive elements favouring the

---

35 B. Wassenberg, B. Reitel & J. Peyrony, *Territorial Cooperation in Europe – A Historical Perspective*, Publications Office of the European Union, Luxemburg, 2015.

36 J. W. Scott, "European Politics of Borders, Border Symbolism and Cross-Border Cooperation", *In* T. M. Wilson & H. Donnan (eds.), *A Companion to Border Studies*. Wiley-Blackwell, London, 2012, pp. 83–99.

37 L. Hooghe & G. Marks, G., *Multi-Level Governance and European Integration*, Rowman & Littlefield, Lanham, Oxford, 2001.

meeting of the actors concerned. Symbolically, it reinforces its position as an established institution and a European-level player. It contributes to a European narrative of the border and of the cross-border area.[38]

A cycle of meetings or a project approach can encourage mutual cultural and administrative learning between the Belgian and French actors, before rebuilding trust. Experience shows that it is effective to first bring together actors who are already convinced about a concrete project. Indeed, the people interviewed are sometimes close to the field and to operational work and will not all be available for governance meetings or intercultural training sessions.

On the basis of successful experiences and committed dynamics, governance is more easily extended to all the actors concerned. European projects can be a co-financed support to move forward on a specific point of the management of non-navigable watercourses. Seeking funding and validation from the European level can have the consequence of making the regional and state levels react. This approach to European recognition initiated by the actors closest to the territory evokes the concept of "European rescaling".[39]

Faced with the various "rebordering" phenomena, good water governance requires a local and collective political dynamic. A strategy of influence can help put the issue on the agenda. It can be useful to mobilise the political will of local actors and to rely on civil society. ENPPSS can provide leadership on this issue as no local actor wants to take the lead.

Conclusion: Opportunities Arising from Obstacles. Start with a Small Project Dynamic

The ENPPSS is a recent EGTC resulting from 40 years of successful cross-border cooperation. Although it has no competence in water policy or river management, it wishes to contribute to building water governance. Indeed, the ENPPSS territory is an inhabited European Natural

---

38 B. Reitel & F. Moullé, "La resémantisation de la ligne frontière dans des régions métropolitaines transfrontalières: Le jardin des 2 Rives à Strasbourg et la place Jacques Delors à Lille", *Belgeo*, 2, 2015, pp. 1–24.

39 P. Pupier, *Recompositions scalaires en Europe occidentale: la construction de régions métropolitaines transfrontalières dans l'Eurorégion Transmanche et le Rhin Supérieur*, Doctoral thesis in geography, University of Artois, Arras, 2021.

Park where numerous watercourses irrigate wetlands and precious ecosystems. The issues related to watercourses are therefore critical. A lack of clarity in the responsibilities for the maintenance of the Canal du Jard has led to overflows and floods in Belgium and France. An industrial accident at the Tereos factory in France led to major pollution with high fish mortality along the French and Walloon Scheldt. A cross-border flood prevention project could not result in the creation of all the flood extension zones due to administrative time differences. The inconsistencies, misunderstandings and tensions identified in cross-border water-related cooperation are mainly due to the asymmetry of competences of waterway managers between France and Belgium. Despite international agreements and, above all, the European harmonisation brought about by the Water Framework Directive, French and Belgian systems diverge on many points. Laws, authorities and procedures changes dramatically at the border. The territorial distribution of competences for river basins in France is compared with a categorical distribution of watercourses in Wallonia. Planning and the need for environmental studies and compensation are more demanding in French law. The ENPPSS can activate several legal and administrative levers to contribute to good water governance. It could take on an active role as a mutual trust builder, facilitator of governance and leader of cross-border projects.

Bibliography

D. Aubin & F. Varone, "La gestion de l'eau en Belgique. Analyse historique des régimes institutionnels (1804–2001)", *Courrier Hebdomadaire Du CRISP*, 26, n° 1731–1732, 2001, pp. 5–75.

M. Commaux, *La gouvernance environnementale de l'eau sur l'Escaut en contexte transfrontalier*, Master's thesis, University of Artois, Arras, 2022.

E. Evrard & T. Chilla, T., "Asymétries institutionnelles et territoriales. Quelles conséquences pour la gouvernance transfrontalière?", *In* J. Beck & B. Wassenberg (eds.), *Vivre et penser la coopération transfrontalière. (Volume 6): Vers une cohésion territoriale?*, Franz Steiner Verlag, Stuttgart, 2014, pp. 63–78.

T. Ghils, *Mémoire des inondations sur le bassin versant de la Scarpe aval. Étude historique pour le Parc naturel régional Scarpe Escaut et le SAGE Scarpe Aval*, Tournai, 2015.

L. Hooghe & G. Marks, *Multi-Level Governance and European Integration*, Rowman & Littlefield, Lanham, Oxford, 2001.

N. Jacquin, *SAGE en France: quelles démarches territoriales de gestion de l'eau dans les autres pays européens? Etude de cas de 6 pays et 5 régions européennes*, International Office for Water, Limoges, 2014.

M. Kamto, "L'entrée en vigueur de la Convention sur le droit relatif à l'utilisation des cours d'eau internationaux à des fins autres que la navigation", *Revue juridique de l'environnement*, 42, 2017, pp. 13–36.

F. Lasserre & Y. Vega Cárdenas, "L'entrée en vigueur de la Convention de New York sur l'utilisation des cours d'eau internationaux: Quel impact sur la gouvernance des bassins internationaux?", *Revue québécoise de droit international*, 29(1), 2018, pp. 85–106.

C. Mir & P. Noualhaguet, *Rupture d'une digue d'un bassin de lagunage de la sucrerie Tereos le 9 avril 2020 à Thun-Saint-Martin (59) et pollution de l'Escaut*, Report for the Ministry of Ecological Transition, General Council for the Environment and Sustainable Development, Paris, 2020.

D. Newman, "The lines that continue to separate us: Borders in our "borderless" world", *Progress in Human Geography*, 30(2), 2006, pp. 143–161.

D. Nordman, "Note critique. Géographie ou histoire d'une frontière: la frontière franco-belge", *Annales. Histoire, Sciences Sociales*, 32(3), 1977, pp. 433–444.

P. Pupier, *Recompositions scalaires en Europe occidentale: la construction de régions métropolitaines transfrontalières dans l'Eurorégion Transmanche et le Rhin Supérieur*, Doctoral thesis in geography, University of Artois, Arras, 2021.

B. Reitel & F. Moullé, "La resémantisation de la ligne frontière dans des régions métropolitaines transfrontalières: Le jardin des 2 Rives à Strasbourg et la place Jacques Delors à Lille", *Belgeo*, 2, 2015, pp. 1–24.

A. Schwabach, A., "The United Nations Convention on the Law of Non navigable Uses of International Watercourses. Customary International Law and the Interest of Developing Upper Riparians", *Texas International Law Journal*, 1998, pp. 257–258.

J. W. Scott, "European politics of borders, border symbolism and cross-border cooperation", *In* T. M. Wilson & H. Donnan (eds.), *A Companion to Border Studies*. Wiley-Blackwell, London, 2012, pp. 83–99.

H. Van Houtum & T. Van Naerssen, "Bordering, ordering, and othering", *Tijdschrift Voor Economische En Sociale Geografie*, 93(2), 2002, pp. 125–136.

B. Wassenberg, B. Reitel & J. Peyrony, *Territorial Cooperation in Europe – A Historical Perspective*, Publications Office of the European Union, Luxemburg, 2015.

A. Wolf, K. Stahl & M. Macomber, "Conflict and cooperation within international river basins: The importance of institutional capacity", *Water Resources Update*, 31, 2003, 125(2), pp. 31–40.

The Franco-Belgian and Franco-German Borders. A Brief Comparative Analysis

Bernard Reitel

From the North Sea to the Swiss border at Basle, three land dyads[1] follow one another and share a common characteristic, that of joining France to another founding state of the EEC, Belgium, Luxembourg and Germany. However, beside the short 73-km stretch separating France from Luxembourg, the Franco-Belgian and Franco-German borders examined below are much longer, the former covering 620 km and the latter 448 km. The former is a typical example of an artificial border, based on no major physical obstacle.[2] The latter is often associated with the Rhine, seen as a physical barrier, whereas the dyad runs only partially alongside the river.[3] These representations, while not erroneous, do hide the fact that these two dyads share a number of similarities, the main one being that they are both internal borders of the European Union (EU). Germany, Belgium and France are all members of this entity, but also of the Schengen area and the Eurogroup, which means that the borders separating these three States have gone through all the stages of devaluation associated with the process of European integration.[4]

One should remember that any border, as a limit established between two sovereign powers and their territories, is above all a convention,

1 A dyad is a section of border separating two neighbouring states.

2 F. Lentacker, *La frontière franco-belge. Étude géographique des effets d'une frontière internationale sur la vie de relations*, Morel & Corduant, Lille, 1974.

3 K. Denni, *Rheinüberschreitungen – Grenzüberwindungen. Die deutsch-französische Grenze und Ihre Rheinbrücken (1861–2006)*, UVK, Konstanz, 2008.

4 M. Foucher, *La République européenne*, Belin, Paris, 2000.

a historical and political construct.[5] However, a border is not a mere limit: it is also a regulatory device, a marker of identity, a discontinuity that forms part of a geographical space and that brings people into contact as much as it separates them.[6] A border has effects on the lives of the populations and neighbouring economic and political actors, on both sides. Consequently, this analysis of the Franco-Belgian and Franco-German borders will focus on the different characteristics mentioned above and take into account cross-border spaces as well as the borders themselves and what they represent in their various dimensions. The aim is to draw up a portrait of those two dyads from a multidisciplinary perspective, emphasising their geographical and political characteristics, and to understand what they represent for territorial players in an integrated European area, but one in which the tensions resulting from power struggles between territorialised powers still persist. This portrait will be completed by a presentation of some of the issues facing these borders in a context where States are taking a renewed interest in their borders, the Covid-19 pandemic and the tensions linked to international migration being a case in point.

Each border is a unique system whose complexity is the effect of a set of characteristics. However, our aim is to go beyond a monographic study by comparing two dyads: this should help to better identify the singularity of each one, which is indeed the objective of the Frontem programme. This comparison, emphasising a geopolitical dimension, will concern two borders for which information has been accumulated on the ground for over ten years.

After briefly recalling how these two dyads have been discussed in the scientific literature and describing the method used for our analysis, we will present the results of the comparison by identifying some key characteristics. Finally, we will attempt to outline a number of possible future developments, based on the challenges that these two dyads may face as a result of developments in the states that border them.

---

5 D. Newman, "On borders and power: A theoretical framework", *Journal of Borderlands Studies*, vol. 18, n°1, 2003, pp. 13–25.

6 B. Reitel, "La frontière internationale, objet sémique, processus multi-dimensionnel, interface signifiante", *LISER Working Paper* 2011/43, Esch-Alzette, 2011. Retrieved from : <https://www.liser.lu/?type=module&id=104&tmp=2753>. Accessed 9 June 2023.

A Brief Historiography and Presentation of the Analytical Method

Two Emblematic Borders in Western Europe

A first description of these two borders show that they have gone through the same kind of changes as other Western European borders since the beginning of the 20th century. Initially, a defensive approach was adopted. Seen from France, these borders are problematic for the defence of the "Pré carré", the national territory.[7] As they do not constitute a "natural frontier", their fragility highlights the vulnerability of the country from a military point of view. The initial project was therefore to circumscribe the territory with a defence system consisting of fortified sites and cities. This conception gradually gave way to a more political approach: the boundary that replaced the defence zone was associated with a regime that emphasised regulation and filtering.[8] The vulnerability of the Franco-Belgian border lies above all in the fact that it is not supported by any natural obstacle, whereas the Franco-German border is mainly associated with conflicts and a limit that has been redefined several times. This view of the Franco-German border is largely shared in German-language publications, even though it is acknowledged that the political and linguistic borders do not coincide.[9] For Belgium, its four land borders are primarily political boundaries: the Franco-Belgian border is therefore not fundamentally different from the other three.[10]

The functional dimension has been very much scrutinised in publications. It is associated with the importance of interactions and life on these borders, but also and above all with the filtering function they perform and the differences it reveals. This is particularly true of the Franco-Belgian border, whose artificial nature, which is regularly mentioned, emphasises

7 R. Dion, *Les frontières de la France,* Hachette, Paris, 1947.

8 M. Foucher, "L'invention des frontières: un modèle géopolitique français", *Hérodote,* n°33, 1986, pp. 54–88; M. Foucher, *Fronts et frontières,* Fayard, Paris, 1991 ; D. Nordman, *Frontières de France. De l'espace au territoire. XVIe-XIXe siècles,* Gallimard, Paris, 1998.

9 D. E. Khan, *Die deutschen Staatsgrenzen: rechtshistorische Grundlagen und offene Rechtsfragen,* Mohr Siebeck, Tübingen, 2004.

10 M.-T. Bitsch, *Histoire de la Belgique : de l'Antiquité à nos jours,* Éditions Complexe, Bruxelles, 2004; F. Lentacker, *La frontière franco-belge, op.cit.*

the role played by customs and their legitimate regulatory function.[11] In both cases, there is a great deal of trade, but smuggling is more particularly mentioned on the Franco-Belgian border. In the post-war period, the western borders of the Federal Republic of Germany (FRG) enjoyed a special dimension, as they ensured the new state's link with the "free world", Western Europe, whereas the eastern border, which was part of the Iron Curtain, separated it from the states under Communist rule, in particular the German Democratic Republic.[12] While the eastern border was associated with division, the western border was seen as a link and a new-found legitimacy. In this context, the border with France was linked to the idea of reconciliation,[13] which gave it a singular character, setting it apart from other borders. The cross-border dimension was mentioned more and more frequently, and this was due in particular to the many cooperation initiatives.[14]

Generally speaking, publications on the Franco-German border abound, which in itself should hardly come as a surprise: it joins the two most powerful states in the EU, whose relations were tormented, to say the least, between 1870 and 1945. The quantity of research on the Franco-Belgian border is much more modest, which is surprising given its length.[15] Is this lesser interest due to the fact that it is of no major

---

11 B. Touchelay, *Fraudes, frontières et territoires (XIIIe-XXIe siècle),* Institut de la gestion publique et du développement économique, Paris, 2020.

12 É. François, K. Seifarth & B. Struck, *Die Grenze als Raum, Erfahrung und Konstruktion. Deutschland, Frankreich und Polen vom 17. Bis zum 20. Jahrhundert,* Campus, Frankfurt-am-Main, 2007.

13 S. Schirmann, "La réconciliation franco-allemande. Éléments d'une problématique", *In* N. Paun & S. Schirmann (eds.), *Borders, Identities, Communities. The road to reconciliation in Central and Eastern Europe,* Nomos, Baden-Baden, 2016, pp. 211–221; B. Wassenberg, "La coopération transfrontalière dans l'espace du Rhin supérieur: un modèle de réconciliation franco-allemande?", *In* N. Paun, S. Schirmann (eds.), *Borders, Identities, Communities, op. cit.*, pp. 295–317.

14 B. Wassenberg, *Vers une eurorégion? La coopération transfrontalière franco-germano-suisse dans l'espace du Rhin supérieur de 1975 à 2000,* Peter Lang, Berne, 2007 ; J. Beck & B. Wassenberg, *Vivre et penser la coopération transfrontalière: les régions frontalières françaises,* Franz Steiner, Stuttgart, 2010 ; P. Hamman, *Sociologie des espaces-frontières. Les relations transfrontalières autour des frontières françaises de l'est,* Presses Universitaires de Strasbourg, Strasbourg, 2013.

15 G. Hamez, *Du transfrontalier au transnational: Approche géographique. L'exemple de la frontière franco-belge,* Doctoral thesis, Université Panthéon-Sorbonne, Paris, 2004; S. Verger, *Les chemins multiples de la coopération transfrontalière franco-belge: analyse des dynamiques institutionnelles,* Doctoral thesis, Université de Grenoble, Grenoble, 2011.

interest to the two States, or because it seems to be a quintessentially peaceful border, devoid of any conflict? Another possible explanation to the lower output of publications is asymmetry. For demographic reasons, there are far fewer social scientists in Belgium than in France, and since the federalisation of 1993 they have been divided between two communities, Flemish and Walloon, each of which has its own research programme. This brief overview provides an initial set of characteristics that the use of a comparative method should make it possible to refine. The comparison is based on the establishment of an analytical grid, so as to identify a number of essential similarities and differences.

A Comparison Based on a Multidimensional Analytical Grid

To compare the Franco-Belgian and Franco-German borders, we shall use an analytical grid whose aim is to provide a diversified overview by integrating a multidisciplinary dimension. The information collected concerns the border itself – the boundary – but more specifically the cross-border area, i.e. an area close enough to the border to be affected by it. The grid includes several categories, whose number should be kept to a minimum in order to facilitate description. The three approaches described by H. Van Houtum (2000) for analysing a border in a cross-border context served as a basis for developing this grid.[16] The first approach, which consists in examining the flows that cross the border, is similar to a socio-economic analysis. A second approach studies the structures of cross-border cooperation: it is based on a political and institutional vision. The third seeks to identify representations: described as a "people's approach", it has a socio-anthropological dimension. These three approaches are reminiscent of, but not really similar to, the registers mentioned by Michel Foucher (1991), namely the real (legal and political), the symbolic (belonging to a political community), and the imaginary (the relationship with others).[17] However, these three dimensions do not take into account the construction of the border over time, and while flows are taken into account, the space in which the border is located is only mentioned indirectly.

---

16 H. Van Houtum, "An overview of european geographical research on borders and border regions", *Journal of Borderlands Studies*, vol. 15, n°1, 2000, pp. 57–83.

17 M. Foucher, *Fronts et frontières: Un tour du monde géopolitique*, Fayard, Paris, 1991.

Our analytical grid is based on those three dimensions, complemented by three others which should help interpretation (See Table 1).[18] First, we retrace the genesis of the border as a boundary. To this temporal and historical approach is added a geographical approach that combines an analysis of the distribution of population and the main centres of spatial organisation. It relies on the idea that every border is part of a geographical space, and that there are links between the two. A third approach, described as functional, consists in identifying interactions and flows. The fourth, institutional, dimension depicts cross-border governance by studying the political systems of each of the states and the cross-border frameworks established. A fifth dimension examines the symbols with which the border is associated through discourse and representations. In addition to these five approaches, there is a sixth which deals with the cultural dimension. Culture conveys a certain conception of the world that influences people's behaviour. However, the objective is modest: it takes into account the linguistic dimension by asking whether or not the political border constitutes a linguistic discontinuity.

Ultimately, each of these six approaches is represented by a category that makes it possible to identify the characteristics of each border, including a multiscalar dimension, where useful. The context of the single European market should not blind us to the fact that, due to the small size of certain States, the distances separating borders may be quite short: for example, less than 100 km lie between the Franco-Belgian and Belgian-Dutch dyads.

---

18 We could have relied on other grids, but these are more general and less suited to a comparison between two dyads, i.e, R. Ratti & M. Schuler, "Typologie des espaces-frontières à l'heure de la globalization", *Belgeo,* n°1, 2013, pp. 1–14 ; V. Kolossov & J. W. Scott, "Theorizing borders. Border studies: Changing perspectives and theoretical approaches", *Belgeo*, n° 1, 2013, pp. 1–19.

Table 1: Main results of the comparison between the two borders[19]

Approach	Franco-Belgian border	Franco-German border
Temporal	Stable	Lively to pacified
Spatial	Alternation of high and low density Dense systems of cities Cross border urban areas	
Functional	Trans-European axis crossing the border	Trans-European axis partly on the border Real economic asymmetry?
Institutional	Unitary/federal Cross-border cooperation ≠ political regimes (conflictual federal, unitary)	Unitary/federal Cross-border cooperation ≠ political regimes (peaceful federal, unitary)
Symbolic	Artificial? Unbalanced / continuity? Small country/"world power" but counterbalanced by the presence of the European institutions	Reconciliation Balance : unbalanced 2 major countries in Europe?
	Economic Asymmetry? Benelux countries	
Cultural	Continuity/discontinuity/continuity	Continuity/discontinuity
Scalarity	Benelux and North Sea	Western Europe with Switzerland / European

Two Internal Borders in the EU: Similarities and Differences

Stability Versus Fluctuation

The genesis of the two borders is markedly different.[20] The Franco-Belgian dyad is the oldest: the border was established bit by bit under

[19] Sources: Mission opérationnelle transfrontalière; bibliographical references mentioned in this chapter, author's work on the ground. The Mission opérationnelle transfrontalière, an agency set up by the French government as an association, which specialises in analysing border areas and advising border local authorities, provides a detailed description of many of the dyads of European states on its website. The table includes the main information gathered for the two dyads under each item. This information has been verified in the scientific literature presented in this article. Retrieved from: <http://www.espaces-transfrontaliers.org>. Accessed: 2 May 2023.

[20] D. Nordman, *Frontières de France, op.cit.*

various agreements, starting from the Treaty of Westphalia in 1648. Following on from the Congress of Vienna (1815), the Treaty of Courtrai between the Kingdoms of France and the Netherlands in 1820 established a boundary that incorporated elements from previous treaties. The originality of this construction is that the line was established before Belgium was founded as a state in 1830: France recognised the boundary in 1839. The stability that has characterised this dyad for more than two centuries differs from the variations experienced by the Franco-German border between 1870 and 1955. This oscillation is the result of the balance of power between France and Germany, but also reflects changes in the political regime of the latter and its evolving position as a geopolitical player in Europe. It was not until 1955, when Saarland became part of the Federal Republic of Germany (FRG), founded six years earlier, that the border was stabilised. The reconciliation that had been implemented between France and Germany since the creation of the European Coal and Steel Community (ECSC) in 1951 undeniably helped to ease tensions. Whereas the Franco-German border is the archetypal border as "Scar of History" that has fluctuated according to the relations between the two States,[21] the Franco-Belgian border is identified as an artificial border, characterised by great stability.

Landscape Continuities and Population Contrasts

These two borders are located in Western Europe, in an area called "Rhenish Europe" in 1968 by Strasbourg geographer Étienne Juillard,[22] and which has subsequently been described as a megalopolis, a European backbone[23] or, more recently, the European Pentagon.[24] Without giving too much credence to these configurations, which are also spatial constructs designed to provide a structured vision of the European area, we may nevertheless focus on a number of geographical and geopolitical

---

21 B. Wassenberg, "La frontière, objet d'intégration ou cicatrice de l'Histoire? L'étude du cas de l'espace du Rhin supérieur", *RECERC*, n°1, online review, article 2, 2007.

22 É. Juillard, *L'Europe rhénane. Géographie d'un grand espace*, Armand Colin, Paris, 1968.

23 R. Brunet, *Les villes 'européennes'*, La documentation française, DATAR-RECLUS, Paris, 1989.

24 Commission Européenne, *Schéma de Développement de l'Espace Communautaire. Vers un développement spatial équilibré et durable du territoire de l'union européenne*, Publication Office of the European Communities, Luxemburg, 1999.

characteristics derived from these descriptions. The area is characterised by a high population density and a dense concentration of cities, as demonstrated by the presence of numerous metropolitan areas that are easily accessible from one to the other, with numerous interactions.[25] This cluster has been identified as a major centre of development and innovation on a global scale. Urban continuums, some of which constitute genuine cross-border conurbations of several hundred thousand inhabitants, are located on these dyads:[26] the West-Vlaanderen-Flanders-Dunkirk-Opal Coast seaside conurbation and Lille-Kortrijk-Tournai between France and Belgium, Saarbrucken-Moselle-Est, Strasbourg-Ortenau between France and Germany. The urban region centred on Luxembourg, which spills over into the two cross-border areas under scrutiny, and the trinational conurbation of Basel, whose centre is in Switzerland but which has suburbs in France and Germany, also feature on this list. Apart from Dunkirk, all the urban centres mentioned structure the cross-border area to a greater or lesser extent through their metropolitan functions. However, this observation needs to be qualified by two observations. On the one hand, low-density zones separate the main urbanised areas: in the Franco-Belgian space, an invisible line linking Saint-Quentin in France to Charleroi in Belgium separates a very densely populated western part (300–400 inhabitants per km^2) from an eastern part characterised by low densities (below 60); in the Franco-German dyad, the Vosges and Palatinate mountains are less densely populated (less than 100 inhabitants per km^2) than either the Moselle urban corridor and the Saarland and Lorraine coalfields in the west or the Upper Rhine in the east (more than 250).

In addition, between the metropolitan centres, certain interstitial areas have experienced – and are still experiencing, in some cases – economic and demographic decline: the Nord-Pas-de-Calais and Borinage coalfields, the Sambre valley, the Ardennes, the Lorraine and Saarland coalfields, and the Lorraine iron and steel basin. Although these industrial areas have sometimes developed in contrasting ways within national territories, their genesis was often based on similar logics in the 19th century: mining, the metal and steel industry, the textile industry.

---

25 É. Juillard, H. Nonn, *Espaces et régions en Europe occidentale*, CNRS, Paris, 1976.

26 P. Pupier, *Recompositions scalaires en Europe occidentale : la construction de régions métropolitaines transfrontalières dans l'Eurorégion Transmanche et le Rhin Supérieur*, Doctoral thesis, Université d'Artois, Arras, 2021.

More recently, the development of leisure activities and tourism has contributed to the growth of the mid-mountain ranges (Ardennes, Vosges, Palatinate) and the coastal areas around the North Sea. Apart from a few exceptions such as this coastline – the French part looks very industrial, while the Belgian coast is dominated by seaside and leisure activities – the landscapes show cross-border continuity. The differences are only perceptible to informed observers, which reinforces the artificial character attributed to the two boundaries. Similarly, although the Rhine constitutes a discontinuity on the Franco-German border, as evidenced by the need to build infrastructure in order to bridge the river, crossing the border does not give the feeling of a genuine landscape discontinuity.

Connectedness and Trans-European Traffic Routes

An examination of flows reveals that the boundaries under consideration are far removed from the image associated with the margins or the periphery: they are characterised by a high degree of connectivity, i.e. the existence of significant traffic flows for both goods and people.[27] This is reflected in the fact that Germany is France's leading trading partner, while Belgium is its second. These two borders are crossed by major trans-European routes, some of which are supported by waterways (Rhine, Moselle, Scheldt). The western part of the Franco-Belgian dyad includes seaside areas that form part of the Northern Range, a coastal interface that is connected to all regions of the world by a network of seaports with substantial traffic, between Hamburg and Le Havre. The hinterland of Belgian (Antwerp, Ghent, Zeebrugge) and Dutch ports (Vlissingen, Rotterdam, Terneuzen) extends right across Northern France.[28] The Benelux-Paris link is an essential component of a trans-European axis linking the North Sea, the Atlantic Ocean and the Mediterranean Sea. In the Franco-German dyad, the cross-border Rhine corridor is a key route for overland trade

---

27 European Commission, Directorate-General for Mobility and Transport, *White paper on transport: roadmap to a single European transport area: towards a competitive and resource efficient transport system*, Publications Office, Luxemburg, 2011. Retrieved from; <https://data.europa.eu/doi/10.2832/30955>. Accessed 23 May 2023.

28 A. Beyer, "La Belgique à la croisée des chemins. Une géohistoire des grandes infrastructures de transport du plat pays à l'aune de ses frontières", *Annales de Géographie,* Vol. 681, n°5, 2011, pp. 465–485.

between northern and southern Europe, and is also linked to the Moselle axis between Luxembourg, Germany and France. The Rhine corridor is also connected to the major ports of the Northern Range, especially Rotterdam and Antwerp.[29] The meridian organisation of these routes is part of the Trans-European Transport Network (TEN-T),[30] an infrastructure development programme launched by the EU in 1996. The intensity of flows largely reflects the organisation of population densities and urbanisation, with sparsely populated areas generally located away from the main routes. Long-distance exchanges are intermingled with short-distance cross-border flows of various kinds, which are particularly intense and diverse. Although the number of cross-border workers is lower than in other dyads (France-Luxembourg or France-Switzerland), it is still significant. However, the unilateral nature of the flows – they are mainly directed from France to neighbouring territories – shows a persistent socio-economic differential. Of the approximately 40,000 cross-border commuters in the Franco-Belgian dyad, some 30,000 worked in Belgium and just over 10,000 in France in 2017.[31] The asymmetry is even more obvious in the German-French dyad: 46,000 cross-border commuters go to Germany, while the reverse flow only represents 5,000 people in 2018.[32]

Multiple Forms of Cooperation between a Federal State and a Unitary State

At institutional level, each dyad separates a unitary political system, France, from a federal system, Belgium and Germany.[33] This marker is

29 A. Beyer, "Nœuds de transport et frontières. L'invention de la métropole bâloise", *Annales de géographie*, n° 657, sept.-octobre, 2007, pp. 451–369.

30 European Commission. Retrieved from: <https://transport.ec.europa.eu/transport-themes/infrastructure-and-investment/trans-european-transport-network-ten-t_en>. Accessed 9 June 2023.

31 E. Crocquey, F. Mille & A. Roszak, "L'emploi frontalier des Hauts-de-France vers la Belgique : une attractivité modérée et contrastée", *INSEE Analyses Hauts-de-France,* n°94, 2019. Retrieved from : <https://www.insee.fr/fr/statistiques/3899294>. Accessed 23 May 2023.

32 T. Martini & S. Villaume," Le travail frontalier dans le Grand Est : en 2018, 182 000 habitants de la région partent travailler dans un pays voisin", *INSEE Analyses Grand Est,* n° 144, 2022. Retrieved from : <https://www.insee.fr/fr/statistiques/6444588>. Accessed 23 May 2023.

33 R. Pasquier, *Organisation territoriale et démocratie locale en Europe,* La Documentation française, Paris, 2022.

essential, as it leads to a very different distribution of powers and competences, and is sometimes seen as an obstacle to cross-border cooperation. These differences are expressed in particular in political cultures. In France, although the central State is often described as an omnipresent authority, the *mille-feuille* approach seems to be dominate, in reality: decision-making and funding require a partnership between different local authorities, with which the prefectures are associated, as decentralised authorities of the central State. In Belgium and Germany, the federated States, Regions and Länder have extensive powers, which puts them on an equal footing with the French central State in transnational cooperation.

However, this apparent similarity between the two federal states conceals real differences resulting from the origins of these political systems.[34] In Germany, federalism was instituted by the three allied forces – the United States, France and Great Britain – occupying the western part of Germany in 1949.[35] This division of powers between two territorial levels was designed to prevent the resurgence of an imperialist power in Germany. Federalism was widely adopted by the new state as a means of promoting democracy in opposition to the National Socialist regime, whose totalitarian nature was associated with excessive centralisation of power. During German reunification, federalism became a key feature, with the creation of new *Länder* (*Neue Länder*) in the newly integrated territories in 1990.

In Belgium, federalism is a much more recent phenomenon.[36] Officially implemented in 1993, it was a response to the tensions between the two main language communities, Flemish and Walloon. To put it simply, there are two elements that distinguish Belgian federalism from German federalism. On the one hand, the former is undeniably more conflictual than the latter, even if in both cases compromise is established as a value between the political forces. On the other hand, the institutional organisation is paradoxically simpler in Germany than in Belgium. Germany is divided into 16 *Länder*, all being granted the same powers.

---

34 M. Uyttendaele (ed.), *Fédéralisme et relations internationales : étude comparée de la pratique en Allemagne et en Belgique : Actes du colloque, 20 Septembre 1996*, Bruylant, Bruxelles, 1998.

35 É. Dufeil, "Allemagne : l'avenir du fédéralisme", *Futuribles*, vol. 420, no. 5, 2017, pp. 19–32.

36 C. Xhardez (ed.), *50 Ans de fédéralisation de l'État belge : institutions, acteurs, politiques publiques et particularités du fédéralisme belge*, Académia, Louvain-la-Neuve, 2020.

In Belgium, there exist two institutions, the region and the community, which only overlap geographically in the case of Flanders. The three communities – Flemish, French and German – have extensive powers in the fields of culture and education, while the three regions – Flanders, Wallonia and Brussels – bring together many state powers (agriculture, regional planning, the economy, energy, the environment, housing, etc.).

Cross-border cooperation initiatives have nevertheless been taken by public authorities at regional and local levels on both borders since the 1960s and 1970s. The originality of Franco-German cooperation lies in the fact that it is often established with a third partner at regional level – Switzerland in the Upper Rhine, Luxembourg in the case of Sar-Lor-Lux. Concerning Luxembourg, the creation of the Grande Région, which was initiated in 1995 and established in 2007, also involves Belgian institutions. This multi-national dimension is also manifested in the "Interreg A" European territorial cooperation programmes. For the period 2021–2027, the Franco-German border is involved in two programmes: the Upper Rhine programme (Germany-France-Switzerland) and the Grande Région programme, which also includes the Franco-Belgian border (France-Belgium-Germany-Luxembourg). Another programme exists for the latter: Wallonie-Vlaanderen-France. It reflects a particular feature of Franco-Belgian cross-border cooperation. As in the Upper Rhine programme, it is in fact a multilateral partnership, between three actors: the two regions, Wallonia and Flanders, cooperate with French public authorities, which is notably the case in the Lille-Kortrijk-Tournai Eurometropolis.

However, cooperation on the Franco-Belgian border was formalised in the 1990s, later than on the Franco-German border. Since the beginning of the 21st century, cooperation has become institutional in both dyads with the creation of European Groupings of Territorial Cooperation (EGTC West Vlaanderen-Flandres-Dunkerque-Côte d'Opale, Lille-Kortrijk-Tournai, Grande Région, Strasbourg-Ortenau) and sectoral cooperations that are often considered as models for other cross-border regions: in health for the Franco-Belgian dyad, in training, public transport and environmental management for the Franco-German dyad.

Artificiality Without an Emblem vs. Infrastructure as a Symbol of Reconciliation

Despite the artificial nature attributed to the Franco-Belgian border, there exist significant differences that are not immediately perceptible.

This idea of a border that "cannot be seen", but which is there nonetheless, may lead to downplay certain differences, particularly in political terms, which paradoxically could hinder cross-border cooperation.[37] What is the point in setting up cross-border projects if you feel that your neighbour is similar to you, or if you are completely unaware of their specificity? This idea of continuity is perhaps more perceptible in France, for which Belgium merely seems to extend the national territory towards Northern Europe, in certain representations. As a result, it is difficult to find an emblem that would symbolise the border, apart from the Lille-Kortrijk-Tournai Eurometropolis, which claims to be the first European Grouping of Territorial Cooperation (EGTC), but to which it is difficult to associate a representation. The absence of an emblem is a factor in the invisibility of the border, which might affect cross-border cooperation and render it illegible or superfluous.

The Franco-German border is based on an entirely different symbolism, that of reconciliation,[38] which is the driving force behind powerful cooperation at all levels. The privileged partnerships that have been established between the states and between local authorities are therefore seen as a model. This is the sign of a strong commitment from both countries, cross-border cooperation being somehow the local or regional counterpart of the "Franco-German couple" (couple franco-allemand) or the "German-French friendship" (*Die Deutsch-französische Freundschaft*), depending on which side of the Rhine you stand.[39] This transition from a conflictual to a peaceful border that allows for all kinds of cooperation seems to be reflected in the Treaty of Aachen (2019),[40]

---

37 P. Pupier, *Recompositions scalaires en Europe occidentale. op.cit.*

38 S. Schirmann, "La réconciliation franco-allemande. Éléments d'une problématique", *op.cit.*; B. Wassenberg, "La coopération transfrontalière dans l'espace du Rhin supérieur: un modèle de réconciliation franco-allemande?", *op.cit.*

39 P.-E. Thomann, *Le couple franco-allemand et le projet européen : Représentations géopolitiques, unité et rivalités*, L'Harmattan, Paris, 2015.

40 J. Beck (ed.), *Grenzüberschreitende Zusammenarbeit nach der Pandemie La coopération transfrontalière après la pandémie. Rechtlich-institutionelle Flexibilisierung im Kontext des Aachener Vertrags Flexibilisation juridique et institutionnelle dans le cadre du Traité d'Aix-la-Chapelle*, Peter Lang, Bruxelles, 2021 ; See also MOT , France-Allemagne : Retrieved from : <http://www.espaces-transfrontaliers.org/ressources/territoires/frontieres/frontieres-en-europe/frontiere-france-allemagne/frontiere-france-allemagne-1/>. Accessed, 12 May 2023.

which opened up new prospects, and the creation of the *Collectivité européenne d'Alsace* (CEA) (2021), one of whose key competences is cross-border cooperation.[41] However, this idyllic image was questioned by lockdown during the coronavirus pandemic. On the one hand, the border was closed unilaterally at the beginning of the first containment period, without consultation between the states, which for a time weakened the rhetoric on cross-border cooperation and the associated arrangements. The authorities subsequently tried to demonstrate the reality of cooperation by transferring patients, but the damage had been done. On the other hand, attitudes of rejection of neighbours were again identified in public spaces when border restrictions were lifted, whereas this type of behaviour seemed to have disappeared. Undeniably, one should not overestimate these tensions and restrictions, but they do raise questions about the scope of this model cooperation, with the Rhine as its main emblem. The river, which forms the border between France and Germany for around 190 km, is a powerful symbol of the existing link. On the one hand, it is an essential transport route, as mentioned above. In fact, following the Congress of Vienna in 1815, the first international cooperation organisation, the Central Commission for Navigation on the Rhine (CCNR), was created precisely to develop this transnational river. Since the millennium, a number of bridges have been built across the river. The construction of infrastructure to improve traffic flow by rail (train, tram) and local transport (pedestrians, cyclists) is one way of challenging the image of the Rhine as a natural border between France and Germany. The cooperation resulting from reconciliation is therefore reflected in the creation of infrastructure. Finally, it should be remembered that although the Saar river also forms a border along some 20 km of its course between Saarland and the Moselle department, the rest of the Franco-German border does not rest on any real natural obstacle, including in its mountainous part (the Vosges and Palatinate uplands).

---

[41] O. Vergne, *La Collectivité européenne d'Alsace. Différenciation territoriale, identité régionale et enjeux transfrontaliers*, doctoral thesis, Université Paris 8, Saint-Denis, 2021.

Linguistic Practices between Continuity and Discontinuity

Generally speaking, the cultural and linguistic borders that are particularly numerous in Europe do not necessarily overlap with political borders.[42] The learning of a *lingua franca* that enables all the inhabitants of a given territory to understand each other can paradoxically lead to the emergence of a linguistic discontinuity on a political border, but in different ways, due to a gradual change in practices.[43] In the Franco-Belgian dyad, the linguistic continuity between France and Wallonia may be related to the high permeability of the border. However, the linguistic continuity that still existed between Belgian and French Flanders in the 1950s has largely disappeared, although it can still be seen in the place names of localities.[44] A similar process of distancing in linguistic practices can be observed to a lesser extent on the Franco-German border, albeit with a time lag. The use of Germanic dialects in Moselle and Alsace has declined sharply since the 1980s.[45] However, policies to support the learning of the "neighbour's language" or bilingual classes in

---

42 C. Bromberger & A. Morel, *Limites floues, frontières vives,* Maison des Sciences de l'Homme, Paris, 2000 ; V. Konrad, "Borders and Culture: Zones of Transition, Interaction and Identity", *Eurasia Border Review,* vol. 5, n°1, 2014, pp. 41–57.

43 C. Raffestin, *Pour une géographie du pouvoir,* Litec, Paris, 1981; E. Gellner, *Culture, identity and Politics,* Cambridge University Press, Cambridge, 1987; A. Raasch, *L'Europe, les frontières et les langues. Guide pour l'élaboration des politiques linguistiques éducatives en Europe – De la diversité linguistique à l'éducation plurilingue,* Council of Europe, Strasbourg, 2002 ; I. Moira-Cavaion, "Contiguous or Border Languages and Their Teaching in Europe", *In* S. Bratoz, A. Pirih & A. Kocbek (eds.), *Pathways to Plurilingual Education*, University of Primorska Press, Koper, 2020, pp. 47–59.

44 P. Tourret, "Les identités culturelles et leurs territoires en Belgique", *Hommes et Terres du Nord,* vol. 4, n°1, 2001, pp. 191–197; S. Dubois, "The Making of Nations in Belgium and Western Europe in Historical Perspective (Fifteenth-Twentieth Century): National Ideology, Ethnicity, Language and Politics", *In* N. Slocum-Bradley, *Promoting Conflict or Peace through Identity,* Routledge, London, 2009, chapter 2.

45 P. Erhart, "Les effets de la frontière sur les pratiques linguistiques dans le Rhin supérieur", *Cahiers du plurilinguisme européen*, vol. 9, 2017. Retrieved from : <https://doi.org/10.57086/cpe.971>. Accessed 23 May 2023; C. Polzin-Haumann & C. Reissner, "Traverser les frontières et les langues dans la Grande Région", pp. 117–135 *In* G. Hamez & J.-M. Defays (eds.), *Réalités, perceptions et représentations des frontières. L'espace transfrontalier de la Grande Région Sarre-Lor-Lux*, EME Éditions, Louvain-la-Neuve, 2020 ; C. Faucompré, "L'enseignement de la langue du voisin: contribuer à la cohésion interrégionale et interfrontalière dans le Rhin Supérieur", *In* N. Dziub, *Le Transfrontalier. Pratiques et représentations*, Presses Universitaires de Reims, Reims, 2020, pp. 115–143.

Alsace and Moselle have not compensated for the decline in cross-border comprehension skills, especially as on the opposite side of the border, the decrease in the learning of French in Flanders and in the German Länder bordering France has also contributed to a feeling of declining intercomprehension. In both cases, the political and linguistic borders are becoming similar, with the notable exception of the section between Wallonia and Northern France. Only a strong political will in favour of language learning could counteract this trend or at least reduce it.

Two Borders Characterised by Multilateral Cooperation

The use of the analytical grid shows that the two dyads, although born of different origins and associated with different symbols, are part of comparable spaces, characterised by contrasting populations and intense interactions. In some respects, the border seems to be getting stronger despite European integration, and this has been observable for some decades in the use of languages, with the exception of the part of the Franco-Belgian border between France and Wallonia. In both cases, the asymmetry of powers between a federal system and a decentralised unitary system does not prevent close and multiple forms of cross-border cooperation, which cannot, however, be reduced to bilateral agreements. Instead, cooperation is tripartite (France-Wallonia-Flanders on one side, Germany-France-Switzerland on the other), or even quadripartite around the Grand Duchy of Luxembourg (France-Belgium-Germany-Luxembourg). This means that each border, considered as a link between two territorial powers, is likely to go through changes depending on all the states in its immediate vicinity. Our objective is now to complete the portrait of the two borders by identifying a number of issues that are both common and distinct.

Two Borders and Cross-Border Spaces in Western Europe: Some Trends and Issues

Our analysis of the two dyads occurs in a specific temporal and spatial context. First of all, Europe experiences a growing tension, as was strikingly demonstrated during the pandemic. On the one hand, the trend toward integration is characterised by the existence of the single market, which reinforces all kinds of interdependence between Member States, and by decisions taken at supranational level in a growing number of fields. On the other hand, there is a nationalist temptation, which considers the national level to be the most relevant for responding to all the challenges facing a society, and which very often seems to call into question the European dimension. This form of nationalism tinged with Euroscepticism is propagated by political parties, certain media and some sections of civil society; it envisages the state as the only legitimate organisation to meet the needs of the national community. This trend is also in line with the states' growing preoccupation, considering territorial security as a systematic priority over other issues. We contend that this tension between Europeanisation and nationalism can have significant repercussions on cross-border regions, because it encourages national authorities to see their territory as a coherent whole, without taking into account the specific characteristics of border areas. This process does not mean that the international dimension is ignored, or that the cross-border dimension, i.e. the local international dimension, becomes a secondary concern. As a result, the autonomy of local or regional public authorities in the field of territorial diplomacy could be disregarded or even ignored. The experience of the first few weeks of the covid-19 pandemic, during which each state decided to manage the crisis unilaterally, neglecting the cross-border dimension, is an admittedly extreme example of what could happen.

In this context, the persistence of socio-economic asymmetries deserves to be highlighted. Indeed, the EU's cohesion policy is designed to reduce socio-economic disparities between Member States. However, national borders continue to show often significant gaps between the wealth produced in two neighbouring states. The existence of socio-economic discontinuities at borders was already identified by the European Commission with the abolition of customs duties between Member States in 1968. Although the gaps have narrowed at EU level, they may persist locally, or even widen, which can give rise to tensions. The specialisation of border areas that is indicative of cross-border integration – to put it simply, a strong employment dynamic on one side and a strengthening of the residential function on the other – is likely to be interpreted as the formalisation of a domination and risks being expressed through frustrations and resentments[46]. Cross-border cooperation should

---

[46] R. Ysebaert et al., *A socio-economic typology on cross-border regions at European scale (2000–2012), Key findings*, UMS RIATE, Paris, 2016; A. Decoville & F. Durand,

therefore be aware of the differences on either side of a border, whether socio-economic or cultural and linguistic.

In addition, the two borders here described are part of Western Europe, a particular region of the continent. Some of the specific geographical features of this area have already been mentioned, such as population density and the proximity of urban areas. Historically, this part of Europe was also seen as a centre for the emergence of political ideologies and geopolitical concepts such as nation states.[47] Finally, several of the EU's major decision-making centres are located within this area: alongside the EU's political and symbolic centres of power, such as Brussels, Luxembourg, Strasbourg and Frankfurt am Main, one finds the capitals of the EU's two most influential states. Of course, only Paris is really located in this area, while Berlin is geographically quite far away. This observation needs to be qualified, however, as Bonn, the former federal capital, still retains a number of ministries and federal institutions, and many companies have their head offices in Rhineland and Southern Germany[48]. The Benelux countries should also be taken into account: even if this territorial configuration is part of the EU and its role appears more symbolic than real, it signals the existence of close links between the governments of the three states that make it up, Belgium,

"Challenges and obstacles in the production of cross border territorial strategies: the example of the greater region". *Transactions of the Association of European Schools of Planning*, vol. 1, n°1, 2017, pp. 65–78; B. Reitel, B. Wassenberg & J. Peyrony, "The INTERREG Experience in Bridging European Territories. A 30-Year Summary", *In* E. Medeiros (ed.), *European Territorial Cooperation. Theoretical and Empirical Approaches to the Process and Impacts of Cross-Border and Transnational Cooperation in Europe,* Springer, Dordrecht, 2018, pp. 7–24; S. Bourdin, "Analyse spatiale de l'efficacité des Fonds structurels européens sur la croissance régionale", *Revue d'Économie Régionale & Urbaine*, no. 2, 2019, pp. 243–269 ; E. Nadaud & O. Bouba-Olga, "Au-delà du PIB par habitant : la diversité des contextes régionaux européens", *Revue d'Économie Régionale & Urbaine*, no. 4, 2019 pp. 727–747 ; C. Grasland, M. Guerois & R. Ysebaert, "Border Discontinuities", *In* B. Wassenberg & B. Reitel (eds.), *Critical Dictionary of Borders, Cross-border Cooperation and European Integration*, Peter Lang, Brussels, 2020, pp. 123–126.

47 R. Brunet, *Les villes 'européennes', op.cit.* ; R. Brunet, D. Pumain, T. Saint-Julien & R. Ferras (eds), *Géographie universelle, France, Europe du Sud,* Belin-Reclus, Paris, 1990 ; R. Brunet, J.-P. Marchand & P. Riquet (eds.), *Géographie universelle, Europe du Nord, Europe médiane*, Belin-Reclus, Paris, 1996.

48 A. Laporte, *De Bonn à Berlin – Le transfert d'une capitale (1990–2010),* Presses Universitaires du Mirail, Toulouse, 2015.

Luxembourg and the Netherlands.[49] Finally, the two neighbouring states of the two dyads should also be considered, as their policies may have repercussions on the borders under study. The importance of Brexit, the United Kingdom's withdrawal from the EU, goes without saying. However, another country deserves attention: Switzerland. In the 1990s, it signed bilateral agreements that have been renegotiated since 2015. Here again, these new developments may have consequences.

Our analysis examines two aspects: on the one hand, we will see that certain internal characteristics of Belgium, France and Germany are likely to influence the evolution of these two borders. On the other hand, we aim to understand how the states located in the vicinity of the borders studied (United Kingdom, Luxembourg, Switzerland) could affect them.

Borders Shaped by Neighbouring States

Neil Brenner (2004) has shown that, far from being immutable structures, states are subject to numerous changes. In the present case, we have identified for each of the three states a marker that distinguishes the evolution of the public sphere in recent years and whose characteristics we believe may influence the borders: the process of federalisation in Belgium, the merger of regions in France, the strategy of border areas in Germany.

States Reconsidering Their Cross-Border Spaces

The "Regionalisation" of Belgium

Belgium has been a federal state since 1993, but federalisation is a process that does not appear to be completed yet, as demonstrated by the persistent tensions between the Flemish and Walloon regions[50]. Cross-border cooperation is actually based on the existence of a balance between the actors present in a framework involving three partners from Flanders, Wallonia and France: such is the case of the Interreg

---

[49] B. Wassenberg, "Benelux States: Belgium, Luxembourg and the Netherlands", *In* B. Wassenberg & B. Reitel (eds.), *Critical Dictionary of Borders, op. cit.*, pp. 101–105.

[50] C. Xhardez (ed.), *50 Ans de fédéralisation de l'État belge, op.cit.*; R. Pasquier, *Organisation territoriale et démocratie locale en Europe, op. cit.*

A 2021–2027 "France-Wallonie-Vlaanderen" programme and the Lille-Kortrijk-Tournai Eurometropolis. Could this balance be endangered by an increase in tensions? What role would the French partner be able, compelled, or willing to play? Depending on the case, the French partner can act as a mediator or, on the contrary, take a back seat. In the first situation, cooperation would continue, whereas in the second, the ability to set up cross-border projects would certainly be affected. Another possibility would be for bilateral cooperation to be favoured between Flanders and France on the one hand and Wallonia and France on the other, to the detriment of trilateral cooperation, but with an uncertainty about the future. Indeed, the linguistic proximity between Wallonia and France seems to favour cooperation between these two sides to the detriment of cooperation between Flanders and France. Paradoxically, the relationship between Flanders and Wallonia shows the characteristics of a real border, but this is not taken into account in the Interreg programmes, which only consider international borders and not internal borders within federal states. Belgium's unique situation is therefore not really taken into account.

One way of overcoming these difficulties would be to define a strategic vision of the cross-border area over the next few years, and to establish a framework for action on a contractual basis, provided that there is a willingness on the part of each political authority to do so. At the same time, the actors involved on both the Belgian and French sides should be encouraged to continue their training so as to acquire a superior knowledge of the institutional and cultural systems of the neighbouring country, and in particular to invest in language learning.

Changes to the Boundaries of the Regions in France

France is a unitary state, but the process of decentralisation has profoundly altered the roles of local authorities since 1982. In 2014 and 2015, three laws (Modernisation de l'action publique territoriale et d'affirmation des métropoles, MAPTAM ;[51] loi relative à la délimitation

[51] République française, Vie publique. Retireved from: <https://www.vie-publique.fr/loi/20674-loi-maptam-action-publique-territoriale-et-daffirmation-des-metropoles>. Accessed 23 May 2023.

des régions; Nouvelle organisation territoriale des Régions,[52] NOTRe[53]) redefined the powers and scope for action of local public bodies. At first sight, this process should result in simplification.[54] Only one region now borders the Franco-German dyad, the Grand Est, instead of two previously, Alsace and Lorraine. Two are adjacent to the Franco-Belgian border: Hauts-de-France for the western part and Grand Est for the eastern part, instead of three, Nord-Pas-de-Calais, Champagne-Ardennes and Lorraine. However, once established, the new merged regions devoted much of their energy to implementing internal integration policies and mechanisms. Attention to cross-border areas might well be supplanted by other issues considered to be more significant. One should therefore look closely at the ambitions of the political authorities. In Hauts-de-France, the new regional perimeter now borders on the Île-de-France region, which has a strong power of attraction. As a result, the cross-border relationship with Belgium has lost some of its importance, due to the influence of the Paris conurbation throughout the southern part of the Region, but also because of Brexit. Paradoxically, the creation of the Grand Est region allows for a single global vision of the Franco-German dyad, something unprecedented. However, the two pre-existing cross-border cooperation frameworks, the Upper Rhine and the Grande Région, both have Interreg programmes based on a development strategy, the Trinational Metropolitan Region of the Upper Rhine since 2007 for the former and the Grande Région in 2020 for the latter. The CEA, whose creation in 2021 was based on the idea that the former Alsace Region has a long history of close cooperation with neighbouring German and Swiss regions within the framework of the Upper Rhine, seems to reflect a status quo between the two frameworks. The challenge now is to ensure coordination between the two cross-border perimeters in order to avoid contradictions. However, the 3DS law (territorial differentiation, decentralisation, deconcentration, simplification) passed in

---

52 République française, Vie publique. Retireved from: <https://www.vie-publique.fr/loi/20720-elections-regionales-elections-departementales-calendrier>. Accessed 23 May 2023.

53 République française, Vie publique. Retireved from: <https://www.vie-publique.fr/loi/20721-loi-notre-loi-du-7-aout-2015-nouvelle-organisation-territoriale-de-la>. Accessed 23 May 2023.

54 É. Négrier, V. Simoulin, *La fusion des régions : le laboratoire d'Occitanie*, Presses Universitaires de Grenoble, Grenoble, 2021 ; J.-C. Némery, *Quelle organisation pour les grandes régions en France et en Europe?*, L'Harmattan, Paris, 2015.

2022 opens up new prospects for cross-border cooperation, as it enables local authorities that so request, and subject to obtaining authorisation from the central state, to carry out experiments and, in particular, transfers of competence. Some French local authorities seem ready to embark on experiments to strengthen their cross-border cooperation.

A Federal State Strategy for Cross-Border Regions in Germany

The reunification of Germany and the eastward enlargement of the EU have strengthened this country's geopolitical centrality in Europe, but this has also prompted the Federal State to reconsider its spatial planning policy. One of the objectives is to reduce the effects of territorial boundaries, particularly those of the Länder, by encouraging the emergence of inter-territorial governance structures. The creation of metropolitan regions in 1995, several of which straddle several Länder, is a case in point.[55] Secondly, since 2008 the role of border areas has been the subject of a specific strategy as part of model projects (*Modellvorhaben der Raumordung* (MORO)): border areas characterised by the existence of strong interactions with neighbouring national territories, with which specific cooperation can be undertaken, have thus been identified.[56] Without calling into question the powers of the *Länder*, which have jurisdiction over their territories, this means that the federal state is encouraging public authorities to coordinate with a view to strengthening cross-border cooperation in functional areas considered relevant. These frameworks obviously concern both the Upper Rhine and the Grande Région, meaning that it is part of a multilateral approach involving Luxembourg and Switzerland.

Besides, Franco-German cooperation, which was reaffirmed by the Treaty of Aachen in 2019, undeniably opens up new prospects. However, one may wonder whether these prospects are reflected in the policies adopted by the three Länder bordering France. Since 2015, Saarland has

---

55 C. Schulz, "La politique des 'Metropoleregionen' en Allemagne", *Bulletin de l'Asscoiation des Géographes Français*, n°4, 2012, pp. 515–522.

56 T. Chilla & F. Sielker, "Cross-border spatial development in Bavaria – Dynamics in Cooperation – Potentials of Integration", *Arbeitsberichte der ARL*, n°34, 2022. Retrieved from: <https://www.bbsr.bund.de/BBSR/DE/forschung/programme/moro/studien/2020/deutsch-franzoesische-planspiele/01-start.html>. Accessed 23 May 2023.

adopted a strategy called "Frankreichstrategie", which aims to develop a special partnership with neighbouring French regions[57]. Bilingualism is one of the proposed actions. But Saarland is one of Germany's smallest *Länder*. In contrast, Baden-Württemberg, one of the most powerful Länder in the Federal Republic of Germany, shares a long border with both France and Switzerland. For decades, this Land has prioritised integration into globalisation, the cross-border dimension being somewhat neglected. Moreover, within the federal framework, it is often in competition with neighbouring Bavaria as a location for strategic investments.[58] Seen from Stuttgart, the Land capital, cooperation with the Swiss cantons might seem more attractive than with the neighbouring French regions, but this would need to be explored in greater depth. As for Rhineland-Palatinate, its border with France is very short and its main urban centres are quite far away. In addition, it lies on the periphery of several very dynamic Länder, but this specificity undeniably brings it closer to the neighbouring French regions.[59] Ultimately, in this very favourable context for cross-border cooperation at federal level, future directions will depend on the policies pursued by each of the three Länder, which are not necessarily concerted. Ultimately, public players capable of driving forward cross-border initiatives are also found in towns close to the borders, like Saarbrücken, Karlsruhe and Freiburg im Breisgau.

The Proximity of Neighbouring States and the Effects on Cooperation

The Impact of Brexit

The United Kingdom (UK)'s withdrawal from the EU has undeniable consequences, both for neighbouring territories and for the EU

---

57 Saarland, "Frankreichstrategie", 2023. Retrieved from: <https://www.saarland.de/stk/DE/portale/europa/frankreichstrategie/frankreichstrategie_node.html>, Accessed 12 May 2023. Accessed 23 May 2023.

58 Baden-Würtemberg, "Innovationsstrategie Baden-Württemberg", 2020. Retrieved from! <https://2021-27.efre-bw.de/wp-content/uploads/5422_MinBw_Innovationsstrategie_2020_WEB.pdf>. Accessed 12 May 2023.

59 Rheinland-Pfalz. Zukunftsrat. Nachhaltige Entwicklung, 2023. Retrieved from:<https://www.rlp.de/fileadmin/02/Service/Pressemitteilungen/Arbeitsprogramm_Zukunftsrat_Nachhaltige_Entwicklung_RLP_2023.pdf>. Accessed 12 May 2023.

as a whole, not all of which have yet been identified.[60] Indeed, despite the signing of the Windsor Framework between the EU and the UK in February 2023, many adjustments are still possible in the coming years.[61] The Channel and the North Sea are discontinuities, but above all an area whose management is shared between neighbouring states. The challenge here is to manage maritime space, the movement of ships and container ships, the associated risks, but also the resources, while limiting conflicts.

Two risks are currently apparent. The first is the existence of a Franco-British bias, which stems from the fact that the United Kingdom is not part of the Schengen area and that the Strait of Pas-de-Calais is the shortest passage between the two countries. Brexit has heightened tensions between the two states by systematically associating the issue of migration with that of security, which tends to overshadow other issues that concern the states around the North Sea. The second is that Belgium's concerns are not necessarily the same as those of France. Beyond those differences, however, Brexit offers an opportunity to reconsider cross-border cooperation between the two shores, which was not always very active when the United Kingdom was a member of the EU: the management of the maritime space in all its dimensions is a common issue for all neighbouring states. From this point of view, France and Belgium could be a joint force for proposals – with or without the Netherlands – to initiate new forms of partnership.

Luxembourg, a Cross-Border Metropolis

A process of metropolisation has been going on for several decades in the Grand Duchy of Luxembourg, hence strong growth in the number of highly skilled jobs in its main urban centres.[62] The areas bordering Belgian Luxembourg, Lorraine in France, and Saarland and Rhineland-Palatinate in Germany, are all part of this process, which

---

[60] B. Wassenberg & N. Suzuki (eds.), *Origins and Consequences of European Crises: Global Views on Brexit*, Peter Lang, Brussels, 2020.

[61] UK Government, "The Windsor Framework". Retrieved from: <https://www.gov.uk/government/publications/the-windsor-framework>. Accessed 23 May 2023.

[62] C. Sohn, *Luxembourg: An Emerging Cross-border Metropolitan Region*, Peter Lang, Brussels, 2010.

has substantially raised the standard of living in the cross-border region.[63] However, the effects of metropolisation are not all positive: it has become difficult to recruit labour due to competition from salaries offered in the Grand Duchy, and the cost of land has sharply increased.[64] Local authorities on their own do not have the necessary resources to solve these problems, which can be perceived as a form of domination by the local population. An ambitious cross-border cooperation policy aimed at regulating asymmetries or at least limiting the effects of differentials would open up some prospects and perhaps make it possible to limit or at least manage tensions at the regional level. Luxembourg undeniably plays a leading role in cross-border cooperation, but nothing prevents the public authorities of the other three countries from initiating bilateral or trilateral cooperation so as to respond together to common challenges or to constitute a single force of proposition vis-à-vis the Luxembourg metropolis. The excessive focus on Luxembourg and the size of the Grande Région might relegate the concerns of other territories to a secondary position and weaken bilateral relations at local level: this could be the case in the Ardennes mountains, between French and Belgian local authorities.

Switzerland, between European Integration and Autonomy

Switzerland has signed two bilateral agreements with the European Union, which came into force in 2002 and between 2007 and 2009 respectively. The Confederation is thus associated with certain EU policies.[65] This à la carte integration was the subject of talks from 2014 onwards, which were interrupted in 2021 by the Swiss government. Negotiations resumed in 2023 a few months after a vote in which Swiss citizens approved increased funding for Frontex, the European border and coastguard agency, meaning that the Confederation would remain

---

63 Interreg Grande Région, “Eine grenzüberschreitende operative Strategie für die Großregion”, 7 October 2020. Retrieved from : <https://www.sig-gr.eu/content/dam/sig/publications/sdt-gr/20210112-Grenzuberschreitende-operative-Strategie-REKGR.pdf>. Accessed 12 May 2023.

64 L. Diop, *La production de l’espace résidentiel dans l’aire métropolitaine transfrontalière de Luxembourg*, Doctoral thesis, University Paris 10, Nanterre, 2013.

65 Confédération Suisse, “Politique européenne de la Suisse”. Retrieved from: <https://www.eda.admin.ch/europa/fr/home.html>. Accessed 12 May 2023.

part of the Schengen area.[66] A resolution to draw up new agreements is expected in 2024, but whatever the outcome of these negotiations, the involvement of the Swiss public authorities in the dynamics of cross-border cooperation in the Upper Rhine is being questioned. Since the 1960s, the region's cooperation model has been based on a trinational dimension. The canton of Basel-Stadt plays a leading role in this area in north-western Switzerland, and for a long time it dragged the other four cantons in its wake. For some years now, however, Basel-Stadt has seemed more concerned with cooperation at the level of the trinational conurbation than at the level of the Upper Rhine, and the other cantons, with the notable exception of Jura, seem less interested in cross-border issues.[67] As a result, the weakening of the participation of Swiss players, even if only partial, would require the establishment of a new balance between the French and German public authorities. If this were the case, leadership could not be provided by a player from only one side, as it might be interpreted as a form of domination. However, two factors mitigate this risk. On the one hand, the Treaty of Aachen signed in 2019 opens up new prospects for cross-border cooperation between France and Germany, in particular by providing an innovative legal framework for public authorities. On the other hand, the Collectivité européenne d'Alsace created in 2021 within the *Grand Est* region, following the desire of the elected representatives of this former region, as a reaction to the regional merger of 2014, could provide leadership for the French part.[68] In this context, it raises questions about the correspondent that the CEA will have to find in Germany for cross-border cooperation in the Upper Rhine. On the Franco-German border, the two cross-border cooperation frameworks, the Upper Rhine and the Grande Région, therefore continue to have different perspectives.

66 Confédération Suisse, "Le Conseil fédéral. Le Portail du Gouvernmaent suisse". Retrieved from: <https://www.admin.ch/gov/fr/accueil/documentation/communiques.msg-id-95910.html>. Accessed 12 May 2023.

67 P. Pupier, *Recompositions scalaires en Europe occidentale. op.cit.*

68 O. Vergne, *La Collectivité européenne d'Alsace, op.cit.*

Conclusion

In the end, this study of the Franco-German and Franco-Belgian dyads shows some similarities, but also some notable differences.

In both cases, the border is above all a political distinction between a federal system and a decentralised unitary system. At local level, cross-border cooperation brings together a French side made up of several partners, including the central state, and a Belgian or German side which, because of subsidiarity, has only one public player, usually the region or Land. The regional cooperation frameworks are most often based not on a bilateral logic, but on a trilateral one, involving France, Wallonia and Flanders on the Belgian border, Germany, France and Switzerland on the German border, or even a quadripartite one in the case of the *Grande Région*, which concerns the two dyads studied. The political dimension is all the more significant given that the boundaries do not present any marked differences in terms of landscape, topography or population. The main discontinuities appear perpendicular to the borders. The connectivity that characterises the two borders is based on the presence of major trans-European traffic routes, the Rhine axis completed by the Moselle axis for the Franco-German border and the Netherlands-Paris axis for the Franco-Belgian border. Finally, linguistic differences are tending to increase between France and Germany and on part of the Franco-Belgian border between France and Flanders, which hampers cross-border cooperation.

The uniqueness of the Franco-Belgian border, however, lies in the linguistic continuity over a large part of its length, between France and Wallonia, and above all in the absence of a marker. Paradoxically, continuity seems to be the main characteristic of this border, but Belgian federalism is another important feature, even if it is less perceptible. Flanders and Wallonia are invited to participate, but the Belgian state does not systematically take part in the bilateral cooperation in which the French central state is involved, which can be disconcerting for the decentralised authorities, depending on the case. In the cross-border cooperation frameworks themselves, the presence of the Grand Duchy of Luxembourg has become essential in the context of the Grande Région with German public-sector players who are not adjacent, which clearly reflects the polarisation of Luxembourg. Interest in cross-border cooperation between France and Belgium could wane as a result of a number of factors: Brexit, which increases the attention paid by each of the two states, France and Belgium, to the United Kingdom; the change in the boundaries of the regions as part of the regional mergers in France, with one region, Hauts-de-France, likely to look more towards Paris than Brussels Gent and Antwerp; and tensions between Wallonia and Flanders in Belgium, at a time when a number of areas require increasing attention, such as transport and security.

The uniqueness of the Franco-German border is much more obvious than that of the Franco-Belgian border. On the one hand, it is similar to a "scar of history", which, thanks to Franco-German reconciliation, has become a peaceful border. This symbolic dimension of reconciliation is omnipresent in all areas of cross-border cooperation and is undeniably a driving force at local and regional level. But this dimension can also be seen as a burden, to be mitigated by the presence of the public authorities of the neighbouring states – Switzerland for the Upper Rhine, Luxembourg and Belgium for the Grande Région: cooperation should focus on objectives that are more tangible than symbolic. The Rhine is undeniably a symbol of bi-national cooperation, as it is reflected in the infrastructure built over the last two decades to reduce the effects of discontinuities in the river, which is also managed by an international organisation, the CCNR. Franco-German cooperation could be given new impetus if the public authorities on both sides of the river make use of some of the possibilities offered by the Treaty of Aachen. On the French side, the 3DS law also offers local and regional authorities the possibility of carrying out experiments. Two entities could play a leading role, the Collectivité européenne d'Alsace and *Saarland*, but each is located in another cross-border region, the Upper Rhine for the former and the *Grande Région* for the latter, so it is necessary to find a partner on the other side of the border within the same cross-border region.

The introduction of a bilateral treaty between France and Belgium along the lines of what has been achieved on the German border could certainly provide a useful framework for any future cooperation, as it would enable the main public players, and in particular the Belgian regions, to coordinate their activities.

The two cross-border areas are characterised by intense interactions, but also by a possible heightening of tensions linked to the persistence of socio-economic differences and the existence of often contradictory dynamics: reactivation of more or less marked forms of nationalism, increased restrictions on border crossings for reasons of territorial security by the States, accentuation of metropolisation with both an effect of concentration and economic structuring, but also a risk of social polarisation and functional specialisation. In all cases, the strengthening of partnerships with different configurations, bilateral or multilateral, whatever the scale, is an option that is widely envisaged. This argues in favour of maintaining the capacity of regional and local authorities to design, develop and implement public cross-border cooperation policies with the help of government departments.

In spite of multiple challenges, the most serious one being the exacerbation of nationalisms, cross-border cooperation may be expected to continue to thrive across the Franco-German and Franco-Belgian open borders.

Bibliography

J. Beck (ed.), *Grenzüberschreitende Zusammenarbeit nach der Pandemie La coopération transfrontalière après la pandémie. Rechtlich-institutionelle Flexibilisierung im Kontext des Aachener Vertrags Flexibilisation juridique et institutionnelle dans le cadre du Traité d'Aix-la-Chapelle*, Peter Lang, Bruxelles, 2021.

A. Beyer, "Nœuds de transport et frontières. L'invention de la métropole bâloise", *Annales de géographie*, n° 657, sept.–octobre, 2007, pp. 451–369.

A. Beyer, "La Belgique à la croisée des chemins. Une géohistoire des grandes infrastructures de transport du plat pays à l'aune de ses frontières", *Annales de Géographie,* vol. 681, n°5, 2011, pp. 465–485.

M.-T. Bitsch, *Histoire de la Belgique : de l'Antiquité à nos jours*, Éditions Complexe, Bruxelles, 2004.

S. Bourdin, "Analyse spatiale de l'efficacité des Fonds structurels européens sur la croissance régionale", *Revue d'Économie Régionale & Urbaine*, no. 2, 2019, pp. 243–269.

C. Bromberger & A. Morel, *Limites floues, frontières vives,* Maison des Sciences de l'Homme, Paris, 2000.

R. Brunet, *Les villes 'européennes'*, La documentation française, DATAR-RECLUS, Paris, 1989.

R. Brunet, D. Pumain, T. Saint-Julien & R. Ferras (eds.), *Géographie universelle, France, Europe du Sud*, Belin-Reclus, Paris, 1990 ; R. Brunet, J.-P. Marchand & P. Riquet (eds.), *Géographie universelle, Europe du Nord, Europe médiane*, Belin-Reclus, Paris, 1996.

T. Chilla & F. Sielker, "Cross-border spatial development in Bavaria – Dynamics in cooperation – Potentials of integration", *Arbeitsberichte der ARL*, n°34, 2022.

Commission Européenne, *Schéma de Développement de l'Espace Communautaire. Vers un développement spatial équilibré et durable du territoire de l'union européenne*, Publication Office of the European Communities, Luxemburg, 1999.

E. Crocquey, F. Mille & A. Roszak, "L'emploi frontalier des Hauts-de-France vers la Belgique : une attractivité modérée et contrastée", *INSEE Analyses Hauts-de-France*, n°94, 2019.

A. Decoville & F. Durand, "Challenges and obstacles in the production of cross-border territorial strategies: the example of the greater region". *Transactions of the Association of European Schools of Planning*, vol. 1, n°1, 2017, pp. 65–78.

K. Denni, *Rheinüberschreitungen – Grenzüberwindungen. Die deutsch-französische Grenze und Ihre Rheinbrücken (1861–2006)*, UVK, Konstanz, 2008.

R. Dion, *Les frontières de la France*, Hachette, Paris, 1947.

L. Diop, *La production de l'espace résidentiel dans l'aire métropolitaine transfrontalière de Luxembourg*, Doctoral thesis, University Paris 10, Nanterre, 2013.

S. Dubois, "The making of nations in Belgium and western Europe in historical perspective (fifteenth-twentieth century): National ideology, ethnicity, language and politics", *In* N. Slocum-Bradley, *Promoting Conflict or Peace through Identity*, Routledge, London, 2009, chapter 2.

É. Dufeil, "Allemagne : l'avenir du fédéralisme", *Futuribles*, vol. 420, no. 5, 2017, pp. 19–32.

P. Erhart, "Les effets de la frontière sur les pratiques linguistiques dans le Rhin supérieur", *Cahiers du plurilinguisme européen*, vol. 9, 2017.

European Commission, Directorate-General for Mobility and Transport, *White Paper on Transport: Roadmap to a Single European Transport Area: Towards a Competitive and Resource Efficient Transport System*, Publications Office, Luxemburg, 2011.

C. Faucompré, "L'enseignement de la langue du voisin: contribuer à la cohésion interrégionale et interfrontalière dans le Rhin Supérieur", *In* N. Dziub, *Le Transfrontalier. Pratiques et représentations*, Presses Universitaires de Reims, Reims, 2020, pp. 115–143.

M. Foucher, "L'invention des frontières: un modèle géopolitique français", *Hérodote*, n°33, 1986, pp. 54–88.

M. Foucher, *Fronts et frontières: Un tour du monde géopolitique,* Fayard, Paris, 1991.

M. Foucher, *Fronts et frontières,* Fayard, Paris, 1991.

M. Foucher, *La République européenne*, Belin, Paris, 2000.

É. François, K. Seifarth & B. Struck, *Die Grenze als Raum, Erfahrung und Konstruktion. Deutschland, Frankreich und Polen vom 17. Bis zum 20. Jahrhundert,* Campus, Frankfurt-am-Main, 2007.

E. Gellner, *Culture, Identity and Politics*, Cambridge University Press, Cambridge, 1987.

C. Grasland, M. Guerois & R. Ysebaert, "Border discontinuities", *In* B. Wassenberg & B. Reitel (eds.), *Critical Dictionary of Borders, Cross-border Cooperation and European Integration*, Peter Lang, Brussels, 2020, pp. 123–126.

G. Hamez, *Du transfrontalier au transnational: Approche géographique. L'exemple de la frontière franco-belge*, Doctoral thesis, Université Panthéon-Sorbonne, Paris, 2004.

P. Hamman, *Sociologie des espaces-frontières. Les relations transfrontalières autour des frontières françaises de l'est,* Presses Universitaires de Strasbourg, Strasbourg, 2013.

É. Juillard, *L'Europe rhénane. Géographie d'un grand espace*, Armand Colin, Paris, 1968.

É. Juillard & H. Nonn, *Espaces et régions en Europe occidentale*, CNRS, Paris, 1976.

D. E. Khan, *Die deutschen Staatsgrenzen: rechtshistorische Grundlagen und offene Rechtsfragen,* Mohr Siebeck, Tübingen, 2004.

V. Kolossov & J. W. Scott, "Theorizing borders. Border studies: Changing perspectives and theoretical approaches", *Belgeo*, n° 1, 2013, pp. 1–19.

V. Konrad, "Borders and culture: Zones of transition, interaction and identity", *Eurasia Border Review,* vol. 5, n°1, 2014, pp. 41–57.

A. Laporte, *De Bonn à Berlin – Le transfert d'une capitale (1990–2010),* Presses Universitaires du Mirail, Toulouse, 2015.

F. Lentacker, *La frontière franco-belge. Étude géographique des effets d'une frontière internationale sur la vie de relations*, Morel & Corduant, Lille, 1974.

T. Martini & S. Villaume," Le travail frontalier dans le Grand Est : en 2018, 182 000 habitants de la région partent travailler dans un pays voisin", *INSEE Analyses Grand Est,* n° 144, 2022.

I. Moira-Cavaion, "Contiguous or border languages and their teaching in Europe", *In* S. Bratoz, A. Pirih & A. Kocbek (eds.), *Pathways to Plurilingual Education*, University of Primorska Press, Koper, 2020, pp. 47–59.

E. Nadaud & O. Bouba-Olga, "Au-delà du PIB par habitant : la diversité des contextes régionaux européens", *Revue d'Économie Régionale & Urbaine*, no. 4, 2019, pp. 727–747.

É. Négrier & V. Simoulin, *La fusion des régions : le laboratoire d'Occitanie*, Presses Universitaires de Grenoble, Grenoble, 2021.

J.-C. Némery, *Quelle organisation pour les grandes régions en France et en Europe?,* L'Harmattan, Paris, 2015.

D. Newman, "On borders and power: A theoretical framework", *Journal of Borderlands Studies*, vol. 18, n°1, 2003, pp. 13–25.

D. Nordman, *Frontières de France. De l'espace au territoire. XVIe-XIXe siècles,* Gallimard, Paris, 1998.

R. Pasquier, *Organisation territoriale et démocratie locale en Europe,* La Documentation française, Paris, 2022.

C. Polzin-Haumann & C. Reissner, "Traverser les frontières et les langues dans la Grande Région", pp. 117–135. *In* G. Hamez & J.-M. Defays (eds.), *Réalités, perceptions et représentations des frontières. L'espace transfrontalier de la Grande Région Sarre-Lor-Lux*, EME Éditions, Louvain-la-Neuve, 2020.

P. Pupier, *Recompositions scalaires en Europe occidentale : la construction de régions métropolitaines transfrontalières dans l'Eurorégion Transmanche et le Rhin Supérieur,* Doctoral thesis, Université d'Artois, Arras, 2021.

A. Raasch, *L'Europe, les frontières et les langues. Guide pour l'élaboration des politiques linguistiques éducatives en Europe – De la diversité linguistique à l'éducation plurilingue*, Council of Europe, Strasbourg, 2002.

C. Raffestin, *Pour une géographie du pouvoir,* Litec, Paris, 1981.

R. Ratti & M. Schuler, "Typologie des espaces-frontières à l'heure de la globalization", *Belgeo*, n°1, 2013, pp. 1–14.

B. Reitel, "La frontière internationale, objet sémique, processus multidimensionnel, interface signifiante", *LISER Working Paper* 2011/43, Esch-Alzette, 2011.

B. Reitel, B. Wassenberg & J. Peyrony, "The INTERREG experience in bridging European territories. A 30-year summary", *In* E. Medeiros (ed.), *European Territorial Cooperation. Theoretical and Empirical Approaches to the Process and Impacts of Cross-Border and Transnational Cooperation in Europe,* Springer, Dordrecht, 2018, pp. 7–24.

S. Schirmann, "La réconciliation franco-allemande. Éléments d'une problématique", *In* N. Paun & S. Schirmann (eds.), *Borders, Identities, Communities. The Road to Reconciliation in Central and Eastern Europe,* Nomos, Baden-Baden, 2016, pp. 211–221.

C. Schulz, "La politique des 'Metropoleregionen' en Allemagne", *Bulletin de l'Asscoiation des Géographes Français*, n°4, 2012, pp. 515–522.

C. Sohn, *Luxembourg: An Emerging Cross-border Metropolitan Region,* Peter Lang, Brussels, 2010.

P.-E. Thomann, *Le couple franco-allemand et le projet européen : Représentations géopolitiques, unité et rivalités,* L'Harmattan, Paris, 2015.

B. Touchelay, *Fraudes, frontières et territoires (XIIIe-XXIe siècle),* Institut de la gestion publique et du développement économique, Paris, 2020.

P. Tourret, "Les identités culturelles et leurs territoires en Belgique", *Hommes et Terres du Nord*, vol. 4, n°1, 2001, pp. 191–197.

M. Uyttendaele (ed.), *Fédéralisme et relations internationales : étude comparée de la pratique en Allemagne et en Belgique : Actes du colloque, 20 Septembre 1996,* Bruylant, Bruxelles, 1998.

H. Van Houtum, "An overview of European geographical research on borders and border regions", *Journal of Borderlands studies*, vol. 15, n°1, 2000, pp. 57–83.

S. Verger, *Les chemins multiples de la coopération transfrontalière franco-belge: analyse des dynamiques institutionnelles,* Doctoral thesis, Université de Grenoble, Grenoble, 2011.

O. Vergne, *La Collectivité européenne d'Alsace. Différenciation territoriale, identité régionale et enjeux transfrontaliers*, doctoral thesis, Université Paris 8, Saint-Denis, 2021.

B. Wassenberg, "La frontière, objet d'intégration ou cicatrice de l'Histoire? L'étude du cas de l'espace du Rhin supérieur", *RECERC*, n°1, online review, article 2, 2007.

B. Wassenberg, *Vers une eurorégion? La coopération transfrontalière franco-germano-suisse dans l'espace du Rhin supérieur de 1975 à 2000*, Peter Lang,

Berne, 2007 ; J. Beck & B. Wassenberg, *Vivre et penser la coopération transfrontalière: les régions frontalières françaises*, Franz Steiner, Stuttgart, 2010.

B. Wassenberg, "La coopération transfrontalière dans l'espace du Rhin supérieur: un modèle de réconciliation franco-allemande?", *In* N. Paun & S. Schirmann (eds.), *Borders, Identities, Communities. The Road to Reconciliation in Central and Eastern Europe,* Nomos, Baden-Baden, 2016, pp. 295–317.

B. Wassenberg, "Benelux States: Belgium, Luxembourg and the Netherlands", *In* B. Wassenberg & B. Reitel (eds.), *Critical Dictionary of Borders, Cross-border Cooperation and European Integration*, Peter Lang, Brussels, 2020, pp. 101–105.

C. Xhardez (ed.), *50 Ans de fédéralisation de l'État belge : institutions, acteurs, politiques publiques et particularités du fédéralisme belge,* Académia, Louvain-la-Neuve, 2020.

R. Ysebaert et al., *A Socio-economic Typology on Cross-border Regions at European Scale (2000–2012), Key Findings*, UMS RIATE, Paris, 2016.

Part 3

Cross-Border Cooperation on the Isle of Ireland: A (Still) Invisible Border?

Resisting Re-bordering. Cross-Border Mobilities in an Irish border Region in Anticipation of Brexit

Milena Komarova & Katy Hayward

Introduction

As with other borders born out of the retreat of empire, the Irish border has historically been associated with structural violence and social division.[1] Habits of political contestation, economic stagnation and social estrangement marked the border region for much of the 20th century. However, common United Kingdom (UK) and Irish European Union (EU) membership gradually helped form "soft spaces" for integration across the region, facilitated by the common EU corpus of law, spirit of cooperation and financial support. The metamorphosis of the border is illustrated by the mobility across it. Traversed by over 177,000 trucks per month[2] and by over 30,000 people per day[3] for services, shopping, work, leisure, tourism, or family, everyday mobilities demonstrate

1 The meandering UK-Ireland land border runs for just under 500 km across the northern part of the island of Ireland. It divides the independent state of the Republic of Ireland from Northern Ireland, a region of the United Kingdom. The island was first divided into two separate jurisdictions by The Government of Ireland Act (1920) which envisaged partition as an internal United Kingdom matter, and as a temporary answer to the thorny question of contested sovereignty across the island. However, in 1922, the unionist-dominated government of Northern Ireland exercised its right not to be included in the Irish Free State, and the border officially became an international frontier. Over time, it steadily took a more concrete form as a customs and security barrier as well as a political divide.

2 S. Leheny, "Making Northern Ireland a bridge between the UK and the EU", *UK in a changing Europe* [Website], 2018. Retrieved from: <https://ukandeu.ac.uk/making-northern-ireland-a-bridge-between-the-uk-and-the-eu/>. Accessed 27 April 2023.

3 UK Parliament, Northern Ireland Affairs Committee, "The land border between Northern Ireland and Ireland", 2018. Retrieved from: <https://publications.parliament.uk/pa/cm201719/cmselect/cmniaf/329/32902.htm>. Accessed 27 April 2023.

a "waning" of the border as a barrier over time. When Prime Minister May announced[4] that the UK would interpret "Brexit" to mean exit from both the EU Single Market and its customs union, it was inevitable that there would be more friction for movement across the UK-EU borders. This would be felt most profoundly at the Irish border as the part of the UK-EU border that was not only the most deeply integrated, but also the most socially and politically sensitive.[5]

Over a period of four years, between 2017 and 2021, we conducted a series of research studies to capture the impact of Brexit on those living in the central border region of Ireland/Northern Ireland as it unfolded.[6] By adopting a theoretical perspective that draws on the intersections between mobility and border studies, we here analyse some of those experiences. People's accounts of their mundane movements revealed the extent to which everyday life for many in the region is genuinely cross-border. They also show how significant cross-border mobility is for them in symbolic terms as well as practical and personal ones. It is predominantly through the lens of its anticipated effects on cross-border mobility

---

4 T. May, "The government's negotiating objectives for exiting the EU: PM speech", 2017. Retrieved from: <https://www.gov.uk/government/speeches/the-governments-negotiating-objectives-for-exiting-the-eu-pm-speech>. Accessed 27 April 2023.

5 K. Hayward, "The origins of the Irish border", UK in a changing Europe, 2017. Retrieved from: <https://ukandeu.ac.uk/explainers/the-origins-of-the-irish-border/>. Accessed 27 April 2023.

6 K. Hayward, "Bordering on Brexit: Views from local communities in the Central Border Region of Ireland/Northern Ireland", Queen' University Belfast, Report for the Irish Central Border Area Network, 2017. Retrieved from: <https://pureadmin.qub.ac.uk/ws/portalfiles/portal/137861639/Bordering_On_Brexit_full_report_Nov17.pdf >. Accessed 27 April 2023.

K. Hayward, "Brexit at the border: Voices of local communities in the Central Border Region of Ireland/Northern Ireland", Irish Central Border Area Network (ICBAN), 2018. Retrieved from: <https://www.qub.ac.uk/brexit/Brexitfilestore/Filetoupload,824444,en.pdf>. Accessed 27 April 2023.

K. Hayward & M. Komarova, "The border into Brexit. Perspectives from local communities in the Central Border Region of Ireland / Northern Ireland", Irish Central Border Area Network (ICBAN), 2019. Retrieved from: <https://ukandeu.ac.uk/wp-content/uploads/2019/12/The-Border-into-Brexit-perspectives-from-local-communities-in-the-central-border-region-of-Ireland-and-Northern-Ireland.pdf>. Accessed 27 April 2023; K. Hayward & M. Komarova, "The border after Brexit: Experiences of local communities in the Central Border Region of Ireland / Northern Ireland", Irish Central Border Area Network, 2021. Retrieved from: <https://qpol.qub.ac.uk/wp-content/uploads/2021/10/The-Border-After-Brexit-published.pdf>. Accessed 27 April 2023.

that Brexit – and with it, the possibility of a "hard border" – was judged by those in the border region.

In what follows we first discuss the conceptual framework that helps us understand the relationship between accounts of everyday movements across the border and the multiple layers and scales of de- and re-bordering that have occurred in the 21st century. We then turn to a brief historical summary of the mobility regimes variously characterising the Irish border before outlining the methodology of our research. Following this discussion, we present our analysis of research findings. Finally, we consider the conceptual and practical implications of this analysis and draw conclusions about the political effects of mundane cross-border mobilities on the island of Ireland and beyond.

Borders and Mobilities

> "Both empirically and conceptually, research on borders and bordering raises questions that are intricately entwined with movement and mobilities. Similarly, it is difficult to conceive of mobilities without confronting the ways in which [they] are constrained and regulated by borders and bordering practices".[7]

It is in the very character and purpose of a border to affect mobility. Although historically seen as the territorial container line of a state and its sovereignty, the reality of any border is best captured in terms of its consequences for those who wish to cross it.[8] It is in this sense that Rumford has argued, "to theorize mobilities and networks is at the same time to theorize borders".[9] Importantly, ease of traversing a border, especially in

[7] T. Richardson, "Borders and mobilities: Introduction to the special issue", *Mobilities*, vol. 8, n°1, 2013, p. 1.

[8] S. Pickering & L. Weber, "Borders, mobility and technologies of control", *In* S. Pickering and L. Weber (eds.), *Borders, Mobility and Technologies of Control*, Springer, New York, 2006, pp. 1–19; A. Mountz, K. Coddington & R. T. Catania, "Conceptualizing detention: Mobility, containment, bordering, and exclusion", *Progress in Human Geography*, vol. 37, n°4, 2012, pp. 522–541.

[9] C. Rumford, "Theorizing borders", in *European Journal of Social Theory*, vol. 9, n°2, 2006, p. 156.

the context of globalisation[10] is not to imply that borders are diminishing in significance. The securitisation of borders and border practices is a global phenomenon that has largely come in response to growing mobility in the form of migration "crises".[11] Such acts of securitisation conceive of borders as lines of division at the same time as treating them as points of connection which can be obstructed or narrowed. Of course, borders can be hardened for some types of movement and categories of persons at the same time as being opened for others.[12] As such, borders are manifested in the processes of identifying, categorizing, and managing the multiple mobilities across them.[13] Such processes increasingly occur away from official state borderlines,[14] and are also enacted "from below" by a multitude of non-state actors and individuals.[15]

---

10 V. Kolossov & J. Scott, "Selected conceptual issues in border studies", *EUROBORDERSCAPES Working Paper*, n° 4, 2013. Retrieved from: <http://www.euborderscapes.eu/index.php?id=working_papers>. Accessed 27 April 2023.

11 S. Lèonard, "EU border security and migration into the European Union: FRONTEX and securitization through practices", *European Security*, vol. 19, n°2, 2010, pp. 231–254; R. Jones, *Violent borders: Refugees and the right to move*, Verso, London, 2016; R. Jones & C. Johnson, "Border militarization and the re-articulation of sovereignty", *Transactions of the Institute of British Geographers*, vol. 41, n°2, 2016, pp. 187–200; C. Minca & A. Rijke, "Walls! Walls! Walls!", *Society and Space*, vol. 40, n°2, 2017. Retrieved from: <https://www.societyandspace.org/articles/walls-walls-walls>. Accessed 27 April 2023.

12 M. Sheller & J. Urry, "The new mobilities paradigm", *Environment and Planning A: Economy and Space*, vol. 38, n°2, 2006, pp. 207–226.

P. Adey, *Mobility*, Routledge, London and New York, 2010; T. Cresswell, "Towards a politics of mobility", *Environment and Planning D: Society and Space*, vol. 28, n°1, 2010, pp. 17–31; K. Buscher, J. Urry & K. Witchger, "Introduction: Mobile methods", *In* K. Buscher, J. Urry & K. Witchger, K. (eds.) *Mobile methods*, Routledge, London and New York, 2011, pp. 1–20.

13 N. Vaughan-Williams, *Vernacular border security. Citizens' narratives of Europe's 'migration crisis'*, Oxford University Press, Oxford, 2021.

14 C. Johnson, R. Jones, A. Paasi, L. Amoore, A. Mountz, M. Salter & C. Rumford, "Interventions on rethinking 'the border' in border studies", *Political Geography*, vol. 30, n°2, 2011, pp. 61–69; D. Andersen & M. Sandberg, "Introduction", *In* D. Andersen, M. Klatt, and M. Sandberg (eds.), *The border multiple*, Ashgate, Farnham, 2012, pp. 1–22; M. Lemberg-Pedersen, "Forcing flows of migrants: European externalization and border-induced displacement", *In* D. Andersen, M. Klatt, and M. Sandberg (eds.), *The border multiple*, Ashgate, Farnham, 2012, chapter 2.

15 M. Doevenspeck, "Constructing the border from below: Narratives from the Congolese–Rwandan state boundary", *Political Geography*, vol. 30, n°3, 2011, pp. 129–142; C. Rumford, "Introduction: Citizens and borderwork in Europe", *Space and Polity*, vol. 12, n°1, 2008, pp. 1–12.

Importantly, Cooper and Rumford stress an understanding of the term "boundary" as a social act that "helps to highlight the ways in which borders ... need not be restricted to territorial peripheries".[16] Rather, they constitute "an envelope of action within which people from different 'sides' act in relation to one another – but the sides in question need not be local".[17] As social and political boundaries, state borders both represent and enact the inextricable link between national territory and social or political identity that we are all socialised in, but they also reflect the "complexities of difference, hybridity and commonality that co-exist across borders".[18] Borders may invoke not only official and national but also a myriad of other social and personal identifications, memories and understandings of place.[19] Indeed, the case of the Irish border reminds us of the importance of the territorial border as a "place", the physical crossing of which entwines various scales and layers of meaning and connection. The symbolism and benefits of this actualised mobility are encapsulated well in the narratives and experiences of people in the border region reflecting the unfurling of the Brexit process.

Bordering and Everyday Moves

To understand borders as everyday practices requires us to appreciate everyday mobilities which are central to such practices and, more broadly, to the reproduction of social life at large.[20] The importance of physical

K. Cassidy, G. Wemyss & N. Yuval-Davis, "Debordering and everyday (re)bordering in and of Dover: Post-borderland borderscapes", *Political Geography*, vol. 66, 2018, pp. 171–179.

16 A. Cooper & C. Rumford, "Monumentalising the border: Bordering through connectivity", *Mobilities*, vol. 8, n°1, 2013, p. 112.

17 *Ibid.*

18 K. Cassidy et al., "Debordering and everyday (re)bordering in and of Dover", *op.cit.*, p. 173.

19 M. Degen, C. DeSilvey & G. Rose, "Experiencing visualities in designed urban environments: Learning from Milton Keynes", *Environment and Planning A: Economy and Spaces*, vol. 40, n°8, 2008, pp. 1901–1920; S. Green, "Performing border in the Aegean: On relocating political, economic and social relations", *Journal of Cultural Economy*, vol. 3, n°2, 2010, pp. 261–278.

20 J. Binnie, T. Edensor, J. Holloway, S. Millington & C. Young, "Mundane mobilities, banal travels", *Social & Cultural Geography*, vol. 8, n°2, 2007, pp. 165–174; K. Manderscheid, "Criticising the solitary mobile subject: Researching relational mobilities and reflecting on mobile methods", *Mobilities*, vol. 9, n°2, 2014,

mobility for social experience is underlined by Binnie et al who stress the profound extent to which "[a]ll of us are caught up in banal or mundane mobilities", the "commonplace and regular occurrence" of which is deeply "enmeshed with the familiar worlds we inhabit".[21] Yet, those working at the intersection of mobility and border studies have often, and for good reason, been drawn to the more extraordinary or spectacular forms of border crossing, either by focusing on forced migration,[22] or on the cosmopolitan movements of internationally-mobile, privileged populations.[23] Similarly, scholarly attention is often directed at "extensive movements across the planet rather than the usually more modest journeys of the everyday through familiar and 'homely' space".[24] But what of cases where such homely space straddles a state borderline? With some notable exceptions[25] borders and mobility researchers have not devoted too much attention to how everyday lives are lived across state borders.

---

pp. 188–219; K. Doughty & L. Murray, "Discourses of mobility: Institutions, everyday lives and embodiment", *Mobilities*, vol. 11, n°2, 2016, pp. 303–322.

21 J. Binnie, T. Edensor, J. Holloway, S. Millington & C. Young, "Mundane mobilities", *op.cit.*, p. 165.

22 J. Darling, "Becoming bare life: Asylum, hospitality and the politics of encampment", *Environment and Planning D: Society and Space*, vol. 27, n°4, 2009, pp. 649–665; D. Bigo, "The (in)securitization practices of the three universes of EU border control: Military/navy – border guards/police – database analysts", *Security Dialogue*, vol. 45, n°3, 2014, pp. 209–225; V. Squire, "Desert 'trash': Posthumanism, border struggles, and humanitarian politics", *Political Geography*, vol. 39, 2014, pp. 11–21; R. J. Topinka, "'Wandering the settled tribes': Biopolitics, citizenship and the racialized migrant", *Citizenship Studies*, vol. 20, n°3/4, 2016, pp. 444–456; V. C. Tran, "The Borders around us: Forced migration and the politics of border control", *City & Community*, vol. 19, n°2, 2020, pp. 323–329.

23 S. Croucher, "Migrants of privilege: The political transnationalism of Americans in Mexico", *Identities: Global Studies in Culture and Power*, vol. 19, n°4, 2009, pp. 463–491; S.Y. Koh & B. Wissinik, "Enabling, structuring and creating elite transnational lifestyles: Intermediaries of the super-rich and the elite mobilities industry", *Journal of Ethnic and Migration Studies*, vol. 44, n°4, 2018, pp. 592–609; E. Spence, "Unraveling the politics of super-rich mobility: A study of crew and guest on board luxury yachts", *Mobilities*, vol.9, n°3, 2014, pp. 401–413.

24 J. Binnie, T. Edensor, J. Holloway, S. Millington & C. Young, "Mundane mobilities", *op.cit.*, p. 166.

25 C. Wille & U. Roos, "Cross-border everyday lives on the Luxembourg border? An empirical approach: the example of cross-border commuters and residential migrants", *In* C. Wille & B. Nienaber (eds.), *Border Experiences in Europe*, Nomos, Baden-Baden, 2018, pp. 101–126; S. Ghosh, "Cross-border activities in everyday life: The Bengal borderland", *Contemporary South Asia*, vol. 19, n°1, 2011, pp. 49–60.

One upshot is that the social and political effects and potential of ordinary mundane mobilities across state borders may sometimes go unnoticed. Such effects are rooted in two characteristics of mobilities at large, as well as in the specificity of different border contexts. First, as people do not move equally smoothly through space, even banal mundane journeys raise and highlight questions over mobility (in)justices.[26] Who gets to cross borders and how, becomes a question of enacting citizenship and of the associated rights and (in)equalities. In this sense, given its history as a customs frontier and a violently contested security barrier, the very mundaneness and unremarkability of the present-day practices of crossing the Irish border may be seen as an extraordinary achievement, addressing an injustice over those border residents whose mobilities had, for generations, been profoundly restricted by the mere presence of the border.

Second, human mobility is not a technical process but inevitably carries fundamental meaning and value to us; places are made meaningful through it, and "who we are" is experienced through the connection to place that movements produce.[27] Mobility, in other words, is a relational practice, shaped by and itself shaping places, identities and social relationships. Manderscheid sees this relationality as captured by the Foucauldian notion of *dispositif*[28] (1977) and by Bourdieu's understanding of socially structured practices (2000). Patterns of movement "emerge through social and spatial relations and cannot be sufficiently understood as the function of rational decisions made by autonomous subjects".[29] Again, this is particularly significant in border regions, not least due to the historical violence usually inherent in their creation. In our case study the symbolic importance and political contestation of the Irish border remain "live" and open because of the political salience of the "constitutional status of Northern Ireland" (and thus the future of the

26 M. Sheller, *Mobility Justice. The Politics of Movement in an Age of Extremes*, Verso, London, 2018.

27 H. Easthope, "Fixed identities in a mobile world? The relationship between mobility, place, and identity", *Identities: Global Studies in Culture and Power*, vol.19, n°1, 2009, pp. 61–82; O. Jensen, "Negotiation in motion: Unpacking a geography of mobility", *Space and Culture*, vol.13, n°4, 2010, pp. 389–402.

28 A concept that draws attention to the multiple interrelations between discourses, material landscapes, practices and subjectivities that shape, and are shaped by, movement and mobilities.

29 K. Manderscheid, "Criticising the solitary mobile subject", *op.cit.*, p. 193.

border). As revealed in our data, local people construct and enact their relationship to the border, and to those on either side of it, politically and in terms of home and belonging, through their movements across it. Here, the history, the materiality and rurality of the border landscape, as well as patterns of inter-communal relationships, and attitudes to (state) authority are all constituted via, and constitute, mobilities.

"Telling Stories from the Ground": The Vernacular Approach to Borders

Methodological shifts in the study of borders have reflected a recognition that as social categories borders are not stable but "socially situated, culturally mediated and discursively constructed".[30] Cassidy et al for instance, examine the "post-borderland borderscape" of Dover in South East England, identifying narratives that elucidate "local consciousness(es) or border imaginaries … not reducible to national-level contextual discourses in the media and politics".[31] The authors thus see the border as produced "in the situated imaginations"[32] of the borderlanders, the meanings of which may either align with or challenge other emergent processes of re- or de-bordering. Similarly, Newman and Paasi argue that it is the local sphere of everyday life experience "where the meanings of (state) boundaries are ultimately reproduced and contested".[33] More recently, Vaughan-Williams[34] seeks to apply a "vernacular" approach to border security to grasp better the relationship between the macro-level nation-state bordering and border anxieties at the micro-level of the citizen. What is more, everyday processes of de- or re-bordering are patchy, multi-scalar and multi-faceted, with their vernacular dimensions tied strongly to different social positionings, identifications and histories. As

---

30 N. Bubandt, "Vernacular security: The politics of feeling safe in local, national and global worlds", *Security Dialogue*, vol. 36, n°6, 2005, pp. 275–296, quoted by N. Vaughan-Williams, *Vernacular border security*, *op.cit.*, p. 8.

31 K. Cassidy et al., "Debordering and everyday (re)bordering in and of Dover", *op.cit.*, p. 172.

32 M. Stoetzler & N. Yuval-Davis, "Standpoint theory, situated knowledge and the situated imagination", *Feminist Theory*, vol. 3, n° 3, 2002, pp. 315–333.

33 D. Newman & A. Paasi, "Fences and neighbours in the postmodern world", *Progress in Human Geography*, vol. 22, 1998, p. 197.

34 N. Vaughan-Williams, *Vernacular border security*, *op.cit.*, p. 3.

Vaughan-Williams suggests, a vernacular border studies approach must focus on "how diverse publics construct and narrate personal experiences" of bordering and "how in so doing they express and reproduce notions of self-identity, community and belonging in contradistinction with notional 'others' ".[35]

Applied to the context of the Irish border, this proposition echoes Donnan's writings[36] which depict how the waning of the border as a result of the peace process induced counter-narratives among border Protestants. Whereas for many the post-1998 Agreement, "Irish cultural borderscape provide[s] opportunities for escaping the territorial cage of conflict in Northern Ireland, thus underpinning and advancing conflict transformation",[37] the narratives of border Protestants by contrast are ones of heightened border anxiety. They remind us that "though borders and frontiers may be zones of flux and blurring, they are also places where identities can harden as well as change".[38]

The Irish Border's Regimes of Im/mobility

From the very beginning of partition, everyday life along the border was shaped profoundly and above all by the ways in which border mobility was now officially regulated. The border had not been negotiated in any specific geographical detail and a number of towns, villages and their hinterlands were divided by it.[39] Such arbitrariness has often characterised the creation of many state borders but, unlike other cases,[40] in Ireland this was not accompanied by state attempts to homogenise

35 *Ibid.*, p. 9.

36 H. Donnan, "Cold war along the emerald curtain: Rural boundaries in a contested border zone", *Social Anthropology*, vol. 18, n°3, 2010, pp. 253–266.

37 C. McCall, *Border Ireland: From partition to Brexit*, Routledge, London, New York, 2021, p. 51.

38 H. Donnan, "Material identities: Fixing ethnicity in the Irish borderlands", *Identities: Global Studies in Culture and Power*, vol. 12, n°1, 2005, pp. 69–105.

39 P. Leary, *Unapproved routes: Histories of the Irish border, 1922–72*, Oxford University Press, Oxford, 2016;

P. Leary, "A house divided: The Murrays of the border and the rise and decline of a small Irish house", *History Workshop Journal*, vol. 86, 2018, pp. 269–290.

40 Such as the Balkans in the earlier part of the 20th century. See for instance T. Dragostinova, *Between two motherlands. Nationality and emigration among the Greeks of Bulgaria, 1900–1949*, Cornell University Press, Ithaca, 2011.

the populations on either side of it, whether by forceful expulsions or by population exchanges. As such, the natural social, economic, and cultural connections, familial relations, and sense of home for many border residents remained decidedly cross-border.

A contradictory regime of controlling and managing human and non-human movement was established. While crossing the border never became subject to immigration control,[41] a strong classificatory regime over taxation, currency, and the flow of goods was swiftly put in place.[42] When in 1923 the Free State imposed a customs barrier at the border, the vast majority of the hundreds of cross-border roads were rendered "unapproved" for carrying dutiable goods across. But whether smuggling or not, those not using "unapproved" roads (even only for simple convenience) were immediately subject to some suspicion, "making the very act of crossing the border potentially disruptive and rebellious".[43] As such, border residents' response to cross-border mobility arrangements – albeit often based on pragmatic considerations entwined with the necessities of social and familial life – profoundly affected their relationship to the states either side of the border. These were in turn shaped by how cross-border movement (especially of goods) was governed and controlled.[44]

The rise of violent conflict marked the start of "the troubles" at the end of the 1960s, resulting in the swift militarisation and securitisation of the whole region. This included the erection of army checkpoints, military bases and watchtowers along the border, and the physical blocking of

---

41 As the centuries-long history of free movement of people on the island of Ireland and between Britain and Ireland continued under what became known as the Common Travel Area (CTA): A gradually established regime of a series of semiformal arrangements between the British and Irish governments, with the cumulative effect that British and Irish citizens were to be *de facto* treated as "nationals" in each other's territories.

42 L. Cañás Bottos, "Assemblages of sovereignty and anti-sovereign effects on the Irish border", *Focaal: Journal of Global and Historical Anthropology*, vol. 71, n°1, 2015, pp. 86–99.

43 K. Hayward, P. Leary & M. Komarova, "The Irish border as sign and source of British-Irish tensions", *In* Ribas-Mateos, N. & Dunn, T. (eds.), *Handbook on Human Security, Borders and Migration*, Edward Elgar, Cheltenham, 2021, p. 361.

44 C. Nash, L. Dennis & B. Graham, "Putting the border in place: Customs regulation in the making of the Irish border, 1921–1945", *Journal of Historical Geography*, vol. 36, 2010, pp. 421–443.

border crossings by the British Army.[45] Continuing violence, the presence of heavily militarised infrastructure and the inconvenience of navigating security checks translated for many into effective immobility. Segregation and alienation increased between unionist and nationalist communities. A multitude of familial, social, business and economic cross-border connections were strangled, although not severed. Thus, border (im)mobilities continued to be central to political and social relationships in the border region – between and within communities, and between citizens and the state. In terms of border symbolism, Northern Ireland unionists considered the border as a means of preserving their culture and protecting them against republican violence.[46] Irish nationalists, at the same time, detested it as a manifestation of British colonialism and a reminder of their role in the "northern state" as "second class citizens".[47]

The "Soft Spaces" of European Integration and Peace

By opening both states to the processes and practices of European integration, the UK and Ireland's joint membership of the then European Economic Community (EEC) from 1973 gradually countered the local structures, practices, and experiences of border fixity. The economic and cross-border cooperation it facilitated would become integral aspects of both peace and of regional devolution. The introduction of EEC regulations on customs declarations in 1987 already eased the movement of goods across the border, while the creation of the Single Market (operational from January 1993) steadily removed many obstacles to the mobility of goods, services, capital and people. Cross-border cooperation between the north and south of the Irish border was given legitimacy and financial support through a series of EU regional development and peace programmes. At intergovernmental level, this had the effect of practically and symbolically recasting the border from a barrier to movement to an enabler of cooperation and partnership.[48]

---

45 P. Mulroe, *Bombs, bullets and the border: Policing Ireland's frontier: Irish security policy, 1969–1978*, Irish Academic Press, Newbridge, 2017.

46 H. Patterson, *Ireland's Violent Frontier: The border and Anglo-Irish relations during the Troubles*, Palgrave MacMillan, New York, 2013.

47 E. Moloney, *A Secret History of the IRA*, Penguin, London, 2007.

48 J. Coakley & L. O'Dowd, "The transformation of the Irish border", *Political Geography*, vol. 26, 2007, pp. 877–885; K. Hayward & M. Murphy, "The (soft)

The 1998 Good Friday (Belfast) Agreement further extended and reinforced this transformation through the institutions for cross-border governance it created. Importantly, our research in the border region shows that locals credit the 1998 Agreement, and not EU membership, for the openness of the border.[49] Specifically, the Agreement created and formalised the mechanisms for north-south and east-west cooperation. North-south cooperation has included cross-border implementation bodies, such as those for the promotion of food safety and all-island trade. Together, these multi-level arrangements constitute some of the most advanced forms of cross-border governance and cooperation in Europe. Significantly, their practical operation through over 140 areas of cross-border cooperation on the ground was largely enabled by EU law and policies.[50]

Yet the 1998 Agreement also left the symbolic significance of the Irish border intact. By affirming that no change to the constitutional status of Northern Ireland could occur without a majority referendum vote in the region (a so-called "border poll"), the Agreement ostensibly anchored the border's legitimacy in the exercise of democracy itself. Our research confirmed the extent to which the symbolic importance of the border to both nationalists and unionists has remained strong.

Questions over the symbolic and practical importance of the Irish border were given a new significance by the United Kingdom's withdrawal from the European Union. The EU-UK withdrawal negotiations centred upon the matter of how to avoid a hard border on the island of Ireland after Brexit. The Protocol on Ireland/Northern Ireland in the Withdrawal Agreement addressed this in a peculiar fashion. Northern Ireland was to *de facto* remain in the single market for goods and under the Union Customs Code, thus necessitating border checks and controls on sea and air (rather than land) border entry points, including from the rest of the United Kingdom. These arrangements were unknown until the very last

---

power of commitment: The EU and conflict resolution in Northern Ireland", *Ethnopolitics*, vol.11, n°4, 2012, pp. 439–452; K. Hayward, "The future of the Irish border", *Renewal: A Journal of Social Democracy*, vol. 26, n°4, 2018, pp. 12–22.

49 K. Hayward, "Bordering on Brexit", *op.cit.*, p. 46.

50 L. C. Whitten, "North-south cooperation and the Protocol on Ireland/Northern Ireland", Queen's University Belfast, 2022. Retrieved from: <https://drive.google.com/file/d/1bWFFejNg8yC-RhQ4lmxdbm96-XGKs8W6/view>. Accessed 27 April 2023.

round of our research in 2021, and our findings revealed the extent to which the very prospect of Brexit had revived tensions and anxieties in the border region about checks and controls on cross-border movement.

Researching Brexit in the Irish Central Border Region

Our research was conducted in four rounds between 2017 and 2021, each consisting of an online survey of border residents (aged 18+) and a series of focus groups composed of volunteering survey respondents. The latter two years of research also included individual in-depth interviews with representatives from key sectors from the region.[51] The aim was to give people living and working in the region's eight member council areas[52] the chance to share their expectations and experiences of Brexit. It is important to note that survey respondents were self-selecting. This meant that the sample we ended up with in each year was not representative of the general population of the region. Instead, there was over-representation of those with exclusively Irish citizenship and under-representation of those who voted Leave in the 2016 Referendum.[53] Those most motivated to respond to a survey on Brexit were predominantly Remain voters.

Participant accounts of cross-border mobilities, even when not enquired about, appeared central to how the border region was spoken of as a place, and to how the actual or expected effects of Brexit were interpreted. In our analysis, we understand these mobility accounts

[51] Between 2018 and 2021 the research was funded by the Department for Foreign Affairs' (ROI) Reconciliation Fund. Additional financial support was provided in 2019 and 2021 by the UK's Economic and Social Research Council.

[52] The eight member council areas of the Central Border Region include Armagh City, Banbridge and Craigavon; Fermanagh and Omagh; Mid Ulster and the counties of Cavan, Donegal, Leitrim, Monaghan and Sligo. The region had a population of approximately 850,000 in 2011. This is a predominantly rural area, characterised by a dispersed population and distance from major urban centres. Approximately one third of the population live in settlements over 1,500 population. Two-thirds live in smaller settlements and open countryside. The region accounts for 20 % of the land area of the island of Ireland, with a high quality landscapes of coastline, lakes, inland waterways and hills. It further comprises some 12 % of the population of the island.

[53] Since the research straddled the border, we could not rely on the traditional unionist/nationalist distinction as a signifier of belonging to a political community. Instead, we enquired about citizenship and vote cast during the 2016 Referendum as the relevant characteristics to views on the effects of Brexit at the border.

as "spatial stories",[54] constructed within and with reference to specific places and through associated mobile spatial practices. As O'Dowd and Komarova explain,[55] in Simonsen's terms they are to be understood as "performed telling" rather than "decontextualised texts with purely cognitive import".[56] Conversely, once articulated, such spatial stories, constitute places and subjectivities. We find that accounts and performances of mobility are essential to the processes by which the top-down governance regimes of the Irish border are interpreted locally.

"Popular" Geographies: The Place of the Border vs the Border as a Place

While as the UK's only land border with an EU-member state, the Irish border was at first conspicuously absent from Brexit campaigns and debates, by the spring of 2017 the unique circumstances on the island of Ireland were acknowledged through a series of high-level documents. The EU Council's negotiation guidelines stressed the need for "flexible and imaginative solutions" to avoid a hard border "while respecting the integrity of the Union legal order" [emphasis added].[57] The UK government accepted such a commitment in its Position Paper on Ireland/Northern Ireland[58] and proposed addressing the task through four broad legal and policy areas. These included upholding the Belfast/Good Friday Agreement in all its parts, maintaining the Common Travel Area and associated rights, avoiding a hard border for the movement of goods, and the preservation of North-South and East-West cooperation. Such high-level discourses treated borders as sites of administrative policy regimes,

---

54 L. O'Dowd & M. Komarova, "Three narratives in search of a city", *City: Analysis of Urban Trends, Culture, Theory, Policy, Action*, vol.17, n°4, 2013, pp. 526–546.

55 *Ibid.*, p. 528.

56 S. Simonsen, "Spatiality, temporality and the construction of the city", *In* J. O. Baerenholdt & K. Simonsen (eds.), *Space odysseys: Spatiality and social relations in the 21st century*, Ashgate, Farnham, 2004, p. 53.

57 European Council, "European Council (Art. 50) guidelines following the United Kingdom's notification under Article 50 TEU", 29 April 2017. Retrieved from: <https://www.consilium.europa.eu/en/press/press-releases/2017/04/29/euco-brexit-guidelines/>. Accessed 27, April 2023.

58 HM Government, "Northern Ireland and Ireland Position Paper", 2017. Retrieved from: <https://www.gov.uk/government/publications/northern-ireland-and-ireland-a-position-paper>. Accessed 27 April 2023.

and as territorial demarcation lines of state sovereignty. By contrast, borderlanders' accounts in our research tended to focus on the lived experience of everyday life in the border region. As one survey respondent put it:

> "There's always more layers to the politics at the border. … If you're geographically far away from the border, it's probably not something you're going to think about but if the border goes through your farm or your town, then it's part of your everyday reality".[59]

Many pointed out that throughout the years since the end of violent conflict, the border had, for all intents and purposes, ceased to exist, or had become completely irrelevant in their daily lives: "Yes, the decision is to shop in Armagh or Dundalk, but the border doesn't come into it. … Not sure that people are even aware they are crossing the border for shopping or whatever – I don't think about it".[60]

It was also absent from popular geographic imagination: "I don't see a border. So, as far as I can see, at the moment, there isn't a border. I live in Leitrim [Republic of Ireland]. I think Leitrim is just the same as Fermanagh [Northern Ireland]".[61] Indeed, one interviewee neatly summarized the paradoxical juxtaposition between the present-day invisibility of the border in daily life and its continued political significance:

> "I think [you] have to understand that of course the border is a line. It's a line that people have put on a map. That's all it is for many people who live along the border. It's a place where for you maybe is a field or a road but for many border people it doesn't exist. It doesn't exist in terms of where they farm their animals, where they cut their grass, where they bring their milk from… So, for many people their natural affinity, their natural environment is in another jurisdiction".[62]

Our participants' narratives thus demonstrated the popular or "imaginative geographies"[63] of the border, denoting how border residents themselves relate to it as a place. Such words evoked a sense of the border

---

59 Survey respondent, 2019.

60 Focus group 1, 2017.

61 Focus group 3, 2019.

62 Interview 5, 2019.

63 D. Gregory, "Imaginative geographies", *Progress in Human Geography*, vol. 19, n°4, 1995, pp. 447–485.

region as a familiar, lived, homely space, in sharp contrast to the manner in which the UK borders had been represented in Brexit discourses. Here, the particular entwining of physical, political and cultural geography was seen as commanding a certain way of life that constantly criss-crossed the territorial border. What is more, such accounts highlight the artificiality of the sovereign border "in the local worldview".[64]

Crucially, cross-border mobility played a central role in these popular homely geographies. On the one hand, it was almost universally prized by those holding British, Irish, Leave or Remain identities alike, for servicing the convenience of daily life. On the other hand, cross-border mobility was also rhetorically used to explain, justify, construct or contest ideas of home and belonging, identity and citizenship, as well as of state territoriality. As such, the mobility accounts we elicited enacted a "vernacular border" that resonates with the wide-ranging scholarship reflecting upon home as an "intensely political site affected by, and affecting, broader power relations".[65]

Cross-border Ties, Cross-Border Lives: Between Utilitarianism and Belonging

Crossing the border was first and foremost spoken of as an act of necessity and practicality, not so much imbued with political or ideological significance, as rooted in one's personal and social life. As one participant put it: "Live here for a week and see how many times you cross the border! ... I don't think they [British politicians] know that people cross it daily... You wouldn't even know you were crossing the border".[66] A survey respondent concurred: "I live within a mile of the border. I'm constantly in and out of the South for various reasons. I live in a village that the border literally splits in two"![67] Similarly, a focus group participant the following year suggested:

---

64 S. Ghosh, "Cross-border activities in everyday life", *op.cit.*, p. 54.

65 See A. Ní Mhurchú, "Intimately inhabiting borders: Walking in-between belonging and otherness through constructions of home", *Geopolitics*, vol. 26, n°2, 2021, p. 412. The author quotes A. Blunt & R. M. Dowling, *Home*, Routledge, London and New York, 2006.

66 Focus group 2, 2018.

67 Survey respondent, 2018.

> "I cross the border six times on the way to work and six times the way back. So, there's a complete misunderstanding of the day-to-day commute, the use of crossing the border, that the political class have. It's our day-to-day reality... [M]ost people in Pettigo have a long commute, by virtue of where hey live".[68]

Crossing the border, therefore, is for many simply the way of border life. Indeed, data from the 2017 survey revealed that over half of the survey respondents crossed the border at least once a week (some daily or even several times a day). This figure was more or less the same for respondents on either side of the border. Only a handful suggested they crossed the border "rarely", while none shared that they never cross the border at all. A participant in a focus group commented that road networks now meant that one can cross the border several times on a single journey: "For those of us that live near the border, it has to be crossed. They recently closed the roads near where I live for roadworks and I counted that, [following the alternative route] I crossed the border 12 times".[69]

The 2017 survey results also suggested that the main reason people crossed the border "frequently" or "very frequently", were social, with 52 % stating they did so to meet family and friends, many of whom lived on both sides of the border. One survey respondent wrote succinctly about the cross-border geographies of social and familial ties:

> "My family is split across both sides of the border. I have 50 first cousins with approximately half on each side of the border. I am a member of some literature groups and they regularly meet on both sides of the border. I use the libraries in both Derry and Letterkenny. My house is literally 250 metres from the border across the fields. *Everything I do crosses the border.* My bus to Letterkenny crosses the border to go to Derry first, then back across the border to complete its journey to Letterkenny. The border going back up, will affect every single aspect of my life in a negative manner".[70]

In the following year of 2018 the survey specifically enquired about accessing services across the border. Over half of the respondents (57.6 %) reported that they rely "very much" or "quite a lot" on such access, with the use of travel services mentioned most frequently. From

---

68 Focus group 4, 2019.
69 Focus group 2, 2017.
70 Survey respondent 2017.

this perspective, the prospect of Brexit was seen as potentially devastating because of the cross-border dimensions of numerous aspects of daily life:

> "We use the Bus Éireann service to Dublin and usually fly from either Dublin or Knock – even to England. Our local library is across the border. General grocery and clothes shopping. Even to visit Enniskillen (local county town) we use Bus Éireann – Ulster Bus has all but abandoned us. Cinema, meals out, walking for pleasure and health are all done across the border for reasons of proximity and convenience".[71]

> "I work in Derry [Northern Ireland]. My children are educated in Donegal [Republic of Ireland]. When they go to third level, institutions in NI would be my first choice. I access healthcare in ROI for convenience, but if I needed care at a specialist level I would opt to use services in NI due to the distance to access services in Galway or Dublin. … As family, we shop, travel, engage in leisure in NI".[72]

The above accounts reveal the breadth and depth of cross-border entanglements where mobility interweaves the servicing of mundane needs with the inhabiting of familial spaces. Indeed, mundane mobilities across the Irish border are deeply relational practices. At one level, they are framed by present-day transport infrastructure, and the availability of cross-border services. At another level, cross-border movements are driven by social networks and considerations of convenience. As a cumulative result, and whether purposely or not, in the everyday acts of transgressing it, for many residents the border literally and metaphorically ceases to exist. Resonating with other analyses of contested borders, our research thus suggests that practices of cross-border movement have the effect of "overcoming the border"[73] and of "re-imagining" and "re-placing" national space.[74]

Contesting a Hard Border Through Mobility

The previous section suggests that the ways in which the border is experienced in daily life has seriously diminished its practical significance.

---

71 Survey respondent, 2018.

72 Survey respondent, 2018.

73 S. Ghosh, "Cross-border activities in everyday life", *op.cit.*

74 H. Donnan, "Cold war along the emerald curtain", *op.cit.*

This diminution often presents itself as unreflexive – a natural consequence of the political and administrative regime that has allowed for unfettered access across it, and for the dismantling of its visible markers. However, research participants' discussion of the meaning of a "hard border" and their expectations of its effects after Brexit demonstrated that cross-border mobility is not simply an unreflexive, utilitarian practice but also serves as a rhetorical tool in affirming, contesting or reimagining state territoriality.

The anticipation of a hard border intensified in 2018 as UK-EU negotiations stumbled and fears of a "No Deal" exit rose. In the central Irish border region, most respondents understood the term "hard border" to mean "checkpoints", "a physical border", "customs checks" and "the end of freedom of movement". There was common expectation that it would lead to delays and that crossing the border would generally become more difficult. Still, a small portion among our Leave-voting survey respondents were unconcerned about the prospect of a hard border. This was mainly because they did not believe that there would be a "hard border", i.e. that it would be unnecessary, or that the threat of it was merely an EU negotiating tactic. Equally, some expressed the belief that a hard border would lead to "a more secure situation". Such responses reflect the continuing salience of what Donnan terms a "cold war along the emerald curtain"[75] – the contrasting symbolic significance of the border to different communities.

Yet, among Leave respondents we mainly saw a fear (shared with Remain voters) of restrictions to the freedom of movement. Thus, concern with the potential obstruction of the presently seamless connection across the border was the most often shared response. While the practice of cross-border movement itself was not always explicitly present in such responses, a broader understanding of mobility as "connection" and "seamlessness" – sometimes in flows of money, communications or, indeed, in local senses of belonging that straddle the border – was always implied as that which will be negated by a hard border and was, therefore, seen as unacceptable: "[T]hey don't even understand that we're dual currency; they don't understand the issues we have with our mobile phones [roaming charges]... You don't realise how much of a hindrance the border is on a day-to-day basis".[76]

75 *Ibid.*

76 Focus group 2, 2018.

Similarly, a number of participants (particularly those who self-identified as Irish or nationalist) elaborated on the positives of the border's present-day openness by stressing that the principle of "soft" territoriality that it embodies means the imaginative geography of the nation is now not seen to conflict with the official geography of the state: "[T]he … border being seamless … allowed nationalists north of the border to feel connected to Dublin, and allowed to concede that NI is British because they had that security of knowing Dublin was a seamless connection away".[77]

As another participant explained, Brexit challenges this because it "has now suddenly forced choice back: Who are you? Where do you belong? Where do you want to be? What is your future? Where is your allegiance"?[78] But not all accounts of the value of "soft" territoriality came from an Irish nationalist perspective, as demonstrated by a participant from a unionist background south of the border:

> "Why does the border matter to people like myself? – it's because I come from a unionist family in County Monaghan, which you know was left in the south during partition and on the whole, people like us were not happy to be left out of the six counties. [We also] felt a very strong economic impact of partition…. I felt, as someone who is a British citizen living in the Republic of Ireland, that this referendum was a terrible mistake".[79]

For the majority of respondents thus no hardening of the border was acceptable because of the manner in which it threatened to sever social relations across a variety of scales – from the immediacy of the border region as a homely space, to that of the state and jurisdiction on its other side. Additionally, many claimed that the use of technology to manage border controls would change the current experience of crossing the border and could thus undermine the 1998 Agreement, creating more tension. This idea was built upon in the following extract where peaceful, good cross-community relationships were constructed as directly underpinned by the removal of the immediacy, visibility and materiality of the border in everyday life, and as a symbol of state territoriality:

77 Survey respondent, 2018.

78 Focus group 5, 2019.

79 Focus group 1, 2019.

> "I believe any form of a hard border will have an adverse effect on relations here between Unionists and Nationalists... As someone who lives in a border area I have witnessed how contentious simply erecting KM/H [kilometres per hour road speed signs, used in Republic of Ireland] and "Welcome to Northern Ireland" signage has been in particular rural areas. These are always quickly defaced or removed. I cannot for the life of me imagine what [effect] erecting physical signs of a border [or] establishing border infrastructure... would have on community cohesion and the continued acceptance by Nationalists of the current constitutional position [given that they] believe their Irishness is underpinned by membership of the EU and an invisible border".[80]

The words of this respondent confirm and resonate with those of Donnan and Wilson's emphasis[81] on how symbolism at the frontiers of a state is meant to reiterate its presence, and as such evokes both powerful territorial identifications and, in cases of contested state territories, is often ignored or actively attacked and defaced by borderlanders. Instead, as Ghosh suggests, border residents "continue to scale their world in a way that their mental maps do not coincide with state borders".[82] In their words below, another survey respondent literally weaved together all the layers of border contestation mentioned in previous quotes, while specifically connecting the presently soft, invisible and open border to the 1998 Peace Agreement, including for identity:

> "I am Irish, born and bred in the north. I voted for peace in 1998. The GFA [Good Friday Agreement] recognises that I am Irish. In my heart, in my mind and in reality, Brexit and a hard border or any new border equipment or physical presence takes that away from me again. All I hear in relation to the border is trade. Well for many people along the border, it is family, social, cultural; our heritage, our way of life that come before this".[83]

In the multitude of such accounts, therefore, a hard border was seen not simply as contravening cultural and political belonging but also the associated rights of cross-border identity and mobility on the post-Agreement island.

---

80 Survey respondent, 2017.

81 H. Donnan & T. Wilson, *Borders: Frontiers of Identity, Nation and State*, Berg, Oxford, 1999.

82 S. Ghosh, "Cross-border activities in everyday life", *op.cit.*, p. 55.

83 Survey respondent 2018.

Conclusions

Few among scholars and politicians, or the people living and working in the Irish border region today contest the impression that over the past 25 years the region has transformed remarkably. A former security barrier, the border is now the epitome of the "soft spaces" of integration reflected in the discourses of the 1998 Belfast/Good Friday Agreement and in the EU's strife for internal de-bordering.[84] This remarkable transformation of the border is best illustrated through the mundane mobilities it now enables and invites. In the everyday, cross-border movements are a necessity and a fact of life, valued by many for enacting the Good Friday Agreement's recognition of the complex and hybrid senses of citizenship and national belonging. As such, freedom of movement across the border has helped entrench its everyday "waning".

However, by challenging the established regimes of governance of the Irish border, Brexit has provoked deep uncertainty over the future of cross-border mobilities, and revived the social and political tensions centering on the border's legitimacy. Our participants' accounts of their cross-border movements revealed a clash between official and local discourses over the future of the Irish border after Brexit. They demonstrated the extent to which, like in other borderlands, the Irish border region continues to be "[a site] of anxiety for the modern state … where state authority is contested, interrupted and qualified in everyday life".[85] Mundane mobilities are an essential element of such processes of contestation. They are not always purposeful acts attempting to consciously subvert various border mobility regimes, and may simply be seen as representing "competing forms of social constructions of space".[86] Yet as such, cross-border movements reflect and affect "cognitive maps of imagined communities and everyday affiliations",[87] producing and enacting the vernacular border as an "intensely political site" affected by, and affecting, broader power relations.[88]

---

84 C. Walsh, "Brexit geographies: Spatial imaginaries and relational territorialities on the island of Ireland", *Irish Geography*, vol. 52, n°2, 2019, pp. 137–152.

85 S. Ghosh, "Cross-border activities in everyday life", *op.cit.*, p. 58.

86 *Ibid.*, p. 58.

87 *Ibid.*, p. 50.

88 A. Ní Mhurchú, "Intimately inhabiting borders", *op.cit.*, p. 412.

Bibliography

P. Adey, *Mobility*, Routledge, London and New York, 2010.

D. Andersen & M. Sandberg, "Introduction", *In* D. Andersen, M. Klatt & M. Sandberg (eds.), *The Border Multiple*, Ashgate, Farnham, 2012, pp. 1–22.

D. Bigo, "The (in)securitization practices of the three universes of EU border control: Military/navy – border guards/police – Database analysts", *Security Dialogue*, vol. 45, n°3, 2014, pp. 209–225.

J. Binnie, T. Edensor, J. Holloway, S. Millington & C. Young, "Mundane mobilities, banal travels", *Social & Cultural Geography*, vol. 8, n°2, 2007, pp. 165–174.

N. Bubandt, "Vernacular security: The politics of feeling safe in local, national and global worlds", *Security Dialogue*, vol. 36, n°6, 2005, pp. 275–296.

K. Buscher, J. Urry, & K. Witchger, "Introduction: Mobile methods", *In* K. Buscher, J. Urry & K. Witchger, K. (eds.), *Mobile Methods*, Routledge, London and New York, 2011, pp. 1–20.

L. Cañás Bottos, "Assemblages of sovereignty and anti-sovereign effects on the Irish border", *Focaal: Journal of Global and Historical Anthropology*, vol. 71, n°1, 2015, pp. 86–99.

K. Cassidy, G. Wemyss & N. Yuval-Davis, "Debordering and everyday (re)bordering in and of Dover: Post-borderland borderscapes", *Political Geography*, vol. 66, 2018, pp. 171–179.

J. Coakley & L. O'Dowd, "The transformation of the Irish border", *Political Geography*, vol. 26, 2007, pp. 877–885.

A. Cooper & C. Rumford, "Monumentalising the border: Bordering through connectivity", *Mobilities*, vol. 8, n°1, 2013, p. 112.

T. Cresswell, "Towards a politics of mobility", *Environment and Planning D: Society and Space*, vol. 28, n°1, 2010, pp. 17–31.

S. Croucher, "Migrants of privilege: The political transnationalism of Americans in Mexico", *Identities: Global Studies in Culture and Power*, vol. 19, n°4, 2009, pp. 463–491.

J. Darling, "Becoming bare life: Asylum, hospitality and the politics of encampment", *Environment and Planning D: Society and Space*, vol. 27, n°4, 2009, pp. 649–665.

K. Doughty & L. Murray, "Discourses of mobility: Institutions, everyday lives and embodiment", *Mobilities*, vol.11, n°2, 2016, pp. 303–322.

M. Degen, C. DeSilvey & G. Rose, "Experiencing visualities in designed urban environments: Learning from Milton Keynes", *Environment and Planning A: Economy and Spaces*, vol. 40, n°8, 2008, pp. 1901–1920.

M. Doevenspeck, "Constructing the border from below: Narratives from the Congolese–Rwandan state boundary", *Political Geography*, vol. 30, n°3, 2011, pp. 129–142.

H. Donnan, "Material identities: Fixing ethnicity in the Irish borderlands", *Identities: Global Studies in Culture and Power*, vol. 12, n°1, 2005, pp. 69–105.

H. Donnan, "Cold war along the emerald curtain: Rural boundaries in a contested border zone", *Social Anthropology*, vol.18, n°3, 2010, pp. 253–266.

H. Donnan & T. Wilson, *Borders: Frontiers of Identity, Nation and State*, Berg, Oxford, 1999.

T. Dragostinova, *Between Two Motherlands. Nationality and Emigration among the Greeks of Bulgaria, 1900–1949*, Cornell University Press, Ithaca, 2011.

H. Easthope, "Fixed identities in a mobile world? The relationship between mobility, place, and identity", *Identities: Global Studies in Culture and Power*, vol. 19, n°1, 2009, pp. 61–82; O. Jensen, "Negotiation in motion: Unpacking a geography of mobility", *Space and Culture*, vol. 13, n°4, 2010, pp. 389–402.

S. Ghosh, "Cross-border activities in everyday life: The Bengal borderland", *Contemporary South Asia*, vol. 19, n°1, 2011, pp. 49–60.

S. Green, "Performing border in the Aegean: On relocating political, economic and social relations", *Journal of Cultural Economy*, vol. 3, n°2, 2010, pp. 261–278.

D. Gregory, "Imaginative geographies", *Progress in Human Geography*, vol. 19, n°4, 1995, pp. 447–485.

K. Hayward, "Bordering on Brexit: Views from local communities in the Central Border Region of Ireland/Northern Ireland", Queen' University Belfast, Report for the Irish Central Border Area Network, 2017.

K. Hayward, "Brexit at the border: Voices of local communities in the Central Border Region of Ireland/Northern Ireland", Irish Central Border Area Network (ICBAN), 2018.

K. Hayward, "The future of the Irish border", *Renewal: A Journal of Social Democracy*, vol. 26, n°4, 2018, pp. 12–22.

K. Hayward & M. Komarova, "The border into Brexit. Perspectives from local communities in the Central Border Region of Ireland / Northern Ireland", Irish Central Border Area Network (ICBAN), 2019.

K. Hayward & M. Komarova, "The border after Brexit: Experiences of local communities in the Central Border Region of Ireland / Northern Ireland", Irish Central Border Area Network, 2021.

K. Hayward, P. Leary & M. Komarova, "The Irish border as sign and source of British-Irish tensions", *In* Ribas-Mateos, N. & Dunn, T. (eds.), *Handbook on Human Security, Borders and Migration*, Edward Elgar, Cheltenham, 2021.

K. Hayward & M. Murphy, "The (soft) power of commitment: The EU and conflict resolution in Northern Ireland", *Ethnopolitics*, vol. 11, n°4, 2012, pp. 439–452.

C. Johnson, R. Jones, A. Paasi, L. Amoore, A. Mountz, M. Salter & C. Rumford, "Interventions on rethinking 'the border' in border studies", *Political Geography*, vol. 30, n°2, 2011, pp. 61–69.

R. Jones, *Violent Borders: Refugees and the Right to Move*, Verso, London, 2016; R. Jones & C. Johnson, "Border militarization and the re-articulation of sovereignty", *Transactions of the Institute of British Geographers*, vol. 41, n°2, 2016, pp. 187–200.

S. Y. Koh & B. Wissinik, "Enabling, structuring and creating elite transnational lifestyles: Intermediaries of the super-rich and the elite mobilities industry", *Journal of Ethnic and Migration Studies*, vol. 44, n°4, 2018, pp. 592–609.

V. Kolossov & J. Scott, "Selected conceptual issues in border studies", *EUROBORDERSCAPES Working Paper*, n° 4, 2013.

P. Leary, *Unapproved Routes: Histories of the Irish Border, 1922–72*, Oxford University Press, Oxford, 2016.

P. Leary, "A house divided: The Murrays of the border and the rise and decline of a small Irish house", *History Workshop Journal*, vol. 86, 2018, pp. 269–290.

S. Leheny, "Making Northern Ireland a bridge between the UK and the EU", *UK in a changing Europe* [Website], 2018.

M. Lemberg-Pedersen, "Forcing flows of migrants: European externalization and border-induced displacement", *In* D. Andersen, M. Klatt & M. Sandberg (eds.), *The border multiple*, Ashgate, Farnham, 2012, chapter 2.

S. Lèonard, "EU border security and migration into the European Union: FRONTEX and securitization through practices", *European Security*, vol. 19, n°2, 2010, pp. 231–254.

K. Manderscheid, "Criticising the solitary mobile subject: Researching relational mobilities and reflecting on mobile methods", *Mobilities*, vol. 9, n°2, 2014, pp. 188–219.

C. McCall, *Border Ireland: From Partition to Brexit*, Routledge, London and New York, 2021, p. 51.

C. Minca & A. Rijke, "Walls! Walls! Walls!", *Society and Space*, vol. 40, n°2, 2017.

E. Moloney, *A Secret History of the IRA*, Penguin, London, 2007.

A. Mountz, K. Coddington & R. T. Catania, "Conceptualizing detention: Mobility, containment, bordering, and exclusion", *Progress in Human Geography*, vol. 37, n°4, 2012, pp. 522–541.

P. Mulroe, *Bombs, Bullets and the Border: Policing Ireland's Frontier: Irish Security Policy, 1969–1978*, Irish Academic Press, Newbridge, 2017.

C. Nash, L. Dennis, & B. Graham, "Putting the border in place: Customs regulation in the making of the Irish border, 1921–1945", *Journal of Historical Geography*, vol. 36, 2010, pp. 421–443.

D. Newman & A. Paasi, "Fences and neighbours in the postmodern world", *Progress in Human Geography*, vol. 22, 1998, p. 197.

A. Ní Mhurchú, "Intimately inhabiting borders: Walking in-between belonging and otherness through constructions of home", *Geopolitics*, vol. 26, n°2, 2021, p. 412. The author quotes A. Blunt and R. M. Dowling, *Home*, Routledge, London and New York, 2006.

L. O'Dowd & M. Komarova, "Three narratives in search of a city", *City: Analysis of Urban Trends, Culture, Theory, Policy, Action*, vol. 17, n°4, 2013, pp. 526–546.

H. Patterson, *Ireland's Violent Frontier: The Border and Anglo-Irish Relations during the Troubles*, Palgrave MacMillan, New York, 2013.

S. Pickering & L. Weber, "Borders, mobility and technologies of control", *In* S. Pickering and L. Weber (eds.), *Borders, Mobility and Technologies of Control*, Springer, New York, 2006, pp. 1–19.

T. Richardson, "Borders and mobilities: Introduction to the special issue", *Mobilities*, vol. 8, °1, 2013, pp. 1–6.

C. Rumford, "Theorizing borders", *European Journal of Social Theory*, vol. 9, n°2, 2006, p. 156.

C. Rumford, "Introduction: Citizens and borderwork in Europe", *Space and Polity*, vol. 12, n°1, 2008, pp. 1–12.

M. Sheller, *Mobility Justice. The Politics of Movement in an Age of Extremes*, Verso, London, 2018.

M. Sheller & J. Urry, "The new mobilities paradigm", *Environment and Planning A: Economy and Space*, vol. 38, n°2, 2006, pp. 207–226.

S. Simonsen, "Spatiality, temporality and the construction of the city", *In* J. O. Baerenholdt & K. Simonsen (eds.), *Space Odysseys: Spatiality and Social Relations in the 21st Century*, Ashgate, Farnham, 2004, p. 53.

E. Spence, "Unraveling the politics of super-rich mobility: A study of crew and guest on board luxury yachts", *Mobilities*, vol.9, n°3, 2014, pp. 401–413.

M. Stoetzler & N. Yuval-Davis, "Standpoint theory, situated knowledge and the situated imagination", *Feminist Theory*, vol. 3, n° 3, 2002, pp. 315–333.

V. Squire, "Desert 'trash': Posthumanism, border struggles, and humanitarian politics", *Political Geography*, vol. 39, 2014, pp. 11–21.

R. J. Topinka, "'Wandering the settled tribes': Biopolitics, citizenship and the racialized migrant", *Citizenship Studies*, vol. 20, n°3/4, 2016, pp. 444–456.

V. C. Tran, "The borders around us: Forced migration and the politics of border control", *City & Community*, vol. 19, n°2, 2020, pp. 323–329.

N. Vaughan-Williams, *Vernacular Border Security. Citizens' Narratives of Europe's 'Migration Crisis'*, Oxford University Press, Oxford, 2021.

C. Walsh, "Brexit geographies: Spatial imaginaries and relational territorialities on the island of Ireland", *Irish Geography*, vol. 52, n°2, 2019, pp. 137–152.

L. C. Whitten, "North-south cooperation and the Protocol on Ireland/ Northern Ireland", Queen's University Belfast, 2022.

C. Wille & U. Roos, "Cross-border everyday lives on the Luxembourg border? An empirical approach: the example of cross-border commuters and residential migrants", *In* C. Wille & B. Nienaber (eds.), *Border Experiences in Europe*, Nomos, Baden-Baden, 2018, pp. 101–126.

The Visible Invisible Border: Reflections on the Frontem Focus Group Discussions on Mobility across the Ireland-Northern Ireland Border

Anthony Soares

Introduction

The ritual pilgrimage to the border between Ireland and Northern Ireland that became a practice for countless political leaders from across Europe and beyond in the run-up to the 2016 referendum on the United Kingdom (UK)'s membership of the European Union (EU), and during the tortuous period of negotiations that followed, captured some of the essence of the border's duality. These visits, accompanied by the world's press, produced numerous images of various dignitaries and their entourages looking with grave intent at, essentially, nothingness. Occasionally, that nothingness was disturbed as groups of citizens, some of them dressed to resemble officials or police, erected sham customs posts and barriers to make clear that they did not want Brexit to result in a "hard" border on the island of Ireland – a form of protest that made, even if only temporarily, what is invisible visible. More recently, and in what is now a post-Brexit context, protests have been organised to raise awareness of "an invisible – racist – hard border stopping people from moving freely to work, to access healthcare or to visit friends",[1] which was already in existence before the UK's departure from the EU.

In the discussions that took place during two focus group discussions organised in March 2022 as part of the Frontem network's activities relating to the Ireland-Northern Ireland border, some of that duality

[1] North West Migrants Forum, "No racist border on the border of Ireland", 2021. Retrieved from: <https://www.nwmf.org.uk/common-travel-area-petition-north-west-migrants-forum/>. Accessed 21 May 2023.

manifested itself as participants stressed its present invisibility while fearing its visible return. However, the ways in which the border meandered between invisibility and visibility differed between the two focus groups. This difference resulted not only from the slightly different purposes behind the constitution of each focus group, but also perhaps from the relation each of them had to cross-border mobility.

The first focus group brought together several key actors with oversight of or involved in a range of areas relevant to cross-border mobility,[2] namely: tourism, trade, health, labour, education, culture, transport and the green economy. A core purpose for organising this focus group was to gather their views on and experience of managing and monitoring cross-border mobility. The second focus group on the other hand was composed of representatives from a range of community groups, as well as civic society organisations whose work includes addressing issues related to citizens' cross-border mobility.[3] Four of the seven participating organisations were based in Northern Ireland, three in the Republic of Ireland, with four located within the border region and three outside of it. A central purpose of constituting this focus group was to discover their

---

2 Participants in this focus group were drawn from the following institutions and organisations: East Border Region (a cross-border network of six local authorities from the east coast of Ireland and Northern Ireland); EURES Cross Border Partnership (an alliance of representatives from the employment services in Ireland and Northern Ireland, trade unions from both sides of the border, chambers of commerce, and business and employers' federations; this partnership has now reconstituted itself as the Cross Border Partnership Employment Services); the EU & North/South Unit of the (Irish) Health Service Executive; Irish Central Border Area Network (a cross-border network of eight local authorities from the central border area); InterTradeIreland (established as one of the implementation bodies under Strand Two of the 1998 Belfast/Good Friday Agreement, its role is to help small businesses in Ireland and Northern Ireland to explore new cross-border markets); Middletown Centre for Autism (established in 2007 by Ireland's Department of Education and Skills and Northern Ireland's Department of Education to support the promotion of excellence in the education of children and young people with Autistic Spectrum Disorders); and the North South Ministerial Council (NSMC) Joint Secretariat (established under Strand Two of the Good Friday Agreement, the NSMC brings together the governments of Ireland and Northern Ireland to develop consultation, cooperation and action within the island of Ireland, supported by a Joint Secretariat drawn from the Irish Department of Foreign Affairs and Northern Ireland's Executive Office).

3 The organisations participating in this focus group were: Arts as Exchange, Bolster Community, Committee on the Administration of Justice (CAJ), Creative Spark, Cross Border Workers' Coalition, Longford Women's Link, and Women's TEC.

perceptions of the border and the importance of citizens' cross-border mobility.

In advance of each focus group meeting all participants completed a questionnaire. Reflecting the differing purposes of the focus groups, there were two different questionnaires with different sets of questions. The questionnaire aimed at key actors with oversight of or involved in areas relevant to cross-border mobility included questions on monitoring and reporting of cross-border mobility, the implementation of measures addressing changes to rates of mobility, impacts of Covid-19 on mobility, and organisational collaboration on mobility. The questionnaire completed by the second focus group included questions on perceptions of the border, the importance citizens place on cross-border mobility, monitoring and restricting cross-border mobility, and closing the border. The responses were then used to inform the focus group discussions, allowing them to explore in greater depth some of the issues raised by the questionnaires.

The Cross-Border Framework

Before reflecting on the content of the focus group discussions, it is important to recall some of the underlying framework for cross-border mobility and cooperation applying to the island of Ireland, and the fact that as a result of Brexit the border between Northern Ireland and the Republic of Ireland is now an EU external border. That change of circumstance has given rise to considerable challenges in accommodating the institutional structures and bodies established by the 1998 Good Friday Agreement, including those charged with cooperation between Northern Ireland and the Republic of Ireland.

The second of the 1998 Agreement's three core strands put in place a formal structure for dialogue and cooperation between the Irish Government and the devolved administration in Northern Ireland, as well as setting out the creation of a number of bodies (known as implementation bodies) to undertake practical cross-border and all-island cooperation in a range of specific areas. The formal platform for cooperation between the Dublin and Belfast governments is the North South Ministerial Council, whose Joint Secretariat, staffed by personnel from the Irish Civil Service and Northern Ireland Civil Service, is located in Armagh. The six implementation bodies that would result from the

1998 Agreement are: the Food Safety Promotion Board (known as safefood); the Foyle, Carlingford and Irish Lights Commission (consisting of two agencies – the Loughs Agency, and the Lights Agency, although the latter is yet to be established); the Language Body (consisting of two agencies – Foras na Gaeilge and Tha Boord o Ulster-Scotch/Ulster-Scots Agency); the Special European Union Programmes Body (SEUPB); the Trade and Business Development Body (known as InterTradeIreland) and Waterways Ireland.

Strand 2 of the Agreement consists of 19 paragraphs and an Annex, setting out what the structures for North-South cooperation and their functions would be. The first paragraph confirms the establishment of the North South Ministerial Council:

> "Under a new British/Irish Agreement dealing with the totality of relationships, and related legislation at Westminster and in the Oireachtas, a North/South Ministerial Council to be established to bring together those with executive responsibilities in Northern Ireland and the Irish Government, to develop consultation, co-operation and action within the island of Ireland – including through implementation on an all-island and cross-border basis – on matters of mutual interest within the competence of the Administrations, North and South".[4]

Two points should be highlighted here. These formal structures for cooperation were created as part of a Good Friday Agreement that references the fact that for the British and Irish governments it was seen as a means "to develop still further the unique relationship between their peoples and the close co-operation between their countries as friendly neighbours and as partners in the European Union". Therefore, the approaches to cross-border mobility and cooperation were undertaken through the prism of a shared EU framework and where, specifically in terms of cross-border mobility, answers to resolving obstacles to mobility could be sought within EU regulatory frameworks. In the absence of those frameworks, the pre-existing Common Trave Area (CTA) has been cast by the UK and Irish governments as the bilateral regulatory regime that supports the free mobility of British and Irish citizens between their territories. Established in 1922, the CTA is not only a legal travel regime,

---

4 UK Government, "Good Friday Agreement", 10 April 1998, 1rst paragraph. Retrieved from: <https://www.gov.uk/government/publications/the-belfast-agreement>. Accessed 23 May 2023.

but also a regime of citizens' rights that includes access to public services, to healthcare and social benefits, and some voting rights in each other's territories. Crucially, of course, it does not apply to citizens who are neither Irish or British, nor does it have the same legal stature as the EU treaties that guarantee citizens' mobility and rights within the EU.[5]

The second point to be noted is that the Good Friday Agreement made clear that the administrations of Ireland and Northern Ireland can only cooperate on matters that are within their competence. In the case of Northern Ireland, as a devolved administration within the United Kingdom, this means that it cannot enter into discussions on cooperation with the Republic of Ireland involving areas that are the sole competence of the government in London, which include international relations, nationality, immigration and asylum, UK-wide taxation, import and export controls, and financial services and pensions regulation.

This in turn leads to two further considerations. Where policy and legislation is developed in London on matters where Northern Ireland does not have competence, it is essential that the border between Northern Ireland and Ireland does not become invisible during the process. If the border becomes invisible to policy-makers and legislators in London, it is more likely that this will make it more visible to citizens on the island of Ireland as they are faced with obstacles to their cross-border mobility and cooperation. Where, on the other hand, Northern Ireland is in a position to cooperate with the Republic of Ireland on matters within its competence, its ability to do so in the post-Brexit context will rely to an extent on the UK Government's putting into practice its desire to diverge from the EU. That desire is expressed in the UK Government's Command Paper on the Windsor Framework, which refers to how there will be "now considerable underlying regulatory divergence North and South", and goes on to state:

> "The Government recognises that these changes do create a different legal and practical context on the island of Ireland, with substantial and likely increasing divergence between Northern Ireland and Ireland over time – building,

---

5 In this regard, the 2019 Memorandum of Understanding between the Government of the United Kingdom of Great Britain and Northern Ireland and the Government of Ireland concerning the Common Travel Area and associated reciprocal rights and privileges states in its final provisions: "The foregoing record represents the common understanding of the Participants upon the matters referred to therein. It is not of itself intended to create legally binding obligations".

of course on the in-built capacity for divergence in the vast majority of areas outside the Protocol including environmental law, professional qualifications, employment law, procurement, immigration, banking, data and a wide range of services and other rules".[6]

Bearing in mind that some areas in which the UK indicates it wishes to diverge are actually ones where Northern Ireland has competence, such as environmental and employment law, the question of managing cooperation and mobility across the Ireland-Northern Ireland border has become loaded with political tensions.

Added to the tensions surrounding the border between Ireland and Northern Ireland arising from Brexit, approaches to the border are also always haunted by the past. It is important to recall that while the creation of the EU Customs Union and Single Market led to the removal of border infrastructure to control the movement of goods, the Northern Ireland-Ireland border remained entirely visible. The security infrastructure that continued to closely control and monitor the movement of people would only begin to be dismantled following the 1998 Good Friday Agreement, which also led to the re-opening of border crossings that had been closed during the period of the conflict. This haunting memory of the pre-1998 border, along with the disquiet caused by the more recent change to an external border of the EU, were simultaneously present during the Frontem focus group discussions.

Imagining the Border: Where Does Cross-Border Mobility Begin?

Participants in both of the Frontem focus groups on the island of Ireland were asked to indicate where they thought the limits of the Ireland-Northern Ireland cross-border region were located: their mental map of the region. The results are represented here (Figure 1).

6 UK Government, "The Windsor Framework: A new way forward", 2023, paragraph 55. Retrieved from: <https://www.gov.uk/government/publications/the-windsor-framework>.

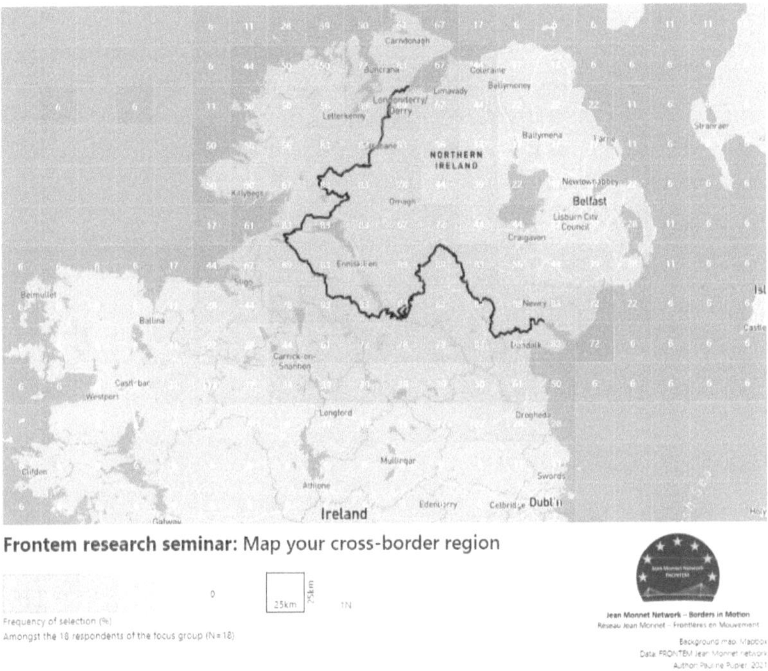

Figure 1: Mental map made during Frontem's focus-groups in Ireland.[7]

Their mental map is clearly focused on locations that span the six border counties of the Republic of Ireland (Cavan, Donegal, Leitrim, Louth, Monaghan and Sligo), and parts of the Local Government Districts in Northern Ireland immediately abutting the border. They correspond in large part to the flows of cross-border mobility related to employment and cross-border shopping, but are narrower than the cross-border region as delimited in terms of the eligible area for the EU's Interreg A and PEACE programmes (now combined into the new PEACE PLUS programme for the 2021–2027 period), which is constituted by all of Northern Ireland

7 Source: University of Strasbourg/Euro-Institute, *Toolkit on Models of Border Management and Perception in the EU. 5 Case Studies*, Strasbourg, 2023, p. 150. Retrieved from; <https://centre-jean-monnet.unistra.fr/2023/09/05/toolkit-modeles-de-gestion-de-la-frontiere-et-de-la-perception-sur-lunion-europeenne/>. Accessed 18 December 2023.

and the six border counties of the Republic of Ireland. Indeed, on the Northern Ireland side, the mental map shows how the focus groups' participants imagined the cross-border region to not encompass to a significant extent the entirety of the Local Government Districts adjacent to the border, which would have resulted in an area between the red lines shown in the map below (Figure 2).

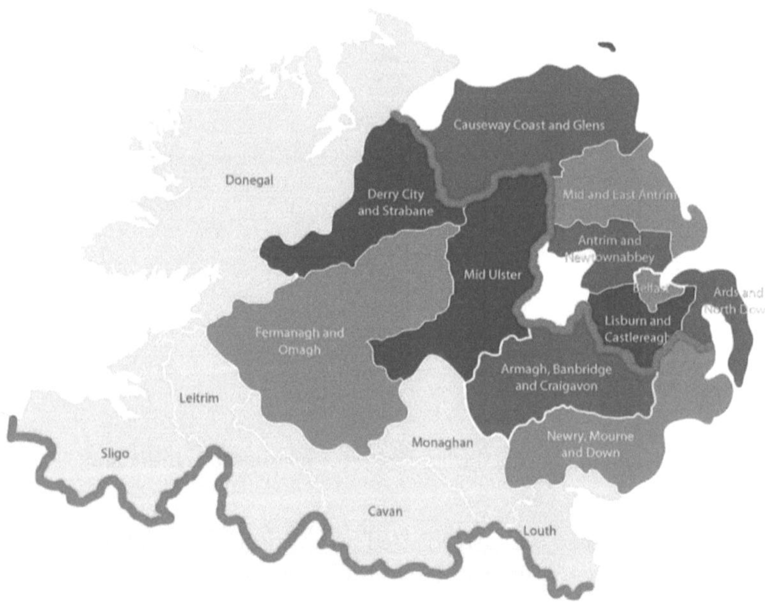

Figure 2: Area eligible for Interreg A Financing in Ireland (counties from the Republic of Ireland in yellow, counties of Northern Ireland in blue).[8]

This narrower mental map of the focus groups' participants of the cross-border region appears to be more attuned to the lives and concerns of communities residing more closely to the Ireland-Northern Ireland border, rather than to concepts designed to address the concerns or perspectives of entities located outside the region, such as the delineation of a programming area. In turn, this could mean that the focus groups'

8 Source: Peace Programmes. Retrieved from: <https://peaceplatform.seupb.eu/en/story-of-peace/key-themes/local-authorities/>. Accessed 24 November 2023.

participants are more conscious of the perceptions of the border held by those who live in its immediate vicinity, which includes their valuing of its invisibility not only as a physical presence, but also as a dividing line between two jurisdictions with different administrative and legal regimes. Crucially, this would take into account how communities living within this mental map would react to any divergence between the two jurisdictions that would create obstacles for their cross-border lives, returning them to a past they believed had been forsaken following the 1998 Good Friday Agreement.

"Managing" and Monitoring Cross-Border Mobility

The View from Key Actors

There was a very mixed response among key actors with oversight of or involved in cross-border mobility in relation to the question of the management of such mobility (See Figure 3). This was immediately apparent from the responses to the questionnaire sent to participants before the focus group discussions, and the reasons for these differing attitudes were explored during the focus group meeting.

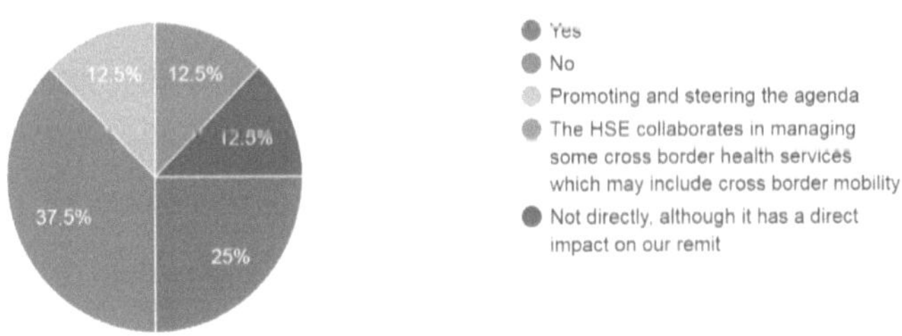

Figure 3: Percentages of responses to the question "Do you see your organisation as being involved in the 'managing' of cross-border mobility?"[9]

Most respondents (37.5 %) did not see themselves as "managing" mobility, with some (25 %) saying they did, while others indicated that they managed specific aspects restricted to their particular area of operation, or that they did so indirectly. Similarly, any monitoring of

[9] Source: Centre for Cross-Border Studies, Armagh, Frontem survey, 2022.

cross-border mobility was largely indirect, with insights drawn from the volume of services delivered, all-island business monitors and general stakeholder engagement. A number of responses to a question on the measures and methods these key actors had to monitor cross-border mobility are revealing of why monitoring might take place, and the extent to which responsibility is seen as lying elsewhere: "We don't hold a policy role as such so monitoring would be carried out elsewhere"; "We do not formally monitor cross border mobility but rely on others to provide the statistics"; "Completed 4 Brexit community impacts surveys [...] going back to 2017. The latest study in 2021 also examined the impacts of the pandemic on mobilities".[10]

Although the question of who might have responsibility for monitoring cross-border mobility will be picked up again later, some of the responses to the questionnaire are suggestive of a reactive approach to such monitoring. In other words, monitoring is seen as important when it is prompted by a particular event, such as Brexit or the Covid-19 pandemic, and that it may not necessarily be a systematic and ongoing process.

It is important to note, however, that a significant proportion of responses to the questionnaire (37.5 %) pointed to an increase in cross-border mobility over the last five years, with half reporting that levels of mobility had remained relatively stable (See Figure 4).

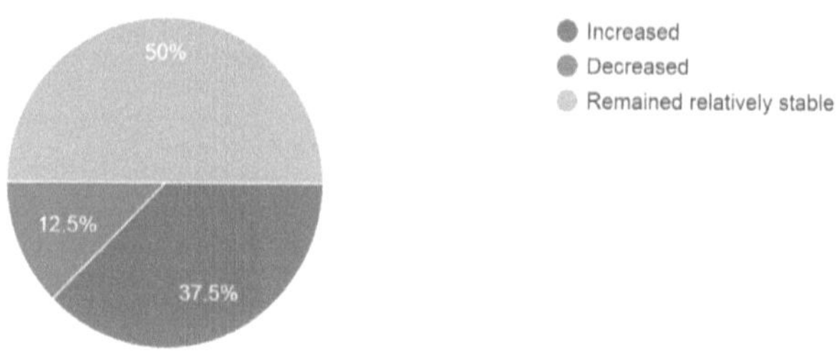

Figure 4: Percentages of responses to the question "In the last 5 years, has cross-border mobility in the area for which your organisation has responsibility…"[11]

[10] Frontem survey respondents, Centre for Cross-Border Studies, 2022.

[11] Source: Centre for Cross-Border Studies, Armagh, Frontem survey, 2022.

Additionally, half of the respondents to the questionnaire indicated that their organisation reported levels of cross-border mobility in the area for which they were responsible (See Figure 5).

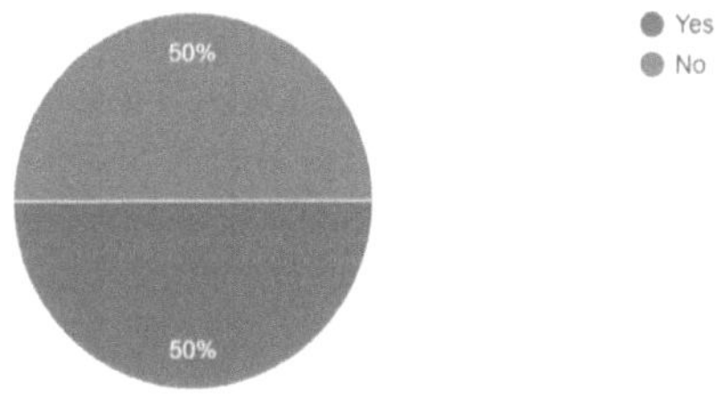

Figure 5: Percentages of responses to the question "Does your organisation report levels of cross-border mobility in the area for which it is responsible?"[12]

Moreover, all of those respondents who said they reported levels of cross-border mobility also stated that they reported those levels to national authorities.

Initially, there may seem to be a degree of contradiction between the responses from key actors to the questionnaire, whereby they appear to indicate that they do not necessarily have responsibility for managing or monitoring cross-border mobility, and yet they are in a position to state whether such mobility has increased or not in the last five years, and half state that they report on levels of mobility. Indeed, in the focus group discussion with key actors, one of the comments made was illustrative of the general position of participants, which was that they did not see themselves as having any formal responsibility for the monitoring or management of the border: "Organisations on this island have a tendency to look to legislation to dictate their remit and function, because there's safety in what's been agreed and written down".[13]

If the legislation on which these key actors' organisations or institutions is based does not stipulate monitoring or managing of cross-border mobility, then they are reluctant to do so. This is especially true for organisations established under Strand 2 of the Good Friday Agreement, where policy is set by agreement between the Dublin and Belfast administrations.

12 Source: Centre for Cross-Border Studies, Armagh, Frontem survey, 2022.

13 Focus group, 2022.

Therefore, to stray beyond what has been agreed risks censure from one or other government or from political representatives.

For example, while the North South Ministerial Council (NSMC) can focus attention on specific mobility issues, it can only do so where the issue in question relates to a power devolved to the Northern Ireland Assembly. Thus, in the area of immigration, a matter reserved to the UK Government, while the development of policy in this area is likely to impact on cross-border mobility on the island of Ireland, it does not fall within the competency of the Northern Ireland Executive and, therefore, of the NSMC.

The focus group discussion revealed that although certain amounts of data are gathered on a sectoral basis, there is no dedicated mechanism by which to get a "panoramic" view of cross-border mobility on the island of Ireland. Participants agreed that while multiple organisations offer support to facilitate cross-border mobility, it is not within the gift of any to "manage" it. This means that information on mobility is gathered indirectly and according to specific sectoral needs and objectives. For example, the Cross Border Partnership for Employment Services gathers information on labour and mobility based on its advice service, which is shared with both statistics agencies on the island of Ireland. Surveys it conducts are also shared with academics, government departments, policymakers, the European Commission, and used by the partnership to focus services and delivery.

Nevertheless, the absence of a more systematic approach to the monitoring of cross-border mobility became all the more obvious once the UK had voted to leave the European Union. Prompted by Brexit and the need to gain an understanding of its potential impacts, a report published by Northern Ireland's Department for the Economy noted that "there is no one complete and definitive data source that records the number and purpose of all cross border movements between Northern Ireland, Great Britain and ROI [the Republic of Ireland]".[14] However, referring to the existence of different estimates as to how many people commute across

---

[14] Department of the Economy, "Background Evidence on the Movement of People across the Northern Ireland-Ireland Border", 2018, p. 4. Retrieved from: <https://www.economy-ni.gov.uk/sites/default/files/publications/economy/movement-people-northern-ireland-ireland-border.pdf>. Accessed 23 May 2023.

the Northern Ireland-Ireland border, the same report provides its own estimate:

> "The exact number of people that cross the border to work or study is difficult to quantify. The Centre for Cross Border Studies has used a variety of studies to provide an estimate that there are between 23,000 and 29,000 people who commute across the border. Based on the available information, an estimate of 25,000 trips (including daily and less regular commuters) across the border every day for work or study appears reasonable".[15]

Indeed, participants in the key stakeholders' focus group reflected on the change of circumstances Brexit had brought about in terms of cross-border mobility, and doubts as to the extent to which the Common Travel Area could maintain things as they were pre-Brexit. As one participant put it:

> "A lack of freedom of movement cross-border was not a perceived circumstance when the remits of cross-border groups were established under the Good Friday Agreement. We thought the CTA [Common Travel Area] would provide us with more protection than it is".[16]

In some sense there may have been a belief that the lack of a panoramic view of cross-border mobility on the island of Ireland was not a critical issue since common membership of the EU guaranteed that mobility. Brexit has changed that belief.

The View from Civic Society

There was a very mixed reaction from representatives of civic society when asked in a questionnaire whether citizens in the border region thought it important to monitor levels of cross-border mobility (See Figure 6).

[15] *Ibid.*, p. 18.

[16] Focus group, 2022.

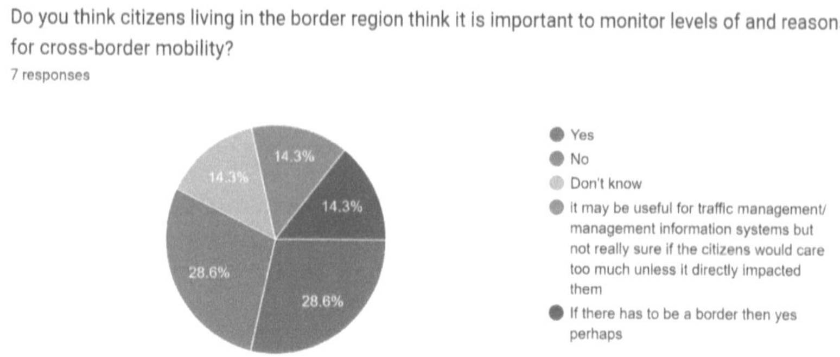

Figure 6: Percentages of responses to the question "Do you think citizens living in the border region think it is important to monitor levels of and reasons of cross-border mobility?"[17]

Respondents were equally divided between those who thought monitoring was important to citizens in the border region and those who were of the opposite view, with some respondents preferring to offer more detailed but nevertheless tentative answers. However, there was a greater degree of certainty when asked whether citizens from outside the border region thought monitoring of cross-border mobility was important (See Figure 7).

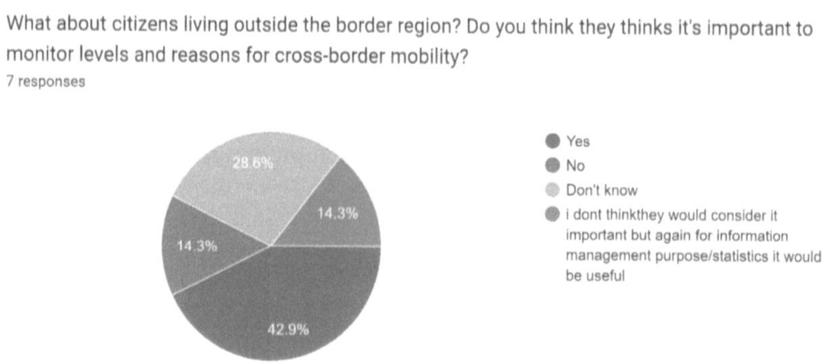

Figure 7: Percentages of responses to the question "What about citizens living outside the border region? Do you think they thinks it's important to monitor levels and reasons for cross-border mobility?"[18]

17 Source: Centre for Cross-Border Studies, Armagh, Frontem survey, 2022.

18 Source: Centre for Cross-Border Studies, Armagh, Frontem survey, 2022.

Some of the respondents took up the opportunity to offer comments on their answers to these questions, and the following perhaps indicate the ambiguous attitude to the issue of monitoring cross-border mobility: "To underpin improved understanding of economics, to underpin dual responsibility for improving infrastructure, for taxation purposes, for analysis of economics – yes"; "A successful all-island economy depends upon cross-border mobility. That impacts everyone."; "I suppose some people may believe that the movement of people across the border should be controlled as ROI [Republic of Ireland] opens its borders and UK closes their borders."; "To monitor the volume of cross-border mobility implies that it's not actually free/allowed/encouraged/welcome".[19]

While the comments provided point to a certain acknowledgement of the benefits of monitoring of cross-border mobility for specific purposes, they also illustrate a perception that such monitoring would come as a result of differing attitudes to immigration or that it would go against the grain of what is generally seen as an open – an invisible – border.

However, the focus group discussion with representatives of civic society revealed how the border has not been invisible to all, and that some citizens may not necessarily wish it to be open for certain people. This sentiment was captured in the following intervention by one of the participants:

> "We have encountered examples where people are less open to cross border mobility based on the nationality or immigration status of the person crossing the border, particularly when it concerns access to healthcare, welfare and employment. It is common to see this promoted as something that should only apply to British/Irish citizens. This is often linked to false narratives and negative perceptions of the migrant community".[20]

It is important to recognise that even before Brexit there were people on the island of Ireland who could not cross the border with the freedom available to others. Brexit simply expanded the categories of people potentially having to see the border as an immigration obstacle, and its openness gradually being reserved for those of Irish or British nationality.

Nevertheless, there was overwhelming opposition within the focus group to the idea of "monitoring" cross-border mobility, with concerns

---

19 Frontem survey respondents, Centre for Cross-Border Studies, 2022.

20 Focus group, 2022.

expressed that "you don't want your car counted or checked every time it crosses the border".[21] This was explicitly linked to a psychological legacy of historic border checks during the period of conflict, which should not be returned to. The discussions frequently returned to this memory of the heavily fortified and controlled border whenever the issue of monitoring of cross-border mobility was raised. At the same time, there was an acceptance that a lack of data can lead to poor policy making:

> "We don't want to encourage checks, but there are frustrating issues such as the Frontier Workers Scheme which was introduced with no data to support it.[22] [There was a] very low uptake [of the Scheme] on this island. We know frontier workers are out there and it's only when something goes wrong that it'll be noticed".[23]

In the context of supporting improved public policy it was suggested that: "A degree of study is needed to ascertain real data to know where problems are. Perhaps there is a place for the Shared Island Unit to do all island study. But it doesn't have to be invasive".[24]

The focus group agreed on the use of terms like "research into" and "studies of" cross-border mobility, but remained deeply opposed to the term "monitoring". In some senses, therefore, the attitudes displayed by this focus group towards the idea of the monitoring of cross-border mobility echoes what was revealed in the focus group of key stakeholders: that some monitoring is undertaken on a sectoral basis, but that there is no "panoramic" view of cross-border mobility on the island of Ireland.

Managing and Monitoring, and No Return to a "Hard Border"

Article 1 of the Protocol on Ireland/Northern Ireland states that it "sets out arrangements necessary to address the unique circumstances on the island of Ireland, to maintain the necessary conditions for continued North-South cooperation, to avoid a hard border and to protect the 1998

---

21 *Ibid.*

22 The Frontier Workers Scheme was introduced by the UK Government under the Withdrawal Agreement, and is aimed at safeguarding the rights of frontier workers working in the UK on or before 31st December 2020.

23 Focus group, 2022.

24 *Ibid.*

Agreement in all its dimensions".[25] Arguably, the attitudes of the key stakeholders focus group towards border management and monitoring could be interpreted as implicit rejections of being implicated in actions that could be seen as a "hardening" of the border and in conflict with the Good Friday Agreement. A reluctance to introduce novel modes of border management and monitoring may be simultaneously motivated by a desire not only to avoid bringing back into the present aspects of the pre-1998 border, but also to evade the consequences of the border's new status as an external EU border and to enshrine as far and as long as possible the current nature of cross-border mobility. Capturing the anxieties being felt, one of the key stakeholders noted in the focus group discussion: "It has gone from a contested border to peace and free movement, and we have become used to that. Now that is questioned again suddenly".[26]

Another reflected on how, even without local stakeholders implementing new forms of border management and monitoring, North-South cooperation would suffer a setback as a consequence of Brexit: "there will be a period of time over the next five to ten years where we are all playing catch-up to deliver policy and legislation to get us back to where we were in 2016".[27]

There was a similar double undercurrent shaping the discussions in the focus group with representatives from civic society. While, as has been noted before, their strong rejection of the concept of border monitoring was motivated by a refusal to return to the "hard" border of the conflict, it was also informed by a desire not to lose the benefits of an island of Ireland within the EU space. In terms of the latter, this focus group was much more explicit in linking the introduction of any form of border management and monitoring to a reversal of what they regarded as commitments made during the Brexit negotiations and embedded within the Protocol. One of the focus group participants commented:

---

25 UK Government, "Protocol on Ireland and Northern Ireland", 31 January 2020, Article 1. Retrieved from: <https://assets.publishing.service.gov.uk/media/5da863ab40f0b659847e0184/Revised_Protocol_to_the_Withdrawal_Agreement.pdf>. Accessed 23 November 2023.

26 Focus group, 2022.

27 *Ibid.*

> "We [in Northern Ireland] voted against Brexit, and Ireland didn't have a say. We didn't ask for any of this, and we were promised that things wouldn't change on the island of Ireland, that the Good Friday Agreement would be protected – all of it. If anyone finds it more difficult to cross the border, or to do the things they've been able to do since 1998, then we have been lied to".[28]

What can be characterised as a "light-touch", unobtrusive and largely uncoordinated approach to border management and monitoring by key stakeholders since 1998 has been accepted (if not even unknown) to citizens, and both groups appear to be resistant to any change to this relationship to the border even if the context created by Brexit would seem to make change inevitable. Indeed, those changes are already taking place, not only due to UK changes to immigration policy, as has been noted before, but also because Northern Ireland is no longer within the EU Single Market for services.

The Covid Moment: Managing the Border at Times of Crisis

During the Covid-19 pandemic there were no closures of the border between Northern Ireland and the Republic of Ireland, unlike what was seen elsewhere in Europe. As the imposition of border closures would be politically divisive given the context on the island of Ireland, as well as being extremely difficult if not impossible given the number of border crossing points, other means had to be found to prevent the spread of Covid-19. Essentially, success in preventing the spread of the pandemic within and between the jurisdictions was predicated on the willingness of citizens to voluntarily abide by the public health restrictions put in place by their respective authorities, with enforcement – which the closure of the border would represent – not employed as a general practice.

In its approach to the imposition of travel restrictions in the Republic of Ireland, for example, the policy of the Garda (Irish police) was "to engage, explain, encourage and only if necessary enforce the emergency legislation".[29] A similar approach was taken by the Police Service of Northern Ireland. Although attempts were made by the Irish police to discourage Northern Ireland citizens from crossing the border into the

---

28 *Ibid.*

29 *Ibid.*

Republic of Ireland at traditionally intensive periods for cross-border leisure transit, and random checks were made to confirm the purpose of journeys being made by citizens travelling from the Republic of Ireland into Northern Ireland, these cannot be seen as border closures.

The responses from key actors with oversight of or involved in a range of areas relevant to cross-border mobility to one of the questions in the questionnaire circulated prior to the focus group meeting is of significance in relation to this topic (See Figure 8).

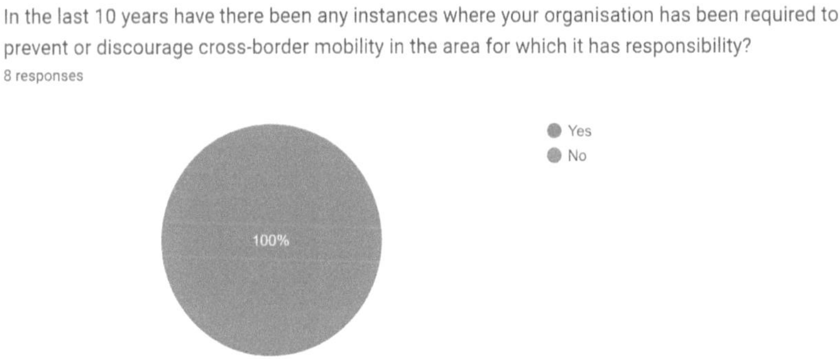

Figure 8: Unanimity in the responses to the question "In the last 10 years, have there been any instances where your organisation has been required to prevent or discourage cross-border mobility in the area for which it has responsibility?"[30]

None of the respondents indicated that their organisation had been required to prevent or discourage cross-border mobility in the last ten years, which includes the period of the Covid-19 pandemic. Nevertheless, the majority (75 %) of these respondents also pointed to interruptions to cross-border mobility, which would have been a result of the public health restrictions put in place in both jurisdictions rather than because of any border closures. Such interruptions, whether occasioned by a pandemic or for any other reason, were considered to be detrimental to cross-border cooperation, as illustrated by one the comments offered by a respondent:

> "Obviously interrupting [cross-border] mobility has potential to directly impact [North-South] cooperation, which we aim to foster and promote.

---

[30] Source: Centre for Cross-Border Studies, Armagh, Frontem survey, 2022.

When barriers arise citizens look at other options which could potentially be to the detriment of [North-South] cooperation".[31]

Interestingly, one of the comments offered by a respondent to the questionnaire aimed at representatives from civic society organisations referred to citizens' recourse to alternative options when faced with restrictions to their cross-border mobility: "During Covid people understood the implications of the government advice. In the past people found a way around restrictions by accessing alternative routes or taking business elsewhere".[32]

However, this comment once again highlights how the pre-1998 border is the ever-present ghost in people's memories. The Covid pandemic is immediately positioned here in a past of security controls and cross-border road closures, where those living in the border region were forced to find other crossing sites or abandon their plans and remain within their own jurisdiction.

The past was also a significant presence when the representatives of civic society organisations were asked about periods of restrictions to cross-border mobility (See Figure 9).

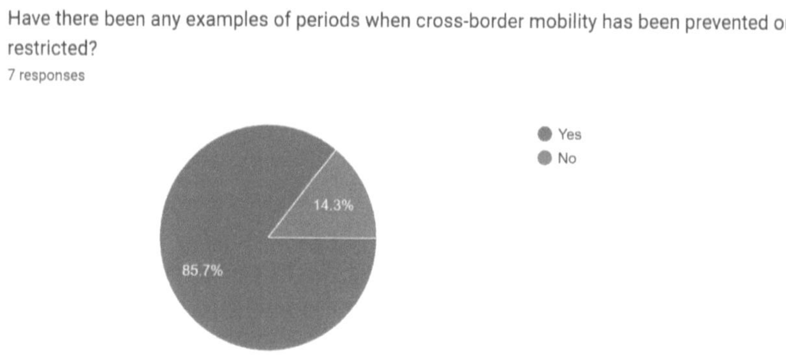

Figure 9: Percentage in the responses to the question "Have there been any examples of periods when cross-border mobility has been prevented or restrictes?"[33]

[31] Frontem survey respondents, Centre for Cross-Border Studies, 2022.

[32] *Ibid.*

[33] Source: Centre for Cross-Border Studies, Armagh, Frontem survey, 2022.

With the majority of them saying there had been periods when cross-border mobility had been prevented or restricted, the additional comments offered frequently intertwined the Covid-19 pandemic and other recent developments with the period of the conflict, as in these examples: "The troubles in the North. Soldiers, checkpoints. Roads closed due to unsafe conditions."; "Recently the Covid-19 pandemic impacted on cross-border travel, yet work continued well online. In my youth cross-border travel was restricted due to the onerous searches and customs points at the border and security crossings."; "When there were border checkpoints, road closures or visa restrictions"; "Checks during the troubles and currently for visa nationals"; "Border controls pre-1990s".[34]

When asked whether there were any circumstances where restricting or preventing cross-border mobility would be justifiable, the responses from the civic society representatives suggested that public health imperatives could justifiably lead to such restrictions, but with a high degree of reluctance: "During global pandemics, in the absence of harmonised all-island approaches"; "Potentially on public health grounds, but not if contravening the GFA [Good Friday Agreement]"; "Only in the event of medical/national health scenarios. But even then I don't feel it should be restricted or prevented – impact limitation actions should be put in place".[35]

However, the responses were much more vociferous when asked what they thought would be the reaction of citizens on one side of the border if the authorities in the other jurisdiction introduced border closures or border controls: "It would be unacceptable."; "Depends – using Covid as an example, outrage from some quarters, others would be more favourable."; "There would be outrage. People on the island of Ireland have family, friends, jobs on both sides of the border. This would not be accepted by anyone."; "Absolutely deplorable, unnecessary and a backwards step"; "Back to troubles – don't think it would be accepted."[36]

These responses, which were echoed again in the focus group discussion, are suggestive of how civic society is not readily accepting of restrictions of or interruptions to cross-border mobility, even when faced with a major crisis. The focus group discussion in particular revealed

---

34 Frontem survey respondents, Centre for Cross-Border Studies, 2022.

35 *Ibid.*

36 *Ibid.*

how during the Covid-19 pandemic citizens were in favour of all-island responses that would obviate the need for border restrictions, and that any such restrictions were indicative of a failure of the authorities to cooperate effectively. And again, restrictions on cross-border mobility were interpreted as signs of a return to a past that had been rejected by the Good Friday Agreement.

Skipping the Border: Cross-Border Cooperation

Although the key actors with oversight of or involved in areas relevant to cross-border mobility appeared to be reluctant to assume significant roles in terms of border management or monitoring, they were much more comfortable in indicating how they collaborated with others in relation to cross-border mobility and other areas of cooperation. When asked in the questionnaire circulated before the focus group meeting whether their organisations collaborated with others in relation to cross-border mobility, 87.5 % of respondents indicated that they did. They reported that their collaboration was principally with government departments and public bodies in both jurisdictions, the formal bodies for North-South cooperation established under Strand 2 of the Good Friday Agreement, health authorities, local councils, chambers of commerce, universities, and NGOs.

The focus group discussion revealed how, faced with the uncertainties resulting from Brexit, key actors were making efforts to maintain their stakeholders' commitment to cross-border cooperation. In the case of those involved in local authority cross-border networks, for example, they responded to these challenges by revisiting the core concerns and commitments of their member councils. In doing so, they drew on their past experience of securing buy-in from stakeholders during the period of the Belfast/Good Friday Agreement. They underlined how the two core motivating factors for increased cross-border cooperation that existed during that period remain now, namely the combination of Interreg and PEACE funding from the EU (which have been integrated into a new programme, PEACE PLUS, for the 2021–2027 period), and the hard evidence that cooperation for cross-border development benefits everyone in the region, regardless of political persuasion. Local authority networks faced the reality that half of their membership remains in the EU and that politicised discussions of the consequences of Brexit were not constructive. In this context, these networks asked local authorities to recommit to cross-border cooperation and, in one instance, to agree a renewed charter for ongoing strategic cross border cooperation.

There was unanimous agreement among the focus group participants on the existence of current and potential challenges to cross-border mobility as a result of Brexit. These included current or potential UK divergence in procurement, General Data Protection Regulation (GDPR), the recognition of professional and technical qualifications, and standards.

Within the healthcare sector, organisations which previously employed medical professionals on an all-island basis reported impacts on workforce planning due to a lack of mutual recognition of qualifications, accentuated by differential wage rates North and South. It was reported that particular impacts were being felt in the north-west cross-border region, where many healthcare services had been established on a dedicated cross-border basis.

Recalling that the majority of the key actors participating in this Frontem focus group did not see themselves as having responsibility for managing or monitoring cross-border mobility, what appears to be emerging here is that these key actors' focus is on continuing cross-border cooperation despite the changing circumstances brought about by Brexit. Their attention is on what lies beyond the border, and not on the border itself. Moreover, the challenges to cooperation they identified are the result of actual or potential UK divergence from the EU and, therefore, the Republic of Ireland. It is London, and not Belfast or Dublin, that is driving divergence, and therefore the policy levers to maintain the conditions for cross-border cooperation and mobility are not in the hands of these key actors.

Conclusion: Managing and Monitoring the Ireland-Northern Ireland Border Remotely

The UK Government has repeatedly been at pains to point out that there will be no immigration controls whatsoever at the Ireland-Northern Ireland border, and that in terms of the movement of goods there will be no "new checks or controls at the international border between Northern Ireland and Ireland".[37] Effectively, this means that management and

[37] UK Government, *The Windsor Framework*, *op.cit.*

monitoring of the Ireland-Northern Ireland border will take place away from the border.

In terms of the movement of people, this is nothing new. For some time, public service providers, employers, landlords and educational institutions, among others, have been responsible for acting as proxy border managers, ensuring that those accessing their services, those they employ or rent their properties have the required immigration status. What *is* new for those in Northern Ireland is that a new category of people will need to prove their right to work or access services: non-Irish EU citizens. Indeed, once the system is in place, such citizens if arriving on the island of Ireland as visitors will have to apply for an Electronic Travel Authorisation before entering the UK, according to the new UK Nationality and Borders Act 2022. This will include those who may have arrived in the Republic of Ireland, but wish to visit Northern Ireland. However, again, it is important to recall that there will be no immigration controls at the Ireland-Northern Ireland border, meaning that enforcement of this new system will have to take place away from the border.

There is little appetite among key actors on the island of Ireland with oversight of or involved in areas relevant to cross-border mobility to be responsible for such management or monitoring of the border. It appears antithetical to the principle of maintaining the necessary conditions for continued North-South cooperation, changing the underlying context that has existed since the 1998 Good Friday Agreement.

Crucially, whereas management of the Ireland-Northern Ireland border may take place remotely, that is not the case for the maritime border between the Republic of Ireland and Great Britain, nor indeed for the movement of goods between Great Britain and Northern Ireland. If no new infrastructure has appeared at the land border, it is emerging at Northern Ireland ports, where monitoring of transit from Great Britain is taking place. The becoming visible of the Irish Sea border has led to the collapse of government in Northern Ireland, with the Democratic Unionist Party refusing to form an Executive until this border between Northern Ireland and the rest of the United Kingdom is removed. However, the abandonment of border management and monitoring at Northern Ireland ports would immediately require it to take place at the land border. This would be to go back to the pre-1998 border, and to fail to comply with the commitment made by the UK and EU to avoid a "hard" border between Ireland and Northern Ireland.

The question of managing and monitoring the land border on the island of Ireland carries with it a heavy load. As was seen in the Frontem focus groups, the concept of border management and monitoring raises the ghosts of the past and the memories of a border amid a conflict that it was hoped had been left behind. It is a concept that also reminds those in Northern Ireland who voted against Brexit that they are no longer part of a space that brought them beyond the constrictive and divisive space they had been largely limited to. The idea of managing and monitoring the border makes the invisible visible, at least psychologically. However, by keeping the land border invisible, it has made visible a maritime border that is being fiercely contested by unionists who view it as an unacceptable fault-line between Northern Ireland and the rest of the United Kingdom, and thereby as an assault on their British identity. At the heart of the question of managing and monitoring of borders in the current context of the island of Ireland is whether it can ever fulfil the commitment to protect the 1998 Agreement in *all* its dimensions. This is the dilemma created by Brexit.

Bibliography

Case Studies, Strasbourg, 2023.

Centre for Cross-Border Studies, Armagh, Frontem survey, 2022.

Department of the Economy, "Background Evidence on the Movement of People across the Northern Ireland-Ireland Border", 2018.

UK Government, "Good Friday Agreement", 10 April 1998.

UK Government, "Protocol on Ireland and Northern Ireland", 31 January 2020.

UK Government, "The Windsor Framework: A new way forward", 2023.

University of Strasbourg/Euro-Institute, *Toolkit on Models of Border Management and Perception in the EU. 5*

Practices of Cross-Border Mobility. The Standing Conference on Teacher Education North and South SCoTENS as a Learning Cross-Border Organization

Kathy Hall

Introduction

This short chapter is based on a presentation at Frontem's event, "The Northern Ireland-Ireland border region: Policies and practices" at Queens University Belfast in March 2022. This piece draws on various papers presented at the American Educational Research Association Conference in New York in April 2018 for which Kathy Hall was a co-presenter along with Marie Campbell, Linda Clarke, Conor Galvin, Teresa O'Doherty and Noel Purdy. The chapter outlines the background and rationale of SCoTENS as a learning cross-border organization and then goes on to describe its objectives and main activities. There is reference to its impact before drawing a conclusion.

Origin, Context and Rationale

The Standing Conference on Teacher Education North and South (SCoTENS) is a unique network of teacher educators from north and south of the border which separates Northern Ireland from the Republic of Ireland. SCoTENS was first conceived following the Belfast /Good Friday Agreement in 1998[1] and was formally established in 2003. While

1 J. Coolahan, J., "SCoTENS: How teacher educators took the lead in North-South cooperation", *Journal of Cross Border Studies*, 3, 2008, pp. 50–55.

its *raison d'être* lies in peace building, SCoTENS' objectives are focused primarily on cross-border co-operation for the enhancement of teacher education on the island of Ireland.

It is a network of 40 colleges of education, university education departments, teaching councils, curriculum councils, education trade unions and education centres on the island of Ireland with a responsibility for and interest in teacher education in the broadest sense. Thus, the membership is drawn from the full range of stakeholder institutions for teacher education across the island. A part of the broader peace dynamic, SCoTENS has always been rooted in the deepest commitment to quality teaching and learning for all. Teacher education was perhaps an unlikely vehicle for the promotion of peace on the island, yet it has been posited that SCoTENS, an organisation dedicated to the promotion of teacher education, has been one of the most effective cross-border initiatives in Ireland. SCoTENS is the only network of its kind in the world, operating across a contested border.

While teacher education on the island of Ireland had a shared history and structure until partition in 1921, thereafter a divergence emerged in the respective policies and practices of the two jurisdictions- north and south. However, a shared passion for and commitment to promoting research in education contributed to the establishment of the Education Studies Association of Ireland (ESAI) in 1976 as an all-Ireland organisation. In the early years following, the annual conferences of the ESAI were held in Northern Ireland in 1978 and again a decade later in 1988, while it revisited Belfast in 1998 and in 2003. For many involved in Irish education, these trips to the ESAI conferences were their first forays across the border into Northern Ireland. Travel to the North was considered to be a risk, and the return South from the 1998 conference where the train carrying many of the southern delegates was terminated some short distance from Belfast, due to the identification of an incendiary device on the train line, served to underscore how politics and violence were encountered in everyday life for colleagues who lived and worked in Northern Ireland.

Building on a deep commitment to quality teaching and learning for all, and a shared tradition where teaching is a valued profession, SCoTENS was established as a non-political and non-sectarian organization working across higher education institutions on the island of Ireland in support of the peace process. As will be described below this was achieved though the development of a number of shared learning activities that specifically support collaborative research and professional activities.

The growth in the peace process paralleled a growing interest in teacher education policy renewal and development. While teacher education policy had experienced extended periods of policy drift and inertia in the South,[2] the first decade of the 21st century witnessed a renewed interest by national and international agencies in teaching and teachers (OECD, 2005).[3] Consequently the platform provided by SCoTENS for shared policy discussion and the professional development of teacher educators became increasingly important and valuable as the years progressed. The work of SCoTENS has gathered momentum since its establishment, mirroring the enhanced level of engagement by policy makers in the area of teacher education, not only for pre-service teachers, but for all teachers across the continuum.

The emergence of the SCoTENS initiative was characterised by a new cautious political optimism in North-South relations and the establishment of new north-south political institutions such as the North/South Ministerial Council which aimed to develop consultation, cooperation and action in areas of mutual interest. Amid the optimism, there was still much political caution and even resistance to the Good Friday Agreement and opposition to the idea of cross-border institutions of any kind. For instance some 29 % of the electorate had voted "NO" in the referendum on the Agreement on 22 May 1998.

Key individuals, meetings, happenstance, serendipity conspired to enable the birth of SCoTENS. In an interview carried out in September 2017 with the two principal founding fathers of SCoTENS, Professor John Coolahan (Republic of Ireland) and Professor Harry McMahon (Northern Ireland), John Coolahan talks of the initial caution among civil service departments in Northern Ireland who were "a little bit hesitant" when first approached for financial support following the initial North/South teacher education conference in Belfast in May 2000. The turning point came when Coolahan's participation in a government review of

2 T. O'Doherty, "Defining moments in policy development, direction, and implementation in Irish initial teacher education policy", *Center for Educational Policy Studies Journal* 4, 2014, p. 38.

3 OECD, *Teachers matter: Attracting, developing and retaining effective teachers*, OECD, Paris, 2005.

post-primary schooling in Northern Ireland (*The Burns Report*, 2001)[4] gave him the opportunity to have a "good chat" over dinner with Martin McGuinness (then Sinn Fein Minister of Education in the Northern Ireland Assembly, and former IRA leader). Coolahan recalls explaining his "plan" for a cross-border teacher education body within the "Good Friday Agreement climate" of the time. He recounts that McGuinness "listened very carefully" and asked for a follow-up briefing paper to be sent to him, after which "he signalled to the others to play ball with this." As Coolahan adds: "there were no other problems then after that" and support was quickly gained from both the Department of Education and the Department of Employment and Learning (responsible for Higher Education in Northern Ireland). He concludes that "having made the break", the previously hesitant government departments in Northern Ireland did buy into the project. Indeed, Coolahan notes that Ministers from both sides of the Irish border "honoured" SCoTENS by frequently coming together to open the Annual Conferences, a practice which has continued, diaries permitting, right up to the present day.

The intervention by Martin McGuinness sparked a series of developments which led to the decision in July and September 2002 by the Department of Education and the Department of Employment and Learning in Northern Ireland and the Department of Education and Science in the Republic of Ireland to financially support the work of SCoTENS, initially committing £40,000 from each jurisdiction for a two-year pilot which would run from February 2003 to February 2005. The committee met in September and November 2002 to prepare a programme of work to be submitted in December 2002 to the respective Departments for approval, and negotiated arrangements for the secretariat to be provided by the Centre for Cross Border Studies in Armagh. From the outset the constitution of the SCoTENS Committee was carefully balanced: five members from Initial Teacher Education (ITE) institutions in Northern Ireland, and five from the Republic of Ireland as well as one member from the Centre for Cross Border Studies. The committee, from the outset, had two co-chairs, one from the North and the other from the South.

---

4 Department of Education, Northern Ireland, "Education for the 21st Century – Report by the Post-Primary Review Body [The Burns Report]", 2001. Retrieved from <https://www.education-ni.gov.uk/publications/report-review-body-post-primary-education-burns-report>. Accessed 23 November 2023.

Since 2006 the contribution from officialdom in both jurisdictions is extended considerably through additional funding from subscription fees from member institutions, thus reducing the dependence on governmental funding alone. This fee in the case of teacher education providers at college and university level is based on the number of students registered for the study of teacher education in the respective institutions.

Objectives, Practices, Activities

The objectives of SCoTENS are set out on its website as follows: to provide a supportive framework for collaborative research and professional activities in teacher education, North and South; to hold invitational conferences on themes of mutual interest to teacher educators, North and South; to provide seed funding for North-South research projects on teacher education issues; to promote position papers on issues of mutual concern to teacher educators, North and South; to use its good offices to assist in obtaining funding for approved research activities; to support North-South exchanges between teacher educators for approved purposes as part of its concern to strengthen inter-professional and inter-institutional linkages; to act as an agency for advice and consultation by policy makers in the Education Departments of both jurisdictions; and to establish and maintain a website which will provide information and resources of interest to teacher educators and researchers, North and South.[5]

In terms of activities and practices to achieve these objectives SCoTENS organises several events and practices on a yearly basis. An annual conference (including roundtables for doctoral researchers) and a cross-border student–teacher exchange are among key activities. Crucially, it also provides seed-funding which supports collaboration between groups of teacher educators in designing, developing and implementing small-scale, north/south research projects. The scale of achievement relating to these activities are well documented.[6]

---

5 SCoTENS: <https://scotens.org/>. Accessed 21 November 2023.

6 L. Clarke & M. Campbell, "Reimagining SCoTENS for the Future: observations from an on-going project in organisational self-renewal", Paper presented at the American Educational Research Association (AERA) Conference, New York, 2018; L. Clarke et al., "Assessing the value of SCOTENS", *op.cit.*; C. Galvin, "The standing conference on teacher education", *op.cit.*

The SCoTENS annual conference is a key fixture in the education calendar on the island of Ireland. It provides a forum where teacher educators, policy makers, teacher practitioners, and education leaders can engage in open, critical and constructive analysis of current issues in education with a view to promoting a collaborative response to these issues bridging research, policy, practice. International scholars are invited to give keynote addresses while delegates are invited from the north and south to work on joint presentations related to the theme of the conference. A doctoral conference within the main conference is also featured where current doctoral students are invited and funded to present work in progress pertaining to the broad theme of mutual understanding across both jurisdictions. Among the conference themes in recent years are the following: education as a common good; education for social justice; educational innovation: the challenge of evidence-informed change; leadership across boundaries: challenges for educators and teacher educator; communities of practice: learning together to teach together; and reflecting, reconnecting and re-engaging with the core purpose(s) of education.

The establishment of the SCoTENS north-south exchange programme for student teachers has been particularly well received politically and described as an "unequivocal success" in terms of professional action.[7] Designed to enable students undertake a teaching placement in the neighbouring jurisdiction, this process has promoted cultural sensitivity by engaging students educationally and socially with each other, teachers and pupils in a cross-border, cross-community context on the island of Ireland. This innovative cross-border exchange project is interested in learning how to overcome barriers, reduce prejudice and lead to increasing co-operation between colleges of education and their students on the island.

The provision of high quality research and resource materials through the SCoTENS website is another important facet of SCoTENS work. It offers valuable, contextually relevant and timely resources to students and teachers in their classrooms – the materials, reports, examples of good practice, have been generated by Irish teachers and researchers and are culturally responsive to the needs of schools on the island. Working with

7 SCoTENS, Annual Report, 2014, p. 52. Retrieved from: <https://scotens.org/>. Accessed 21 November 2023.

and through materials which are relevant and appropriate to the needs of teachers is an excellent method of ensuring that policy becomes part of the pedagogy and practice within classrooms.[8]

Impact

Evidence from commissioned independent evaluations together with a major values-oriented investigation conducted by several committee members (all volunteers) shows the positive impact of SCoTENS as an organisation seeking to bridge boundaries on the island.[9] Educationalists, both North and South, who had engaged with aspects of the SCoTENS project, indicate interesting and encouraging value-gain and benefit. In sum, research participants attribute significant professional development to their engagement with SCoTENS and align the extent of this development with the level of interaction SCoTENS facilitated with other professionals from across the North/South divide. Themes highlighted in reviews and evaluations note the value of all of the following: mentoring, opportunities to collaborate with colleagues from the "other" jurisdiction, exposure to difference, the stimulation of critical thinking and empowerment to move beyond "comfort zones".

It is clear that those who engage with the research process have benefited professionally from their participation in SCoTENS and attribute this to the North/South dimension as evidenced by the theme of leadership together which was prevalent throughout the findings presented in Clarke et al. (2021).[10] Student teachers on the exchange programmes feel challenged and supported in adjusting to different curricula and teaching cultures. The doctoral students appreciated the opportunity to "go public" with their work in a different jurisdiction.

The challenging of taken-for-granted assumptions about "the other" and the embracing of opportunities to do things a little differently feature in several of the conversations among the SCoTENS seed-fund recipients.[11] They see the funding, though small in real terms, not only as a previously unavailable impetus to conduct cross-border educational

8 See SCoTENS: <https://scotens.org/>. Accessed 21 November 2023.

9 L. Clarke et al., "Assessing the value of SCOTENS", *op.cit.*

10 *Ibid.*

11 *Ibid.*

research with immediate and longer-term benefits (such as engaging in professional dialogue, learning from each other, finding common interests, and instigating subsequent funding applications) but also as facilitating strongly transformative experiences in itself because of the nature of the research topics and the openness of SCoTENS to a rich variety of activities SCoTENS-funded research is seen as a catalyst for a related shift in teaching, research interest, and professional understandings.

Teacher educators within the SCoTENS community appear to have benefited tremendously from its research and networking opportunities. Of note in this respect is that SCoTENS would appear to support two agendas – the personal and collective. Academics value the forum afforded for competitively generated income (albeit small-scale), networking, research activity, and publication – a theme increasingly relevant for higher education staff on both sides of the Irish border in terms of professional and academic respect and promotional prospects. However, this personal benefit completely aligns with the collective – in that only projects that promote mutual understanding across both jurisdictions are funded. Hence, the corpus of work now available on the SCoTENS website and published variously in respected professional and academic journals bearing on this broad principle linked to teacher education.[12]

SCoTENS is now recognised at home and abroad e.g. by OECD and UNESCO as a remarkable experiment in teacher education and cross border co-operation. The first Southern Joint Secretary of the North South Ministerial Council (Tom O'Connor) noted that SCoTENS is an outstanding example of co-operation between professional bodies that is an example for others to follow while the independent evaluator John Furlong observed that "there's a broader agenda … greater mutual understanding, reducing barriers. It's a long-term process, takes several generations. If there is a range of cross border activities, a sense of normality sets in – people get used to working together. Over time, a generation, a different climate emerges".[13]

12 See SCoTENS: <https://scotens.org/>. Accessed 21 November 2023.

13 J. Furlong, A. Pendry & P. Mertova, *SCoTENS: An Evaluation of Its First 8 Years*, Oxford University, Oxford, 2011, p. 21.

Conclusion

The future of SCoTENS looks promising based on the experience to date. However, in a post-Brexit world the articulation of its value as a unique, all-island professional learning network will have to continue to be assessed and demonstrated using critical concepts and evidence. By focusing systematically on what SCoTENS has done to date and its activities – through the voices and insights of those involved, it will be important to conduct timely evaluations of its culture and practice.

Readers outside Ireland could be forgiven for thinking that the account offered of an initiative designed to encourage collaboration and sharing among teachers, teacher educators and student teachers is somewhat unremarkable. After all, on the face of it, is it not the positive account one might expect when peers and professional colleagues come together to work on projects pertaining to their professional lives, common interests, and settings? However, it is in fact quite remarkable, significant and inspiring. Prior to SCoTENS, there has not been a tradition of formal academic and professional engagement across teacher education on this island. While there were undoubtedly examples of sharing and co-operation among academics, a case in point being the cross-jurisdictions ESAI with its well-regarded academic journal and annual conference, mentioned earlier, there was no state/official support mechanism that enabled the two parts of the island to engage around teacher education, and there was no official mechanism for bringing teachers, teacher educators and student teachers into dialogue together. Indeed, it is far more likely that colleagues, north and south, have typically had more involvement with their counterparts in England, Scotland or Wales or elsewhere, than with their peers across the two parts of Ireland itself. Thus, the comparative lived experiences, educational practices and policies of those in our nearest jurisdiction remained mostly hidden until the late 1990s.

SCoTENS has helped the teacher education community on the island of Ireland to move on from the political turmoil and violence which led to the loss of over 3,000 lives in the years leading up to the relative peace which developed following the signing of the Belfast /Good Friday Agreement in 1998. The opening-up of conversation among "the geographically close" if "culturally different" is generating awareness of both difference and commonality, but, most importantly, fostering trust, familiarity, and mutual under-standing.

Bibliography

L. Clarke & M. Campbell, "Reimagining SCoTENS for the future: Observations from an on-going project in organisational self-renewal", Paper presented at the American Educational Research Association (AERA) Conference, New York, 2018.

J. Coolahan, J., "SCoTENS: How teacher educators took the lead in North-South cooperation", *Journal of Cross Border Studies*, 3, 2008, pp. 50–55.

J. Furlong, A. Pendry & P. Mertova, *SCoTENS: An Evaluation of Its First 8 Years*, Oxford University, Oxford, 2011.

T. O'Doherty, "Defining moments in policy development, direction, and implementation in Irish initial teacher education policy", *Center for Educational Policy Studies Journal* 4, 2014, pp. 29–49.

OECD, *Teachers Matter: Attracting, Developing and Retaining Effective Teachers*, OECD, Paris, 2005.

Department of Education, Northern Ireland, "Education for the 21st Century – Report by the Post-Primary Review Body [The Burns Report]", 2001.

SCoTENS, Annual Report, 2014.

Changing Perspectives on the Irish Border, from the Southern Border Counties. Negotiating New Regional Frames and Constitutional Debate[1]

Jennifer Todd

From the end of the 20th century, we have become used to seeing borders as bridges as well as boundaries, as cultural units of interaction and fluidity and trans-state governance even more than as areas of state contest and state sovereignty. The importance of border regions has been emphasised practically and discursively by the European Union (EU). Meanwhile, in Ireland, the period beginning with the ceasefires of 1994 and continuing through the Good Friday Agreement (GFA) of 1998 coincided with increasing EU economic support for the border area.[2] The GFA itself provided for all-island and cross border institutions and increased linkages and economic flows, to be instituted by the British and Irish governments with key Northern Irish ministerial participation.[3] This was supported not simply by the EU but also by the International Fund for Ireland.

Of course, the Irish border is also a line of state sovereignty, it was for long an arena of smuggling and, for over a quarter century after 1969, a military resource for the paramilitaries who crossed it, and for the state forces who closed it and guarded it.[4] Since Brexit, the Irish border is

---

1 I acknowledge funding from the Irish Research Council New Foundations Scheme; the Shared Island Unit, the Irish Department of the Taoiseach; and the Irish Department of Foreign Affairs, Reconciliation Fund.

2 B. Laffan & D. Payne, "Creating living institutions: EU cross border cooperation after the Good Friday Agreement", *Centre for Cross-Border Studies*, Report, Armagh, 2001.

3 J. Coakley & L. O'Dowd (eds.), *Crossing the Border: New Relations between Northern Ireland and the Republic of Ireland*, Irish Academic Press, Newbridge, 2007.

4 H. Patterson, *Ireland's Violent Frontier: The Border and Anglo-Irish Relations during the Troubles*, Palgrave MacMillan, New York, 2013.

again problematised and much political and social effort has focussed on ensuring that it does not again become a barrier to movement, trade or security. Whether and how the emerging debate on Irish unification will affect this remains unclear.

In this paper, I look at the changing everyday perceptions of the border area and of North-South linkages from the period immediately after the GFA to the present. I trace how respondents in the border counties use their local expertise to attempt to steer constitutional discussion along safe and constructive paths. I focus on perceptions within the Southern border counties (particularly the counties of Louth, Monaghan and Cavan) mentioning other counties only occasionally. These Southern border counties are important not just because they are less well studied than the North, but also because they give the clue to how Southerners respond to Northerners, and how far they will move to accommodate them politically.

My data are primarily from interviews and focus groups undertaken in research projects in 2004–2006, just after the GFA (55 interviews in Louth/Monaghan border area); in 2014, just before Brexit (17 interviews, Louth, Cavan, Donegal); and after Brexit in 2020–2022 (focus groups and deliberative cafes, 45 participants Louth/Monaghan/Cavan/Donegal).[5] The samples are not representative, but they are meaningful, grasping a range of perspectives in the local areas. The research was undertaken as part of wider projects that focussed on diverse perspectives on the island of Ireland: the changing border focus was brought by the participants, it is therefore one of the research findings, not a product of our research agenda.

The interviews were open-ended, focussing on how participants situated themselves in a changing political context. The focus groups were

[5] I have many acknowledgements, first to those interviewed, second to colleagues who worked as co-PIs on the projects, Orla Muldoon, Karen Trew, Joanne McEvoy, Dawn Walsh, and to research fellows Lorenzo Cañás Bottos, Nathalie Rougier, Susan McDermott, and Oisin O'Malley-Daly. Method and findings are summarised in other publications, in particular J. Todd, *Identity Change after Conflict: Ethnicity, Boundaries and Belonging in the Two Irelands*, Springer International, New York, 2018 (see especially. pp. 235–264) and J. McEvoy, J. Todd & D. Walsh, "Participatory constitutionalism and the agenda for constitutional change: Socio-economic issues in Irish constitutional debates", *Irish Studies in International Affairs*, vol. 33, n°2, 2022, pp. 140–171. In this chapter, I draw on already published analyses of these research projects and compare findings over time.

loosely-semi-structured, starting with a political issue (constitutional change) and asking participants how it was important (or not) to them, and how they would want to frame an agenda around the issue. The deliberative cafés followed the suggestions of the focus groups by beginning with a practical policy issue (health provision in the border area), providing expert information, and asking the participants to help define policy and political priorities.

Contextualising the Changing Irish Border

The Irish border area has fluctuated in extent, depending at once on the activity and the period.[6] It incorporated large towns in narratives of Irish Republican Army (IRA) camaraderie in the 1940s and 1950s. It might be condensed to a few roads in areas of military contestation in the 1970s and 1980s.[7] During the troubles, it was described in terms of penumbra of danger radiating from core areas through to normally safe havens. More recently, it has become at once a liminal zone, where state boundaries are not seen until it is clear that one has passed some miles beyond them, and a territorially delimited area for European funding.[8]

In all of these guises, the border area is more important politically than the numbers who live there might suggest. Identity formation in "frontier" areas, where state-ness is routinely problematised or challenged, throws light on national identity and national distinctions more generally.[9] In Ireland, the economic problems and demographic decline of the area, and the importance of EU funding, suggest that it may also be an area where transgressive, hybrid, cross-border and European identities are likely to emerge faster than in other parts of the island.[10]

---

[6] J. Coakley & L. O'Dowd "The transformation of the Irish border", *Political Geography*, 2007, vol. 26, pp. 877–885.

[7] H. Patterson, *Ireland's Violent Frontier*, *op.cit.*

[8] K. Hayward, *What Do We Know and What Should We Do about the Irish Border?*, Sage, Thousand Oaks, 2021.

[9] J. A. Agnew, "A world that knows no boundaries? The geopolitics of globalization and the myth of a borderless world", CIBR working papers in Border Studies, Queens University of Belfast, 2003; H. Donnan & T. M. Wilson, *Borders: Frontiers of Identity, Nation and State*, Berg, Oxford, 1999, pp. 45–50.

[10] K. Hayward, "From barriers to bridges: The europeanisation of Ireland's borders", CIBR Working Papers in Border Studies, Queens University of Belfast, 2004.

My research in the border area suggests that changes took place in a patterned sequence, affecting not just North-South interactions but also internal religious and political oppositions in the South, and perceptions of all of these interrelations. Although this article focusses primarily on the changing views of those on the Southern side of the border, there is much evidence that views on the Northern side also changed, although in different phases.[11]

2004–2006: A Cautious Thaw[12]

The border region for its older inhabitants was a unit of long familial residence, with family relationships spanning the border area.[13] The border itself was explicitly ignored as irrelevant to identity and to linkages. In small Southern towns, it was common to answer questions about the border by saying "Border, there is no border here" or to deny real distinctions between North and South "It's all the same, we're all the same." (Female second generation Catholic). These were, however, sensitive issues about which people were slow to speak, especially in border towns that had experienced violence. The almost ritual denial of the existence

Retrieved from: <https://www.researchgate.net/publication/258862043_The_moral_boundaries_of_the_nation_subverting_state_borders_and_constituting_national_identity_in_the_South_Eastern_border_counties_of_the_Irish_state>. Accessed 27 April 2023; B. Laffan & D. Payne, "Creating living institutions: EU cross border cooperation after the Good Friday Agreement", *op.cit.*; E. Tannam, "Cross-Border co-operation between Northern Ireland and the Republic of Ireland: Neo-functionalism revisited", *The British Journal of Politics and International Relations*, vol. 8, n°2, 2006, pp. 256–276.

11 Between 2003 and 2006, I found that people in Northern Ireland were twice as likely to move significantly away from traditional identities as those in the Republic of Ireland, although by 2014 there was reversal of change in the North, and acceleration of it in the South. See J. Todd, *Identity Change after Conflict: Ethnicity, Boundaries and Belonging in the Two Irelands*, *op.cit.*, pp. 97–122.

12 Quotations and analysis are largely drawn from J. Todd, O. Muldoon, K. Trew, L. Cañás Bottos, N. Rougier & K. McLaughlin, "The moral boundaries of the nation: Nation, state and boundaries in the Southern Irish border counties", *Ethnopolitics*, vol. 5, n°4, 2006, pp. 365–382. See also J. Todd, *Identity Change after Conflict*, *op.cit.*

13 L. Cañás Bottos & N. Rougier, "Generations on the border: Changes in ethno-national identity in the Irish Border area" *In* J. Todd, L. Cañás Bottos & N. Rougier (eds.), *Political Transformation and National Identity Change: Comparative Perspectives*, Routledge, London and New York, 2008, pp. 303–328.

of the border was a way at once of asserting a deeper identity than state boundaries defined, and of avoiding contentious discussion.

And yet state distinctions were evident. One older respondent from Louth remembered encountering the Northern security forces in the 1950s: "you were amazed, probably even shocked by [...] the colour of the uniforms [...] the dark black uniform [...] the first thing you noticed was the guns and you found that quite overpowering really" (male, first generation, Catholic). And the border, albeit liminal and denied, was also crucial to the conduct of everyday life: "County Monaghan has no natural resources, well, the border is its natural resource, and its inhabitants have always mined it" (male second generation Catholic).

When pressed, respondents in the Southern part of the border area described the difference between North and South as one of values, not of nationality: speaking of "refugees" from the violence in Northern Ireland, one respondent said: "They'd even take over the whole conversation. We were placid, quiet, but they were all gun blazing" (male, second generation, Catholic). "You could get beaten up in the pub for saying something, if you said the wrong thing they'd look at you and you might get beaten up for saying something" (male, second generation, Catholic). We concluded that border dwellers in the Republic of Ireland constituted the Irish nation and its boundaries morally, and that the North was included only as its people accepted the moral standards of the South.[14]

The border was also a zone where a significant minority of Protestants lived. Into the 2000s, they remained largely endogamous. Many of the Protestants we interviewed were critical of state narratives and public practices: even the fact that Gaelic Athletic Association (GAA)[15] games were played on Sundays meant that they and their children stayed separate from this important sphere of rural social life. They were Irish but, they themselves said, "I suppose we are probably not traditionally Irish at all" (male, second generation Protestant). They distanced themselves from local Catholics in their understandings of history and in their attitudes to key symbols of the state, like the national anthem: "I just couldn't walk into my local village [...] and [...] tell them there what I'm telling you, I don't think that would go down too well" (male second generation Protestant). But they also distanced themselves from Protestants

---

14 J. Todd et al., "The moral boundaries of the nation", *op.cit.*

15 GAA games include Gaelic football, hurling and rounders.

across the border: "I think em just the hard line, the […] southern protestants look at the north and see the very very hard line that Ulster says no to everything, you know" (female, first generation Protestant). "I'm a Protestant, I'm proud to be, you know, but I don't agree with what the loyalists, you know, the whole thing in the North, fighting over […]" (female, second generation Protestant).

2014: Relaxation[16]

Ten years later, in 2014, we conducted a new set of interviews. At this time, respondents had relaxed into the peaceful and open post Good Friday interaction. Cross border trade, training, sport had increased: "Yes, well I would say, you know, the big change has been perhaps in people's minds: there is no border" (male first generation Catholic). A GAA enthusiast spoke of the new diversity in the GAA and a new ease in cross-border interaction. His older Protestant neighbour agreed that tensions had eased and that religious distinction did not matter much to the young, although she noted that in the past "you wouldn't want to give your mind at all". She suggested this hadn't totally gone away "It's like the elections coming up, you just have to be careful. They're a little bit hot and bothered now, and I don't blame them. […] And the best way is to keep quiet."

In general, however, political sensitivities and caution had relaxed. In Dundalk, both older conservative Catholics and younger republicans felt the atmosphere had become "less harsh, less entrenched, more willing to reach out the hand and see the other point of view" (third generation female Catholic background). People are now "more tolerant of other people's views […] there's an openness and people will talk an awful lot more and not afraid to express the view that they may have" (first generation male Catholic). Farther west, in Donegal, young people on both sides of the border spoke of the benefits of mixed schooling and freely crossed the border to socialise and to go to university.

Was there a shared sense of being part of a border-region? Perhaps, but that was not what our respondents emphasised: they did not define

---

16 Quotations and analysis are drawn primrily from J. Todd, *Identity Change after Conflict*, *op.cit.*, pp. 201–205.

a new territorial unit, but rather focussed on the easing of constraints on interaction and movement.

2020–2022: After Brexit[17]

By 2020, the narratives were different. In focus groups that I conducted with Joanne McEvoy and Dawn Walsh, women recounted the great change they had experienced with the GFA. Their security fears were lifted and they no longer had to worry that their relatives would be arrested for no reason. The end of violence and the normalisation of the border region was a great relief. These women identified as "border dwellers", with a clear strong regional identification with the border region. Some lived in one jurisdiction and worked in the other, and their organisations were tasked to span the border, collaborating with sister organisations on the other side. They were very aware of EU funding, and they shared the belief that they should emphasise shared experience not divisive politics. There was also a regionalist distancing from Dublin. The women said that politicians "don't care" about the border area. One noted the misrecognition she systematically faced: she went to meetings in Dublin, and said that anyone North of Drogheda was taken to be from Northern Ireland. The border region, our respondents said, had a definite identity and a set of distinct social problems, and its residents had to be listened to.

The women also feared the return of violence. One respondent expressed "the fear and threat" of constitutional discussion, advising that "there is huge fear in Northern Ireland and the border communities, north and south", largely due to Brexit. Another suggested that heightened constitutional discussion is "not a good thing […] it's almost as if we're speaking about division again". Many of them had a long history of cross-border peace building and they spoke of their fear of the remaining

17 The focus groups were conducted with Joanne McEvoy and Dawn Walsh and facilitated by the NCCWN (National Collective of Community Based Women's Networks, now WCI Womens Collective Ireland). Most quotations are from J. McEvoy, J. Todd & D. Walsh, "Participatory constitutionalism and the agenda for constitutional change", *op.cit.* Some quotations are from J. McEvoy & J. Todd, "Constitutional inclusion in divided societies: Conceptual choices, practical dilemmas, and the contribution of the grassroots in Northern Ireland and the Republic of Ireland", *Cooperation and Conflict*, 2023. Retrieved from: <https://journals.sagepub.com/doi/full/10.1177/00108367221147790>. Accessed 27 April 2023.

"sensitivities". One woman noted that when you got Northerners and Southerners together, Northerners would ask the Southerners "where were you during the war?"

They brought this experience to bear in discussing the emerging debate on Irish unification. As we wrote in a recent paper, they wanted to refocus the conversation from ideology to shared experience. For example, one focus group participant said that "united Ireland" and "unification" "are very loaded terms" that can cause a "huge reaction in people, either pro or anti". Constitutional politics was seen as far removed from people's priorities. One focus group participant commented: "when people talk about a united Ireland, there is no sense of what the reality or the practicality of the lived experience is like [...] or how those conversations about or those statements affect the lived reality of people in Northern Ireland and the border regions".

This was not simply a retreat to particular interests. Participants interrelated their particular lived experience with more general social issues relevant to constitutional division. There was an important logic in their views that narrow constitutionalism misses. There are different levels of socio-economic provision and different rights in each jurisdiction on the island of Ireland.[18] The GFA provided for harmonisation of rights and equality across both parts of the island, but this has not been fulfilled. These differences in provisions tap into differences in legal practices, constitutional traditions and cultural expectations in each jurisdiction. Thus, discussion of the different welfare and rights provisions leads directly into properly constitutional issues – the foundations of social order, the basic assumptions about authority – providing an entry point to discussion about the relative merits of British and Irish modes of governance that can be relevant to everyone, and allow analysis of the prerequisites, costs and benefits of unification.

To summarise, our border focus group participants were disengaged from political parties not from constitutional discussion. There was widespread support for an inclusive process of discussion around North-South relations. There were multiple reasons why they had not participated in such discussion to date but above all, the reason most

---

18 M. P. Murphy "A new welfare imaginary for the island of Ireland", *Irish Studies in International Affairs*, vol. 32, n°2, 2021, pp. 532–557; C. Fitzpatrick & C. O'Sullivan, "Comparing social security provision north and south of Ireland: Past developments and future challenges", *Irish Studies in International Affairs*, vol. 32, n°2, pp. 283–313.

frequently mentioned was their desire to change the language of debate. They wanted to talk about real issues, bread and butter issues, how best to foster gender rights and a vibrant democracy. They wanted to short-circuit polarising political meanings through "experiential politics". They were adamant that the border experience is important, that it shows up the problems on the island in a really clear way.

In part as a response to this, Joanne McEvoy and I began a new phase of research in early 2022, using a deliberative café method, and focussed on issues of transborder socio-economic concern. We began with the topic of public health in the border area, and organised 3 cafés in a local hotel, each lasting 3 hours, facilitated by our partners in a local womens community organisation.[19] Again, the participants were women and the smallest café had 8 participants and the largest around 20. We prepared information sheets, organised zoom based Q/A sessions with an expert in public health, and provided tea and snacks.[20] During the first hour we invited participants to share experiences of health provision in the region; the second hour centred on policy discussion and questions for our expert; after the Q/A the final half hour concerned wider political issues. Analysis is still ongoing, but we have several provisional findings.[21]

Participants shared many personal narratives of dysfunctions of the health system in the border region, and moved on to credible analysis of regional problems of public health, which were not being tackled by the requisite agencies. Problems, which they said had got worse, included lack of transmission of information from hospitals in one jurisdiction to health – care providers in the other – patients themselves have to copy files and take them to their General Practioner; ambulance services may take a critically ill patient to a hospital an hour away on the same side of the border rather than to one 5 minutes away on the other side; and there is a lack of efficacy of the relevant agencies, which are "overloaded with bureaucracy".

They went farther to analyse the systemic regional problems, for example lack of digitisation that would allow seamless transfer of information,

---

19 Thanks to the organisers in Women's Collective Ireland.

20 Particular thanks to Dr Ann Nolan, TCD, who was our expert on public health and answered questions by zoom in all 3 cafés.

21 The discussion that follows draws on the draft paper by J. McEvoy & J. Todd presented at the ECPR conference on gender and politics in Llubjana, July 2022.

and inadequate clarification of roles and responsibilities across different health agencies. And they were clear that lack of harmonisation of policies on each side of the border had bad social and political effects, retriggering North-South tensions. For example, the different Covid regimes caused resentment – one participant mentioned the "cowboys" from the North, at the height of a Covid surge, who came South without masks, laughing at the people on the Southern side of the border.

But when it came to policy and politics, there were few answers; the problems outlined were too big, they existed on each side of the border, and coordination between two problematic systems did not seem possible: for one woman living south of the border "until we get our system right how can we collaborate with another system and get that to integrate with ours when it's such a mess?" This was presented as a trap: these were regional problems that could not be resolved regionally, but had to be resolved in each state first, and where politicians had no will to resolve them. There was a clear sense of frustration among our participants.

We tried to introduce wider political discussion. The participants noted more coordination in policing than in health, and they noted that the GFA had brought benefits to the region. We suggested that in both cases this was a function of top-down political will and determination, from London, Dublin and Belfast, but they did not follow up. There was little discussion of political initiatives or of constitutional change.

Why then did the fluency of discussion end so abruptly, after the participants had given an expert analysis of regional problems? Why did they not look further to political causes and remedies? One possibility is that at last the regional unit had embedded in people's perceptions, but that there were no discursive repertoires to connect the regional level with contentious politics of statehood and constitutions. In the context of an emerging debate on constitutional change, these repertoires need to be developed if the expert experience of border dwellers is to inform the debate.

Conclusion

There is little doubt but that the border region has evolved, not simply in institutions and linkages, but also in the perceptions of its inhabitants. Within 60 years, the border changed from a liminal area between states which Southerners both recognised and denied, to a highly fortified barrier systematically breached by smugglers and paramilitaries, to a region that was first cautiously negotiated, then opened seemingly without barriers, then more clearly defined as a regional unit, with its own practices, identity, problems and expertise.

The participants in our recent focus groups and deliberative cafés had a regional experience and regional identity. They were border dwellers and women, quite as much as they were Northerners, Southerners, nationalists, unionists, Protestants or Catholics or other. They wanted to use their experience to help navigate the new difficult phase of Brexit and constitutional discussion. They saw their "transversal" identity – as border dwellers, as women – as an antidote to polarisation. They saw their experience as important, providing lessons for those who had not, as yet at least, had to navigate different jurisdictions and different administrative cultures and public expectations on a daily basis.

This signals a delayed but important embedding of a regional frame of perception and analysis that has many benefits for people in the region and on the island more generally. It offers a capability of region-level analysis of problems and dysfunctions and a "transversal" approach to contentious constitutional politics that emphasises overarching umbrella values that can overcome group opposition and divisive identity politics; values of a vibrant democracy, socio-economic and gender rights, sustainability, respect for difference.

The question, however, remains whether and how to join up this perspective with constitutional discussion on the island. At the very least, the problems already existing in the border region, so eloquently discussed by participants in our research, need to be fed into constitutional discussion. It is an important question if and how constitutional change to a united Ireland would impact the dysfunctions of the health service in the border area, and how different institutional models would tackle those dysfunctions. For example, would a devolved Northern Ireland in an Irish sovereign state simply make the problems worse, by allowing differential health care in one state with common taxation? And how might different models work in practice? Certainly, the regional border experience would be a crucial resource to test the arguments of those who want the status quo, and who want change.

There are clearly major challenges ahead for the border region. Grassroots participants highlight socio-economic and rights concerns, but they cannot themselves answer complex legal questions of how rights can best be embedded in a polity or efficacy ensured in all the regions of that polity. Grass-roots deliberation has to be combined with interaction with policy experts and politicians, at once helping participants tease out their priorities and preferences, and facilitating change in the political agenda. Our research suggests that this is likely to be a tricky, challenging yet very rewarding process, which stands to change both regional and constitutional ways of thinking.

Bibliography

J. A. Agnew, "A world that knows no boundaries? The geopolitics of globalization and the myth of a borderless world", CIBR working papers in Border Studies, Queens University of Belfast, 2003.

L. Cañás Bottos & N. Rougier, "Generations on the border: Changes in ethno-national identity in the Irish Border area", *In* J. Todd, L. Cañás Bottos & N. Rougier (eds.), *Political Transformation and National Identity Change: Comparative Perspectives*, Routledge, London and New York, 2008, pp. 303–328.

J. Coakley & L. O'Dowd "The transformation of the Irish border", *Political Geography*, 2007, vol. 26, pp. 877–885.

J. Coakley & L. O'Dowd (eds.), *Crossing the Border: New Relations between Northern Ireland and the Republic of Ireland*, Irish Academic Press, Newbridge, 2007.

H. Donnan & T. M. Wilson, *Borders: Frontiers of Identity, Nation and State,* Berg, Oxford, 1999, pp. 45–50.

C. Fitzpatrick & C. O'Sullivan, "Comparing social security provision north and south of Ireland: Past developments and future challenges", *Irish Studies in International Affairs*, vol. 32, n°2, pp. 283–313.

K. Hayward, "From barriers to bridges: The europeanisation of Ireland's borders", CIBR Working Papers in Border Studies, Queens University of Belfast, 2004.

K. Hayward, *What Do We Know and What Should We Do About the Irish Border?*, Sage, Thousand Oaks, 2021.

B. Laffan & D. Payne, "Creating living institutions: EU cross border cooperation after the Good Friday Agreement", *Centre for Cross-Border Studies,* Report, Armagh, 2001.

J. McEvoy & J. Todd, “Constitutional inclusion in divided societies: Conceptual choices, practical dilemmas, and the contribution of the grassroots in Northern Ireland and the Republic of Ireland”, *Cooperation and Conflict*, 2023.

J. McEvoy, J. Todd & D. Walsh, “Participatory constitutionalism and the agenda for constitutional change: Socio-economic issues in Irish constitutional debates”, *Irish Studies in International Affairs*, vol. 33, n°2, 2022, pp. 140–171.

M. P. Murphy “A new welfare imaginary for the island of Ireland”, *Irish Studies in International Affairs*, vol. 32, n°2, 2021, pp. 532–557.

H. Patterson, *Ireland's Violent Frontier: The Border and Anglo-Irish Relations during the Troubles*, Palgrave MacMillan, New York, 2013.

E. Tannam, “Cross-Border co-operation between Northern Ireland and the Republic of Ireland: Neo-functionalism revisited”, *The British Journal of Politics and International Relations*, vol. 8, n°2, 2006, pp. 256–276.

J. Todd, *Identity Change after Conflict: Ethnicity, Boundaries and Belonging in the Two Irelands*, Springer International, New York, 2018.

J. Todd, O. Muldoon, K. Trew, L. Cañás Bottos, N. Rougier & K. McLaughlin, “The moral boundaries of the nation: Nation, state and boundaries in the Southern Irish border counties”, *Ethnopolitics*, vol. 5, n°4, 2006, pp. 365–382.

Part 4

Danish-German Cross-Border Cooperation: A Model for Minority Management?

The Danish-German Border Region. A Critical Introduction

Steen Bo Frandsen

In the beginning of the 21st century, the Danish-German border region is still very much influenced by the national conflict of the 19th century that made it a border region in the first place. Since then, the dominant conviction of a fundamental national antagonism between Danish and German has never really been challenged or modified, even if the border region has changed from a conflict zone to a place where visiting politicians and heads of state praise the unique development or even hail it as a model region. The discourse of a positive turn and the development of good relations constitute an important and continuous peace work that should not be questioned. Still, there are reasons to mention difficulties, unsolved problems, and an overall lack of visions for this region as a whole.

The Schleswigian border region illustrates the almost perfect fulfilment of national ideologies: here a traditionally mixed culture was effectively broken up and reorganised along a hitherto unknown national cleavage. In this process, Schleswigians were denied the possibility to be just Schleswigians – they had to make a simple choice between "Danish" or "German". Nationalists even told them that they had been archenemies for hundreds of years. The narrative of antagonism was not just imposed upon the past; it effectively created the widespread idea that there were no nuances in between these two national categories. The "nation-state" made the nationalists' dream come true: organising peoples in separate states with as less interaction as possible, hereby creating the reality of parallel lives.

Rather naively, nationalists thought it possible to divide nationalities perfectly along clear lines. In Schleswig, the conflict and the open wars from the 1840s demonstrated how unrealistic this conviction really was. In 1920 a line was finally drawn, dividing the region according to the principle of national self-determination, laid down by the Peace Conference in Paris. However, this was no perfect solution either. Schleswig was divided into two parts with large majorities, but the separation also created national minorities. This imperfection to a national solution laid the ground for a regime of segregation in both halves of the former region that turned out to confirm and further the dominant national way of organising societies.

Uncritically accepting the nationalist arguments of the region being a battleground for the two nationalities since times immemorial, efforts are seldom undertaken to unfold a different past – a past widely unconscious of a fundamental national conflict. In fact, the Schleswigians could boast about a long record of peaceful co-existence within their region. Wars occured, but they were almost exclusively brought to the duchy from the outside. That times and society before nationalism might actually constitute an attractive counter story to division and antagonism is taboo in nationalist historiography. Even today, we still live comfortably in the era of nation-states. All other models of state organisation are rejected as suspicious and less developed.

The Outside View on the Border and the Border Region

The process of separating a Danish from a German Schleswig came to an end with the division of 1920. North and South Schleswig were thenceforth fixed within separate nation-state discourses excluding and ignoring the other part as well as most shared traditions. This marked the beginning of an escalated estrangement among the Schleswigians too. At the same time, national politics and culture established a perspective in dealing with the region that always came from the outside, from the capital or other places in the nation-state, but never from the border region itself.

The triumphant progress of this perspective had been under way ever since Danish and German nationalists chose Schleswig as their main battleground. There was no room for a Schleswigian history, Schleswigian interests, or a Schleswigian identity characterised by heterogeneity, multilingualism, and mixed cultural influences. The look from the border (region) itself was not wanted and almost totally inexistent. However, considering the present-day border region, it is productive to reintroduce this perspective. What became of Schleswig and of the Schleswigians' common experience as this in between region was split up between the nation-states?

This brief introduction includes some of the formative events on the road towards becoming the border region of the 21st century that we know today. Instead of just repeating rather well-known national narratives, we will try to uncover some of the almost forgotten tensions between regional and national approaches to the region's history. We will also address the striking fact that almost every important change in the region has been provoked or initiated from the outside. This reveals a lot about the process of the Schleswigian region being reduced to a rather passive periphery – or rather a double periphery, very different from the much more prominent economic and transitional position Schleswig held before the introduction of national ideologies – and of the nation-states' border.

The End of a Region in between (1848–1864)

Our point of departure is the duchy of Schleswig, a typical European region "in between", as we know it, from other lands between pre-national composite states.[1] The transitional region with its close contacts and infrastructural connections to both the Danish kingdom and the different German neighbours became the victim of nationalist efforts to include the region in one of the competing nation-state projects. During this process, Schleswig became politicised and nationalised, ending up as a theatre of bitter conflict and military interventions, before finally being dismantled and divided.

The history of Schleswig owns a reputation of being complex and labyrinthic. It is still very popular to refer to the quote assigned to Lord Palmerston – one of the many international politicians who took great pains in finding out about this not very prominent but incomprehensible region. According to Palmerston, only three people had ever understood the Schleswig-Holsteinian business – the Prince Consort, who was dead, a German professor, who had gone mad, and himself, who had forgotten

---

1 P. Ther, "Caught in between. Border regions in Modern Europe", *In* O. Bartov & E. D. Weitz (eds.), *Shatter Zones of Empires*, Indiana University Press, New Haven, 2013, pp. 485–502.

all about it.[2] One of the main reasons might be that the past was characterised by phenomena not easily translated into the context of modern times. Compared to many other regions, Schleswig was not particularly complex. Nationalist historians however always welcomed the chance to ridicule former ways of organising society and to reject a past not suitable for their simplistic and antagonistic narrative. They enthusiastically greeted the chance to write their own – recently constructed – national conflict back into history, thereby happily refraining from any effort to understand a seemingly irrational region "in between".

The nineteenth century successfully reduced the Schleswigian situation to two slogans. The Danes gathered behind the claim "Danmark til Eideren" demanding a border along the river Eider River. This river constituted the historic border between the duchies of Schleswig and Holstein, and simultaneously between the lands of the Danish king and the Holy Roman Empire. This clear message was not totally wrong as the whole of Schleswig had "always" been a Danish fief. But since Schleswig was made a separate duchy in the Middle Ages, the significance of "Danish" had been more and more reduced to the meaning that it was ruled by the Danish king. A Danish nation-state ending at the Eider River as the slogan implied was heavily overstretched. Schleswig as a whole, or a unit, was neither culturally nor linguistically a Danish land. People living there did not consider themselves to be like their Danish neighbours in the kingdom. The Schleswigians were generally proud of their ambivalent regional identity where several languages were practiced, and different cultural influences affected daily life. To many the reality of living "in between" was in fact defining their identity. Especially those living in the southernmost third of the duchy rejected any inclusion into a Danish nation-state.

The German-Schleswigian regionalists confronted the Danes with their catching "up ewig ungedeelt", claiming the duchies of Schleswig and Holstein to be an eternally indivisible unity. According to them, the Danish king had promised in 1460 that Schleswig and Holstein should remain together in perpetuity.[3] This reading has been the cause of excited discussions ever since, but it is safe to say that "up ewig ungedeelt" was

---

2 The famous quote is found in several versions, e.g. in L. Strachey, *Queen Victoria*, Brace and Company, New York, 1921, cited from Full Text Archive, p. 199. Retrieved from: <https://www.fulltextarchive.com/pdfs/Queen-Victoriax8761.pdf>. Accessed 6 June 2022.

3 See C. Jahnke, "'dat se bliven ewich tosamende ungedelt.' Neue Überlegungen zu einem alten Schlagwort", *Zeitschrift der Gesellschaft für schleswig-holsteinische Geschichte*,

taken out of its context of the letter of privileges from 1460 by the 19th century historian Friedrich Christoph Dahlmann. He most effectively popularised these words that came to represent the dogmatic position of the Germans that not only Schleswig but also the two duchies could never be divided.[4] Apart from justified doubts that the Danish king nor anybody else should have been able to make eternal promises, also this all or nothing argument cannot stand a closer scrutiny. It foresaw another river, one dividing Schleswig and Northern Jutland, as the border; thereby neglecting that most people living in the northern part of Schleswig were opposed to such a construction and considering themselves closer to Denmark and the Danes.

The slogans represented the antagonistic opponents in the national fight over Schleswig – arguably the first but definitely not the last conflict over a national European border. The exclusiveness of the claims made any understanding impossible; no compromise was found.[5] Although both slogans found supporters within the duchy itself, they represented claims of a unity and a context imposed upon Schleswig from the outside. Neither of the two political slogans had their origin in Schleswig and they simply postulated a kind of unity unknown to most people in the region. Petitions, letters, and diaries of Schleswigians unimpressed by national mobilisations show that many people did not want to choose between the two national alternatives, and their way of defining a regional identity. Instead, they stressed the values of a world "in between" like multilingualism and diversity.[6] These positions are almost totally absent from the historiography. Nationalist historians ignored and rejected them. Their works describe the inevitable victory of the idea of nation-states which retrospectively came to define the past and the

vol. 128, 2003, n°2, pp. 45–59, and T. Riis, "Up ewig ungedeelt. Ein Schlagwort und sein Hintergrund", *In* T. Stang-Kuhlmann et al. (eds.), *Geschichtsbilder. Festschrift für Michael Salewski*, Stuttgart, Franz Steiner, 2003, pp. 158–167.

4 W. Bleek, *Friedrich Christoph Dahlmann: Eine Biographie*, C.H. Beck, Munich, 2010.

5 S. B. Frandsen, "Die deutsch-dänische Grenze im Zeitalter der nationalen Gegensätze", *In* M. Krieger, F. Lubowitz & S. B. Frandsen (eds.), *1200 Jahre deutsch-dänische Grenze. Aspekte einer Nachbarschaft*, Wachholtz, Neumünster, 2013, pp. 225–234, esp. p. 230 and 232; I. Adriansen, "Dänemark bis an die Eider! Die deutsch-dänischen Grenzen als Erinnerungsorte im 19. Und 20. Jahrhundert", *In* M. Krieger et al., *1200 Jahre deutsch-dänische Grenze., op. cit.*, pp. 235–247.

6 H. Schultz Hansen, "Schleswigsche Identität in den 1840er Jahren – ein historischer Begriff wird wieder aktuell", *Grenzfriedenshefte*, vol. 2, 1997, pp. 239–252.

present of Schleswig. Histories about this region are so dominated by national narratives that regional questions and perspectives are seldom even mentioned. Instead, national historians found what they wanted and expected: a region caught up in deep national antagonisms. This national focus has constructed the region as we know it.

In the first half of the 19th century a Schleswigian – or rather a Schleswig-Holsteinian – regionalism developed in opposition to the Danish dominated composite state. This regionalism would be considered "German" due to its background in the German-speaking part of the population. Consequently, a regional history dimension is easier to be found on the German side, whereas it is almost totally neglected in a Danish context that include Schleswig as a part of Danish history without problematising "regionality" at all. Although the border region today is partly located in a regional Schleswig-Holsteinian state, it has been difficult for the Germans as well too to reach a regional narrative that is not caught up in a national discourse.

National histories successfully created their *lieux de mémoires* almost exclusively defined by the national conflict. Their hagiographies of heroic nationalists consist of the provocateurs and activists that tried to create tension, conflicts, or generally strived to construct obstacles to a life in the region as it had been before. Moderate voices and bridge-builders are mostly ignored. National histories avoid all things unclear, ambivalent, shifting, unsteady, and changing – although exactly these elements should be promoted and supported in a post-national culture or in a border region reality that hardly reflects the antagonism of the nationalist tradition. Still, the silence about all the "in betweens" cannot be taken as a proof of their nonexistence.

There is a need for a history of Schleswig seeing things from the border and taking a critical approach to all that happened to a region that had not been a border region between national communities in the past. It could be curious about regional interests. It would be aware of those many Schleswigians that did not consider themselves a hundred percent one nationality or the other. As if there were no mixed marriages, no crossing borders nor divided families. No loyalties and no nostalgia. Such a history would offer space for local protagonists like the successful merchant Jens Wulff of Bredebro who – in the middle of the national conflict of the 1840s – wrote a diary, criticising the nationalists for spreading their ideas and their hate among the local people, for promoting the language as an exclusive identity marker instead of a means for communication.[7] Or, for the last mayor of Flensburg loyal to the Danish king, who naively believed it possible to be German speaking and a Danish patriot at the same time.[8] This would make us more aware of all those people who did not feel well at all by the prospect of nation-states, where they would have to make a simple choice and where national identity was reduced to Danish or German without any *in-betweens*. They were, of course, left without the alternative of the "terrible" possibility that someone might want to be a Schleswigian, accepting the specific mix as the real marker of identity. We will never know how many people thought that way. They could not vote for it, and they even never existed according to the national narratives.

The Division of Schleswig 1920

A division of the region had already been proposed and discussed as a solution in the middle of the 19th century. Some dividing lines that were considered came very close to the outcome of the 1920 plebiscites. Retrospectively, one might wonder why the antagonists did not manage to compromise about a division already then. The answer is, quite simple, that neither part was prepared to give up their dogmatic convictions about the unity of the region. Ironically, it was even the Danes that took a particularly rigid position in this matter. Instead, Prussia occupied and later annexed the entire duchy as a result of the Second Schleswigian War of 1864.

During the 1850s the Danish government had launched a cultural and linguistic offensive. It wanted to secure Danish control over the entire duchy and a future border at the Eider River once and for all. In the southern parts of the region, the German-minded population was mobilised against the "Danisation" and even many indecisive Schleswigians reacted and now chose a German position. The offensive backfired and made it clear that a Danish Schleswig would never happen. But neither did the Prussians learn this lesson. After their victory in 1864, they began

---

7 P. K. Iversen, "Kniplingskræmmer Jens Wulffs Dagbog. Uddrag", *Sønderjyske Aarbøger*, vol. 66, pp. 60–130, vol. 67, n°1, pp. 55–152, vol. 67, n°2, pp. 184–227, 1954–1956.

8 H. F. Schütt, "Das loyale Flensburg im nationalliberalen Konflikt," *Flensburger Bilderbogen*, vol. 31, 1998.

oppressing Danish-minded Schleswigians even more heavily, thereby provoking that the then inhabitants of Schleswig's northern parts felt even stronger about their Danish preferences.

In this way, both sides strengthened their adversaries and paved the way for an increasing separation along lines of language and national orientation. If Schleswig had been – generally – rather uncertain of its national preferences in 1848 or 1864, the Danes in the north of the region in effect managed to carry through something like a nation-building process within the Prussian state. A dividing line across the peninsula became distinct even before being introduced with the 1920 plebiscites.

Following the principles of national self-determination, the Paris Peace Conference decided to let plebiscites define a border in Schleswig. The Danes in the north were well prepared for this after almost sixty years of nation-building. Their dream of joining the Danish nation-state was fulfilled with their vote in February 1920 – a vote that buried the idea of an undivided Schleswigian region. The plebiscites of 1920 produced very clear and indisputable results with majorities for Denmark and Germany respectively of about 75 % in each zone. The fact that this line was so overwhelmingly confirmed contributed decisively to the lasting national border, which is still valid today.

However, it is also an important chapter in the story of the border that feelings of disappointment and injustice on either side would last for many years. The 1920 border represented the southernmost line the Danes could claim according to the principle of self-determination, but still, to many Danes, it seemed a defeat that the remaining parts of Schleswig and especially Flensburg had been lost. The Germans as well were dissatisfied with the modalities of the border drawing and the plebiscites and argued that some areas north to the new border had German majorities.[9]

With the border defined along national preferences, the new question of national minorities came up on both sides. The roles and the rights of these groups would be one of the biggest problems in the new order and it would take many years to reach acceptable solutions.[10] Hopes for a

---

[9] T. Fink, *Da Slesvigs blev delt*, Institut for Grænseregionsforskning, Aabenraa, 1979; H. F. Schütt, "Das loyale Flensburg im nationalliberalen Konflikt," *op. cit.*

[10] J. P. Noack, *Det danske mindretal i Sydslesvig 1920–1945*, Institut for Grænseregionsforsking, Aabenraa, 1989; H. Becker-Christensen, *Det tyske mindretal i Nordslesvig 1920–1932*, Odense Universitetsforlag, Odense, 1990.

border revision were kept alive on both sides, irredentism lingered on for decades, and the minority question would always work as a tool for those eager to keep up a tense atmosphere. The argument for a different border was sometimes stronger in Denmark, sometimes in Germany, although neither had particularly convincing reasons given the clear votes of 1920. One could even bring up the argument in favour of the new line that the nationalists on both sides were disappointed and rejected the solution as insufficient.

The highly mobilised and nationalised region would not be the place to introduce any initiatives towards a more relaxed and forward-looking development. The governments in Copenhagen and Berlin were not eager to reanimate the old conflict, which on the other hand always remained as a threat and a weapon that could be used in bilateral relations. The border region was a mere object, a ball that could be played in one direction or the other.

During the 1920s, the German government still hoped for adjustments to the Versailles Treaty, and it did in fact not officially recognise the new border, as it hoped the international balance would change, so that possibilities would come up to put Denmark under pressure for a revision. Although, the German government never questioned the line, and even the national-socialist government took no action after coming to power in 1933. It even respected the 1920 border as a dividing line during the occupation 1940–1945. This could in fact be interpreted as a silent acceptance.[11]

The Peace Treaties after the First World War represented the victory of the nation-state and the defeat of the empires of the past. The new European state system meant the end of historical regions *in between* in Europe, and although 1864 had already marked the formation of the nation-state order in Schleswig, the large Danish-minded minority in its northern part represented a clear imperfection. In 1920 however, the role of an indecisive Schleswig *in between* came definitively to an end. From

---

11 H. Skov Kristensen, "Den 9. april 1940 og grænseflytningsrygterne," *In* H. Schultz Hansen & H. Skov Kristensen (eds.), *Sønderjylland under krig og besættelse 1940–1945*, Historisk Samfund for Sønderjylland, Aabenraa, vol. 91, 2003, pp. 8–88, esp. pp. 16–20; K. C. Lammers, "Die neue dänische-deutsche Grenze als 'Versailles-Grenze'. Die Grenzfrage in den dänisch-deutschen Beziehungen 1933–1955," *In* M. Krieger et al. (eds.), *1200 Jahre deutsch-dänische Grenzen, op. cit.*, pp. 259–268.

this point north and south were integrated in their respective states, and all Schleswigian common tradition died.

The border turned Schleswig into a border region – or two regions as both nation-states now focused almost exclusively on their own side. The division caused not least economic difficulties on both sides, and Schleswig experienced considerable economic problems as the border now stopped the flow and traditional contacts within the region.[12] Early celebrations on the Danish side were soon followed by a much more sceptic and somewhat disillusioned atmosphere. "Reunion", or "Genforening" as it was ideologically called, was a challenge in a region with a past of its own.

The Declarations of Bonn and Copenhagen (1955)

The border region had not seen a lot of action during the Second World War, but with its end and the cessation of Denmark's German occupation, the border was questioned once again. Now in the superior position again, the Danes tended to forget that the border was "definitive", as they had claimed themselves a few years earlier. Growing tensions were provoked by the high numbers of refugees that came to Schleswig mainly from the former German lands in East Prussia and Silesia. The composition of the population changed drastically and the old continuity of a Schleswigian identity was broken. A much more heterogeneous population would not be equally informed about the old conflict.

As the German Federal Republic wanted to join the North Atlantic Treaty Organization (NATO) in 1955, the Danish government linked its acceptance to the question about the border and the role of the minorities. The declarations between the governments of Denmark and the Federal Republic of Germany, separately signed in Copenhagen and Bonn, have become the symbol of a new understanding between the neighbours.[13] Ever since they have been considered a basis for detente and a peaceful development of the borderland, so that they are now treated as almost

---

[12] M. Andersen, *Den følte grænse. Slesvigs deling og genopbygning 1918–1933*, Historisk Samfund for Sønderjylland, Aabenraa, 2008.

[13] J. Kühl, *København-Bonn Erklæringerne 1955–2005. De dansk-tyske mindretalserklæringers baggrund, tilblivelse og virkning*, Institut for Grænseregionsforskning, Aabenraa, 2005.

sacrosanct. They contributed to relieve tensions, but they did not mark an immediate beginning of a close relationship.

This process was difficult. It reflected the complex relations between Danes and Germans at the time. The usual practice of signing the treaty jointly was rejected by the Danes. Therefore, the declarations were adopted separately by both parliaments. A common ceremony with exchange of signatures was never held, which must be seen in the light of the past conflict. The Danes still remembered the difficulties caused by the treaties of the 1850s with the German powers enabling Austria and Prussia to interfere in what was considered "inner affairs". They would not again end up in a situation where the bigger neighbour could put them under pressure by referring to a treaty. It did not reduce the importance of the declarations, but it is significant that the declarations did not include any promises or any roadmap for future cooperation and cross-border normality. They constituted the perfect foundation for a peaceful relationship of parallel lives and as such they can be seen as an impressive achievement.

The importance of the 1955 declarations seems to have grown over time; and returning to the regional perspective of the Schleswigian border, they also reflect the dominant outside perspective of the border. Once more, an important decision was taken in other places, caused by other events – in this case the German access to NATO – than anything happening in the borderland itself. There were no initiatives for a closer relation or less border from a local level, and there were no strong demands from the region itself for a different border regime. Both sides had submitted to the national ideologies, and it would have been exceedingly difficult to find signs of a Schleswigian identity or tradition that would keep the two parts together. In the nation-states the idea of the border region was built on the past, on the narratives of conflict and national tensions.

Events happening in the region mostly did not challenge these ideas, but as demonstrated by Tobias Haimin Wung-Sung we should be more cautious in believing the national narratives of closed communities. In the post-war decades youth culture for instance began to be influenced and inspired from other sources than the respective national ideologies. On both sides of the border but also very much in both the minorities young people were attracted by fashion, music and lifestyle coming from the outside world, not least the United States, and grew quite distant from the culture of the traditional national communities. For the first time since the introduction of nationalism, both sides were exposed to the same foreign influences that challenged the old narratives.[14]

Denmark Joins the Common Market (1973)

Cold War and youth culture were contesting the dominance of traditional nationalist ideas in the borderland, but neither part of Schleswig would profile itself as a particular open or dynamic part of their respective nation-states. National traditions were strong; the regions developed back-to-back.

European integration offered Germans a possibility to free themselves from the frustrations of the immediate past in the post-war period. The Danes never felt such needs. Denmark developed strong economic relations to Great Britain, and only the British rapprochement towards the Six made the Danes contemplate a future within the European Community from the early 1960s. It was however only in the early 1970s that a membership became a real option. In the referendum of October 1972, a surprisingly clear majority decided to follow Britain into the European Economic Community (EEC).

The prospect of a membership had reanimated traditional Nordic-nationalist positions in a fierce campaign against the EEC. Since the conflict of the 19th century Schleswig held a prominent position in the national narrative as a bulwark against Germany (and Europe), and now it was expected to have a significant say in the political conflict over a Danish membership.

Support for the Danish opposition to EEC came from the Danish minority in Germany that represented a strong anti-European position.[15] The minority did not consider a Danish membership as a possibility for themselves to come closer to their kinstate. It remained stuck in a traditional nationalist idea of Nordic culture as an alternative to German or European culture. In general, the arguments of the Danish minority did not find much sympathy north of the border. The German counterpart clearly advocated a Yes to a Danish membership, hoping this could pave

---

14 T. Haimin Wung-Sung, *Beyond the Border. Young Minorities in the Danish-German Borderlands 1955–1971*, Berghahn, Oxford, New York, 2019.

15 M. Rostgaard Nissen, *Karl Otto Meyer: politiker, publicist, polemiker*, Syddansk Universitetsforlag, Odense, 2018.

the way for easier contact and movement across the border. Interestingly, the majority population of North Schleswig also took a clear stand for a European membership. In the end the region had the highest percentage of votes in favour of a Danish entry in the entire kingdom. Although nationalists had believed this region a dam against Europe it actually turned out strongly in favour of a Yes.

The Danish membership of the EEC from 1973 signalled the beginning of a stronger formalised cooperation between the neighbours now both being part of the same European project, but at the same time it hardly had immediate effects on border region politics. Initiatives would not come from the region itself, and the governments showed no hurry when it came to stepping up integration. A prominent example was the absence of an infrastructural modernisation that would have brought the nation-states closer to each other and could have strengthened the contacts within the border region. Fifty years later, the railway connection is still slow and inefficient, and has hardly seen any noticeable improvements. Another in some respectcompeting infrastructural project was the construction of a motorway from Hamburg to Kolding. Denmark and Germany agreed upon this connection in 1964, but it took twenty years before it was finished. As the Danish part was opened in 1984 the German Autobahn had been waiting for six years to be connected.[16] For those who passed the new border station in Ellund it would be immediately clear that Europe looked different on the Danish side. The Danish motorway was organised differently with signs, exits and traffic rules, in order to make it clear to everyone that the border signalled a change.

The failing enthusiasm in developing the border region cannot be explained alone by the overall reluctant Danish politics towards European integration. The border region itself did not show a lot of interest in developing a common space, and there were hardly any challenges to the reality of parallel lives that had been established in the decades before. A forum for regional affairs did not exist after the division and both halves of the former duchy played a rather discreet role in their new respective entities.

In general, it is remarkable how modest the effects of the Danish decision to join the EEC turned out to be in the border region. Apart from

---

[16] Vejdirektoratet, "Den jyske motorvej Skovby – Christiansfeld," *Dansk vejtidsskrift*, vol. 9, 1984.

chilly Danish feelings towards European integration and historic reservations towards the southern neighbour, the reason could also be found in the fact that the border region experienced a diminishing status after the end of the conflict. In a certain respect, Schleswig now was even further away from the capital, and it even lost its role as the Danish border region at the turn of the millennium – as the Øresund region was discovered as a "new" border region just outside the capital. Now it was remarkable to witness how cross-border investments, cross-border regional planning, cross-border border rhetoric, and even a European dimension in regional politics (and economy) were discovered by the political elite and the government.

The European turn did not unravel a new interest in the history or any efforts to re-write or re-discover regional history in Schleswig. Both sides remained in their national categories and continued to celebrate their nationalised histories separately. The most prominent example is the battlefield at Dybbøl where the decisive battle of 1864 had been fought. There were no initiatives to open up the celebrations or to develop a common memory culture. Even at the beginning of the millennium, German presence was still considered to be disturbing by the Danes. They preferred to celebrate their national history alone.

The Euroregion and the Schengen Border

The positive narrative about the border region would stress the development from a conflict zone to a tranquilised borderland. The declarations of 1955 had successfully created overall structures that made it possible for majorities and minorities on both sides of the border to organise themselves in more or less separate and parallel lives with their own institutions. The Danish entry into Europe in 1973 also marked the beginning of a slow and cautious change in the border region.

Two developments illustrate how impacts from the outside once again disturbed the border region. The North Schleswigians submitted to the centralist structures of the nation-state without demonstrating any serious ambitions of using their position at the border as a means to strengthen their region. This was made noticeably clear with the initiative to create a Euroregion. It clearly exposed the different positions with respect to cooperation and cross-border regional structures in Denmark and Germany, but the discussion also revealed old idiosyncrasies and prejudices. In general though, the plan was only met with heavy opposition on the Danish side.

Euro regions did of course already exist across many European borders demonstrating the European Union (EU)'s wish for a stronger and deeper cross-border cooperation. In North Schleswig the proposal was met with strong opposition – even the obvious "Schleswig" was rejected as a name for the common region although it historically comprised both sides of the border. Danish opponents now saw Schleswig as a symbol of German dominance as they had acquainted themselves with "Sønderjylland" as the name used by the Danish nationalists. The discussion was an emotional one and showed how far away the border region was from being an ordinary place. The historic conflict could still be a weapon in the political debate if necessary. In the end a much less ambitious regional structure for cooperation was created, and even if some mostly pragmatic issues are dealt with quite effectively, the scope and the ambitions of the cooperation are very modest indeed.[17]

The successful resistance against the Euroregion was combined with a less successful mobilisation against the Schengen agreement that Denmark finally decided to introduce in 2001. This would turn out to be the biggest single event in the history of the border since 1920, but the decision and the debate about this step were not only border region issues, but very much national politics.

Still, the border region played a prominent role in the discussion. In the minds of many Danes the land border in Schleswig is "the border" – and also the place where the history of the wars with Germany and the establishment of the 1920 border were used to dramatise the scope of the announced decision to join the Schengen zone. Nationalist activists organised themselves in a so-called Sønderjyllandskomite – using the national name for Northern Schleswig in their effort to mobilise old resentments. From a Danish perspective, the region had once again to be protected from being taken over by the aggressive neighbours, and once more the border became a prominent symbol for national consciousness and demarcation towards a threatening outside world.

Trying to copy the impressive freedom movements of the Baltic States, a human chain across the peninsula was supposed to show that

---

[17] S. B. Frandsen, "J. Andresen, Grænseoverskridende samarbejde i den dansk-tyske grænseregion", *Økonomi og Politik*, vol. 2, 2020, pp. 69–77.

the border should remain closed and under control. It ended in a debacle because only a few thousand and not the expected 25,000 people turned up. Prominent newspapers like *Politiken* and *Jyllandsposten* criticised the "professional EU opponents" for instrumentalising Schleswig and the history without considering the present-day relations to the neighbour.[18] Following the introduction of the Schengen border Denmark experienced at least some years within the mainstream of EU border politics before the government used the refugees of 2015 to introduce the provisional border control.[19]

Since then, the border has again become more and more visible. Where the major traffic corridors cross the border, a growing number of control barracks have erased the unwanted impression of an open Schengen border. Demonstratively the visitors are greeted by the Danish flag and an ensemble of Nordic States' flags. A European flag is nowhere to be seen, and the symbol of the EU remains a very uncommon sight in the rest of the kingdom. Since 2019, 70 km of metal fence are marking the border across the peninsula, raised in order to keep out infected wild boars and protect the Danish pork production. To the Danes it is about economic interests, to the Germans and many other foreigners the fence is more of reminder of a past of walls and fences that some Europeans thought gone since 1989.

The prominence of the border region in the discussions about the Euroregion and the Schengen agreement could not hide the fact that also the Danish part of Schleswig experienced economic difficulties as a periphery. If the North Schleswigians had demonstrated their will and interest in a stronger cooperation with the neighbour in the referendum of 1972, regional self-consciousness was dwindling in the decades that followed. The problems of being a periphery were growing, and the region reacted quite passively in this development, which ended with its liquidation in 2004. Becoming a part of a new artificial administrative region called "Southern Denmark" meant that North Schleswig lost the last bit of

---

18 L. Ejlskov Røhrig, *Sønderjylland – hvad kan det bruges til? En erindringspolitisk undersøgelse af den politiske og ideologiske instrumentalisering af Sønderjylland 1955–2006,* Master thesis, University of Southern Denmark, Odense, 2006.

19 This situation still holds as per 11 May 2023. See European Commission, "Temporary Reintroduction of Border Control", retrieved from <https://home-affairs.ec.europa.eu/policies/schengen-borders-and-visa/schengen-area/temporary-reintroduction-border-control_en>. Accessed 11 May 2023.

administrative and political liberty of action on the regional level, which was of course another of the long-time consequences of the division of 1920 that deprived Schleswig of a local and strong natural centre.

The Border Centenary 2020

In 2020, Denmark celebrated the centenary of the border. This happened with the title of "Genforeningen" or the "Re-Unification" as the nationalist concept for the annexation of Northern Schleswig in 1920. As hoped for and expected this reminder of the glorious nationalist narrative and the heyday of the nation-state turned out to be a mobilising factor.

"Genforeningen" celebrated the nationalist narrative of Schleswigian history stressing the "coming back" to Denmark as the culmination of regional history. This perspective is not really questioned, as there is not much of a regional consciousness connected with traditions, *lieux de mémoires*, ideas, and feelings for the old region "in between".

Then, the international pandemic hit the exclusive national program for the centenary celebrations and caused most of the planned events to be postponed or cancelled. This could be seen as a reminder about the necessity to understand borders and border regions in their international dimension. It is not possible to close the door and keep the world outside.

The closing of the border on 14 March 2020, did however expose some of the features of the border region. These stay usually unrecognised in the national discourse.[20] The border does no longer function as the clear demarcation as foreseen by the nation-state. Even in 1920, not many people in the border region could claim to have no contacts, no family relations, economic transactions with "the other side", but, in 2020, it became clear how much the Schengen border has changed border region reality. Ever since the turn of the millennium there is a noteworthy process taking place which gives cause to cautious optimism – largely without any intervention from the officials. Business relations, people commuting across the border, family relations, love affairs, shopping tourism – it all contributes to a border reality quite different from the traditional ideas of the border region.

---

20 B. Henningsen, "'Die deutsch-dänische Grenze ist die glücklichste, die Deutschland hat.' Freundschaft – nicht nur in Zeiten von Corona", *Nordeuropaforum.blog. Kommentare zum Zeitgeschehen in Nordeuropa*, 23 July 2020. Retrieved from: <https://nofoblog.hypotheses.org/tag/daenisch>. Accessed 29 May 2022.

The inhabitants of the border region use their respective hinterland on the other side as a matter of course as belonging to them – for shopping, trips, cultural events or for meeting friends and relatives.[21] Not last the people understanding themselves as minority do not always feel beholden to their own nation but to their minority status which is no longer a factor for segregation as for many generations before but a uniting moment. Those migrating to the border's other side sometimes even switch their minority with their counterpart.[22] One strong uniting factor is now regrettably lacking: young Germans and Danes are not especially interested in learning the other's language as they prefer to use English instead, thus avoiding the dominance of one language, but giving up the acknowledgement of a somewhat shared history.[23] But especially younger people prefer the impartial contact to each other and value almost any diversity instead of nursing obsolete preconceptions. They reacted with incomprehension to the Covid-19 provisions instated by the far-off capitals, which seemed not especially interested in the local circumstances. The impractical and abrupt border closing angered many inhabitants of the region, hereby evidencing the existence of everyday and personal contacts. Almost prophetic is the film *Das unsichtbare Band* (The Invisible Bond) which even before the pandemic declared that shared everyday life is a matter of course and a necessity if the region shall be and stay a living one.[24]

---

21 See for example the survey "Grænseland på tværs. – Identitet, sprog og bevægelser over grænsen". Retrieved from: <https://issuu.com/region-syddanmark/docs/gr_nselandet_p__tv_rs_-_final_tilg__854f29ac06acda>. Accessed 28 May 2022.

22 W Lagler, "Minderheit ist, wer will", *Der Nordschleswiger*, 25 September 2019.

23 Even some teachers do not acknowledge the importance of the German language. There is an ongoing conflict, here represented by T. Rasmussen, interview with Lykke Friis: "Mit Englisch kommt man durch – mit Deutsch kommt man weiter", *Gymnasieskolen*, 3 May 2021. Retrieved from: <https://gymnasieskolen.dk/mit-englisch-kommt-man-durch-mit-deutsch-kommt-man-weiter>. Accessed 30 May 2022, and F. Gross Jakobsen, and C. Jørgensen, "Lærer: Fjern tysk – Det er et levn fra fortiden", *Folkeskolen*, 23 March 2021. Retrieved from: <https://www.folkeskolen.dk/arbejdsliv-besparelser-faglig-fuser/laerer-fjern-tysk-det-er-et-levn-fra-fortiden/1330741>. Accessed 30 May 2022.

24 W. Hauke, *Das unsichtbare Band. Grenzgeschichten von Dänen und Deutschen*, IDA Film/TV Produktion, NDR, 2020, review: C. von Tiedemann, "Die Minderheit im Fernsehen: NDR-Film zeigt Alltag und Geschichte", *Der Nordschleswiger*, 11 March 2020. Retrieved from: <https://www.nordschleswiger.dk/de/nordschleswig-deutschland-suedschleswig/minderheit-im-fernsehen-ndr-film-zeigt-alltag-und>. Accessed 29 May 2022.

This change should not be exaggerated. There is no revival of the historic *in between*, not even if we reduce it to the immediate border region, but we have moved quite far from the nation-state border of the past. Irritations and demands provoked by the closed border should not be overestimated either, but they still represent another, more conscious idea of a border region with specific interests that might very well be different from the course propagated in the distant centres of the nation-states.

Bibliography

I. Adriansen, "Dänemark bis an die Eider! Die deutsch-dänischen Grenzen als Erinnerungsorte im 19. Und 20. Jahrhundert", *In* M. Krieger et al. (eds.), *1200 Jahre deutsch-dänische Grenze. Aspekte einer Nachbarschaft*, Wachholtz, Neumünster, 2013, pp. 235–247.

M. Andersen, *Den følte grænse. Slesvigs deling og genopbygning 1918–1933*, Historisk Samfund for Sønderjylland, Aabenraa, 2008.

H. Becker-Christensen, *Det tyske mindretal i Nordslesvig 1920–1932*, Odense Universitetsforlag, Odense, 1990.

W. Bleek, *Friedrich Christoph Dahlmann: Eine Biographie*, C.H. Beck, Munich, 2010.

L. Ejlskov Røhrig, *Sønderjylland – hvad kan det bruges til? En erindringspolitisk undersøgelse af den politiske og ideologiske instrumentalisering af Sønderjylland 1955–2006*, Master thesis, University of Southern Denmark, Odense, 2006.

T. Fink, *Da Slesvigs blev delt*, Institut for Grænseregionsforskning, Aabenraa, 1979.

S. B. Frandsen, "Die deutsch-dänische Grenze im Zeitalter der nationalen Gegensätze", *In* M. Krieger, F. Lubowitz & S. B. Frandsen (eds.), *1200 Jahre deutsch-dänische Grenze. Aspekte einer Nachbarschaft*, Wachholtz, Neumünster, 2013, pp. 225–234.

S. B. Frandsen, "J. Andresen, Grænseoverskridende samarbejde i den dansk-tyske grænseregion", *Økonomi og Politik*, vol. 2, 2020, pp. 69–77.

T. Haimin Wung-Sung, *Beyond the Border. Young Minorities in the Danish-German Borderlands 1955–1971*, Berghahn, Oxford, New York, 2019.

W. Hauke, *Das unsichtbare Band. Grenzgeschichten von Dänen und Deutschen*, IDA Film/TV Produktion, NDR, 2020.

P. K. Iversen, "Kniplingskræmmer Jens Wulffs Dagbog. Uddrag", *Sønderjyske Aarbøger*, vol. 66, pp. 60–130, vol. 67, n°1, pp. 55–152, vol. 67, n°2, pp. 184–227, 1954–1956.

C. Jahnke, "'dat se bliven ewich tosamende ungedelt.' Neue Überlegungen zu einem alten Schlagwort", *Zeitschrift der Gesellschaft für schleswig-holsteinische Geschichte* vol. 128, 2003, n°2, pp. 45–59.

J. Kühl, *København-Bonn Erklæringerne 1955–2005. De dansk-tyske mindretalserklæringers baggrund, tilblivelse og virkning*, Institut for Grænseregionsforskning, Aabenraa, 2005.

K. C. Lammers, "Die neue dänische-deutsche Grenze als 'Versailles-Grenze'. Die Grenzfrage in den dänisch-deutschen Beziehungen 1933–1955," *In* M. Krieger et al., *1200 Jahre deutsch-dänische Grenzen*, *op. cit.*, pp. 259–268.

J. P. Noack, *Det danske mindretal i Sydslesvig 1920–1945*, Institut for Grænseregionsforsking, Aabenraa, 1989.

T. Riis, "Up ewig ungedeelt. Ein Schlagwort und sein Hintergrund", *In* T. Stang-Kuhlmann et al. (eds.), *Geschichtsbilder. Festschrift für Michael Salewski*, Stuttgart, Franz Steiner, 2003, pp. 158–167.

M. Rostgaard Nissen, *Karl Otto Meyer: politiker, publicist, polemiker*, Syddansk Universitetsforlag, Odense, 2018.

H. F. Schütt, "Das loyale Flensburg im nationalliberalen Konflikt," *Flensburger Bilderbogen*, vol. 31, 1998.

H. Schultz Hansen, "Schleswigsche Identität in den 1840er Jahren – ein historischer Begriff wird wieder aktuell", *Grenzfriedenshefte*, vol. 2, 1997, pp. 239–252.

H. Skov Kristensen, "Den 9. april 1940 og grænseflytningsrygterne," *In* H. Schultz Hansen & H. Skov Kristensen (eds.), *Sønderjylland under krig og besættelse 1940–1945*, Historisk Samfund for Sønderjylland, Aabenraa, vol. 91, 2003, pp. 8–88.

L. Strachey, *Queen Victoria*, Brace and Company, New York, 1921.

P. Ther, "Caught in between. Border regions in Modern Europe", *In* O. Bartov & E. D. Weitz (eds.), *Shatter Zones of Empires*, Indiana University Press, New Haven, 2013, pp. 485–502.

Vejdirektoratet, "Den jyske motorvej Skovby – Christiansfeld," *Dansk vejtidsskrift*, vol. 9, 1984.

Celebrating the Border: The Danish-German Border Region and 100-Year Anniversary of the "Genforening"

Gaia Cardin

2020 was supposed to be a year of great celebrations in Denmark as it marked the 100-year anniversary of the so-called "Genforeningen" (the Reunification), namely the drawing of a new border between Denmark and Germany and the integration of Northern Schleswig to the Kingdom of Denmark. Due to the Covid-19 pandemic, most of the celebrations and arrangements were postponed to the following year, but despite this inconvenience the anniversary was still commemorated in great fashion and arrangements took place all over Denmark.[1]

The present border between Germany and Denmark was established in 1920, and, as set out by the Treaty of Versailles, it was meant to reflect the "wishes of the population".[2] The border was in fact drawn on the basis of two plebiscites in which the inhabitants of Schleswig-Holstein were asked to choose whether they felt like they belonged to Germany or Denmark. The result of the referendums, although not welcomed by everyone, is still in place today and many believe that such resolution of the border issue represents a model to be followed.[3]

---

1 Grænseforeningen, "Genforeningen 2020", 2022. Retrieved from: <https://graenseforeningen.dk/om-graenselandet/graenselandets-historie/genforeningen-2020>. Accessed on 17 of June .2022.

2 W. R. A. Hughes, N. S. Megoran, & M. C. Benwell, "The Emotional Geopolitics of Contested Peace: The Austerity Crisis and the Danish Minority of South Schleswig", *Territory, Politics, Governance*, vol. 10, n°2, 2022, p. 239.

3 S. B. Frandsen, "Schleswig – A Border Region Caught Between Nation-States", *In* K. Stoklosa & G. Besier (eds.), *European Border Regions in Comparison: Overcoming Nationalistic Aspects or Re-Nationalization?*, Routledge Studies in Modern European History 21, Routledge, New York, London, 2014, pp. 79–97.

The delineation of the new border, however, did not mean that the Schleswig-Holstein question was completely solved. Borders are socially constructed and as such always contested. This means that for a border to remain unchanged, it needs to be continuously reproduced through social practice.[4] In this process a major role is played by the State, which often uses borders to create a sense of common identity and to distinguish those belonging to the nation from those who do not through a mechanism of inclusion and exclusion.[5]

Especially in the 19th and 20th century, Germany and the Germans represented the "Other" from a Danish perspective. After the signature of the Bonn-Copenhagen Declaration and the establishment of a Euroregion between the two countries, the narrative about Germany began to change from enemy to partner. As such, the Danish-German border region has been celebrated as an example of peaceful resolution of inter-State conflict.[6]

The aim of this paper is to reconstruct the narrative put forward by the Danish government during the celebrations of the 100-year anniversary of the "Genforening" to investigate whether the border between Denmark and Germany continues to be portrayed as a barrier dividing the "us" from the "them" or if instead it is depicted as a bridge between two neighbouring countries and an opportunity to celebrate together, and how the border is used in the national ideology.

These aspects have been investigated considering the concept of spatial socialisation as defined by Anssi Paasi. Such concept describes the process through which ideas about territory and borders influence the spatial identity of the individuals.[7] This term is to be considered in relation to the one of "social spatialisation", which refers instead to the process

---

4 D. Newman, "On Borders and Power: A Theoretical Framework," *Journal of Borderlands Studies*, vol. 18, n°1, March 2003, pp. 13–25.

5 A. Paasi, "Part One: Territories and Boundaries in Regional Transformation," *In Territories, Boundaries, and Consciousness: The Changing Geographies of the Finnish-Russian Border*, Belhaven Studies in Political Geography, Chichester, England ; J. Wiley & Sons, New York, 1996, pp. 1–76.

6 S. B. Frandsen, "Schleswig – A Border Region Caught Between Nation-States", *op. cit.*; M. Klatt, "Lessons from the Danish-German Border Region for Post 2020 Interreg A – an Alignment with Cross-Border Functional Regions?," *Europa XXI*, vol. 38, 2020, pp. 139–156.

7 A. Paasi, "The Political Geography of Boundaries at the End of the Millennium: The Challenges of the de-Territorializing World," 1999, pp. 9–24.

through which the spatial is socially constructed.[8] Social spatialisation and spatial socialisation go hand in hand because the social construction of the spatial necessarily influences the way in which the spatial shapes identity. In the national sphere, spatialisation of the social occurs in different ways, but is inevitably influenced by power relations. The function of the border as a marker of identity depends on power elites: those who have power can impose their own meaning and their own narrative and thus influence the process of identity formation. In particular through the process of spatial socialisation, boundaries of the State are legitimised, and thus the ones who are to be considered as part of the State and the ones who are not.[9]

Genforening

A first point that must be made before diving into the narrative of the celebrations regards the term *Genforening* itself. The word literally means "reunification", therefore reflecting a national or even nationalist interpretation in so far that the annexed part of Schleswig had not been part of the Kingdom of Denmark since the Middle Age. By using the word *Genforening* the celebrations are presented as a national matter and the focus is put on Denmark gaining its territory back rather than on the successful establishment of a cooperative border region. What is more, the term *Genforening* excludes the German and Danish minorities living on both sides of the borders and therefore presents a one-sided perspective of the homogenous nation-State. It equates territorial identity with national identity and in so doing it ignores the existence of feelings of regional belonging, which for a long time actually characterised the region. Indeed, conflicts between the Danish Kingdom and the region in the 19th century mostly find their roots in Schleswig-Holsteinian regionalism rather than German nationalism.[10]

In general, in the data analysed, which consisted in official speeches and materials from the celebrations, there are many references to how Southern Jutland (Sønderjylland) was reunified with Denmark and rewon

---

8 A. Paasi, "Part One: Territories and Boundaries in Regional Transformation", *op. cit.*

9 *Ibid.*

10 S. B. Frandsen, "Schleswig – A Border Region Caught Between Nation-States", *op. cit.*

by Denmark. In all her speeches, Prime Minister Mette Frederiksen mentioned how inhabitants of Southern Jutland, with the 1920 plebiscites, "came home" and that "the nation was healed" because of such reunification[11]. Furthermore, she linked the fact that the new border was established in a peaceful and democratic way with a specific characteristic of Demark and Danish identity: "we are a country that decides big things together".[12]

What is more the reunification of Southern Jutland with the Kingdom of Denmark is often interpreted as the moment in which Denmark transformed into the modern nation-State, in which the state coincides with the nation. The reunification of the region was seen as marking the reunion of the Danish community within the same borders. That is why it was decided that the celebrations should not be limited only to Southern Jutland but should involve the whole nation, as it was the case in 1920.[13]

Characterisation of Germany and Germans

When inhabitants of the border region were asked to vote between Germany and Denmark in 1920, Danish propaganda was based on an antagonism between Denmark and Germany. While Denmark was presented as an idyllic place where one can feel at home, Germany was presented as a country which cannot offer a good life as all its economic resources needed to be dedicated to war reparations. This was done through the use of posters, which could reach all strata of the population, and symbols such as the Danish flag and Danish landscapes, which could be understood by everyone as they are often used in Danish sociocultural imaginaries.[14]

---

[11] M. Frederiksen, "Mette Frederiksens Tale Ved Genforenings fejring 2020", *Det Kongelige Teater*, 2020;; M. Frederiksen, "Statsminister Mette Frederiksens Tale Ved Genforenings arrangement på Dybbøl Bankeden 15. Juni 2020", *Dybbøl Banke*, 2020. Retrieved from: <https://www.stm.dk/statsministeren/taler/statsminister-mette-frederiksens-tale-ved-genforeningsarrangement-paa-dybboel-banke-den-15-juni-2020/>. Accessed 17 June 2022.

[12] M. Frederiksen, "Statsminister Mette Frederiksens Tale Ved Genforenings arrangement på Dybbøl Bankeden 15. Juni 2020", *op. cit.*

[13] T. Andresen & S. Faber (eds.), *Genforeningen 2020 – 100 året for Danmarks genforening. Jubilæumsbog – 2021*, Sonderborg, 2020.

[14] N. Jebsen, "Dividing Regions? Plebiscites and Their Propaganda – Schleswig and Carinthia 1920", *In* K. Stoklosa & G. Besier (eds.), *European Border Regions in*

Today however, at least in the official narrative, Germany is no longer presented as the negative "Other" or the enemy, as the focus is instead on the good partnership between the two countries. The relationship between Germany and Denmark is described as a friendship and a partnership in all the official materials from the celebrations, as for the speeches given by the Prime Minister, the informational book on the arrangements organised for the occasion and by Danish politicians in general.[15] Furthermore, this partnership is often framed in relation to the European Union (EU): Denmark and Germany are described as partners within the EU and therefore good neighbours.[16]

This change of perception towards the German population is highlighted also by Queen Margrethe II, who believes that new generations can create an even more positive relation with Germany. Indeed, children do not live with the memories of the occupation and the war.[17] In his interview for *Grænsen*, the historian Hans Schultz Hansen also argues that the Danish national identity is no longer built in opposition to the German one like it was in the past, but instead he argues that "And with the Reunification, where after all it was the people themselves who decided where the border had to go, the popular came to be more central as an identity marker".[18]

Comparison: Overcoming Nationalistic Aspects or Re-Nationalization?, Routledge, New York, London, 2014, pp. 58–78; S. B. Frandsen, "Schleswig – A Border Region Caught Between Nation-States", *In* K. Stoklosa & G. Besier (eds.), *European Border Regions in Comparison, op.cit.*, pp. 79–97.

15 Grænsen, "Hvad Betyder Genforeningen?," *Grænsen – Magasinet Om Mindretal, Sprog Og Kultur*, December 2019; T. Andresen & S. Faber, *Genforeningen 2020, op.cit.;* M. Frederiksen, "Statsminister Mette Frederiksens Tale ved genforenings fejring i Alsion i Sønderborg", Alsion, Sønderborg, 2021. Retrieved from: <https://www.stm.dk/statsministeren/taler/statsminister-mette-frederiksens-tale-ved-genforeningsfejring-i-alsion-i-soenderborg/>. Accessed 17 June 2022.

16 T. Andresen & S. Faber, *Genforeningen 2020, op.cit.;* F. A. Petersen, "Erindring og politik: Et blik på historie og kultur i den dansk-tyske udveksling 2017–2020", *In* P. Greiner et al. (eds.), *Dänisch-deutsches Freundschaftsjahr 2020: Debatten und Perspektiven = Tysk-dansk venskabsår 2020: debatter og perspektiver*, Nordeuropa-Institut, Humboldt-Universität zu Berlin, Berlin, 2021.

17 A. Bjerager, "H.M. Dronningen: 'Jeg Håber, at Hele Danmark Vil Fejre 100-Året for Genforeningen' – Interview", *Grænsen – Magasinet Om Mindretal, Sprog Og Kultur*, December 2019.

18 M. Christophersen, "Genforeningen Skabte Nutidens Danmark", *Grænsen – Magasinet Om Mindretal, Sprog Og Kultur*, December 2019, p. 20.

Despite this change in perspective, by describing Germany as the "big neighbour", Germany still implicitly remains the "Other" although not a negative one anymore.[19] Additionally, the celebrations of the *Genforening* rarely involved the German counterpart, and instead remained very much on the national level, perhaps because they were planned by the central Danish government. The choice of using the word "*Genforening*" also reflects the exclusion of the German side as the new border did not constitute a reunification but rather a loss to Germany.[20]

Characterisation of the Border

The distinction between Denmark and Germany is physically drawn by the border which distinguishes between "outside and home, and between yours and ours".[21] *Genforeningen* was all about celebrating the new border established in 1920, which allowed for the Danes to be reunited. In almost all the analysed materials, the new border is praised because it was established democratically and it is seen as the result of people's will.[22] This narrative, however, forgets that the establishment of the new border left many people unhappy and divided many families. As Hinrich Jürgensen, member of the German minority in Denmark, highlights "it is not us that have moved. It is the border".[23]

Nevertheless, the border was always presented in a positive way during the celebrations. It was no longer conceived only as a barrier but also as a bridge and as a unifying element between the two sides of the border: "Think about how far we have come, when the question of the border is no longer about where we should divide – but how we gather".[24]

---

19 T. Andresen & S. Faber, *Genforeningen 2020, op. cit.* p. 62.

20 F. A. Petersen, "Erindring og politik", *op. cit.*

21 H. M. Dronning Margrethe II, "H.M. Dronning Margrethe II's Tale Af 100-Året for Genforeningen 2021", Frederikshøj, 2021. Retrieved from: <https://www.dansketaler.dk/tale/h-m-dronning-margrethe-ii-s-tale-i-anledning-af-100-aret-for-genforeningen>. Accessed 17 June 2022.

22 Grænsen, "Hvad Betyder Genforeningen?", *op. cit.;* T. Andresen & S. Faber, *Genforeningen 2020, op.cit.*

23 M. Christophersen, "Mindretallene i Det Dansk-Tyske Grænseland: 'Det Er Ikke Os, Der Har Flyttet Os. Det Har Grænsen'", *Grænsen – Magasinet Om Mindretal, Sprog Og Kultur*, December 2019, p. 25.

24 M. Frederiksen, "Statsminister Mette Frederiksens Tale Ved Genforenings arrangement på Dybbøl Bankeden 15. Juni 2020", *op. cit.*

Furthermore, the usual openness of the border is underlined in opposition to the forced closure of the border due to the Covid-19 pandemic.[25] For example, Prime Minister Mette Frederiksen, in her speech at the closing ceremony on the 15th of June, states: "The celebration takes place in quite a different way than it was planned. That we were supposed to gather around, we will have to mark at a distance. We figure it out – but we miss the 'nearness' ".[26] Thus, the closed border due to the pandemic is used to underline the fact that the border is usually open.

Yet, Denmark has actually introduced border control at its border with Germany long before the Covid-19 pandemic and continues to keep it even when most countries in the Schengen area have gone back to normality.[27] Thus, despite the narrative used in the celebrations, the border continues to be a barrier and still constitutes an obstacle for local inhabitants, especially for daily commuters and members of the national minorities on both sides.

Another persisting element of the celebrations of the *Genforening* is the fact that the border region is continuously described as an example and a role model for the rest of the world. The fact that the border between Germany and Denmark was established through a plebiscite is used to justify the uniqueness of this border region.[28] This peaceful resolution of the border question is thus seen as a reason of pride, and it is also used to elevate Denmark in comparison to the rest of the world. Yet, this narrative does not reflect the whole history of the border region, which has been characterised by both ups and downs and tensions between the two countries. Commitment to cross-border cooperation has been relatively

---

25 T. Andresen & S. Faber, *Genforeningen 2020, op.cit.*

26 M. Frederiksen, "Statsminister MetteFrederiksens Tale Ved Genforenings arrangement på Dybbøl Bankeden 15. Juni 2020", *op. cit.*

27 M. Klatt, "The So-Called 2015 Migration Crisis and Euroscepticism in Border Regions: Facing Re-Bordering Trends in the Danish–German Borderlands," *Geopolitics*, vol. 25, n°3, May 26, 2020, pp. 567–86; Politi Danmark, "Border Control/Danish Police", 2022. Retrieved from: <https://politi.dk/en/law-and-information/border-control>. Accessed 17 June 2022.

28 M. Frederiksen, "Mette Frederiksens Tale Ved Genforenings fejring 2020", *op. cit.;* T. Andresen & S. Faber, *Genforeningen 2020, op.cit.;* S. Hyldelund, "Unser Grenzraum strahlt weit über Nord- und Südschleswig hinaus/ Vores grænseregion rækker langt ud over Nord- og Sydslesvig", *In* B. Henningsen, C. Räthel & P. Greiner (eds.), *GRENZ\RAUM: dänisch-deutsche Geschichte(n), 1920–2020*, Rombach Wissenschaft, Auflage, Baden-Baden, 2022.

weak, as demonstrated by the initial opposition to the establishment of a Euroregion and recent re-bordering trends. The two sides also often lack a common interpretation and approach over History linked to issues of the region as for example when it comes to whether what took place in 1920 was a reunification or a division.

Past and Present: Historical Events and Denmark

The celebrations of the *Genforening* are of course linked to the plebiscites in 1920, but other historical events were also mentioned. History is often used by politicians to create a sense of continuity and to justify their rule over a population. Historical events and collective memories are also used to create a sense of uniqueness around the nation and a feeling of common belonging.[29]

In the celebrations of its 100-year anniversary, the main focus was on two events: the plebiscite on the 10th of February 1920 and the visit paid by King Christian X took over the former border on the 10th of July 1920. The plebiscite on the 10th of February is recalled because on that day, inhabitants of the region had the opportunity to decide their identity and "could vote themselves home to Denmark".[30] However, those who went to vote were presented with a binary choice between Germany and Denmark and, especially in the Danish propaganda, where there was no space for a regional identity.[31]

The democratic establishment of the border that followed the plebiscites was also mentioned in opposition to the dark years of the war, and thus is seen as a turning point in Danish history.[32] The wars, however, are also mentioned to honour the sacrifice of Danish soldiers, both those who died in the Schleswig wars and those who died during the First World War fighting for the Germans, therefore evoking feelings of national pride: "In the First World War thousands of Danish-minded

---

29 J. Breuilly, "Nationalism and the Making of National Pasts", *In* S. Carvalho & F. Gemenne (eds.), *Nations and Their Histories*, Palgrave Macmillan UK, London, 2009, pp. 7–28.

30 M. Christophersen, "Genforeningen Skabte Nutidens Danmark.", *op. cit.*, p. 21.

31 N. Jebsen, "Dividing Regions? Plebiscites and Their Propaganda", *op. cit.*

32 H. M. Dronning Margrethe II, "H.M. Dronning Margrethe II's Tale Af 100-Året for Genforeningen 2021", *op. cit.*

individuals had to fight in the trenches for a country which they did not feel was theirs".[33]

Regarding King Christian X's ride over the old border, the event was filled with symbolic meaning. It was interpreted as the day in which Southern Jutland officially returned to Denmark and marked the start of the celebration in 1920.[34] The fact that King Christian X was riding a white horse is always underlined, granting grandeur and regalian character for the event. Furthermore, it was at the centre of a myth regarding whether the horse was actually white and thus led to several speculations on that fact. It is also the subject of some memorial stones, occupying a central role in Danish national history and in the national discourse.[35]

The reunification of Sothern Jutland to Denmark in 1920 can also be related to the loss of the other colonies in the West Indies. After the selling of the oversea colonies, in fact, with the Genforening, Denmark regained territory and although it was still rather small, it felt like a national victory.[36] Thus, the Genforening is seen as sort of a win back and as the day in which the Danish nation was reunited within the same border. As such it should be considered as an important event not only for the region of Southern Jutland, but also for the whole Denmark.[37] As Bertel Haarder, a former member of the cabinet, states: "The *Genforening* is a very national celebration. So, if one does not feel anything for his land, the Reunification can be unimportant".[38] This narrative, however, once again did not take the minorities and regional identities into consideration.

---

[33] M. Frederiksen, "Statsminister MetteFrederiksens Tale Ved Genforenings arrangement på Dybbøl Bankeden 15. Juni 2020", *op. cit.*

[34] A. Bjerager, "H.M. Dronningen: 'Jeg Håber, at Hele Danmark Vil Fejre 100-Året for Genforeningen'- Interview", *op. cit.;* T. Andresen & S. Faber, *Genforeningen 2020, op.cit.*

[35] Danmarkshistorien Aarhus Universitet, "MYTE: Var Christian 10.s hest ved Genforeningen kalket hvid?", 2022. Retrieved from: <https://danmarkshistorien.dk/vis/materiale/myte-var-christian-10s-hest-ved-genforeningen-kalket-hvid/>. Accessed 17 June 2022.

[36] T. Andresen & S. Faber, *Genforeningen 2020, op.cit.*

[37] M. Christophersen, "Genforeningen Skabte Nutidens Danmark", *op.cit.*; M. Frederiksen, "Mette Frederiksens Tale Ved Genforeningsfejring 2020", *op. cit.*

[38] Grænsen, "Hvad Betyder Genforeningen?", *op. cit.*, p. 31.

The Narrative in Discussion

In the celebrations of the *Genforening*, the border was always treated as a fixed entity. The establishment of the border in 1920 was presented as the final and successful solution to a longstanding problem, remaining stable since its establishment in 1920. This narrative ignored the numerous times in which the border has been contested since its institution. For instance, just after the establishment of the border, discontent was to be heard on both sides. The situation was especially tense in Denmark where nationalists pressed the King to install a new government in the hope that it would have led to Danish control over the German city of Flensburg.[39] The border was then contested just after the Second World War and then again in the 1990s with the establishment of the Euroregion.[40]

Furthermore, relations between Germany and Denmark have not always been that positive and only with the Bonn-Copenhagen Declaration a shift from a negative peace to a positive one happened. The Bonn-Copenhagen Declaration protected the rights of the minorities on both sides and laid the foundation for a more cooperative relation.[41] Yet, this does not mean that question was solved, since the border, as a social construction, will always change throughout time.

Regarding the border in light of the process of spatial socialisation, it was used to attribute a certain uniqueness to the Kingdom of Denmark and specific characteristics to the Danish nation. In particular, the fact that the border was established democratically is seen as a result of a disposition of the Danish population to solve problems together and without the use of violence. Yet, the possibility of implementing Wilson's principle of self-determination was already brought forward by the German government as an effort to win goodwill, before the Peace Conference in Paris decided to hold plebiscites in Schleswig.[42] Moreover, the border and its history was also used to create a sense of common identity and was

---

39 S. B. Frandsen, "Schleswig – A Border Region Caught Between Nation-States", *op. cit.*

40 C. Yndigegn, "Bordering and Borderwork: Protecting the Danish Border and National Identity," *Nationalism and Ethnic Politics*, vol. 24, n°4, 2018, pp. 411–432.

41 W. R. A. Hughes, N. S. Megoran & M. C. Benwell, "The Emotional Geopolitics of Contested Peace", *op. cit.*

42 S. B. Frandsen, "Schleswig – A Border Region Caught Between Nation-States", *op. cit.*

still infused with nationalist remarks. The role of the State was revealed in imposing a common interpretation of the territory and in using it for nation-building. This also led to the socialisation of individuals in relation to spatial entities that are the nation and the border region.

Yet, the national question was brought in the region from the outside, although it was then internalised and appropriated by locals themselves. The area was characterised by a regional sense of identity and it was only with the plebiscites in 1920 that individuals were forced to identify with either Germany or Denmark. Of course, there were convinced nationalists on both sides of the border, but the issues were first raised by central governments, as they also had an interest in keeping the conflict alive.[43]

The use of historical events can also be linked to Somers' concept of narrativity in the context of identity formation. Somers underlines how in this process of identity formation and of construction of the "us" vs "them", an important role is played by narrative. In that sense, the "narrative" allows people to make sense of their lived experiences by placing them within a wider context. The narrative people use depends not only on the relations among individuals, but also on the cultural and institutional framework the individuals find themselves in, the so-called public narrative. The latter guides people on how to interpret events and how to connect them together, but also tells them which events are relevant for the process of identification, a mechanism which Somers calls "selective appropriation". One important type of public narrative is the one by the State, especially when talking about borders, because it is usually the State which imposes a narrative and dictates how to interpret the border in the wider framework of national history.[44]

In the case of the celebrations of the 100-year anniversary of the "*Genforening*", events used in the official narrative where either instrumental so to present Denmark and the Danish nation as open and peaceful, such as the democratic establishment of the border; or to highlight the courage of the Danish nation, as it was the case for the Schleswig wars and the World Wars. Yet, the narrative completely left out protests and contestation of the border which could have led to an understanding of Denmark as a nationalist and protectionist country.

---

[43] *Ibid.*

[44] M. R. Somers, "The Narrative Constitution of Identity: A Relational and Network Approach," *Theory and Society*, vol. 23, n°5, 1994, pp. 605–649.

Finally, the construction of the Danish identity is still based on a process of "othering", although the "other" is no longer Germany. Today, the border still remains a line of exclusion and inclusion, especially because of the maintenance of border controls despite the Schengen Agreement. However, this line is no longer drawn to keep the Germans away but is instead meant to protect Denmark from foreigners coming from outside Europe. In particular, border controls to the German border are justified as a way to prevent terrorists from coming into the country as they were introduced following the 2015 crisis.[45] The "Other", therefore, is now the non-Christian immigrant coming from the Middle East and Muslim countries and Denmark has even passed new laws to restrict immigration from these countries.[46] Germany is now perceived and identified as an ally and a friend with whom Denmark shares a successful partnership within the EU.

Conclusion

During the celebrations of the 100-anniversary of the *Genforening*, the Danish government carried forward a specific narrative of the event and of the border and its inhabitants. By using the term of *Genforening*, translatable as "reunification", the focus was put on Denmark regaining territory rather than on the establishment of a cooperative relation with Germany. The term also excludes minorities as they were not reunified and were left unsatisfied by the results of the referenda. Furthermore, the focus on reunification instead of cross-border cooperation gives the celebration a more nationalistic character.

One of the functions of the border is to distinguish between insiders and outsiders; Germany constituted the historical foreign enemy from whom Denmark needed to be protected. Today, Germany is instead portrayed as a friend and ally, while the role of the "other" has shifted to non-Christian immigrants (often universalised as Muslim) and non-European foreigners. Therefore, although Germany is no longer the "other", Danish

---

45 Politi Danmark, "Border Control/Danish Police", *op. cit.*

46 C. Ogan et al., "The Rise of Anti-Muslim Prejudice: Media and Islamophobia in Europe and the United States," *The International Communication Gazette*, vol. 76, n°1, 2014, pp. 27–46.

identity is still built on an opposition and on a process of exclusion and inclusion.

The border is no longer perceived only as a barrier, but as bridge between Denmark and Germany. The border then comes to play a role in the process of spatial socialisation because it defines who is part of the Danish nation and because its democratic establishments is seen as the result of the open and peaceful character of the Danish nation. This narrative treats the border as a fixed entity, without considering the fact that ever since its establishment the border has been contested several times. This narrative of openness linked to Danish national characteristics is also in contrast with Denmark's border policy as Denmark reintroduced border controls following the refugee crisis in 2015.

The narrative used during celebrations also made use of historical events to create a sense of continuity and common belonging. Two events became particularly important during celebrations: the plebiscite in February 1920 and King Christian X's ride on a white horse over the old border in June 1920. The plebiscites are used to underline the unique character of the border region as the border issue was solved peacefully and democratically, unlike in most inter-state conflicts. King Christian X's ride however gives a more grand and regal character to the celebrations as the event already occupied an important role in Danish history and was at the centre of myths and speculations. The official narrative also mentioned the Schleswig wars and the World Wars to praise Danish soldiers and underline the courage of the Danish nation.

The use of these historical events or episodes can be connected to the concept of narrative construction of identity. The selection of events is a key element of this process of construction, since the selection of certain events is instrumental for the building of a certain identity. In this case, the public narrative focused on events that highlight the uniqueness and braveness of Denmark but left out the several occasion in which the border was contested which could have led to a negative depiction of the country.

Bibliography

T. Andresen & S. Faber (eds.), *Genforeningen 2020 – 100 året for Danmarks genforening. Jubilæumsbog – 2021*, Sonderborg, 2020.

A. Bjerager, "H.M. Dronningen: 'Jeg Håber, at Hele Danmark Vil Fejre 100-Året for Genforeningen' – Interview", *Grænsen – Magasinet Om Mindretal, Sprog Og Kultur*, December 2019.

J. Breuilly, "Nationalism and the Making of National Pasts", *In* S. Carvalho & F. Gemenne (eds.), *Nations and Their Histories*, Palgrave Macmillan UK, London, 2009, pp. 7–28.

M. Christophersen, "Genforeningen Skabte Nutidens Danmark", *Grænsen – Magasinet Om Mindretal, Sprog Og Kultur*, December 2019.

M. Christophersen, "Mindretallene i Det Dansk-Tyske Grænseland: 'Det Er Ikke Os, Der Har Flyttet Os. Det Har Grænsen' ", *Grænsen – Magasinet Om Mindretal, Sprog Og Kultur*, December 2019.

H. M. Dronning Margrethe II, "H.M. Dronning Margrethe II's Tale Af 100-Året for Genforeningen 2021", Frederikshøj, 2021.

S. B. Frandsen, "Schleswig – A Border Region Caught between Nation-States", *In* K. Stoklosa & G. Besier (eds.), *European Border Regions in Comparison: Overcoming Nationalistic Aspects or Re-Nationalization?*, Routledge Studies in Modern European History 21, Routledge, New York, London, 2014, pp. 79–97.

M. Frederiksen, "Statsminister Mette Frederiksens Tale ved genforenings fejring i Alsion i Sønderborg", Alsion, Sønderborg, 2021.

Grænsen, "Hvad Betyder Genforeningen?," *Grænsen – Magasinet Om Mindretal, Sprog Og Kultur*, December 2019.

W. R. A. Hughes, N. S. Megoran & M. C. Benwell, "The Emotional Geopolitics of Contested Peace: The Austerity Crisis and the Danish Minority of South Schleswig", *Territory, Politics, Governance,* vol. 10, n°2, 2022, p. 237–255.

S. Hyldelund, "Unser Grenzraum strahlt weit über Nord- und Südschleswig hinaus/ Vores grænseregion rækker langt ud over Nord- og Sydslesvig", *In* B. Henningsen, C. Räthel & P. Greiner (eds.), *GRENZ\RAUM: dänisch-deutsche Geschichte(n), 1920–2020*, Rombach Wissenschaft, Auflage, Baden-Baden, 2022.

N. Jebsen, "Dividing Regions? Plebiscites and Their Propaganda – Schleswig and Carinthia 1920", *In* K. Stoklosa & G. Besier (eds.), *European Border Regions in Comparison: Overcoming Nationalistic Aspects or Re-Nationalization?*, Routledge, New York, London, 2014, pp. 58–78.

M. Klatt, "Lessons from the Danish-German Border Region for Post 2020 Interreg A – An Alignment with Cross-Border Functional Regions?," *Europa XXI*, vol. 38, 2020, pp. 139–156.

M. Klatt, "The So-Called 2015 Migration Crisis and Euroscepticism in Border Regions: Facing Re-Bordering Trends in the Danish–German Borderlands," *Geopolitics*, vol. 25, n°3, May 26, 2020, pp. 567–86.

D. Newman, "On Borders and Power: A Theoretical Framework," *Journal of Borderlands Studies*, vol. 18, n°1, March 2003, pp. 13–25.

C. Ogan et al., "The Rise of Anti-Muslim Prejudice: Media and Islamophobia in Europe and the United States," *The International Communication Gazette*, vol. 76, n°1, 2014, pp. 27–46.

A. Paasi, "Part One: Territories and Boundaries in Regional Transformation," *In Territories, Boundaries, and Consciousness: The Changing Geographies of the Finnish-Russian Border*, Belhaven Studies in Political Geography, Chichester, England, J. Wiley & Sons, New York, 1996, pp. 1–76.

A. Paasi, "The Political Geography of Boundaries at the End of the Millennium: The Challenges of the de-Territorializing World," 1999, pp. 9–24.

F. A. Petersen, "Erindring og politik: Et blik på historie og kultur i den dansk-tyske udveksling 2017–2020", *In* P. Greiner et al. (eds.), *Dänisch-deutsches Freundschaftsjahr 2020: Debatten und Perspektiven = Tysk-dansk venskabsår 2020: debatter og perspektiver*, Nordeuropa-Institut, Humboldt-Universität zu Berlin, Berlin, 2021.

M. R. Somers, "The Narrative Constitution of Identity: A Relational and Network Approach," *Theory and Society*, vol. 23, n°5, 1994, pp. 605–649.

C. Yndigegn, "Bordering and Borderwork: Protecting the Danish Border and National Identity," *Nationalism and Ethnic Politics*, vol. 24, n°4, 2018, pp. 411–432.

The Danish-German-Frisian Region – A Best Practice Example?

Martin Klatt

In this article, I scrutinise the Danish-German-Frisian borderland of Schleswig. This region and its majority-minority settlement are often perceived as a best-practice model of how to reconcile a national conflict and accommodate national minorities.[1] The Danish-German minority settlement is characterised by respective recognition by the state of residence and generous kin-state financial support to operate minority educational and cultural institutions, primarily organised as member-driven private associations. In principle, it ensures the possibility to live as a Dane in South Schleswig or as a German in North Schleswig from cradle to grave. It is often forgotten, though, that Frisians but also Sinti and Roma, two of the other four recognised national minorities in Germany, reside in the region. These minorities do not have a kin-state, and their situation is slightly different.

This article is a product of research financed by the Jean Monnet network Frontiers in Motion (Frontem), which states as is objective:

> "The objective of the Frontem network is therefore to propose a critical reading of the "Europe without borders" model. It starts from the observation that the EU has developed a unilateral view on the border which essentially only takes into account its economic character whilst neglecting

---

1 J. Kühl, "The Bonn-Copenhagen Declarations of 1955: Background, Context and Impact of the Danish-German Minority Regulations", *In* J. Kühl & M. Weller (eds.), *Minority Policy in Action: The Bonn-Copenhagen Declarations in a European Context 1955–2005,* Institut for Grænseregionsforskning, Aabenraa, 2005, pp. 31–89; J. Kühl & R. Bohn (eds.), *Ein europäisches Modell? Nationale Minderheiten im deutsch-dänischen Grenzland 1945–2005,* European Centre of Minority Issues, Verlag für Regionalgeschichte, Bielefeld, 2005.

> other aspects, more symbolic or political. The hypothesis put forward is that the border in the political sense has never disappeared and that it should be analyzed by taking into account its ambivalence as an identity limit and an object of separation, but also as a place of contact and exchange".[2]

Here, especially, I will focus on the second part of the hypothesis, to analyse the border by focusing on its ambivalence as an identity limit and a place of contact and exchange. This leads to the research question of whether the Danish-German border region, with its accommodation of national minorities and reconciliation of national conflict, can be understood as a best practice example. Here, of course, it must be asked "a best practice of what?" I will scrutinise six lenses: a best practice of border determination?; a best practice of national minority construction?; a best practice of reconciliation?; a best practice of cross-border cooperation, in politics, people-to-people and business?; a best practice of an integrated cross-border living space (labour market, public services, residential space, creative industries economies); and thus, a best practice of diversity accommodation?

Three theoretical approaches will be used as analytical frameworks: Oscar Martinez concept of national and transnational borderlanders,[3] Emmanuel Brunet-Jailly's four lenses theory of borderland studies,[4] and Bas Spierings' and Martin van der Velde's concept of unfamiliarity.[5] Oscar Martinez has developed a four stages model of borderlands: alienated, coexistent, interdependent, and integrated borderlands.[6] In addition, he examined borderland populations and differentiated between national and transnational borderlanders. When examining contemporary EU borders, Martinez' model's weakness, in the author's

---

2 Retrieved from: https://centre-jean-monnet.unistra.fr/the-project/. Accessed 1 July 2022.

3 O. J. Martinez, *Border People. Life and Society in the U.S.-Mexico Borderlands*, University of Arizona Press, Tucson, London, 1994.

4 E. Brunet-Jailly, "Theorizing Borders: An Interdisciplinary Perspective", *Geopolitics*, vol. 10, 2005, pp. 633–649.

5 B. Spierings & M. van der Velde, "Shopping, Borders and Unfamiliarity: Consumer Mobility in Europe", *Journal for Economic and Social Geography*, vol. 99, 2008, pp. 497–505.

6 O. J. Martinez, *Border People*, *op. cit.*

opinion, lies that all intra-EU borderlands pendulate somewhere in the interdependent-integrated sphere. There are definitely interdependencies in all EU borderlands, but cross-sectoral integration in form of shared public services, integrated labour markets, joint policymaking and political decision-making, etc., is difficult to spot, with few sectoral exceptions.[7] Therefore, I will primarily focus on the national-transnational borderlanders dichotomy, when focusing on people living in the Danish-German border region, with regard to points two and five. Martinez considers national borderlanders as:

> "People who, while subject to foreign economic and cultural influences, have minimal or only superficial contacts with the opposite side of the border, owing either to their indifference to their next-door neighbours or their unwillingness or inability to function in any significant way in another society".[8]

Transnational borderlanders, on the other hand, are:

> "Individuals who maintain significant ties with the neighbouring nation. They seek to overcome obstacles that impede such contact and take advantage of every opportunity to visit, shop, work, study, or even live on the other side. Thus, their lifestyles strongly reflect foreign influences. […] but for those who are deeply immersed in transborder interaction, foreign links govern central parts of their lives".[9]

Brunet-Jailly's model of borderlands uses four lenses to analyse border regions: market forces and trade flows, policy activities of multiple levels of government, local cross-border culture, and local cross-border political clout (Figure 1).

---

7 S. Zillmer, S. Haarich, F. Holstein, C. Lüer, T. Stumm, E. Marques de Costa, N. Da Costa, P. Ockerl, M. Peterlin, A. Berlina, E. Turunen & J. Teräs, "Cross-border Public Services (CPS). Targeted Analysis. Final Report," *In* ESPON (eds.), *Main Report*, 2018, ESPON, Luxembourg.

8 O. J. Martinez, *Border People*, *op. cit.*

9 *Ibid.*

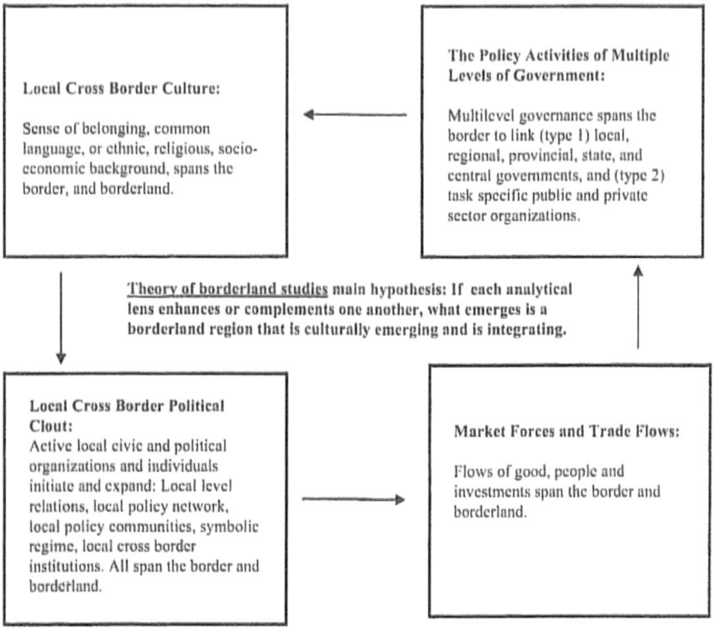

Figure 1: Emmanuel Brunet-Jailly's theory of borderland studies.[10]

These four lenses are interdependent, but there is no clear hierarchy in Brunet-Jailly's model. Empiric evidence, in the author's opinion, demonstrates clearly that market forces and trade flows are a dominant catalyst initiating the activation of the three other lenses.[11] It will be used to measure and analyse points four, five and six.

Spierings' and van der Velde's concept of unfamiliarity[12] explains the reluctance to move across borders even in intra EU borderlands, where

10 Source: E. Brunet-Jailly, "Theorizing Borders", *op. cit.*

11 M. Klatt & H. Herrmann, "Half empty or half full? Over 30 years of regional cross-border cooperation within the EU: Experiences at the Dutch-German and Danish-German border", *Journal of Borderlands Studies*, vol. 26, 2011, pp. 65–87.

12 *Ibid.* B. Spierings & M. van der Velde, "Cross-border differences and unfamiliarity: Shopping mobility in the Dutch-German Rhine-Waal Euroregion", *European Planning Studies*, vol. 21, 2013, pp. 5–23; B. Szytniewski & B. Spierings, "Encounters

no physical barriers or government restrictions to cross borders should exist (Figure 2).

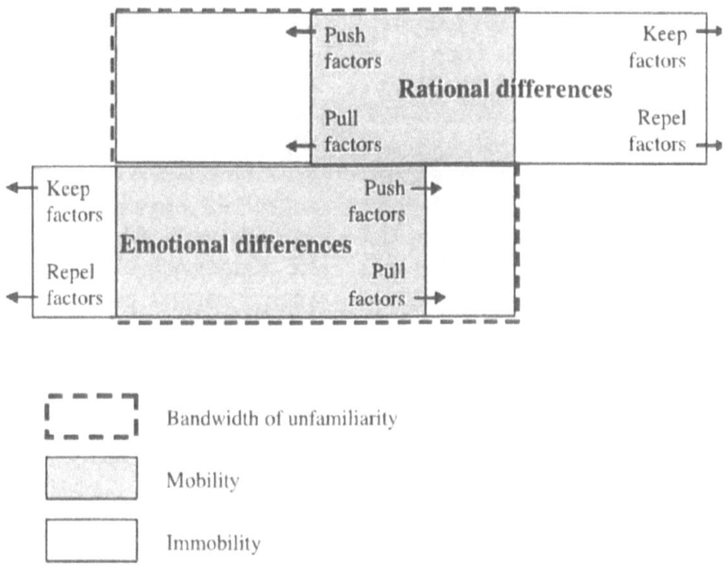

Figure 2: Bandwidth of Unfamiliarity.[13]

Spierings and van der Velde argue that it is not enough to focus on rational differences when analysing motives to cross or not to cross borders. In addition, emotional differences play a role. Perceptions of otherness can both repel or encourage cross-border interaction, creating a bandwidth of unfamiliarity, within which differences are tolerated. In Schleswig, with a majority-minority and a minority-kin-state setting, this approach becomes interesting as it widens the cultural and linguistic scope of borderland populations. We should expect different bandwidths of unfamiliarity here than in other intra EU borderlands. The bandwidth

with otherness: Implications of (un)familiarity for daily life in borderlands", *Journal of Borderlands Studies*, vol. 29, 2014, pp. 339–351.

13 B. Spierings & M. van der Velde, "Shopping, Borders and Unfamiliarity", *op. cit.*

of unfamiliarity may be tied to the share and degree of transnational borderlanders[14] as dependent variable.

A Best Practice of Border Determination

Today's Danish-German border is a product of the post WW I peace order. It divided the former Duchy of Schleswig, a territory having caused a national conflict between Danish and German nation state projects in the 19th century (See Figure 3).[15]

Schleswig had been a Danish fief, but since medieval times connected to Holstein, a fief of the Holy Roman Empire. A Schleswig-Holstein proto-state developed under the Schauenburg dynasty in the 15th century. When the last Schauenburg duke died childless in 1459, the regional nobility decided to elect Danish king Christian I (one of the last duke's three nephews) as their duke. This personal union with the Danish crown lasted until the Danish-German war of 1864, when Prussia overtook Schleswig-Holstein. Germany's defeat in World War I brought the Schleswig-question on the agenda of the Paris Peace Conference, where it was decided to submit the region to a plebiscite.[16]

The exact location of the border was negotiated between Denmark and the Allies at the peace conference, where Denmark insisted on implementation by plebiscite. The terms of this plebiscite, applying two zones voting on 10 February and 12 March 1920, followed a design outlined by the North Schleswig Danes leading political figure H.P. Hanssen. It was based on the "Clausen Line," a line already demarcated around 1900 by a Danish historian as the most southern and best practical line demarcating the part of the lost territory, which Denmark could successfully reintegrate while preserving Denmark as a nation state. This line

---

14 O. J. Martinez, *Border People*, *op. cit.*

15 M. Bregnsbo, "Schleswig(-)Holstein 1848: Legitimism, Nationalism, Constitutionalism and Regionalism in Conflict", *In* M. Bregnsbo & K. Villads Jensen (eds.), *Schleswig-Holstein – Contested Region(s) through History*, University of Southern Denmark Press, Odense, 2016, pp. 177–193.

16 T. Fink, *Da Sønderjylland blev delt*, Institute for Grænseregionsforskning, Aabenraa, 1979.

Figure 3: The Danish-German border region.[17]

became the national border between Denmark and Germany in 1920.[18] The plebiscite modalities involved two zones: a northern zone between the 1864-border at the Kongeå and the Clausen line, and a southern zone stretching from the North Sea islands of Sylt, Amrum and Föhr across the midlands of Schleswig. The Clausen line reflected both national and geographical considerations by going south around the Western towns of Tønder and Højer, which both eventually proved to have significant

---

17 Source: Centre for Border Region Studies. Retrieved from: <https://www.sdu.dk/en/om_sdu/institutter_centre/c_borderregionstudies>. Accessed 21 May 2023.

18 *Ibid.;*T. Fink, "Den første streg på kortet – Clausen linjen", *In* H. Becker Christebseb (ed.), *Grænsen i 75 år,* Institut for Grænseregionsforskning, Aabenraa, 1995, pp. 11–23.

German majorities.[19] Voting in the northern zone was *en-bloc*, while the southern zone voted by parish, opening for local changes of the future border. This provision was especially designed for the city of Flensburg, the region's demographic and economic centre. In Flensburg, a majority (of the male population over 25) had voted for a Danish oriented candidate at the elections to the North German *Reichstag* in 1867, but Danish votes had decreased to less than 5 % in the last pre-war *Reichstag* elections in 1912. Nonetheless, nationalist Danes put a strong campaign focus on the city and hoped to regain it. All men and women over 20 years of age, having resided within the plebiscite area since or before 1 January 1900, were eligible to vote in the plebiscite. This rule was a compromise to account for German in – and Danish outmigration since 1864 and resulted in mobilisation activities of former Schleswigians, stretching all the way to the United States.[20]

The results of the plebiscite confirmed Clausen's prediction: 75 % of the voters in the northern zone voted for Denmark, and 80 % in the southern zone voted for Germany. Flensburg had 25 % Danish votes, but the cities of Tønder (80 %), Aabenraa and Sønderborg of the northern zone had German majorities, as did some rural parishes. Therefore, the post-1920 German minority in Denmark as well as most politicians from Schleswig-Holstein did not acknowledge the plebiscite results and demanded a renegotiation of the 1920 border; demands that became more aggressive with the Nazi takeover in 1933.[21] The border was also challenged by national activists in Denmark during the interwar years,[22] but especially after the surrender of Nazi-Germany in 1945.[23] West Germany's integration into North Atlantic Treaty Organization (NATO) in 1955 implemented acceptance of the 1920-border by both

---

19 *Ibid.*

20 R. Gesme, "Vote for Home: Self-Determination, Wilsonianism, and the Implementation of the 1920 Schleswig Plebiscite", doctoral thesis, University of Tennessee, Knoxville, 2023.

21 P. Hopp & C. Mogensen (eds.), *Ostersturm Påskeblæsten*, Institut für Regionale Foschung und Information im Deutschen Grenzverein e.V., Flensburg, 1993.

22 A. Johnsen, *Dannevirkemænd og Ejderfolk. Den grænsepolitiske opposition i Danmark 1920–1940,* Studieafdelingen ved Dansk Centralbibliotek for Sydslesvig, Flensburg, 2005.

23 J. P. Noack, *Det sydslesvigske grænsespørgsmål 1945–1947*, Institut for Grænseregionsforskning, Aabenraa, 1991.

countries' governments.[24] Nevertheless, in South Schleswig, suspicion on the Danish minority's as well as Danish nationalist circles long-term aim of border revision continued well into the 1960s. This changed when Denmark accessed the European Economic Community (EEC) in 1973, opening for a discourse of de-bordering and cooperation on the German side. This discourse was not fully reciprocated in Denmark until the Millennium.[25] In 2020, when the border's 100th anniversary was celebrated, discourse centred around a "best border of the world" narrative: democratically decided upon, nationally optimal and a meeting point rather than a border dividing people.[26] 100 years after the decision, no serious politician challenges the position of the border. Neither do the members of the reciprocal kin-minorities, nor people in general. In effect, though today presented as exemplary, the model to let people vote on the actual location of a national border, has not been repeated since anywhere in the world.

A Best Practice of National Minority Construction?

The concept of national minority evolved together with the modern concept of the nation as the ideal form to organise a democratic state. I will not go deeper here into the debate between primordialists and social constructivists on the origin of nations.[27] Relevant here is the

---

[24] R. Bohn, "Eine belastete Nachbarschaft. Aspekte der deutsch-dänischen Beziehungen nach 1945", *In* J. Kühl & R. Bohn (eds.) *Ein europäisches Modell?, op. cit.*, pp. 59–76.

[25] M. Klatt, *Fra modspil til medspil?: grænseoverskridende samarbejde i Sønderjylland/Schleswig 1945–2005,* Institut for Grænseregionsforskning, Aabenraa, 2006.

[26] M. Klatt, "1920-grænsen: national optimal, økonomisk katastrofal? Et overset aspekt af den slesvigske model", *Akademisk kvarter,* vol. 21, 2020, pp. 60–73.

[27] B. Anderson & American Council of Learned, *Imagined Communities: Reflections on the Origin and Spread of Nationalism*, Verso, New York, London, 2006; A. Gat, *Nations. The Long History and Deep Roots of Political Ethnicity and Nationalism*, Cambridge University Press, Cambridge, 2013; E. Gellner, *Nations and Nationalism*, Blackwell, Oxford, 1983; A. Hastings, *The Construction of Nationhood: Ethnicity, Religion and Nationalism*, Cambridge University Press, Cambridge, 1997; E. J. Hobsbawm, *Nations and Nationalism Since 1780: Programme, Myth, Reality*, Cambridge University Press, Cambridge, 1992; M. Hroch, *Das Europa der Nationen. Die modern Nationalittätenbildung im europäischen Vergleich*, Vandenhoeck & Ruprecht, Göttingen, 2005; P. M. Judson & M. L. Rozenblit (eds.), *Constructing nationalities in East Central Europe*, Berghahn, New York, Oxford, 2005; A. D. Smith, *The Ethnic Origins of Nations*, Blackwell, Oxford, 1987.

dilemma of aligning territories to nations to create homogeneous, but also sustainable states. Here, the concept of national minority opens for a compromise to secure nationalities cultural, linguistic, and religious rights within a nation state defined by another, dominant nation. This became a core element of the post WW I peace order in the territories of the former Austro-Hungarian, Russian, and German Empires, obliging especially Germany and the newly created or recreated states to enact constitutional minority rights. 70 years later, with the end of the Cold War and several newly independent states on the map of Europe, focus on national minority accommodation experienced a renaissance, with important new legal frameworks as the Framework Convention for the Protection of National Minorities, adopted by the Council of Europe in 1994, and the so-called Copenhagen Criteria of 1993 setting minority rights provisions as a condition to enter the European Union. Hereby, provision for cultural, linguistic, and educational rights to autochthonous national minorities has become a bearing principle of the EU and a universally accepted standard of best practice of diversity accommodation. Attempts to extend these rights to migration-based ethnic minorities so far have not been successful in any of the EU member states, except for the Czech Republic's recognition of the Vietnamese minority. I will here discuss the historical evolvement of national diversity in the Danish-German border region, arguing that the concept of containerised national minorities does not reflect the social reality of borderlands residents and might be a counterproductive factor to Europeanised agendas of cross-border cooperation and cross-border regional development.

Place names in Schleswig suggest a settlement by Nordic language speakers in pre-history, accompanied by colonisation from the South (German and some Slavic speakers) as well as the West (Frisian speakers) from around the turn of the first Millennium.[28] Thus, Schleswig's ethnic composition has been blurred since medieval times, but top-down acculturation into German culture had proceeded slowly, especially since protestant reformation in the 16th century. In effect, a clear national boundary did not exist. Linguistically, the local Danish dialect remained strong in the rather wealthy northern rural areas as well as the rather poor midlands, while German replaced it in the south (17th century) and the wealthy Mideast (19th century). Furthermore, German dominated in the

[28] L. N. Henningsen (ed.), *Sydslesvigs danske historie*, 2005, Studieafdelingen ved Dansk Centralbibliotek for Sydslesvig, Flensburg, 2005.

urban elites. Proto-national and even national identity has not necessarily been tied to language, but remained volatile and multiple, oscillating in times of crises.[29]

The conglomerate state of the Oldenburg monarchy perceived itself as threefold: Danish, Norwegian, and German. Formally, the king ruled over the kingdoms of Denmark, Norway and its dependencies (Iceland, Greenland and the Faeroe Islands), and the duchies of Schleswig and Holstein. In Norway and Denmark, the king ruled as absolute monarch, but the (German speaking) nobility in Schleswig and Holstein enjoyed wide ranging self-rule, including a separate administration. This elevated the status of the German language, which was language of administration and higher education in the duchies, including the northern parts, where the vernacular was a Danish dialect. Result was a gradual shift of the vernacular, where German replaced Danish. The perceived lower social status of the Danish language was a main cause of the national awakening in Schleswig. It resulted in policies to strengthen the Danish language in Schleswig's rural areas after the First Schleswigian War (1848–51). When Schleswig was integrated into Prussia following the war of 1864, strict German language assimilation policies were introduced in the 1880s.

To avoid the 19th century language suppression policies, it was consensus on both sides that the dissenters of the plebiscite (about 25 % in North Schleswig had voted for Germany, about 20 % in South Schleswig had voted for Denmark) should enjoy minority rights and cultural autonomy. This was guaranteed by the Weimar constitution in Germany and specific educational laws in Denmark.[30] Furthermore, when the so-called reunification of Northern Schleswig with Denmark was celebrated in the summer of 1920, the Danish prime minister promised the Danish oriented Schleswigians left behind in Germany that they should not be

---

29 M. Klatt, "Mobile Regions: Competitive Regional Concepts (not only) in the Danish-German Border Region", *In* D. J. Andersen, M. Klatt & M. Sandberg (eds.), *The Border Multiple. The Practice of Borders Between Public Policy and Everyday Life in a Re-Scaling Europe,* Ashgate, Aldershot, 2012, chapter 3; M. Klatt, "Mobilization in crisis – demobilization in peace: protagonists of competing national movements in border regions", *Studies on National Movements,* vol. 4, 2019, p. 30.

30 H. Becker-Christensen, *Dansk Mindretalspolitik i Nordslesvig. Udformningen af den danske politik over for det tyske mindretal 1918–1920,* Institut for Grænseregionsforskning, Aabenraa, 1984; J. P. Noack, *Det danske mindretal i Sydslesvig 1920–1945,* Institut for Grænseregionsforskning, Aabenraa, 1989.

forgotten (*De skal ikke blive glemt*).[31] These words lay ground for the motivation of Danish kin-state support of the Danish minority, and still is its moral base today. Similar pledges were given by German politicians to support the German minority in Denmark.

The interwar years were troublesome: Germany and the German minority did not accept the new border. The terms of the plebiscite were perceived as unjust, because North Schleswig's vote was counted *en-bloc*, ignoring local German majorities in the towns of Tønder, Aabenraa and Sønderborg.[32] Politically, the minority's representatives in the Danish parliament, reiterated their claim for "reunification" of North Schleswig with Germany.[33] Also the Danish minority in Germany struggled, being numerously small and facing the challenge that many of its members and especially their children had no knowledge of the Danish language.[34] Furthermore, there was also considerable opposition to the border within nationalist circles in Denmark, manifesting itself in supporting missionary efforts to convince South Schleswigians of their Danish heritage, roots and souls.[35] The German minority's collaboration with the Nazi occupation of Denmark 1940–45 set a severe strain on Denmark's relation to the minority after liberation,[36] while demands for a border revision to the south raised anxiety in post WWII Germany.[37]

---

31 Prime minister Neergaard's speech [in Danish]. Retrieved from: <https://graenseforeningen.dk/om-graenselandet/leksikon/1920-statsminister-niels-neergaards-tale-paa-dybboel-11-juli>. Accessed 30 June 2022.

32 H. Becker-Christensen, *Det tyske mindretal i Nordslesvig 1920–1932,* Institut for Grænseregionsforskning, Aabenraa, 1990.

33 M. Klatt, "Johannes Schmidt-Wodder und Jens Möller – nordschleswigsche Folketingsmitglieder der Zwischenkriegszeit zwischen Alldeutschtum, Friedensverein und Nationalsozialismus", *In* B. Conrad, C.-C. Manher & J. Kusber (eds.), *Parlamentarier der deutschen Minderheiten im Europa der Zwischenkriegszeit*, Droste, Düsseldorf, 2015, pp. 47–64.

34 J. P. Noack, *Den danske mindretal i Sydslesvig 1920–1945, op.cit.*

35 A. Johnsen, *Dannevirkemænd og Ejderfolk. Den grænsepolitiske opposition i Danmark 1920–1940, op.cit.*

36 H. S. Hansen & H. S. Kristensen, "Mindretal og Flertal i Nordslesvig 1945–1955", *In* J. Kühl (eds.), *København-Bonn Erklæringerne 1955–2005. Mindretalserklæringernes baggrund, tilblivelse og virkning,* Institut for Grænseregionsforskning, Aabenraa, 2005, pp. 125–197.

37 J. P. Noack, *Det sydslesvigske grænsespørgsmål 1945–1947,* Institut for Grænseregionsforskning, Aabenraa, 1991.

Nevertheless, these issues were resolved because of the new post-WWII geopolitical situation, which lead to the abandonment of Danish neutrality and the rearmament and integration of West Germany into NATO. As a bi-product, the Bonn-Copenhagen Declarations of 1955 took geopolitics out of the border region and provided for recognition of the minorities based on subjective identification and the implementation of a wide-ranging model of non-territorial autonomy (NTA).[38] This model of reciprocal, functional minority NTA in Schleswig is based on separate cultural and educational institutions, church congregations, nursing homes, social services, sport clubs and other associations. The Danish-German model of minority institutions financing is characterised by the important role of kin-state support, both material and idealistic. Minority educational institutions in North and South Schleswig operate as private institutions, following the respective state of residence's law on private educational institutions. Nevertheless, they are self-perceived as the minorities' public schools, especially in political discourses equalising them with other private educational institutions.

The model of combined kin-state/state of residence responsibility-sharing is universally accepted. Resource conflicts, though, have been present, usually grounded on the argument of a lack of equalisation. This is understood in such a way that state of residence public funding for minority institutions and associations, usually at municipal level, does not match funding for similar majority institutions and associations. The usual majority narrative when denying equalisation in such resource conflicts is that the Danish minority institutions and associations are generously supported by the kin-state, thus already maintaining higher quality standards than comparable majority institutions and associations. While most of these resource conflicts pass unnoticed for the wider public and media, a one-sided reduction of state of residence

---

38 On NTA see M. Andeva, "Non-Territorial Autonomy in Theory and Practice. A 2020 Report", Report, University American College Skopje, Skopje, 2020; M. Klatt, "The dilemma of responsibility: the role of kin-states and nation-states in implementing non-territorial autonomy models to realize minorities' linguistic, cultural and educational rights", *In* D. J. Smith, I. Dodovski & F. Ghencea (eds.), *Realising Linguistic, Cultural and Educational rights through non-territorial autonomy*, Palgrave Macmillan, Cham, 2022, pp. 43–57; M. Klatt, "Schleswig and Non-Territorial Autonomy – The Territorial Trap?", *Nationalities Papers*, vol. 50, 2022, pp. 906–922; T. Malloy & F. Palermo (eds.), *Minority Accommodation through Territorial and Non-Territorial Autonomy*, Oxford University Press, Oxford, 2015.

financing of minority schools in South Schleswig seriously disturbed the minority peace in 2010–12, creating not just a local crisis but affecting Danish-German bilateral relations as well.[39] It materialised out of negotiations for a post-financial crisis austerity budget in the German state of Schleswig-Holstein, where the ruling government decided to reduce state funding of minority schools from the hitherto 100 % equivalence to the state's public schools funding to just 85 %. The background of this measure demonstrates how global developments as the global financial crisis impact the local: German pressures on EU financial stability and public debt limitations, here especially the Greek bailout, induced similar domestic measures as the constitutionalised "debt brake" for German federal and state budgets – which locally resulted in a severe resource conflict on the financial responsibility towards minority educational institutions. The proposed reduction caught stakeholders by surprise and resulted in a militant resistance within the minority, but also a discourse around the perception of wealthy minority schools – so well equipped that they attracted parents from the majority.[40] Both the Danish and the Federal German government intervened, but the conflict was only resolved after the 2012 parliamentary election in Schleswig-Holstein. The election resulted in a new government coalition including the Danish minority's political party SSW (*Südschleswigscher Wählerverband* – South Schleswigian Voters' Association), which quickly returned to the 100 % equivalence. Since then, the 100 %-principle has remained political consensus. Even though the crisis was solved, it has also led to questions on whether the Schleswig model in fact can be considered a European best practice.[41] It raises questions about acknowledgement and responsibility, but also demonstrated how kin-states' diplomacy could function as an important, functioning conflict resolution mechanism.

---

[39] The following is based on M. Klatt, "Minority Schools Financing in the Danish-German Border Region", *European Yearbook of Minority Issues,* vol. 11, 2014, pp. 359–375; J. Kühl, *Mindretalsmodel i krise,* Dansk Skoleforening for Sydslesvig, Flensburg, 2012.

[40] M. Klatt, "The dilemma of responsibility", *op.cit.*

[41] J. Kühl, *Mindretalsmodel i krise, op. cit.*; W. R. A. Hughes, N. S. Megoran & M. C. Benwell, "The emotional geopolitics of contested peace: the austerity crisis and the Danish minority of South Schleswig", *Territory, Politics, Governance*, vol. 10, n°2, 2020, pp. 237–255; M. Klatt, "Minority Schools Financing in the Danish-German Border Region", *op. cit.;* M. Klatt, "The dilemma of responsibility", *op. cit.*

A Best Practice of Reconciliation?

This section summarises and updates previous research I have done on this issue.[42] Reconciliation presupposes violent conflict and historic traumas, combined with the will of central political and/or civil society actors to actively work for mutual understanding and reconciliation. Prominent examples of active reconciliation policies are the European integration process, but also specifically the Franco-German treaty of friendship and reconciliation of 1963 (Elysée-Treaty) and its German-Polish counterpart, the Treaty on good neighbourhood, of 1991.[43] Also, the South African post-apartheid Truth and Reconciliation Commission (1996) can be mentioned as a best practice approach to reconciliation.

Moving to the Danish-German borderlands, we must constitute that German diplomatic and civil society approaches in the 1960s to negotiate a Danish-German friendship and reconciliation treaty (modelled on the Franco-German example) were met with reservations in Denmark.[44] Local actors had however engaged in reconciliation activities from the 1950s.[45] Albeit, the achievement widely perceived as most important, are the complementary, reciprocal Bonn-Copenhagen governmental declarations of March 1955.[46] They do not concern comprehensive reconciliation

---

42 M. Klatt, *Fra modspil til medspil, op. cit.*; M. Klatt, "Minorities as secondary foreign policy agents in peace-building and reconciliation? The case of Denmark and Germany", *Regional & Federal Studies,* vol. 27, 2017, pp. 239–259.

43 M. Klatt & B. Wassenberg, "Secondary foreign policy: Can local and regional cross-border cooperation function as a tool for peace-building and reconciliation?", *Regional & Federal Studies*, vol. 27, n°3, 2017, pp. 205–218;

E. Opilowska, "Reconciliation through Europeanization: Secondary foreign policy in the German-Polish borderlands", *Regional & Federal Studies*, vol. 27, n°3, pp. 283–304; B. Wassenberg, "Secondary foreign policy as a peace-building tool: A European model? The contribution of cross-border cooperation to reconciliation and stability in Europe", *Regional & Federal Studies*, vol. 27, issue 3, pp. 219–237.

44 M. Klatt, *Fra modspil til medspil, op. cit.*

45 *Ibid.*

46 On the Bonn-Copenhagen Declarations, see M. Klatt, "Von der Abgrenzung zur Grenzüberwindung. Die Minderheiten und die Bonn-Kopenhagener Erklärungen als Wendepunkt der deutsch-dänischen Beziehungen?", *Grenzfriedenshefte*, vol. 62, 2015, pp. 55–64; J. Kühl, "The Bonn-Copenhagen Declarations of 1955", *op. cit.*; European Centre of Minority Issues, J. Kühl, *København-Bonn Erklæringerne 1955–2005. De dansk-tyske mindretalserklæringernes baggrund, tilblivelse og virkning,* Institut for Grænseregionsforskning – Syddansk Universitet, Aabenraa, 2005.

as the bilateral treaties mentioned above, but only a settlement of the minority issues, and, thus, the border issue. The Bonn-Copenhagen Declarations of 1955 are commemorated regularly, where politicians and other stakeholders present narratives of Schleswig being a best-practice example of reconciliation. This is often expressed in the German saying that Danish-German as well as minority-majority relations have moved *vom Gegeneinander zum Miteinander oder gar Füreinander*, which translates in English from living opposed to each other to with or even for each other. The historical context of the two minority declarations is embedded in the post-WWII geopolitics, which required Danish-West German détente. This is often forgotten today. The Cold War gave Denmark a critical role as guardian of the straits to enter the Baltic Sea, as well as to secure NATO's northern flank against a possible Soviet attack. At the same time, it had been central for the Western allies to rearm and integrate West Germany into NATO since the outbreak of the Korean War. West German forces were intended to assist Denmark in guarding NATO's northern flank.[47] Thus, cold war geopolitics had literally taken the Danish-German border issue of the international agenda, with both governments agreeing that a Danish-German national conflict should not resurface. This resulted in West Germany's accession to NATO being side lined by the Danish and the West German government declaring the two reciprocal minorities' rights to cultural autonomy, kin-state connections and kin-state support; minority membership being based on personal decision not to be questioned or verified by state authorities.[48]

The Bonn-Copenhagen Declarations of 1955 hereby laid the ground for building coexistence and cooperation between minorities and majorities, moving from negative to positive peace. The German minority declared that collaboration with Nazi Germany had been a terrible mistake and declared their loyalty to the Danish state and the 1920-border. It redefined their identity as European, and their mission as building bridges between the Danish people and the continent to further European integration.[49] While more sceptical towards Europeanisation

47 P. Villaume, *Allieret med forbehold: Danmark, NATO og den kolde krig en studie i dansk sikkerhedspolitik 1949–1961*, Eirene, Copenhague,1995.

48 J. Kühl, "The Bonn-Copenhagen Declarations of 1955", *op. cit.*

49 M. Klatt, *Fra modspil til medspil?, op. cit.;* F. Lubowitz, "Det tyske mindretal 1955–2005", *In* J. Kühl (ed.), *København-Bonn Erklæringerne 1955–2005. De*

in the 1950s–1980s, of late the Danish minority and their political party also have embraced cross-border cooperation and integration.[50]

In sum, despite the lack of formal reconciliation policies as between Germany and France or Germany and Poland, Denmark and Germany appear to have reconciled. Minority policies have thus contributed to a general reconciliation, even though primarily directed at minority-majority relations.

A Best Practice of Cross-Border Cooperation in Politics, People-to-People and Business?

West Germany's integration into NATO and the accommodation of the two reciprocal national minorities in 1955 did not entail a broader Danish-German rapprochement.[51] Denmark did not follow West Germany into the European Coal and Steel Community or the European Community in the 1950s but remained within the UK-inspired European Free Trade Area. Relations with West Germany developed as between friendly sovereign states – but not more. Paradiplomatic activities of the West German state of Schleswig-Holstein were met with reservations and a clear position that Denmark could not accept a German state as diplomatic partner. This did not change in principle when Denmark joined the EC in 1973. The national balance in the border region continued to be perceived as sensitive, and no institutional cooperation with German partners was desired. Euroregions, established at most inner EEC-borders

---

dansk-tyske mindretalserklæringernes baggrund, tilblivelse og virkning, Institut for Grænseregionsforskning, Aabenraa, 2005, pp. 269–291.

50 M. Klatt, "Minorities as secondary foreign policy agents in peace-building and reconciliation?", *op. cit.*

51 This section is a summary account of my previous research on the topic, see M. Klatt, *Fra modspil til medspil?, op. cit.;* M. Klatt, "Dybbøl 2014. Constructing familiarity by rememberance?", *In* E. Boesen & G. Schuner (eds.), *European Borderlands. Living with bridges and barriers*, Routledge, Milton Park, 2017, pp. 30–46; M. Klatt, "The Danish-German Border Region: Caught between Systemic Differences and Rebordering", *Eurasia Border Review*, vol. 8, 2017, pp. 51–30; M. Klatt & H. Herrmann, "Half empty or half full? Over 30 years of regional cross-border cooperation within the EU: Experiences at the Dutch-German and Danish-German border", *op.cit.*; M. Klatt & I. Winkler, "Lessons from the Danish-German border region for post 2020 Interreg A – an alignment with cross-border functional regions?", *Europa XXVI*, vol. 38, 2020, pp. 139–156.

in the 1970s, were a no-go for the Danish Foreign Ministry as well as local Danish politicians. The only exception was pragmatic cooperation "from case to case": with only one interesting example, a commission for environmental issues in the Flensburg Fjord, an inlet of the Baltic Sea divided by the border. Here, regional and local politicians cooperated and designed a cross-border action plan to improve water quality, which then was implemented by both sides – without ever concluding a formal agreement or a bilateral treaty.

Only when cross-border projects were included in the EU's Cohesion Policy with the Interreg initiative did this change. In 1997, regional Danish politicians were ready to accept continuous German advances to establish a Euroregion. They had underestimated the popular opposition, though, expressing itself in demonstrations, but also vandalism against pro-Euroregion politicians. Also the Germans were taken by surprise. Eventually, the Euroregion agreement was signed in September 1997 – but the Euro – was dropped in the name: it was reduced to Region Sønderjylland-Schleswig. It has operated smoothly since, like other Euroregions, but it is also quite evident that the new political construction never became a political entity integrating Sønderjylland-Schleswig into an operative cross-border region. Rather, it has remained a competence centre for cross-border issues, a forum for ideas, and a project manager for a few cross-border projects, predominantly in the field of culture.

Returning to Brunet-Jailly's borderland theory,[52] it is therefore important to refocus on the market forces and flows. I have earlier argued that economic incitements are an important indicator of successful cross-border activities.[53] Therefore, it makes sense to use the market forces and flows-lens as a starting point to analyse cross-border integration. Data on persons and freight crossing the border in Schleswig reveals a clear picture of an economic hub (See Figures 4 and 5).

52 E. Brunet-Jailly, "Theorizing borders", *op. cit.*

53 M. Klatt & H. Herrmann, "Half empty or half full? Over 30 years of regional cross-border cooperation within the EU: Experiences at the Dutch-German and Danish-German border", *op. cit.*

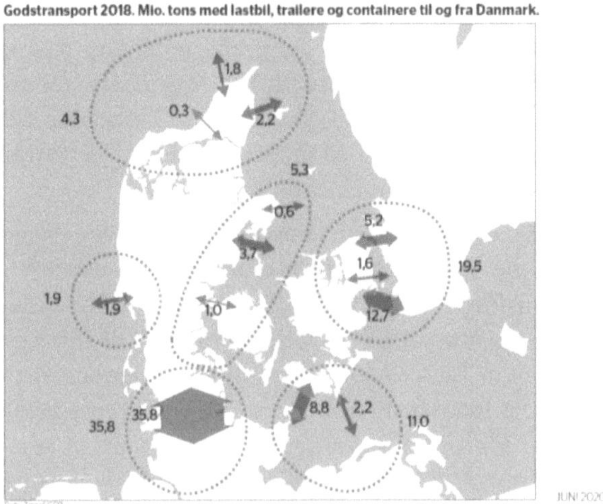

Figure 4: Freight in and out of Denmark, million metric tons (trucks, trailers, containers) 2018.[54]

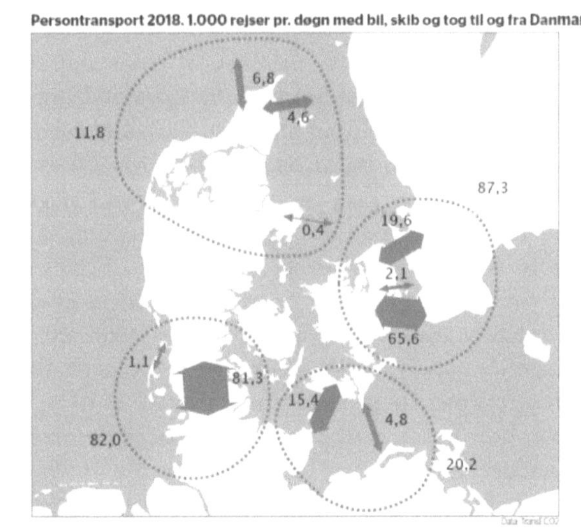

Figure 5: Persons in and out of Denmark (cars, trains, ships): daily, 1000 persons.[55]

---

[54] Source: U. Sonderjylland, "Porten til Europa", Udviklingsråd Sønderjylland, Sonderborg, 2020.

[55] *Ibid.*

This data indicates that the border region is a hub of cross-border freight. Following Brunet-Jailly, a culturally emerging and integrating borderland would require the apparently thriving market forces and trade flows to be enhanced by the three other lenses of his model: a local cross-border culture, policy activities of multiple levels of government and a local cross-border political clout.

I have already demonstrated above that Schleswig lagged behind with regard to the establishment and institutionalisation of a local cross-border political clout. A Euroregion was only established in 1997, and it was accompanied by protests in parts of the regional population, including at least one very prominent member of the Danish minority. Today, it focuses on culture rather than cross-border market forces and flows. Multilevel governance activities are in place. Besides the Euroregion, there is a municipal cooperation in form of the so-called Border Triangle between the cities of Flensburg (DE), Aabenraa and Sønderborg (DK) on the East Coast of Schleswig, and between Tønder (DK) and the Amt Südtondern (DE) on the West Coast. The German state of Schleswig-Holstein claims a close cooperation with the Danish government, it has an observer status in the Nordic Council, and is involved in the STRING network of cities and regional governments stretching from Hamburg to the Greater Copenhagen and Scania region. Schleswig-Holstein and Denmark are also connected in the Baltic Sea Cooperation/Baltic Sea Council, as well as the Interreg C North Sea cooperation. Task specific (Multilevel Governance, type II), there is a Danish-German Transport Commission, involving policymakers from the Danish, Federal German and Schleswig-Holstein ministries of transport, as well as other stakeholders in the transport industry. Therefore, multiple levels of government are involved in cross-border politics. It is however difficult to measure the impact of these different forums. In general, the narrative is of a "better than ever" dialogue, but in practice, prevailing barriers are mentioned, and many best-practice cooperations established in the 1990s have slowly been discontinued or replaced.[56] The lens local cross-border culture and its role in the emergence of an integrating cross-border region is to be discussed. Here, the role of the minorities becomes interesting: how far

[56] M. Klatt & I. Winkler, "Lessons from the Danish-German border region for post 2020 Interreg A", *op. cit.*

do they contribute to the emerging of a cross-border region? In history, as mentioned above, the region of Schleswig was socially and economically integrated. The border of 1920 segregated this integrated region[57] for about 50 years, resulting in severe economic consequences especially for South Schleswig.[58] When Denmark joined the EEC in 1973, in principle, reintegration should have been possible. What had happened to the local cross-border culture in these years? I will here define the local cross-border culture as "Schleswigianism," well aware that this might not be a suitable name (and definitely never has been used in official discourse or official documents). "Schleswigianism" encapsulates a border zone identity in a region characterised by north-south flows, natural input from a rough North Sea tidewater climate in the West, poor, sandy soils in the midlands and fertile soil, mild climate and historic trade relations across the Baltic in the Eastern part. Populated by Jutish tribes (as far as we know), with high interaction with their Saxon and Slavic neighbours, plus Frisians on the West Coast, it became an ethnic melting pot already in medieval times. Cultural influences from the south became stronger after Christianisation (Latinisation), and even more after the Protestant reformation (Germanisation). In early modern times, Schleswig (and the Danish kingdom) were perceived as part of the protestant realm of Central and northern Europe.

Locally, the different regional languages (Danish, Southern Jutlandic, Low German), German and Frisian stood side by side, but were far from equal. German was the undisputed language of administration, courts, and higher education from latest the 15th until the mid-19th century. As such, it also was the language of upward social mobility. State identification was primarily connected to the king and the dynasty, not yet to language.[59] This changed in the 19th century, when national movements gained momentum and challenged the hierarchy of languages that had been established. The anecdote of Hjort Lorenzen "speaking and continuing to speak Danish" in the estate assembly of Schleswig became a

---

57 M. Andersen, *Den følte grænse. Slesvigs deling og genopbygning 1918–1933,* Historisk Samfund for Sønderjylland, Aabenraa, 2008.

58 M. Klatt, "1920-grænsen: national optimal, økonomisk katastrofal? Et overset aspekt af den slesvigske model", *op.cit.*

59 L. N. Henningsen, *Kongen kommer! Kongebesøg og kongeloyalitet i Sydslesvig,* Grænseforeningen, Copenhagen, 2000.

key *lieu de memoire* for the Danish movement,[60] not so much because of the fact that he used Danish, but because of the ensuing debate on language use in this pre-parliament. Here it was agreed to tolerate the use of Danish, if members did not command German that well. A paternalistic attitude reflecting the perceived low social status of the Danish language by the elite in Schleswig.

A narrative of local cross-border culture in Schleswig is contradicted by this inequality of languages even today. When the Euroregion was established in 1997, stakeholders pledged to the aim of mutual comprehension of Danish and German – today, translators are still used at all major cross-border political meetings. Cultural differences might be marginal compared to other intercultural cleavages but are perceived as such by stakeholders and actors in cross-border settings.

A Best Practice of an Integrated Cross-Border Living Space (Labour Market, Public Services, Residential Space, Creative Industries Economics)

The core aim of the four lenses model is to explain emerging integrated cross-border regions. But what are cross-border regions? Oscar Martinez' four stage model of borderlands characterises integrated borderlands:

> "Stability is strong and permanent. The economies of the two countries are functionally merged, and there is unrestricted movement of people and goods across the boundary. Borderlanders perceive themselves as members of one social system".[61]

How far has Schleswig become an integrated cross-border living space? Cross-border living space is here defined in analogy to Martinez' understanding of an integrated borderland: Living spaces is a term derived from German *Aktionsraum* as the space of daily human activities for satisfying basic physical and social needs (*Daseinsgrundfunktionen*). What is important in analysing borderlanders' activities in such a cross-border living space is their perception of the border. Here, the *Aktionsraum* must

---

60 M. Klatt, "Mobilization in crisis – demobilization in peace: protagonists of competing national movements in border regions", *op.cit.*

61 O. J. Martinez, *Border People, op. cit.*

be extended as a space beyond substantial activities. It departs from the understanding that repeatedly shifting space to work or perform other activities on the other side of a state border and returning home for living goes hand in hand with entering and re-entering unfamiliar and familiar symbolic realms. So, it is not only important, whether borderlanders perceive themselves as members of one social system, but rather how their actions are influenced by what the concept of unfamiliarity identifies as emotional differences. Therefore, a cross-border living space is not necessarily depending on borderlanders transformation from national to transnational borderlanders, but rather on how borderlanders use the borderland. Here, the concepts of Regionauts[62] and border surfers[63] might be more useful to characterise cross-border living. They indicate a continuing awareness of the border as a barrier, but also as a source of opportunity.[64]

Here, research shows interesting results. When commuting increased across the Danish-German border in the second half of the 2000s, data reveals a general positive perception of the Danish labour market and working culture, but also prevailing cultural stereotypes about German workers.[65] Newer research has confirmed prevailing or even a strengthening of stereotypes by commuting across the border[66]quite opposite a presumption that cross-border commuting would result in acculturation. The disruption of cross-border activities by the SARS-Cov-2 incited border closure implemented in March 2020, on the other hand, has demonstrated the vast range of existing

62 T. O'Dell, "Øresund and the Regionauts", *European Studies: A Journal of European Culture, History and Politics,* vol. 19, 2003, pp. 31–53; O. Löfgren "Regionauts: the Transformation of Cross-Border Regions in Scandinavia", *European Urban and Regional Studies,* vol. 15, 2008, pp. 195–209.

63 K. Terlouw, "Border Surfers and Euroregions: Unplanned Cross-Border Behaviour and Planned Territorial Structures of Cross-Border Governance", *Planning Practice and Research,* vol. 27, 2012, pp. 351–366.

64 See also Christophe Sohn's conception of the border as a resource: C. Sohn, "The Border as a Resource in the Global Urban Space: A Contribution to the Cross-Border Metropolis Hypothesis", *International Journal of Urban and Regional Research,* vol. 38, 2014, pp. 1697–1711.

65 M. Klatt, "(Un)Familiarity? Labour Related Cross-Border Mobility in Sønderjylland/Schleswig Since Denmark Joined the EC in 1973", *Journal of Borderlands Studies,* vol. 29, 2014, pp. 353–373.

66 D. J. Andersen & I. Winkler, "Grænsearbejde blandt grænsependlere", *Økonomi og Politik,* 2020, pp. 56–68.

cross-border activities beyond commuting, and how many borderlanders had appropriated the idea of a cross-border living space for the design of their personal lives.[67] This has not, though, resulted in functional integration. Agents perceive the border as real and existent – especially when it gets closed. Re-borderings as the "temporary" border controls introduced in January 2016 (and since extended every half year)[68] and the animal health indicated Wild Boar Fence erected in 2019 have an impact on the perception of the borderlands as "open" and the border situation as stable. The latter is not the case, which also reflects on committing cooperation as joint public services. Especially peripheral regions should embrace integrating public services to achieve greater critical mass. For the Danish-German case, a reverse is actually the case, as best-practice examples established in the 1990s have been replaced with national solutions or drastically reduced lately: the cross-border hospital service offering post-surgery radiation treatment to Danish cancer patients in German Flensburg, but also cross-border rescue services in the form of a helicopter and ambulances as well as cross-border study programs offered by the University of Flensburg and University of Southern Denmark.[69]

In fact, Schleswig might not be below European standards (best practice example), as a recent study documented challenges to establish permanent cross-border public services in most European border regions.[70] Still, it is assessed that actual practices to formalise a cross-border living space deviate from political narratives, rhetoric and commitments.

---

67 M. Klatt, "What has happened to our cross border regions? Corona, Unfamiliarity and transnational borderlander activism in the Danish German border region", *In* C. Wille & R. Kanesu (eds.), *Borders in Pandemic Times. Insights into the Covid 19 Lockdown*, University of Luxemburg, University of Trier, Luxembourg, Trier, 2020, pp. 43–47; R. Travet & M. Klatt, "The impact of the Corona crisis on borderland living in the Danish-German border regon with a special focus on the two national minorities", *National Identities*, 2021, pp. 1–18.

68 M. Klatt, "The so-called 2015 migration crisis and Euroscepticism in border regions: Facing re-bordering trends in the Danish-German borderlands", *Geopolitics*, vol. 5, 2020, pp. 567–586.

69 "Dansk-tysk samarbejde skal give merværdi", *I.e. Flensborg Avis*, 22 March 2017.

70 A. Zillmer et al. ""Cross-border Public Services (CPS)", *op. cit.*

A Best Practice of Diversity Accommodation?

Finally, is the Danish-German-Frisian Region a best practice of diversity accommodation? This is the historic challenge after the long 19th century of nationalisation and the 20th century of division and containerisation of borderlanders into a majorities-minorities binarity, which did not reflect the historic social reality, and probably neither has reflected 20th and 21st century social reality. The "best practice" narrative rests on three key assumptions.

First, acknowledgement and accommodation of kin-minorities across the border of 1920, understood as autochthonous minorities "which have always been here – only the border changed." These minorities, in principle understood as coherent and stable groups (autochthonous), are guaranteed cultural and institutional autonomy by legal provision and political goodwill, financed by extensive kin-state support. Second, *Minderheit ist, wer will* (minority is, who wants to be): Subjective identification matters, the state may not register, question or challenge minority membership. This opens for people joining or leaving the minority, raising questions about whether the minorities can still be considered autochthonous – as considerable parts of the membership are de-facto allochthonous. Third, appeasement and reconciliation, also between the majorities. The kin-state based, reciprocal minority model has ensured a general reconciliation, not only between minority and majority, but also Germany and Denmark. Today, no serious politician challenges the position of the Danish-German border, nor the minorities' right to difference as well as the legitimacy of their interests in close cultural connections with their kin-state.

I would argue that there are three components that challenge the best practice narrative. First, hierarchy of minorities: while the Danish minority in Germany (and the German minority in Denmark) can boast of a very comprehensive (and competitive), institutionalised minority system enabling Danish life in South Schleswig/German life in North Schleswig from cradle to grave, this is by far not the case for the two other recognised minorities residing in Schleswig-Holstein, the Frisians and the Sinti and Roma. They do not have a kin-state financing their culture and respective institutions. Here, institutional support of language and culture are very low. In consequence, they are under much stronger assimilation pressure.

Second, containerisation: In principle, the cultural autonomy model applied to the kin-state minorities invites for a parallel society. It is based on the idea of a nation state construct,[71] separating a minority community from the majority community. This has never reflected the social reality in Schleswig: there is absolutely no indication in historical and contemporary sources of a separated Danish community in South Schleswig or a separated German community in North Schleswig. In daily life, both groups interact with society. Furthermore, identification being subjective and based on emotional preferences, intermarriage has been frequent, and most families have members in all four communities (German/Danish majority, German/Danish minority). Recently, minority switching has become more common, as well as cross-minority identification, following increasing political cooperation of the two, in principle antagonistic, kin-minorities.[72] The latter also implies that minority representatives increasingly appropriate a transnational borderlander activist role, proclaiming their key interest in open borders, cross-border cooperation and regional, cross-border development.[73]

Third, kin-state dependency: The Schleswig model of kin-state financed minority institutions has made the minorities' school systems very competitive, attracting parents from outside the minorities. Jørgen Kühl has described this phenomenon by characterising especially the Danish minority as a life-style, "post-national" minority, not really reflecting Danish historical and contemporary perceptions of the Danish minority in South Schleswig, including the Danish motivation for kin-state funding it.[74] This raises issues of responsibility. In the 2010–12 school conflict on the public, Schleswig-Holstein funds for Danish minority schools, it became apparent that kin-state funding of these schools was perceived

---

71 A. Christiansen, *Die Reproduktion der dänischen Minderheit im 21. Jahrhundert zwischen Kultur, Tradition und Gemeinschaftsglauben*, 2008, Tönning, Lübeck.

72 M. Klatt, "Minorities as secondary foreign policy agents in peace-building and reconciliation?", *op. cit.*

M. Klatt, "Genforening af de slesvigske mindretal", *Sønderjysk Månedsskrift*, 2022, pp. 140–147.

R. T. Tarvet, *Re-Imagining Sleswig: Language and Identity in the German-Danish Borderlands*, Syddansk Universitetsforlag, Odense, 2021.

73 R. T. Tarvet, *Re-Imagining Sleswig*, *op. cit.*; M. Klatt, "Minorities as secondary foreign policy agents in peace-building and reconciliation?", *op. cit.*

74 J. Kühl, "'Ihr seid nicht vergessen worden' Königliche Neujahrsgrüsse 2001–2017 und Vorstellungen von der dänischen Minderheit in Südschleswig", *In* J. Kühl (eds.), *Klaar kiming. Festschrift für Thomas Steensen,* Nordfriisk Institut, Bredstedt, Bräist, 2018, pp. 100–116.

as unfair competition by leading conservative politicians in Schleswig-Holstein.[75] The discrepancy between the funding available for Frisian and Sinti/Roma cultural autonomy on the one side and for the Danish and German minorities on the other illustrates the issue. Even though all four minorities are fully acknowledged politically, there are deficiencies in the funds made available to support cultural autonomy.

Conclusion

The best practice paradigm is normative. What is good practice, what is best? From a Danish and German minority perspective, the *status quo* is definitely improved practice, if compared to historical periods of conflict and alienation. Political acknowledgement and social inclusion are best practice with regard to Danish, German and Frisian minorities, while there are clear deficiencies with regard to the Sinti and Roma. Cultural diversity is acknowledged, but the states' titular culture is dominant. The cultural diversity of the region could be made more visible, especially in a cross-border context. Even bilingual road signs, common in many minority regions, are an issue, especially in Denmark. Progress towards an integrated, cross-border region is minimal, when focusing on institutionalisation. This however happens on an actor basis, beyond minority-kin state relations. Recently, German immigration to North Schleswig is a new trend, in fact from all of Germany. Surprisingly, this trend is welcomed by Danish stakeholders in North Schleswig.[76] in the 1960s and 1970s, when Denmark discussed joined the EC, this was perceived as a threat. Having in mind a normative aim of an open border, opening for an inspiring, integrating cross-border living space, there is still room for improvement.

[75] W. R. A. Hughes et al. "The emotional geopolitics of contested peace", *op. cit.;* M. Klatt, "Minority Schools Financing in the Danish-German Border Region", *op. cit.;* J. Kühl, *Mindretalsmodel i krise, op. cit.*

[76] S. Eskilden, "Infoaktionen für potenzielle Zuzügler", *Der Nordschlewiger*, 29 June 2021.

Bibliography

M. Andersen, *Den følte grænse. Slesvigs deling og genopbygning 1918–1933,* Historisk Samfund for Sønderjylland, Aabenraa, 2008.

B. Anderson & American Council of Learned, *Imagined Communities: Reflections on the Origin and Spread of Nationalism*, Verso, New York, London, 2006.

D. J. Andersen & I. Winkler, "Grænsearbejde blandt grænsependlere", *Økonomi og Politik*, 2020, pp. 56–68.

M. Andeva, "Non-Territorial Autonomy in Theory and Practice. A 2020 Report", Report, University American College Skopje, Skopje, 2020.

H. Becker-Christensen, *Dansk Mindretalspolitik i Nordslesvig. Udformningen af den danske politik over for det tyske mindretal 1918–1920,* Institut for Grænseregionsforskning, Aabenraa, 1984; J. P. Noack, *Det danske mindretal i Sydslesvig 1920–1945,* Institut for Grænseregionsforskning, Aabenraa, 1989.

H. Becker-Christensen, *Det tyske mindretal i Nordslesvig 1920–1932,* Institut for Grænseregionsforskning, Aabenraa, 1990.

R. Bohn, "Eine belastete Nachbarschaft. Aspekte der deutsch-dänischen Beziehungen nach 1945", *In* J. Kühl & R. Bohn (eds.), *Ein europäisches Modell? Nationale Minderheiten im deutsch-dänischen Grenzland 1945–2005,* European Centre of Minority Issues, Verlag für Regionalgeschichte, Bielefeld, 2005, pp. 59–76.

M. Bregnsbo, "Schleswig(-)Holstein 1848: Legitimism, nationalism, constitutionalism and regionalism in conflict", *In* M. Bregnsbo & K. Villads Jensen (eds.), *Schleswig-Holstein – Contested Region(s) through History,* University of Southern Denmark Press, Odense, 2016, pp. 177–193.

E. Brunet-Jailly, "Theorizing borders: An interdisciplinary perspective", *Geopolitics,* vol. 10, 2005, pp. 633–649.

A. Christiansen, *Die Reproduktion der dänischen Minderheit im 21. Jahrhundert zwischen Kultur, Tradition und Gemeinschaftsglauben*, 2008, Tönning, Lübeck.

European Centre of Minority Issues, J. Kühl, *København-Bonn Erklæringerne 1955–2005. De dansk-tyske mindretalserklæringernes baggrund, tilblivelse og virkning,* Institut for Grænseregionsforskning – Syddansk Universitet, Aabenraa, 2005.

T. Fink, *Da Sønderjylland blev delt*, Institute for Grænseregionsforskning, Aabenraa, 1979.

T. Fink, "Den første streg på kortet – Clausen linjen", *In* H. Becker Christebseb (ed.) *Grænsen i 75 år,* Institut for Grænseregionsforskning, Aabenraa, 1995, pp. 11–23.

A. Gat, *Nations. The Long History and Deep Roots of Political Ethnicity and Nationalism*, Cambridge University Press, Cambridge, 2013; E. Gellner, *Nations and Nationalism*, Blackwell, Oxford, 1983.

R. Gesme, "Vote for Home: Self-Determination, Wilsonianism, and the Implementation of the 1920 Schleswig Plebiscite", doctoral thesis, University of Tennessee, Knoxville, 2023.

H. S. Hansen & H. S. Kristensen, "Mindretal og Flertal i Nordslesvig 1945–1955", *In* J. Kühl (eds.), *København-Bonn Erklæringerne 1955–2005. Mindretalserklæringernes baggrund, tilblivelse og virkning,* Institut for Grænseregionsforskning, Aabenraa, 2005, pp. 125–197.

A. Hastings, *The Construction of Nationhood: Ethnicity, Religion and Nationalism*, Cambridge University Press, Cambridge, 1997.

L. N. Henningsen, *Kongen kommer! Kongebesøg og kongeloyalitet i Sydslesvig,* Grænseforeningen, Copenhagen, 2000.

L. N. Henningsen (ed.), *Sydslesvigs danske historie,* 2005, Studieafdelingen ved Dansk Centralbibliotek for Sydslesvig, Flensburg, 2005.

E. J. Hobsbawm, *Nations and Nationalism Since 1780: Programme, Myth, Reality*, Cambridge University Press, Cambridge, 1992.

P. Hopp & C. Mogensen (eds.), *Ostersturm Påskeblæsten*, Institut für Regionale Foschung und Information im Deutschen Grenzverein e.V., Flensburg, 1993.

M. Hroch, *Das Europa der Nationen. Die modern Nationalittätenbildung im europäischen Vergleich*, Vandenhoeck & Ruprecht, Göttingen, 2005.

W. R. A. Hughes, N. S. Megoran & M. C. Benwell, "The emotional geopolitics of contested peace: The austerity crisis and the Danish minority of South Schleswig", *Territory, Politics, Governance*, vol. 10, n°2, 2020, pp. 237–255.

P. M. Judson & M. L. Rozenblit (eds.), *Constructing Nationalities in East Central Europe*, Berghahn, New York, Oxford, 2005.

A. Johnsen, *Dannevirkemænd og Ejderfolk. Den grænsepolitiske opposition i Danmark 1920–1940,* Studieafdelingen ved Dansk Centralbibliotek for Sydslesvig, Flensburg, 2005.

J. P. Noack, *Det sydslesvigske grænsespørgsmål 1945–1947*, Institut for Grænseregionsforskning, Aabenraa, 1991.

M. Klatt, *Fra modspil til medspil?: grænseoverskridende samarbejde i Sønderjylland/Schleswig 1945–2005*, Institut for Grænseregionsforskning, Aabenraa, 2006.

M. Klatt, "Mobile regions: Competitive regional concepts (not only) in the Danish-German border region", *In* D. J. Andersen, M. Klatt & M. Sandberg (eds.), *The Border Multiple. The Practice of Borders between Public Policy and Everyday Life in a Re-Scaling Europe*, Ashgate, Aldershot, 2012, chapter 3.

M. Klatt, "Minority schools financing in the Danish-German border region", *European Yearbook of Minority Issues,* vol. 11, 2014, pp. 359–375.

M. Klatt, "(Un)Familiarity? Labour related cross-border mobility in Sønderjylland/Schleswig since Denmark joined the EC in 1973", *Journal of Borderlands Studies*, vol. 29, 2014, pp. 353–373.

M. Klatt, "Johannes Schmidt-Wodder und Jens Möller – nordschleswigsche Folketingsmitglieder der Zwischenkriegszeit zwischen Alldeutschtum, Friedensverein und Nationalsozialismus", *In* B. Conrad, C.-C. Manher & J. Kusber (eds.), *Parlamentarier der deutschen Minderheiten im Europa der Zwischenkriegszeit*, Droste, Düsseldorf, 2015, pp. 47–64.

M. Klatt, "Von der Abgrenzung zur Grenzüberwindung. Die Minderheiten und die Bonn-Kopenhagener Erklärungen als Wendepunkt der deutsch-dänischen Beziehungen?", *Grenzfriedenshefte*, vol. 62, 2015, pp. 55–64.

M. Klatt, "Minorities as secondary foreign policy agents in peace-building and reconciliation? The case of Denmark and Germany", *Regional & Federal Studies,* vol. 27, 2017, pp. 239–259.

M. Klatt, "Dybbøl 2014. Constructing familiarity by rememberance?", *In* E. Boesen & G. Schuner (eds.), *European Borderlands. Living with Bridges and Barriers,* Routledge, Milton Park, 2017, pp. 30–46.

M. Klatt, "The Danish-German border region: Caught between systemic differences and re-bordering", *Eurasia Border Review*, vol. 8, 2017, pp. 51–30.

M. Klatt, "Mobilization in crisis – Demobilization in peace: Protagonists of competing national movements in border regions", *Studies on National Movements,* vol. 4, 2019, p. 30.

M. Klatt, "1920-grænsen: national optimal, økonomisk katastrofal? Et overset aspekt af den slesvigske model", *Akademisk kvarter,* vol. 21, 2020, pp. 60–73.

M. Klatt, "The so-called 2015 migration crisis and Euroscepticism in border regions: Facing re-bordering trends in the Danish-German borderlands", *Geopolitics*, vol. 5, 2020, pp. 567–586.

M. Klatt, "What has happened to our cross border regions? Corona, Unfamiliarity and transnational borderlander activism in the Danish German border region", *In* C. Wille & R. Kanesu (eds.), *Borders in Pandemic Times. Insights into the Covid 19 Lockdown*, University of Luxemburg, University of Trier, Luxembourg, Trier, 2020, pp. 43–47.

M. Klatt, "Genforening af de slesvigske mindretal", *Sønderjysk Månedsskrift*, 2022, pp. 140–147.

M. Klatt, "Schleswig and non-territorial autonomy – The territorial trap?", *Nationalities Papers*, vol. 50, 2022, pp. 906–922.

M. Klatt, "The dilemma of responsibility: The role of kin-states and nation-states in implementing non-territorial autonomy models to realize minorities' linguistic, cultural and educational rights", *In* D. J. Smith, I. Dodovski & F. Ghencea (eds.), *Realising Linguistic, Cultural and Educational Rights through Non-territorial Autonomy*, Palgrave Macmillan, Cham, 2022, pp. 43–57.

M. Klatt & H. Herrmann, "Half empty or half full? Over 30 years of regional cross-border cooperation within the EU: Experiences at the Dutch-German and Danish-German border", *Journal of Borderlands Studies*, vol. 26, 2011, pp. 65–87.

M. Klatt & B. Wassenberg, "Secondary foreign policy: Can local and regional cross-border cooperation function as a tool for peace-building and reconciliation?", *Regional & Federal Studies*, vol. 27, n°3, 2017, pp. 205–218.

M. Klatt & I. Winkler, "Lessons from the Danish-German border region for post 2020 Interreg A – An alignment with cross-border functional regions?", *Europa XXVI*, vol. 38, 2020, pp. 139–156.

J. Kühl, "The Bonn-Copenhagen declarations of 1955: Background, context and impact of the Danish-German minority regulations", *In* J. Kühl & M. Weller (eds.), *Minority Policy in Action: The Bonn-Copenhagen Declarations in a European Context 1955–2005,* Institut for Grænseregionsforskning, Aabenraa, 2005, pp. 31–89.

J. Kühl, *Mindretalsmodel i krise,* Dansk Skoleforening for Sydslesvig, Flensburg, 2012.

J. Kühl, "'Ihr seid nicht vergessen worden' Königliche Neujahrsgrüsse 2001–2017 und Vorstellungen von der dänischen Minderheit in Südschleswig", *In* J. Kühl (eds.), *Klaar kiming. Festschrift für Thomas Steensen,* Nordfriisk Institut, Bredstedt, Bräist, 2018, pp. 100–116.

J. Kühl & R. Bohn (eds.), *Ein europäisches Modell? Nationale Minderheiten im deutsch-dänischen Grenzland 1945–2005,* European Centre of Minority Issues, Verlag für Regionalgeschichte, Bielefeld, 2005.

O. Löfgren "Regionauts: The transformation of cross-border regions in Scandinavia", *European Urban and Regional Studies,* vol. 15, 2008, pp. 195–209.

F. Lubowitz, "Det tyske mindretal 1955–2005", *In* J. Kühl (ed.), *København-Bonn Erklæringerne 1955–2005. De dansk-tyske mindretalserklæringernes baggrund, tilblivelse og virkning,* Institut for Grænseregionsforskning, Aabenraa, 2005, pp. 269–291.

T. Malloy & F. Palermo (eds.), *Minority Accommodation through Territorial and Non-Territorial Autonomy,* Oxford University Press, Oxford, 2015.

O. J. Martinez, *Border People. Life and Society in the U.S.-Mexico Borderlands,* University of Arizona Press, Tucson, London, 1994.

J. P. Noack, *Det sydslesvigske grænsespørgsmål 1945–1947,* Institut for Grænseregionsforskning, Aabenraa, 1991.

T. O'Dell, "Øresund and the Regionauts", *European Studies: A Journal of European Culture, History and Politics,* vol. 19, 2003, pp. 31–53.

E. Opilowska, "Reconciliation through Europeanization: Secondary foreign policy in the German-Polish borderlands", *Regional & Federal Studies,* vol. 27, n°3, pp. 283–304.

A. D. Smith, *The Ethnic Origins of Nations,* Blackwell, Oxford, 1987.

C. Sohn, "The border as a resource in the global urban space: A contribution to the cross-border metropolis hypothesis", *International Journal of Urban and Regional Research,* vol. 38, 2014, pp. 1697–1711.

U. Sønderjylland, "*Porten til Europa*", Udviklingsråd Sønderjylland, Sonderborg, 2020.

B. Spierings & M. van der Velde, "Shopping, borders and unfamiliarity: Consumer mobility in Europe", *Journal for Economic and Social Geography,* vol. 99, 2008, pp. 497–505.

B. Spierings & M. van der Velde, "Cross-border differences and unfamiliarity: Shopping mobility in the Dutch-German Rhine-Waal Euroregion", *European Planning Studies*, vol. 21, 2013, pp. 5–23.

B. Szytniewski & B. Spierings, "Encounters with otherness: Implications of (un)familiarity for daily life in borderlands", *Journal of Borderlands Studies*, vol. 29, 2014, pp. 339–351.

K. Terlouw, "Border surfers and Euroregions: Unplanned cross-border behaviour and planned territorial structures of cross-border governance", *Planning Practice and Research*, vol. 27, 2012, pp. 351–366.

R. T. Tarvet, *Re-Imagining Sleswig: Language and identity in the German-Danish Borderlands*, Syddansk Universitetsforlag, Odense, 2021.

R. Travet & M. Klatt, "The impact of the Corona crisis on borderland living in the Danish-German border regon with a special focus on the two national minorities", *National Identities*, 2021, pp. 1–18.

P. Villaume, *Allieret med forbehold: Danmark, NATO og den kolde krig en studie i dansk sikkerhedspolitik 1949–1961,* Eirene, Copenhagen,1995.

B. Wassenberg, "Secondary foreign policy as a peace-building tool: A European model? The contribution of cross-border cooperation to reconciliation and stability in Europe", *Regional & Federal Studies*, vol. 27, issue 3, pp. 219–237.

S. Zillmer, S. Haarich, F. Holstein, C. Lüer, T. Stumm, E. Marques de Costa, N. Da Costa, P. Ockerl, M. Peterlin, A. Berlina, E. Turunen & J. Teräs, "Cross-border Public Services (CPS). Targeted analysis. Final report," *In* ESPON (eds.), *Main Report*, 2018, ESPON, Luxembourg.

Minorities and Cross-Border Meetings from the Perspective of the Grænseforeningen (Danish Border Association)

Gunvor Vestergaard

The Danish-German border area is often highlighted as a best practice model. People living there went from against each other to with and for each other. Whether you are Danish or German, minority, or majority, living on one side or the other of the border, you are a citizen and an integral part of the local community. It is a story with many positive narratives. But also, with many challenges.

Not all disagreements disappeared. Especially older people still remember the past. Their emotions are easily affected. This was experienced when the German minority in Denmark wanted bilingual city signs, provoking protests. But in the last 30 years people have been brought closer. Many different corporations have arisen. Cooperation adds value to the border area. Cooperations require cross-border meetings.

One of the things affecting cross-border meetings is the interpretation of the border. A closed or strictly controlled border makes meetings more difficult. We are turning our backs to each other. We are focusing on ourselves and our own nation state.

An open border makes us turn around, look at each other, talk to each other. When we're talking, we get to know each other and when knowing each other cooperation becomes much easier. Cross-border meetings are essential not only for the border area but for both nation-sStates.

The Danish Border Association/Grænseforeningen has cross-border meetings as core of its work. Meetings provide cultural knowledgethat can be used in other parts of our lives. A cultural intelligent population benefits both the individual and the society.

The Danish Border Association

The Danish Border Association/Grænseforeningen (GF) grew out of the referendum in 1920. The Schleswig area was divided between Germany and Denmark. A national minority therefore emerged on both sides. In 1920 the focus of the association was on the Danish minority left on the German side of the border. Today the organization is working with a lot of topics connected to the border area. Still in a close connection to and in cooperation with the Danish minority, but also together with the German minority in Denmark and other groups and organisations that have an interest in the border area.

GF is a member organization with local associations all over Denmark. The Danish minority and the border area are topics of interest all around the country.

The slogan of GF is "For an open Danishness". The thought behind is that it must take care of our[1] national values without closing our doors. We may well be equal and different at the same time.

Grænseforeningen is working in the area of public learning (folkeoplysning). A focus point is learning a monocultural Danish population about the multicultural Danish-German border area. It is about understanding the history of the area and its consequences for the Danish society. It is about cultural clashes and how to turn it into something positive. It is about how to jointly strengthen a local area. It is about having focus on what brings us together instead of what tears us apart.

The best way to reach these goals is to let people meet. Therefore, a lot of GF's projects are focusing on bringing people together in Denmark and across the border.

Cross-Border Activities

Cross-border meetings are important for all age groups. GF has chosen to have a special focus on children and youth. Children are mostly openminded and ready to learn. The learning process is important for both minority and majority.

[1] Danish, editor's note.

Most of the cross-border projects are in cooperation with the Danish minority's school association. These projects aim to create meetings between children and young people from the Danish minority and people living in Denmark. In some projects the German minority is also a partner. The aim is to tell about the border area and the minorities.

Elevambassadørerne/ The Student Ambassadors are young people from the three minority high schools in the border area (A P Møller Skolen, Duborg-Skolen, Deutsches Gymnasium für Nordschleswig). For more than 15 years they have traveled around Denmark and visited youth educational institutions. Based on their personal stories they initiate a dialogue about subjects such as national identity, minority issues, the meaning of borders, prejudices, language etc.

They can reach their fellow students in another way than a teacher or a speaker. They are living texts as they make the history of the border area contemporary and relevant.

The ambassadors and their work have been noticed by politicians and other opinion makers and were invited to different kind of arrangements. They have met politicians from both sides of the border several times. In Denmark, this project is supported by the Danish parliament.

The project has also the eyes of the Schleswig-Holstein Landtag. The German association ADS Grenzfriedensbund (ADS-G) have received money to spread the project to high schools in Schleswig-Holstein. The development of the project on the German side of the border is done in close cooperation between GF and ADS-G. Elevambassadørerne became a cross-border project in more than one way.

For more than 10 years it has been possible for children from Denmark and the Danish minority in Germany to be a part of an exchange project. The exchanges are taking part every year in the late autumn. Pupils from the 4th to the 10th grade can take part. Participants are paired based on age, gender, and interests. Because in the project occurs during school time, the pupils' schools must also accept the participation. Concretely, this translates in the fact that during one week, participants will go to school to school together, will practice sports, etc. within one country. The following week, the same happens in the other country.

This is possible despite pupils living in Denmark do not master German proficiently. Going to a Danish school in Germany makes it easier. But it is still in Germany and will give the Danish pupils new experiences. Experiences about being in another country, about being in a border region, about living in a minority. They can even get better at speaking German. For pupils from the Danish minority, the opposite happens. They improve their mastering of the Danish language and get more knowledge about Danish culture and everyday life.

The contact will not only happen between pupils but at the scale of the whole family. Families must however arrange transport by themselves. It forces them to constantly dialogue. Exchanges often endnot just with a single meeting, but long-ending family friendships.

Another way for Danish pupils to learn about the border area and living in another country is to take their 10th grade in a Danish minority school. They can either live at a student home in Flensburg or with a local minority family.

Summer camps became very popular. I is possible during summertime for both minority children and children living in Denmark to be a part of one of six camps arranged on three different locations on both sites of the border. Children are divided in age groups and practice different activities adapted to their age group. This is a joyful way to meet the border area.

Minority Changemaker Programme is a European cross-border project where young people from all over Europe meet at the Danish folk high school (højskole) in Jaruplund in Germany. It is a three-month course combining academic lectures and ECTS credits with non-formal learning. The purpose is to create a space where young people can meet and discuss relevant issues and minority challenges. The wish is to create a stronger Europe united in diversity. The project is a cooperation between Grænseforeningen, Jaruplund Højskole and University of Southern Denmark.

Prospects for the Future

Cross-border projects are constantly evolving. They are being adapted to the reality they are a part of. Developments are taking place on all levels. But we hold on to the basic structure that it is about creating constructive meetings. GF is not the only organization doing cross-border activities in the Danish-German area. All of them are important. All of them are parts of the 2022 narrative.

We are living in a world with many challenges. We can't solve them on our own. It is crucial to create frameworks in which children and young people can meet across borders. Frameworks where they can get to know each other better and together find solutions for the common good. In the Danish-German border area there are useful experiences. The minority youth's narrative about living bilingual and using their cross-cultural possibilities is useful both for nation-states and European cooperation.

Migration Movements of Poles. Comparative Study of the German-Danish Border Region

Katarzyna Stoklosa

The topic of migration has taken on very different dimensions due to the migration wave that came in 2015 and continues to this day. This migration is related to refugees impacted by wars or climate change. The job-seeking migratory movements that took place in Europe from mid-1990 to 2008 were almost forgotten. They also concerned Poland and Poles to a large extent. Many any of them continue to this day, though not as large as in the period mentioned.

This article will analyse the migration movements of Poles in the German-Danish border region that have come due to changes in the European labour market and opportunities to find work abroad. The own research was carried out in the German-Danish border region: in Flensburg, 5 km from the border, and in Sønderborg, 35 km from the border. Both towns became attractive to Polish migrants. Flensburg (90,000 inhabitants) is slightly larger than Sønderborg (75,000 inhabitants), which is the largest city on the Danish side of the border region. As part of a comparative study, working and living conditions and the integration of minorities into the majority on both sides of the Danish-German border region were analysed.

Mobility is one of the important characteristics of today's world. It relates to people, goods, capital and information. Travelling has become cheaper and less complicated. The migrations that occurred after the 2004 European Union (EU) enlargement were not so much related to disasters, wars, or conflicts, but rather to the desire to improve one's standard of living.[1]

---

[1] J. Nakonieczna, *Migracje międzynarodowe a rozwój państwa*, Wydawnictwo Naukowe Scholar, Warsaw, 2007, p. 7.

Poland has significant migration experience. Temporary or permanent moves abroad from Poland took place after the Second World War. In the 1990s, the migration of Poles was mainly caused by the political and economic transformation. Poland metamorphosed from a country of emigration to a country of both emigration and immigration. However, emigration still was predominant and was an important cause of Poland's population loss. After Poland joined the European Union on May 1, 2004, one of the largest waves of migration from Poland in the post-war period occurred. The United Kingdom (UK) topped on the list of destination countries, while the USA dropped to the fifth place.[2]

After May 1, 2004, migration was primarily the domain of men and young people (20–29 years old) with a university degree. At the end of 2007, 2.270.000 Poles temporarily emigrated abroad, including over 1,900,000 in Europe. An increasing number of people were deciding to become permanent residents.[3]

After EU enlargement, one of the largest waves of migration from Poland in the post-war era took place. The exception was a short period in the second half of the 1950s, when a rapid increase in migration was caused by a resolution of the Central Committee of the Polish United Workers' Party (PZPR) on more favourable assessment of applications from German ethnic applicants for emigration.[4] This corresponded to the policy of the Federal Republic of Germany, which, after successful economic development in the 1950s, sought foreign workers (*Gastarbeiter*).[5]

After May 1, 2004, every eleventh Pole between the ages of 20 and 29 decided to leave. The outflow from Poland differed in different regions. Migration of young people covered mainly small and medium-sized towns in Lubusz voivodeship and large cities in Warmian-Masurian and Podlaskie voivodeships. As a consequence, for the first time in the post-war history of Poland, shortages in the labour market appeared. That, in turn, led –under pressure from employers– to changes in Poland's

---

2 K. Iglicka, *Kontrasty migracyjne Polski. Wymiar transatlantycki*, Wydawnictwo Naukowe Scholar, Warsaw, 2008, p. 16.

3 *Ibid.*, p. 18.

4 *Ibid.*, p.20.

5 D. Nouri, *Die Arbeitsmigration nach Deutschland ab 1950. Auswirkungen auf die soziale und wirtschaftliche Entwicklung der Bundesrepublik*, Studylab, München, 2019, pp. 6–7.

immigration laws and a gradual opening of labour markets to its neighbours from the East. The largest group was formed by Ukrainians.[6]

Polish Migrants in the German-Danish Border Region

Apart from Germany and the UK, northern countries have become an important destination of economic migration of Poles after 2004. Iceland was the first in terms of the number of people, followed by Denmark.[7] In later years, Denmark began to dominate. In 2018, 33,889 Poles lived in Denmark. 60 % came for work, 9 % for study.

Border regions became an interesting destination for migrants. Depending on the "passability" of the border, cross-border workers are categorised as border region characteristics. This phenomenon is common in Basel[8] and Øresund regions.[9]

The German-Danish border region has a historical tradition of labour-related migration.[10] The Duchy of Schleswig region had a common labour market before the division based on a referendum in 1920, which decided on today's German-Danish border.[11] On both sides of the border, there are dominant national minorities: the German minority in southern Denmark (Sønderjylland) and the Danish minority in northern Germany (Schleswig-Holstein).[12]

---

6 K. Iglicka, *Kontrasty migracyjne Polski*, *op. cit.*, pp. 15–22.

7 J. H. Friberg & L. Eldring (eds.), *Labour Migrants from Central and Eastern Europe in the Nordic Countries. Patterns of Migration, Working Conditions and Recruitment Practices*, Nordic Council of Ministers, Copenhagen, 2013.

8 Data retrieved from: <https://www.statistik.bs.ch/zahlen/tabellen/11-verkehr-mobilitaet/pendler.html>. Accessed 31 May 2023.

9 A. Wójcicka, "Copenhagen-Malmö. Different sides of integration in a hybrid city?", *In* J. Jańczak (ed.), *Conflict and Cooperation in Divided Cities*, Logos, Berlin, 2009, pp. 167–190.

10 M. Klatt, "(Un)familiarity? Labor related cross-border mobility in Sønderjylland-Schleswig since Denmark joined the EC in 1973", *Journal of Borderlands Studies*, vol. 29, n°3, 2014, pp. 353–373; G. Besier & K. Stokłosa, "Labor market migration in the German-Polish and in the Danish-German border-region", *In* L. Domínguez & I. Pires (eds.), *Cross-border Cooperation Structures in Europe. Learning from the Past, Looking to the Future*, Peter Lang, Berlin, New York, 2014, pp. 169–180.

11 M. Klatt, "(Un)familiarity?", *op. cit.*, p.359.

12 A. Teebken & E. M. Christiansen, *Living Together: The Minorities in the German-Danish Border Regions*, European Centre for Minority Issues, Flensburg, 2001.

Polish migrants in this region are an interesting research topic as they deal with two foreign cultures and languages on a daily basis. After Poland's accession to the EU in 2004, the region also became attractive for Polish people looking for work abroad. For many of them, the German side of the border region was less foreign as they could speak German. Only a few decided to learn Danish. At work, they used mainly English, which is not a problem in Danish workplaces. They quickly discovered how profitable it was to live on the German side and work on the Danish side of the border. This is due to the fact that the prices are lower on the German side, while wages are higher in Denmark. The fact that the Danish labour market became migration-friendly in the border region just after 2004 was a favourable factor.[13]

Compared to Germany, Denmark turned out to be more effective in terms of cultural integration of migrants. In 1999, a three-year integration system was introduced, which included learning Danish language and preparation for the labour market.[14]

The exact number of Polish migrants in the border region cannot be determined. Both on the Danish and German sides, many of them are registered as Danish or German citizens. In Germany, many Polish people are not registered. Most of them move between Poland and Germany. A lot of them are still registered in Poland, even though they have already lived in Germany for a long time.

Methodology

A qualitative study was conducted to investigate the Polish minority in the German-Danish border region. Narrative inquiry, in form of "human stories of experience" seemed to be the most convenient form for this research.[15] This is an inquiry where lives meet in storied ways in which we intentionally come into relation with participants.[16]

---

13 M. Klatt, "(Un)familiarity?", *op. cit.*, p.369.

14 G. Besier & K. Stokłosa, "Labor market migration in the German-Polish and in the Danish-German border-region", *op. cit.*, p.178.

15 P. Mertova & L. Webster, *Using Narrative Inquiry as a Research Method*, Routledge, London, 2020, p. 10.

16 D. J. Clandrin, *Engaging in Narrative Inquiry*, Routledge, London, 2022, p. 12.

I used the method of in-depth individual interviews.[17] They concerned the approach to life, work and a sense of belonging to the region. While the questionnaire distributed to find interviewees contained specific questions (gender, age, occupation, number of years spent in Flensburg or Sønderborg), the interviews were conducted as open conversations. The interlocutors told about the reasons for their emigration from Poland as well as the living and working conditions in the new place.

On the German side (Flensburg) I interviewed Polish members of the Catholic church der Schmerzhafte Mutter and customers of the Polish store "Marek". The Polish priest Kazimierz Sękala, who comes from Kiel (85 km from Flensburg) every second and fourth Saturday to give a mass in Polish in the church in Flensburg, helped me to get important information and interview partners[18]. Before the Saturday Mass before the first Sunday of Advent and on the second day of Christmas, I distributed questionnaires asking for permission to be interviewed. The first time there were about 50 people, including children in the church, the second time there were about 40 people. Father Sękała presented the project as part of pastoral announcements and asked the congregants for help.

I visited the Polish shop "Marek" near the German-Danish border eight times. Marek Borowski, its owner, was interviewed, and the situation in the shop was carefully observed. Each time, Polish customers prevailed. They came from Flensburg and the surrounding area, as well as from more remote regions, such as the island of Sylt. Polish customers usually made large shopping, while Germans bought only single products. After the fourth visit, I left the questionnaires in the shop asking for permission to be interviewed.

On the Danish side (Sønderborg) I managed to reach the Poles thanks to the support of Father Jacek Aniszewski, who once a month commutes from Odense to Sønderborg (165 km) in order to celebrate a Mass in

---

17 J. W. Creswell, *Qualitative Inquiry & Research Design. Choosing Among Five Approaches*, Sage, London, 2007, pp. 132–134; A. M. Cirucci & U. Pruchniewska, *UX Research Methods for Media and Communication Studies*, Routledge, London, 2022, p. 73.

18 My choice of a Catholic church in order to contact the Polish community was based on my previous research experience in the German-Polish border region. In order to examine the cooperation between the inhabitants of the divided cities of Guben and Gubin, I interviewed the congregants of the Catholic Church in Gubin and the Evangelical Church in Guben, see K. Stokłosa, *Grenzstädte in Ostmitteleuropa: Guben und Gubin 1945 bis 1995*, BWV Berliner Wissenschaftsverlag, Berlin, 2003, pp. 32–39.

Saint Pauls Kirke for Poles from this region. About 30 Poles attend the church. In addition, I was helped by the president of the Polish Association in Sønderborg. The questionnaires were distributed after the Mass on Christmas Eve.

Representatives of various social groups were selected for the interviews: from a plumber or an electrician to a designer, a university worker and a music teacher. My own experience of living and working in the German-Danish border region (*autoethnography*) was included in the analysis.

The interviews were carried out two to three times, mostly in respondents' homes. Occasionally there were additional questions or a need to clarify certain points. In such cases, I also contacted by phone. In January 2021, due to the stricter regulations under the deterioration of the COVID-19 situation, only remotely hosted conversations were available. They were conducted via Skype and Zoom.

Initial Research Results

The preliminary research results obtained after the first interviews with Polish inhabitants of the German-Danish border region are presented below.

Five qualitative interviews with Polish residents of Flensburg and four qualitative interviews with Polish residents of Sønderborg were used for this paper (see list of interviews). The oldest person was born in 1976, the youngest one in 1982. Most of the interviewed people were born in the late 1970s. It was not a coincidence. After the enlargement of the European Union and the introduction of labour market liberalisation, it was the generation that decided to emigrate and seek work in a new place. They were still young enough for a fresh start. The purpose of my interviews was to find out how strongly Poles in the region are connected and integrated with Flensburg, Sønderborg or both sides of the border.

For some respondents in the region, such as Adam J. (born 1980), Flensburg was a random choice at the time he decided to emigrate in 2004. His parents and his wife's brother had already been there. First, Adam took a German language course and then found a job working on various types of renovations and construction works, mainly on the German side. After the Covid-19 crisis, he also got work contracts on the Danish side of the border region. Adam employs some Poles (mainly from Szczecin), who help him in carrying out orders. In most cases, he is a linguistic intermediary for the group, helping them with formalities.

In the case of Kasia and Andrzej, the German-Danish border region was consciously chosen as a place to live and work. Andrew B. (born 1978) is a sports footwear designer and got a position at Eco in Tønder. Both Andrzej and his wife Kasia B. (born 1978) spent five years in Nuremberg (2005–2010) and speak fluent German, thus the German side of the border region was less foreign to them. Andrzej did not learn Danish. International workers are the majority at work. English is the main language of conversation. Andrzej and Kasia have lived in Flensburg since 2011, where Kasia works at the municipal music school as a violin and viola teacher. Kasia is a very active member of the cultural life of Flensburg and the region. She takes part in concerts and prepares school children for them, both on the German and Danish sides of the border. Kasia regularly attends Polish masses in Flensburg church. When asked about integration, she emphasised: "One cannot become isolated in one cultural setting or the other"[19].

For Iwona Ż. (born in 1978) and Rafał Ż. (born in 1976), the border region did not play any role in the decision to emigrate from Poland. They moved to Sønderborg from Grodzisk Mazowiecki in 2007 when Rafał was offered to move from Danfoss in Grodzisk to Danfoss headquarters in Nordborg and then in Gråsten. Their place of residence became Sønderborg, located 50 km from the border. Since the beginning, they assumed that they would stay in Denmark. Rafał had worked at Danfoss in Grodzisk Mazowiecki near Warsaw. After being offered a job in Denmark by the Danish boss, the family moved from Poland to Denmark. At work, Rafał uses mainly English. Initially, Rafał worked in Nordborg, and after the closing of the Nordborg branch he was given a job in Gråsten. He was soon required to learn Danish. In order to do so, he was paid for language courses. Iwona has worked in manufacturing for 12 years. Unlike Rafał, she was not required to speak Danish. Iwona took Danish language courses before finding a job and afterwards she attended evening courses. She found mastering the language important to her, in order to communicate with the Danes. The people of Sønderborg and the surrounding area do not speak English as fluently as the people of Copenhagen. Apart from the need to communicate in

19 Interview with Katarzyna B., Flensburg, 12 September 2020.

municipal institutions, Iwona also wanted to be able to interact with her daughters' kindergarten and school teachers without problems[20]. Iwona would really like to find a job in her profession (administration). In Poland, she worked in a company, first in production department, then in administration. Cultivating Polish identity plays an important role for Iwona. When she was questioned over why she was attending a Mass in Polish and not in Danish, I was answered as follows: "To feel a bit like in Poland".

Also in the case of Jacek F. (born in 1981) and Katarzyna J.-F. (born in 1982), the border region played no role in making the decision to live in Denmark. It was Denmark as a country that was attractive as a destination for them. Jacek's first contact with Denmark was in 2006 when, as a PhD student in physics at Jagiellonian University, he went to Odense for a nine-month Erasmus+Staff residency. Jacek was in Sønderborg for the first time in 2008, after his research group from Odense moved to Sønderborg for Alsion[21] and invited Jacek for a short, one–month, residence. After this stay, Jacek's contact with Denmark and Sønderborg was not lost. After completing his PhD at Jagiellonian University in 2010, the research group invited Jacek again for three months. After that, he was offered a postdoc position for two years, then an assistant professor for three years, then an associate professor. Kasia comes from Blachownia near Częstochowa. She studied medical physics at the Jagiellonian University. After graduation, Kasia worked as a bank manager in Kraków. In 2012, when Jacek received an assistant professorship, Kasia decided to move to Denmark with her two-year-old son Maciej to join Jacek. Travelling repeatedly between Sønderborg and Kraków became tiresome for the family. A change was needed and Denmark was financially more attractive than Poland. After Kasia's arrival in Denmark, she and Jacek were not sure if they would stay there permanently. Maciej was only in kindergarten for a few hours a day. Kasia took care of her son, she did not go to work right away. First, she had Danish lessons online. In 2013, Kasia started studying economics (BA Economics and Business Administration) in Sønderborg. Their second son, Nikolaj, was born at the end of 2014 in Aabenraa. In 2015, Kasia started working in Alhmann-Skole in Sønderborg as a teacher of Polish. In 2017, Kasia began her work at the University of Southern Denmark (Centre for

20 Interview with Iwona Ż., Sønderborg, 1 October 2020.

21 Home to part of the University of Southern Denmark.

Industrial Electronics). She attended an on-site Danish language course and speaks Danish at work when required. Nevertheless, communication in Sønderborg is mostly in English since employees are international. At the institute where Kasia works, she is one of the few administrative employees of non-Danish origin. While the older son attends an international school, the younger one is educated at a Danish school.

Jacek and Kasia do not really need the Danish language at work, where the main language of communication is English, but in private life: for communication in kindergarten, school, extracurricular activities and with neighbours. Jacek and Kasia are integrated in Denmark. They cultivate both Polish and Danish traditions. In order to make their older son's life easier, they chose a name for him that is known and used in Denmark. Kasia and Jacek admitted that integration with the Danes is not easy. Despite this, Jacek emphasised: "Many things that once seemed strange to us are normal today".[22]

For Beata Z. (born in 1977) and Lucjan Ż. (born in 1981) the German side of the border region turned out to be more attractive for a family stay than the Danish one. Lucjan comes from Racibórz. In 2007, he started to work for Thansen company in Middelfart[23], near Odense. In 2010, Lucjan met Beata. Half a year later, Lucjan and Beata settled in Denmark (Fredericia) together. However, six months later they moved to Germany. Lucjan commuted from Flensburg to work in Denmark. After being dismissed from a Danish company, Lucjan received unemployment benefits in Germany. After a few months, he found a job at the Nord-Schrott recycling company in Flensburg. However, he was still looking for a job in Denmark ("You earn better in Denmark"[24]). He found a job at Thermo-Transit, a fruit and vegetable shipping company in Padborg (10 km from Flensburg). He worked there for four months. After his period of probation, he was not employed and returned to Nord-Schrott. During their stay in Denmark, Beata and Lucjan did not have the opportunity to learn Danish. In Flensburg they started to learn German, however they do not speak it very well yet. Most of the contacts were made through the children's school. Beata does not work in her profession, in Poland she was a social worker and worked in accounting.

---

22 Interview with Jacek F. and Katarzyna J.-F, Zoom, 5 January 2021.

23 The company sells car parts: Tansen. Retrieved from: <https://www.thansen.dk/>. Accessed 31 May 2023.

24 Interview with Beata Z. and Lucjan Ż., Flensburg (Skype), 21 December 20.

In Flensburg, she works at *minijob*: helping elderly people. Lucjan compared life in Denmark and Germany as follows: "Life in Germany is better than in Denmark. Religion is more Christian, cuisine is more similar to Poland"[25].

A similar experience was shared by Agnieszka Sz. and Piotr Sz. After living in Kassel and Padeborn (Höxster) from 1998 to 2006, they moved to Denmark (Kolding) with their two children where Piotr was offered a job as a welder. Agnieszka worked in a Lidl supermarket for 6 hours a day. After a year, in 2007, they moved to Flensburg. Piotr commutes to work in Sjølund, he is a cross-border worker. Agnieszka found a job in Flensburg, first in Lidl supermarket, and then in Aldi store, where she is a manager. Like Beata and Lucjan, also Agnieszka and Piotr share the opinion that life is better in Germany than in Denmark: "We wanted stability and security, that's why we chose Germany. It was easier for us in Germany than in Denmark." According to Agnieszka, the health system in Denmark is a problem. The word "security" is used again: "I had no sense of security."

Agnieszka and Piotr attend the Catholic Church for masses in both languages: Polish and German. Agnieszka is a member of the parish council of the Catholic Church, she reads sermons both at Polish and German Masses.

Initial Research Results

After the analysis of the interviews with Polish residents on the German and Danish sides of the border, the following results arise.

People who decided to move to the Danish or German side of the border are mostly young people, aged between 25 and 35. In this respect, they do not differ from other migrant groups.

For manual workers, both the German and Danish sides of the border are attractive as a place of seasonal work, but not as a place of permanent migration. Since the way home to Poland is to be made in a half or one day, most respondents (mostly men) decide to travel, earn money and invest money "at home". Language learning is undertaken purely

[25] *Ibid.*

for practical purposes. In most cases, workers master the language just enough to communicate on the job. They often have their intermediaries.

Children are an important reason to decide to stay in a country of temporary emigration. Integration takes place through school and friends. The decision to return to Poland becomes more and more difficult. Many parents decide to learn the language of their new country of residence.

University educated workers mostly choose to migrate with their entire family. This is due to the fact that the purpose of the trip is not only to earn money, but also to become familiar with another culture. For this group, it is important to learn the language of the society in which they are starting a new life and job.

Unlike in the United Kingdom, for highly skilled Polish migrants in the German-Danish border region, Polish centres were not potential workplaces since they are not numerous in the region. This means that Polish migrants had to adapt to the Danish or German (in few cases to both) work model.

Most migrants are bound to one side of the border: German or Danish. Only in few cases do migrants live on one side of the border (especially the German side) and work on the other one (mainly the Danish side). The borderland factor plays only a minor role to most of them.

Similar to Polish migrants in the United Kingdom, Polish people in the German-Danish border region also like to stay in their cultural environments, that is, with other Poles. It does not mean, however, that they do not integrate with their new country of residence.

The willingness to adapt to the culture of the majority is higher on the Danish side than on the German side of the border region. This is related to the Danish policy of integration[26] and the subjective feeling of "being different" and having to adapt to the majority of society. In Germany, the pressure to adapt to the majority is not as high as in Denmark. It is also related to the different sizes of both countries.

For migrants who decide to stay permanently, the Danish side is more favourable for integration purposes. In order to gain access to the labour market, one must have a fairly good command of Danish. The Danish state's assistance in achieving this goal is much more substantial than

---

26 R. Jenkins, *Kontrasty migracyjne Polski. Wymiar transatlantycki*, 2008, Wydawnictwo Naukowe Scholar, Warsaw, pp. 255–281.

that of the German state for learning German language. Participation in Danish language courses is free and compulsory. Only after passing the language exams, access to the labour market or financial assistance in case of unemployment is possible.

While in the case of Danish and German minorities in the Danish-German border region both material and non-material values are exchanged, in the case of the Polish inhabitants of the region it is mainly a development of their new place of living in terms of material aspects. The pragmatism is very widespread among Polish migrants on the German side of the border.

The Covid-19 restrictions meant for Polish migrants in the Danish-German border region a limitation not because the closed German-Danish border but because it was not possible to them to travel to Poland during the pandemic. This fact shows that the German-Danish border region does not have a great importance to Polish migrants living there.

Bibliography

G. Besier & K. Stokłosa, "Labor market migration in the German-Polish and in the Danish-German border-region", *In* L. Domínguez & I. Pires (eds.), *Cross-border Cooperation Structures in Europe. Learning from the Past, Looking to the Future*, Peter Lang, Berlin, New York, 2014, pp. 169–180.

A. M. Cirucci & U. Pruchniewska, *UX Research Methods for Media and Communication Studies*, Routledge, London, 2022.

D. J. Clandrin, *Engaging in Narrative Inquiry*, Routledge, London, 2022.

J. W. Creswell, *Qualitative Inquiry & Research Design. Choosing Among Five Approaches*, Sage, London, 2007.

J. H. Friberg & L. Eldring (eds.), *Labour Migrants from Central and Eastern Europe in the Nordic Countries. Patterns of Migration, Working Conditions and Recruitment Practices*, Nordic Council of Ministers, Copenhagen, 2013.

K. Iglicka, *Kontrasty migracyjne Polski. Wymiar transatlantycki*, Wydawnictwo Naukowe Scholar, Warsaw, 2008.

R. Jenkins, *Kontrasty migracyjne Polski. Wymiar transatlantycki*, Wydawnictwo Naukowe Scholar, Warsaw, 2008.

M. Klatt, "(Un)familiarity? Labor related cross-border mobility in Sønderjylland-Schleswig since Denmark joined the EC in 1973", *Journal of Borderlands Studies*, vol. 29, n°3, 2014, pp. 353–373.

P. Mertova & L. Webster, *Using Narrative Inquiry as a Research Method*, Routledge, London, 2020.

J. Nakonieczna, *Migracje międzynarodowe a rozwój państwa*, Wydawnictwo Naukowe Scholar, Warsaw, 2007.

D. Nouri, *Die Arbeitsmigration nach Deutschland ab 1950. Auswirkungen auf die soziale und wirtschaftliche Entwicklung der Bundesrepublik*, Studylab, München, 2019.

K. Stokłosa, *Grenzstädte in Ostmitteleuropa: Guben und Gubin 1945 bis 1995*, BWV Berliner Wissenschaftsverlag, Berlin, 2003.

A. Teebken & E. M. Christiansen, *Living Together: The Minorities in the German-Danish Border Regions*, European Centre for Minority Issues, Flensburg, 2001.

A. Wójcicka, "Copenhagen-Malmö. Different sides of integration in a hybrid city?", *In* J. Jańczak (ed.), *Conflict and Cooperation in Divided Cities*, Logos, Berlin, 2009, pp. 167–190.

Achieving (Im)mobility. Resilience at the German-Danish Border during the Covid-19 Pandemic

Dorte J. Andersen & Ingo Winkler

Lately it has become somewhat problematic to live an everyday life on both sides of an internal EU border, in our case the German-Danish land-border. What used to be a borderland without stress, as Donnan and Wilson[1] would call it, a smooth space, where most border crossings were unproblematic, is now a strained space[2] with mobilities directed away from the State border, leaving a strange vacuum and emptiness around the border control posts. Even for cross-border commuters, who seem to have a legitimate reason to cross the border, general restrictions on mobility and the almost permanent border control upon entry into Denmark has made it difficult to physically cross the border. In the beginning of the Covid-19 pandemic, with workplaces locking down on both sides of the border, many commuters were not even allowed to cross the land-border and in the beginning of 2021, crossing the border for work began including a constant testing, hence, making sure one goes to the test station regularly and before crossing the border, causing much confusion, and upset among cross-border commuters. For many commuters in this borderland, with more than 20 years of open Schengen borders, these restrictions are first-time experiences.

In this article, we shed light on experiences of cross-border commuters in the German-Danish borderlands during the first years of the pandemic. We contribute to an emerging field of work on experiences

1 H. Donnan & T. M. Wilson, *Borders: Frontiers of Identity, Nation and State.* Bloomsbury, London, 1999.

2 G. Deleuze & F. Guattari, *A Thousand Plateaus*, Trans. B. Massumi, University of Minnesota Press, Minneapolis, 1987.

during and of the Covid-19 pandemic in the European borderlands[3] as we problematise the cross-border commuters' dealings with the measures imposed due to the pandemic in a new way by applying the concept of "borderlands resilience".[4] We do this because when studying the effects of the pandemic in the context of borderlands, we find it necessary to mobilise approaches that are sensitive to the agency of the borderlands. In our ethnographic investigations, it seems as if people in the borderlands are not just conforming to and passively adopting the operation of closure. Rather, they actively deal with, and even resist it, a response also indicated in recent research into the pandemic.[5] Border transitions, including border closures, often work against the everyday dealings of people living in borderlands. Resilience would therefore emerge from the stimulus of those who can impact the trajectory of change, by establishing and finding creative ways to deal with border transitions. Hence, in borderlands, measures implemented to control a pandemic are not just implemented by the state apparatus in and for itself but just as much by people who translate the measures into workable doings in their everyday lives. Cross-border commuters in the German-Danish borderland constitute a real-life example of such resilience.

The article begins by sketching out what we mean by borderlands resilience and how we mobilise the concept analytically to assess the way cross-border commuters working in the German-Danish borderlands have responded to the border closures during the pandemic. The article then continues by clarifying the methodological consideration guiding the empirical research, focusing on the various ethnographic methods, which have been put into play in the investigations. Presenting

---

3 C. Wille & R. Kanesu (eds.), "Bordering in Pandemic Times. Insights into the COVID-19 Lockdown", *UniGR-CBS Thematic Issue*, Bd. 4, 2020; E. Opiłowska, "The Covid-19 crisis: the end of a borderless Europe?", *European Societies*, vol. 23 (sup1), 2021, pp. 589–600; R. Tarvet & M. Klatt, "The impact of the Corona crisis on borderland living in the Danish-German border region with a special focus on the two national minorities", *National Identities*, 2021, pp. 35–52.

4 D. J. Andersen & E. K. Prokkola, "Introduction: Embedding borderlands resilience", *In* D. J. Andersen & E. K. Prokkola (eds.), *Borderlands Resilience: Transitions, Adaption and Resistance at Borders*, Routledge, London, 2022, pp. 1–18.

5 See for instance M. Lois, H. Cairo & M. García de las Heras, "The Politics of Resilience ... politics of borders? In-mobility, insecurity and Schengen 'exceptional circumstances' in the time of COVID-19 at the Spanish-Portuguese border", *In* D. J. Andersen & E. K. Prokkola (eds.), *Borderlands Resilience: Transitions, Adaption and Resistance at Borders*, Routledge, London, 2022, pp. 54–68.

our findings, we show how resilience is practiced, not uniformly, but in multiple ways in the borderlands. We thus problematise resilience in different versions. "Simply just getting there" illustrates the cross-border commuters' response in terms of being able to continue their everyday (commuter) life. "Bending the rules" addresses how the unsettled border restrictions open possibilities of interpreting rules in creative ways. "Resistance" involves the critical distancing from the rules, questioning their normativity and value, including frustrations and annoyance with the constant change of rules and regulations. Finally, "Re-bordering" shows how (memories and imaginations of) national differences are exposed as ways to practice borders. Together these practices constitute the cross-border commuters' manifold abilities to independently upset and settle borders, abilities constituting local resilience towards the stress caused by the pandemic.

Resilience at the Border

Resilience research and not the least its discursive developments in policymaking and public management, has aimed to create awareness and practical tools for local populations at global scale to deal with various stress events. The discourse has been widely promoted in recent years due to general and globally relevant experiences of increased insecurity and continuous crisis[6] and it is today much talked about in relation to so-called "crisis management".[7] Local communities in multiple shapes and forms are uniformly encouraged to develop resilience mechanisms in order to find their own ways to "bounce back" from major stress events caused by armed conflicts, climate crisis, migration crisis and not the least border transitions.[8]

---

6 R. Hassink, "Regional resilience: A promising concept to explain differences in regional economic adaptability?", *Cambridge Journal of Regions Economy and Society*, vol. 3, n°1, 2010, pp. 45–58.

7 See P. Rogers, J. Bohland & J. Lawrence, "Resilience and values: Global perspectives on the values and worldviews underpinning the resilience concept", *Political Geography*, vol. 83, 2020, pp. 1–9.

8 United Nations, "Resilient people, resilient planet: A future worth choosing", Secretary-General's High-level Panel on Global Sustainability, United Nations, New York, 2012; OECD, *Guidelines for resilience systems analysis*, OECD, Paris, 2014.

When border transitions are spoken about in resilience research and management, they are mostly looked upon "externally", so to speak, as policy driven changes related to territorial conflict but also to the liberalisation of state border controls related to globalisation processes.[9] Stress events, transformations, and crises in relation to borders are in other words dealt with as caused by "the systemic level" and all that ordinary people can do in their everyday lives is to "bounce back" and cope with transition, otherwise the ordinary plays no role in the dealings. Agency has, in relation to re-bordering processes, been limited to the level of geopolitics and the supporting national and international institutions, and there has been little to no understanding of the way ordinary people are involved in bordering practices.

The understanding of borders and border dynamics promoted in mainstream resilience research and management as always being imposed externally is rather un-nuanced, as it fails to consider the results of border research done over the last decades where borders have been problematised as practices performed by multiple actors, at many scales and levels and in multiple contexts.[10] This research shows that ordinary people most certainly have the will and power to border –also in the territorial sense of the term – and their resilience in the face of border transitions is always-already related to such abilities, it does not just stem from processes located outside the ordinary. Thus, ordinary and everyday life is ripe with agents and actants who are, in their own ways, both creating and dealing with the stress of border transformation. Zooming in on these everyday dealings will tell us more about how to handle the stress than when we merely focus on geopolitical impositions.

---

9 E. K. Prokkola, "Border security interventions and borderlands resilience", *In* D. J. Andersen & E. K. Prokkola (eds.), *Borderlands Resilience: Transitions, Adaption and Resistance at Borders*, Routledge, London, 2022, pp. 21–35.

10 H. van Houtum, O. Kramsch & W. Zierhofer, "Prologue B/ordering Space", *In* H. van Houtum, O. Kramsch, & W. Zierhofer (eds.), *B/ordering Space*. Aldershot, Ashgate, 2005, pp. 1–13; C. Rumford, "Introduction: Citizens and borderwork in Europe", *Space and Polity*, vol. 12, 2008, pp. 1–12; D. J. Andersen & M. Sandberg, "Introduction", *In* D. Andersen, M. Klatt & M. Sandberg (eds.), *The Border Multiple: The Practicing of Borders Between Public Policy and Everyday Life in a Rescaling Europe*, Routledge, London, 2012, pp. 1–19; C. Brambilla, "Exploring the Critical Potential of the Borderscapes Concept", *Geopolitics*, vol. 20, 2015, pp. 14–34; S. Green, "Lines, Traces and Tidemarks", *In* O. Demetriou & R. Dimova (eds.), *The Political Materialities of Borders: New Theoretical Directions*, Manchester University Press, Manchester, 2018, pp. 67–83.

Still, it is rarely asked in border research exactly how strengthened border securitisation in the Schengen context incites border communities and mobile people to deal with manmade material border infrastructures and how renewal and resistance may emerge as responses to border transitions. Considering a trend away from open borders scenarios in Europe towards more and more border securitisation,[11] these questions become highly pertinent. States and other types of administrations do not always imagine what "the good life" is in ways that match with what people living in borderlands find a life worth living. In understanding how borderland people deal with stress situations caused by increased securitisation, it is important to pay attention to their ways of making sense of different shocks, and how interpretations and articulations of them influence their responses.[12] Our study therefore follows the proposal of Andersen and Prokkola to focus on "[T]he situational practices, experiences and narratives of borders, border crossing and belongings from the perspective of adaptation and resistance among the borderland peoples", and that we are particularly interested in "the ways through which people reorient their practices and relations with regards to border transitions".[13] Thus, we understand resilience intrinsically as the borderland people's responses to their interpretation of the stress they experience and how that impacts their understanding of "the good life" in the borderland: "People's stories of their cultural goals lead us to matters that elucidate their resilience."[14]

Our Ethnographic Access Points

Our research builds on the findings of our now five year and ongoing ethnographic study of the mundane borderwork of cross-border commuters in the German-Danish borderlands.[15] For this manuscript, we mainly draw upon insights generated during the last two years, i.e.

---

11 E. K. Prokkola, "Border security interventions and borderlands resilience", *op. cit.*

12 G. Wandji, "Rethinking the time and space of resilience beyond the West: An example of the post-colonial border", *Resilience: International Policies, Practices and Discourses*, vol. 7, 2019, pp. 288–303.

13 D. J. Andersen & E. K. Prokkola, "Introduction: Embedding Borderlands Resilience", *op. cit.*, p. 6.

14 *Ibid.*, p. 10.

15 D. J. Andersen & I. Winkler, "Grænsearbejde blandt grænsependlere (Borderwork among cross-border commuters)", *Oekonomi & Politik*, vol. 93, 2020, pp. 56–68.

the period of the pandemic. However, we find it an advantage that our research extends beyond the time of the pandemic, thus also touching upon the experiences of cross-border commuting in more "normal" circumstances as well as the experience with border closures in relation to the so-called "refugee crisis" in 2015. This gives us more insight into how the resilience during the pandemic is exceptional in itself and only linked to the measures imposed by the pandemic and how much of this has already become normality for the commuters.

We primarily engage with cross-border commuters, who reside in Germany and commute to Denmark for work because this is by far the most usual direction of cross-border commuting in the context.[16] During the past two years, we mainly focused our investigations on how the commuters have been interpreting and responding to the various and constantly changing regulations that Denmark and Germany issued regarding the possibility to cross the border between the States. We paid particular attention to the commuters' reactions and how they managed to continue their lives as commuters. To get close to the commuters' numerous ways of making sense of and dealing with the situations that the pandemic created for their mobility, we entered into both formal and informal conversations with them as well as experts working in the *Regionskontor & Infocenter*, a German-Danish institution providing advice and support for cross-border commuters. Accessing the commuters proved to be a challenge during the pandemic due to social distancing restrictions imposed upon the populations in both countries. To mitigate this problem, we decided to access commuters we already had close contact with, such as neighbours and colleagues. We also approached commuters at the Corona-test facilities. In addition to these ethnographic interviews, i.e. everyday conversations conducted with the purpose to learn more about the field,[17] we kept a close record of the respective

---

16 Øresundsinstitut, "Grænsependlernes samfundsmæssige værdi samt indkomst i den dansk-tyske grænseregion", October 2016. Retrieved from: <https://www.region.dk/downloads/Publikationen/DK/Graensependlernesamfundsmaessigevaerdi20161116.pdf?m=1549473628&>. Accessed 14 September 2021. Judging from our research, the reasons for this are higher wage levels in Denmark, the social benefits that the commuters are eligible to, the superior working conditions and better working climate, language issues as well as how the availability of jobs has traditionally been greater at the Danish side of the border (this is only partly the case today though).

17 J. P. Spradley, *The ethnographic Interview*, Wadsworth, Belmont (CA), 1979.

border-related restrictions valid on each side of the border, enabling us to connect our material to these measures.

As the pandemic made it harder to conduct traditional ethnographic research, we also spend more time on using netnography as a methodology for the ethnographic study of online communities and cultures.[18] We followed three Facebook groups established to enable the commuters to exchange information, to help each other regarding all sorts of commuting related issues, to express their feelings and thoughts regarding their life as commuter, and to organise themselves. These groups are *Arbeiten in Dänemark* (Working in Denmark), *Grænsependleriet* (Cross-border commuting) and *Flensbook – for danskere i Flensborg* (Flensbook – for Danes in Flensburg). Our prolonged immersion in the online communities that these groups create helped us to further reveal the reality of the cross-border commuters and their various responses to the issues the pandemic created for them.

As both authors are themselves cross-border commuters in this context, we have been affected by the situation ourselves. Therefore, we also used our own experiences of being cross-border commuters, who seek to deal with the strengthened border securitisation – thus integrating an autoethnographic methodology. According to Ellis and Bochner autoethnography constitutes a "genre of writing and research that displays multiple layers of consciousness, connecting the personal to the cultural".[19] We particularly created autoethnographic vignettes[20] where we engage in personal reflections on how we and our mobility has been affected as well as what actions we took to respond to that.

Combining the findings created through the above-described ways of generating empirical material, we sought for patterns in the commuters' ways of dealing with this exceptional situation. Analysing our material we paid particular attention to the commuters' intrinsic resilience, hence, their situational practices, experiences and narratives related to the strengthened border securitisation as expressions of their adaptation and resistance as borderland peoples.

---

18 R. V. Kozinets, *Netnography, Doing Ethnographic Research Online*, Sage, London, 2010.

19 C. Ellis, & A. P. Bochner, "Autoethnography, personal narrative, reflexivity: Researcher as subject", *In* N. K. Denzin & Y. S. Lincoln (eds.), *Handbook of Qualitative Research*, Sage, Thousand Oaks (CA), 2000, p. 733.

20 M. Humphreys, "Getting Personal: Reflexivity and Autoethnographic Vignettes", *Qualitative Inquiry*, vol. 11, 2005, pp. 840–860.

Resilience at the German-Danish Border

Border transitions experienced during the Covid-19 pandemic constitute the perfect scenario to research the ways European borderlanders adapt to, interpret and deal with border transitions in their everyday lives. The Covid-19 pandemic may be categorised as a public health crisis of proportions, and it has brought about measures substantially questioning the foundations of "normality", "rearranging daily life in the form of quarantines, isolation, mobility restrictions, and closure of borders" resulting "from the deployment of exceptional spaces and times destined to mitigate the effects of the pandemic".[21] Despite the European attempts at practicing a more liberal border regime, the closure of borders has –at least since the long summer of migration 2015– turned into a common tool used by European governments in such "exceptional circumstances". Being product of political decisions shaping a social construction of exceptionality, it is important to reflect on people's responses to such security measures, thereby questioning their taken for grantedness as means to maintain public security.

In our first readings of the gathered material related to the pandemic, we agreed that it did not really seem as if there was much resilience among cross-border commuters in the German-Danish borderland. Rather, responses indicated a passive-aggressive attitude, simply just demanding State authorities the abolition of the border closures to make everything return to normal. However, in our second and third readings we paid closer attention to the plurality in the commuters' stories, which enabled us to recognise various forms of resilience. On the backdrop of our theoretical approach, we thus realised that there are multiple versions of resilience at play, reaching from less aggressive forms of everyday resilience to more confrontational strategies of resistance, also mobilising established politics. In the following, we present these versions of resilience.

"Simply Just Getting There"

The first version, which is also the most usual, expresses people's attempts to just navigate the exceptional circumstances created by the authorities and their immediate responses to the pandemic. We call it

[21] M. Lois, H. Cairo, & M. García de las Heras, "The Politics of Resilience", *op. cit.*, p. 55.

"simply just getting there", indicating how the commuters are just interested in and working towards getting to work, or more precisely, in just living their everyday lives as cross-border commuters. As a commuter aptly expressed it in a Facebook post: "But we actually just want to work".[22] This constitutes a form of pragmatic resilience where the commuters try their best to adapt the measures imposed by the governments on both sides of the border, a struggle to accomplish mobility in difficult times. One example of this struggle is exemplified in the following fieldnote excerpt:

> "I met my neighbour the other day. I know that he is a cross-border commuter working for a transport company in Denmark. I asked him how it is to cross the border every day. He said that it is not a problem at all. He has his work contract, which he shows at the checkpoint, in addition to his ID card and that is enough to get into Denmark. Returning to Germany, the German border control only asks for the ID. In the beginning, he learned that there is a specific form for cross-border commuters that could be downloaded from the internet. It was said that showing this form at the border control would smoothen the check as it indicates that one is a cross-border commuter. So, he downloaded the form, completed the necessary information, and got it signed by the employer. However, he quickly realised that in addition to the form he still had to present his ID card at the border. So, there was not really an advantage compared to showing the employment contract. Hence, this is what he does since, showing the employment contract and the ID card works fine".[23]

For this and for other commuters it is important that their mobility is not obstructed by measures that prove to be insuperable. As we learned from a talk with another commuter, "For him the only thing that matters is that the commute is not hindered too much at the border, and this is the case. A couple of minutes waiting and he is on the other side"[24] Talking to various commuters and following the discussions in the Facebook groups, it is apparent that the commuters have been able to cross the border during the pandemic despite new, stricter, and ever-changing restrictions.

In relation to pragmatic resilience, Facebook groups became a means to clarify everyday questions such as, "What documents do I need to

---

22 Facebook group *Arbeiten in Dänemark*, post 1 (authors' translation).

23 Fieldnote excerpt 1.

24 Fieldnote excerpt 2.

present at the border?", "Do I need a Covid19 test?", "What type of test do I need?", "Where can I get the test?", "How often should I get tested?" Furthermore, it is used for reporting on the immediate situation experienced at the border: "How long is the queue?", "What about commuting northbound?", "and southbound?", "Where is it best (easiest) to cross the border?" People were in other words using Facebook to clarify issues concerning what to do to keep on going –and not the least "what to do exactly now", considering the speed with which the restrictions developed at times.

One of the things, which made life difficult for the commuters was, in other words, the constant changes in regulations. Sometimes new measures were introduced literally the day before they entered into force. The confusion was continuously great, yet, peaking in the beginning of the pandemic, and when a testing regime was introduced as a requirement to travel to work in the beginning of 2021, something that demanded quite a lot of adjustments by people who still had to travel to work physically. On Friday the 8th of January 2021 in a press conference, the Danish government announced stricter regulations for the entry into Denmark, especially relevant for cross-border commuters. These regulations specified that persons residing in the border region, who have a recognisable reason for crossing the border into Denmark must be able to present a negative test at entry, which must not be older than one week. It was also revealed that the measures would take effect from the 10th of January, i.e. the following Sunday.

This rather unexpected announcement generated a huge discussion on the Facebook group *Arbeiten in Dänemark*. Commuters were asking "Do I need a test as daily commuter?" and "What about those who live in the border region?" as they had previously been exempt from some restrictions and the information about this was apparently confusing. One commuter complained that he asked at the border and got the wrong information from a police officer that those living in the region would not need to present a negative test from the 10th onwards. The confusion was enormous, and the most urgent question asked on Facebook was simply "What to do when needing a test to be allowed to cross the border for work?" At that time, the test infrastructure on the German side was poorly developed and tests came with costs (30–40 Euro). Some commuters therefore immediately drove to Denmark to get a test. Others made appointments with one of the test stations on the German side and spend hours (and much money) to get a test, however these appointments where soon unavailable for the weekend. Some weekly commuters considered commuting already on the 9th. Still, others booked a test in Denmark for the coming week and wrote that they will try to cross the border by showing the appointment and their work contract. Finally, some just informed their workplace about the situation announcing that they cannot get a test on the weekend and therefore might not be able to come to work on the next Monday.

The following months, testing continued to constitute a concern among the commuters as it took additional time and costs. However, talking to some of them we learned that they swiftly accommodated this new requirement and accepted it as an additional element of their commuter life. As one commuter puts it: "But we should also be grateful that we can still go to work. There are enough who have to stagnate at home with a net salary of 67 % (the short-time allowance from the German governance, authors). That's why I take it easy. Today testing took 12 minutes, including driving to the test station."[25] Commuters quickly learned to navigate the situation by making it into a new routine, as we for example learned from a talk to a commuter at a test station: "He gets tested twice a week. The test facility is often crowded but around 9 a.m. is a good time, he said. In the morning around 5 and 6, there are many because commuters seek to get a test before they cross the border to get to work."[26]

Overall, we learned that simply just getting there and successfully coping with the situation required one to let go of established commuting routines and to create new ones. These routines included for example getting regularly tested (or updating one's Corona passport), continuously updating oneself on the current regulations, and being routinely controlled on the border (associated with impact on travel time). Even though the logic behind many new regulations was difficult to understand and to follow for commuters, and that the constant change of regulations constitutes an ongoing concern as they required continuous effort, commuters have by and large been able to adapt. However, such adaptation is not simple rule following. Rather, the pragmatic resilience of simply just getting there requires (constant) creative responses to the situation in order to fulfil the promise of "just wanting to work".

---

[25] Facebook group *Arbeiten in Dänemark*, post 2 (authors' translation).
[26] Fieldnote excerpt 3 (authors' translation).

Bending the Rules

Some of "the achievements" are less pragmatic and adaptive, but instead interpret into a different form of creativity, a form that is aimed at finding ways around and through the measures and rules imposed. The bending of rules, however, is carefully adopted, so that it would not trouble authorities in a way that would challenge the imposed measures, that is, rules are not bend with the purpose nor with the means to form a resistance[27] but simply just warranted by a will to "survive" in the changed circumstances. While speaking into the "simply just getting there", "bending the rules", in this sense, is more than a creative adaptation to rules. It constitutes a more playful approach towards testing the ground and pushing the limits imposed by restrictions – thereby enlarging the possibilities of having agency. Moreover, bending the rules has a dual connotation, as we observed this form of resilience on the side of the commuters, yet, also on the side of border officials.

Some commuters quickly realised that at the time of day when they crossed the border into Denmark –usually in the early morning– there was not anyone to control them as the controls usually would start later in the day. Hence, for them it seemed to be less relevant to get a negative Corona test and to have all the necessary documents ready for inspection when crossing. As one of them reported in the Facebook group *Arbeiten in Dänemark*, when the commuter crossed the border at 5 a.m. there is no control as the border officials start their work at 7 a.m. and the commuter claimed to not being affected by any stricter border regulations. Similarly, upon return to Germany, commuters also quickly realised that it was rather unproblematic to re-enter Germany as more regular border control was never established. At points, there were different restrictions imposed from the German and regional authorities, such as commuters needing a negative Corona test to enter. However, many commuters simply did not bother because of the lack of control. As one of the commuters aptly put it: "You do not need to have a valid negative test when re-entering Germany as there isn't any control"[28] and as far as we understand the community, many commuters showed this attitude.

---

27 See also G. Wandji, "Rethinking the time and space of resilience beyond the West", *op. cit.*

28 Facebook group *Arbeiten in Dänemark*, post 3 (authors' translation).

Bending the rules therefore included crossing the border without valid legitimation despite the controls. In those cases, some commuters directly confronted the border officials with the fact that they would try to enter Denmark without fulfilling the requirements, what from a legal standpoint can be considered an infringement of entry requirements causing a monetary penalty. However, as one commuter reported in the Facebook group: "I did not have a test but were allowed to cross the border to Denmark anyway".[29] Speaking to one of the commuters, we were able the learn more about the circumstances of such a situation. In this case, the commuter was unable to get a test due to the long waiting time at the test centre. As he had to go to work, he therefore decided to try his luck:

> "This Thursday he tried to cross the border without a test and of course he was checked. He had to pull over at the border crossing and he explained to the border official that he didn't have a valid test. He also showed all his test slips from January until now. Additionally, he offered to leave his ID card at the border control, quickly drive to Bov (the nearest test station in Denmark) for testing and then come back. The border guards then let him in without a test".[30]

As we can see from this example, it was not only the commuters who became creative in bending the rules. Border officials in some cases became complicit as they did not enforce existing regulations but enabled the commuters to slip through, so to speak. On the 6th of February 2021, the local Danish newspaper *Der Nordschleswiger* issued an article entitled "*Über 200 Autofahrer überqueren deutsch-dänische Grenze ohne Kontrolle*"[31] (More than 200 car drivers cross the German-Danish border without being checked). According to the article, the border officials just left one of the border control posts in the late evening thereby abandoning the control until their return on the next morning. We do not know the exact reason for that but according to the article, border officials did work from the assumption that controls (especially controls of tests) are still temporary. What we learned from our fieldwork on the

---

29 *Ibid.*, post 4 (authors' translation).

30 Fieldnote excerpt 4.

31 "Über 200 Autofahrer überqueren deutsch-dänische Grenze ohne Kontrolle", *Der Nordschleswiger*, 6 February 2021. Retrieved from: <https://www.nordschleswiger.dk/de/nordschleswig/ueber-200-autofahrer-ueberqueren-deutsch-daenische-grenze-ohne-kontrolle>. Accessed 14 September 2021.

Facebook groups was that it was a more general experience that commuters were able to cross the border without any control: "They just wave you through" is what some of them write to the extent of complaining that the border officials do not even bother to control them.

Coming across such stories, a question occurred as to who actually makes the rules that are of relevance for the cross-border commuters. Various commuters shared their observation of an apparent mismatch between the government's official regulations and how they are implemented at the border. For some it seemed as if the drivers of the local busses crossing the border became more of an enforcement of controls than the border police,[32] while others experienced arbitrary interpretations of regulations among the police. Regarding the requirement to show a valid negative Corona test when entering Denmark, one commuter narrated the following on Facebook: "I was just refused to cross the Danish border because I have a Danish Corona test and the Danish police said that I, as a German, need a German Corona test?? Since when?? I work in Denmark, pay taxes there, have a CPR number and NemID. They said it had to be a German test and if I live in Germany I have to be tested here as well!!??"[33]. Clearly, the border officials were to blame here for having made a mistake. However, for the commuter their rule making constituted what the commuter experiences to be the reality, an arbitrariness of rule implementation affecting the commuter's mobility.

As we have learned, both the cross-border commuters and the border officers engage in the practice of bending the rule. For the former this form of resilience enabled the commuters to have agency beyond merely complying to new and continuously changing regulation. Even though bending the rules does not constitute a form of resistance, it does express itself as creative disobedience while responding to tightened border regimes, disobediences that allowed the commuters to continue their life, even in situations when rule-following was impossible. The reasons behind the border officers' engagement in bending the rules are difficult to determine, yet, misinformation, miscommunication as well as work-to-rule serve as interpretation from the perspective of commuters. In any case, resilience as bending the rules contributed to the maintenance of mobility in difficult times.

---

32 Fieldnote excerpt 5.

33 Facebook group *Arbeiten in Dänemark*, post 5 (authors' translation).

Resistance

"We know best when it is necessary to control the Danish borders"[34]. These were the words of the Danish foreigner – and integration minister, Mattias Tesfaye, when asked in the Danish Border Association's magazine *Grænsen* about the future of the Schengen cooperation. At this point in time, cross-border commuters in the German-Danish borderlands had been confronted with changing border restrictions for more than a year, and they had to respond to constantly shifting demands for access at the border control posts. The minister's statement was, by many cross-border commuters, read as a denial of the value of their everyday life – and to an extend a denial of their very existence, including the insecurity they experienced because of not knowing what potential restrictions would be imposed on the next day. Considering the Danish government's declarations and attitudes during the peaks of the pandemic in 2020 and 2021, it is not surprising that there has been signs of resistance towards border closures among cross-border commuters, resistance especially targeted at the border security measures imposed by the Danish government.

The Facebook group *Grænsependleriet* was formed as a more articulated resistance towards the security measures imposed at the German-Danish border. In its purpose declaration it says: "For those who commute across the German-Danish border; with the purpose of reducing the queue that has formed at the German-Danish border since June 15".[35] The date mentioned in the declaration refers to the point in 2020 when border controls were lifted by the Danish authorities, which allowed tourists to again travel to Denmark provided they could present a 6-day booking receipt at a hotel or a summerhouse. Consequently, long queues developed at the soon overloaded checkpoints as numerous tourists (most of them German) had to be checked before they were allowed to enter Denmark. Establishing this group is one of the clear, albeit few signs of (active) resistance as some of the commuters unite, creating a platform for exchanging their experiences of how the increased border securitisation interferes with their need for free mobility.

---

[34] M. Tesfaye, "Vi ved bedst, hvornår der er brug for grænsekontrol i Danmark", *Grænsen*, vol. 3, 2021. Retrieved from: <https://graenseforeningen.dk/magasinet-graensen-nr-3-juni-2021/mattias-tesfaye-vi-i-danmark-ved-bedst-hvornaar-der-er-brug>. Accessed 03 June 2021.

[35] Facebook group *Grænsependleriet*, post 1 (authors' translation).

Another attempt to challenge border regulations that made it more difficult to commute was to involve the media (e.g. local newspapers, local and national radio / television) in order to make the public aware of the difficult situation at the border. As a result, in September 2020, the Danish national television, *Dansk Radio*, on its website, launched an article critically asking why it took the government so long to re-open the border after the society shutdown.[36] Furthermore, some cross-border commuters wrote protest letters to politicians and to the parliament trying to inform policy makers about the problematic effects of their decisions.[37]

Moreover, cross-border commuters did not stand alone in their resistance. Also, local politicians became active, such as the mayor of Flensburg, the largest city in the border region. In an interview with the local Danish television, *TV2 Syd*, she declared: "We will travel to both Copenhagen and Berlin and talk to the governments there to ensure that in the future we have a right to be listened to."[38] A couple of days earlier, the Danish MP Bertel Haarder, on his Facebook site expressed his discontent with the current border regulations by asking what the people living in the Danish border regions have done for them to be required to regularly show a negative Covid19 test when crossing the border.[39] He argues that this requirement is disproportionate because crossing the border for work is unrelated to infection rates.

Another, less active and more emotional kind of resistance also informed the resilience that we observed among the cross-border commuters. We observed numerous expressions of anger, frustration, annoyance, yet also cynicism and sarcasm both when talking to commuters and when reading the posts in the Facebook groups. We read that as the more emotional part of protest and resistance, which may create less

---

36 "Grænselandet kritiserer regeringen: Hvorfor åbner I først grænsen nu?", *Danish Radio*, 4 September 2020. Retrieved from: <https://www.dr.dk/nyheder/regionale/syd/graenselandet-kritiserer-regeringen-hvorfor-aabner-i-foerst-graensen-nu?fbclid=IwAR2OagDh-3b-QbU9CgdYvuKjnH4sT2nyDm_72gKE6Vh8BJBGSAHUlOx2MFc>. (authors' translation)

37 Facebook group *Grænsependleriet.*

38 "Politikere vil sikre, at grænsen aldrig lukker igen", *TVS*, 21 June 2020. Retrieved from: <https://www.tvsyd.dk/syd-og-soenderjylland/de-vil-sikre-graensen-aldrig-lukker-igen?fbclid=IwAR234vTLQaFMfvMGtcjr53bp9UTIbKMZRyYWwEAmjAnqe-XF8JhbSfOuXwE>. Accessed 14 September 2021.

39 Facebook post 10.

of a sense of agency in terms of mitigating problems. However, it provides agency in terms of interpreting the situation and thereby creates a counter-discourse denying the legitimacy of border controls, which in the perception of many cross-border commuters had no other effect than to complicate their life. Despite this type of resistance having even less effect on those who issued the stricter regulations, it nevertheless enables the commuters to articulate their dismissive stance towards the measures they had to adapt to:

> "Oh my god, another queue at the border. And another border control officer, who believes that intensively checking the passport and the negative SarsCov2 certificate of each and every person, who wants to enter into Denmark, is very necessary, even checking those who come from the region and are obviously on their way to work. What is the point of all this? In the beginning of the more restricted border controls implemented to hinder the spread of the virus (I think this was the official parlance) my immediate reaction at the border was (and still is) annoyance. Annoyance about the measures – I mean how should the documentation that you have a legitimate reason to cross the border prevent the virus from spreading? But also, annoyance about the obstacle that it caused to my mobility. So, what is the point?"[40]

Letting go of their emotions, commuters attacked regulations themselves, yet also the ones who had to enforce them at the border. On the one hand, enforcing the regulations obstructs the commuters' mobility as some of them rant about: "Took me 'only' one and a half hour to get from Flensburg to Aabenraa on Saturday afternoon."[41] or "We have to cross the border 5–10 times a day! It has become a disaster!"[42] On the other hand, not enforcing the regulations makes the commuters wonder why they have to make additional effort to get the documents necessary for crossing the border: "So we queued for over seven hours yesterday to get a test. This morning at the border I showed my documents, but they didn't even want to see anything. They just wished me a nice day!? Kidding??!!"[43] Various commuters wondered what the point was of imposing such regulations when they are actually not enforced. Either

40 Autoethnographic fieldnotes 1.

41 Facebook group *Grænsependleriet,* post 2 (authors' translation): the usual travel time is about half an hour.

42 *Ibid.*, post 3 (authors' translation).

43 Facebook group *Arbeiten in Dänemark,* post 6 (authors' translation).

they are in place or not. For many commuters these regulations constitute the new reality of cross-border commuting, a reality that commuters tend to adapt to yet also a reality that they despise: "To be honest, no matter where you commute from, you should all stop working. Standing in line for hours in your free time to be allowed to work does not correspond to any definition of free time and rest time."[44]

Resilience as more openly enacted, and emotionally articulated resistance constitutes a more forceful response while dealing with the disturbance the new border regulations formed for the cross-border commuters. This form of resilience aims at openly and discursively challenging tightened border regimes with the explicit goal to remove regulations complicating the commuters' life. The resistance that we observed expresses the commuters' attempts to take their fate in their own hands. It is hard to estimate the actual effects of this type of resilience, yet it enabled commuters to, at least, gain a sense of being in control of their situation and thereby a sense of agency.

Re-bordering

> "It has always been possible for me to cross the border –also in the beginning of the pandemic– and my tests have never been checked. In the beginning of the pandemic, when I was actually controlled at the border, I just told the police that I had to go to the doctor or to get my medication in Denmark and they let me cross. Then the Danish police began letting those with Danish passport through without further notice. As a Danish citizen in possession of a Danish passport, going into Denmark has in other words not been a major issue for me, only the possible queue at the border posed/poses problems. And the German police is never there, or at least I have not experienced them being there more than twice in one and a half year, upon return to Germany. A case of 'Much ado about nothing'?"[45]

Introducing border closures in the internal Schengen context, even when it is in exceptional circumstances, there is clearly a risk of reinforcing already existing or even generating new national othering and stereotypes. In the German-Danish case, the asymmetry between the two States' way of handling the closures, indicated in the above vignette, has

---

44 *Ibid.*, post 7 (authors' translation).

45 Autoethnographic fieldnotes 2.

become a factor in this re-bordering. The fact that Danish citizens very soon experienced no problems entering Denmark, at the same time as the Danish authorities are the ones to make most trouble at the border for commuters, articulates a prioritisation of own citizens and thus also a superiority of some commuters as against others.

This can be seen as a provocation by cross-border commuters who are not Danish citizens, but have been part of the Danish labour market for years, and thus contributing to Danish society. When the requirement for non-Danish cross-border commuters to present a valid negative Covid-19 test at control posts was introduced in January 2021, one commuter shared the following outburst in Facebook: "Yes of course!! Now everything is so pointless and totally without sense... The Danes can move freely, and officials only come down hard on us. Really annoying."[46] Also, during February, when the test centre on the German side that most of the commuters used as it offered free-of charge tests for them, was again very crowded so that the commuters had to wait for several hours, another commuter wrote: "In (name of the border control point) the Danes can just cross the border ... ☹ as always, with them no one gives a shit about any order. They don't need to show anything if you want to go to Germany or to Denmark. Just great."[47]

> "Perhaps paradoxically, even I, as a Danish citizen living in Germany felt this resentment. When going to the shopping centre in Flensburg and looking at all the Danish number plates at the parking lot during the periods of the pandemic where the border was more strictly controlled when going into Denmark than the other way – it did incite some resentment of 'the Danes' from my side as well; they close off their own territory, yet, they still come here as if they cannot themselves be harmful in any way – or maybe they just do not care about people on the other side, we are worthless and hence it is no problem that we might be infected by them?"[48]

What is interesting to see is that many of the commuters in their comments targeted "the Danes", who did however nothing else than following the regulations valid for them. Yet, in the commuters' world, the cross-border mobility of Danish citizens was not or only less obstructed, and this created annoyance, anger and a sense of envy among those

46 Facebook group *Arbeiten in Daenemark*, post 8 (authors' translation).

47 *Ibid.*, post 9 (authors' translation).

48 Autoethnographic fieldnotes 3.

commuters without Danish passports. Why is it only us – the non-Danish commuters, who live in Germany and work in Denmark – who have to face difficulties and not Danish citizens living in Germany, nor those living north of the border, who just continue crossing the border almost effortlessly in order to shop south of it? The re-bordering that we observed in this sense emerged along the creation of otherness in terms of opportunities related to cross-border mobility. While these othering processes had become less and less of an explicit issue before the strengthened border securitisation, perceptions of unfairness established by the different regulations for those coming from the Danish and German side of the border region served to (re-)create a sense of a divided border region where difficulties involved in dealing with the consequences of the pandemic for cross-border mobility seem to depend on one's nationality. Borderland resilience, in this sense, takes the form of creating notions of asymmetrical difference fuelled by the different regulations imposed north and south of the border without taking the borderland's needs into account.

Conclusions

What we show in this article is how the cross-border commuters' experiences and stories reveal how and why the tightened border regime matters for them and how they at once live with, attempt to resist, and also openly resent the stress they experience. Andersen and Prokkola argue that "(p)eople's stories of their cultural goals lead us to matters that elucidate their resilience."[49] As it turns out, the German-Danish borderland is surely not lacking in resilience; rather, there are multiple versions of resilience at play. A majority seem to find pragmatic ways to continue working on the other side of the border, as they understand the pandemic and the associated difficulties of the governments' measures as a temporary infringement on their mobility and as a challenge that can be mastered. They are most inclined to make the situation bearable and manageable, and the majority is successful at doing that. Hence, they continue focusing on "just getting there", crossing the border for work their interpretation of "having a good life" under the specific circumstances. Yet, our

---

[49] D. J. Andersen & E. K. Prokkola, "Introduction: Embedding borderlands resilience", *op. cit.*, p. 10.

ethnographic investigations illustrate that people do not just conform or adapt heedlessly to the measures imposed "from above" but they find their own ways of coping with border transitions, and they do this in more than one way.

One thing which is revealed, by reflecting on pre-pandemic experiences, is that already existing forms of resilience become amplified during the time of the pandemic with the related strengthened border securitisation. Cross-border commuting always entailed making efforts to "simply just get there", that is, to be able to work on the other side of the border.[50] However, during the pandemic, commuters had to invest much more time and energy into creating and sustaining mobility, efforts that were most often unpaid and detracted from their spare time. In this sense, simply just getting there as an integral part of cross-border commuting intensified. By contrast to this everyday form of resilience that already existed prior to the pandemic, the rule bending and the resistance that we were able to observe during the pandemic, constituted more specific forms of resilience emerging in response to pandemic-related tightened border controls. They would be more exceptional, indicating how borderlands can mobilise creative agency in relation to new forms of stress.

The versions of resilience exposed also lead us to the conclusion that resilience practices may sometimes clash, on the one hand, in their relation to geopolitical decision-making, and, on the other, create conflict in the borderlands themselves because social groups cope differently with the stress of border closure. Some bordering practices clearly aim at depoliticising border controls, whereas the version of resilience exhibiting resistance aims at articulating mobilities as were they a normative, universal right of European citizens, thus constituting "open borders" less than a way of life for the commuters than as a political project. There is thus a risk of antagonisms developing between the more pragmatic forms of resilience and the confrontational, normative, political one. At the same time, the simultaneous existence of the re-bordering and the open border movement might also be called antagonistic, thereby indicating the Janus-face of the overarching border logics at stake.

Altogether, the versions of resilience in their multiplicity articulate the inherent complexities and ambiguities generated by current European borders logics where multiple possibilities are at play in the same space.

---

50 D. J. Andersen & I. Winkler, "Grænsearbejde blandt grænsependlere", *op. cit.*

The borderland resilience in the German-Danish context indicates that there are different ways of living life as borderlander, and that they depend on perceptions of the border and related consequences of closures. Hence, to understand the variety of the borderlanders' resilience it is essential to understand their situated interpretations and how they influence experiences and ways of dealing with border restrictions. In this sense, we have indeed been able to demonstrate in this article the necessity and value of understanding resilience intrinsically.

Bibliography

D. J. Andersen & E. K. Prokkola, "Introduction: Embedding borderlands resilience", *In* D. J. Andersen & E. K. Prokkola (eds.), *Borderlands Resilience: Transitions, Adaption and Resistance at Borders*, Routledge, London, 2022, pp. 1–18.

D. J. Andersen & M. Sandberg, "Introduction", *In* D. Andersen, M. Klatt & M. Sandberg (eds.), *The Border Multiple: The Practicing of Borders between Public Policy and Everyday Life in a Re-scaling Europe*, Routledge, London, 2012, pp. 1–19.

D. J. Andersen & I. Winkler, "Grænsearbejde blandt grænsependlere (Borderwork among cross-border commuters)", *Oekonomi & Politik*, vol. 93, 2020, pp. 56–68.

C. Brambilla, "Exploring the Critical Potential of the Borderscapes Concept", *Geopolitics*, vol. 20, 2015, pp. 14–34.

G. Deleuze & F. Guattari, *A Thousand Plateaus*, Trans. B. Massumi, University of Minnesota Press, Minneapolis, 1987.

H. Donnan & T. M. Wilson, *Borders: Frontiers of Identity, Nation and State*. Bloomsbury, London, 1999.

C. Ellis & A. P. Bochner, "Autoethnography, personal narrative, reflexivity: Researcher as subject", *In* N. K. Denzin & Y. S. Lincoln (eds.), *Handbook of Qualitative Research*, Sage, Thousand Oaks (CA), 2000, p. 733.

M. Humphreys, "Getting personal: Reflexivity and autoethnographic vignettes", *Qualitative Inquiry*, vol. 11, 2005, pp. 840–860.

S. Green, "Lines, traces and tidemarks", *In* O. Demetriou & R. Dimova (eds.), *The Political Materialities of Borders: New Theoretical Directions*, Manchester University Press, Manchester, 2018, pp. 67–83.

R. Hassink, "Regional resilience: A promising concept to explain differences in regional economic adaptability?", *Cambridge Journal of Regions Economy and Society*, vol. 3, n°1, 2010, pp. 45–58.

R. V. Kozinets, *Netnography, Doing Ethnographic Research Online*, Sage, London, 2010.

M. Lois, H. Cairo & M. García de las Heras, "The Politics of Resilience … politics of borders? In-mobility, insecurity and Schengen 'exceptional circumstances' in the time of COVID-19 at the Spanish-Portuguese border", *In* D. J. Andersen & E. K. Prokkola (eds.), *Borderlands Resilience: Transitions, Adaption and Resistance at Borders*, Routledge, London, 2022, pp. 54–68.

E. Opiłowska, "The Covid-19 crisis: The end of a borderless Europe?", *European Societies*, vol. 23 (sup1), 2021, pp. 589–600.

E. K. Prokkola, "Border security interventions and borderlands resilience", *In* D. J. Andersen & E. K. Prokkola (eds.), *Borderlands Resilience: Transitions, Adaption and Resistance at Borders*, Routledge, London, 2022, pp. 21–35.

P. Rogers, J. Bohland & J. Lawrence, "Resilience and values: Global perspectives on the values and worldviews underpinning the resilience concept", *Political Geography*, vol. 83, 2020, pp. 1–9.

C. Rumford, "Introduction: Citizens and borderwork in Europe", *Space and Polity*, vol. 12, 2008 pp. 1–12.

J. P. Spradley, *The Ethnographic Interview*, Wadsworth, Belmont (CA), 1979.

M. Tesfaye, "Vi ved bedst, hvornår der er brug for grænsekontrol i Danmark", *Grænsen*, vol. 3, 2021.

R. Tarvet & M. Klatt, "The impact of the Corona crisis on borderland living in the Danish-German border region with a special focus on the two national minorities", *National Identities*, 2021, pp. 35–52.

United Nations, "Resilient people, resilient planet: A future worth choosing", Secretary-General's High-level Panel on Global Sustainability, United Nations, New York, 2012; OECD, *Guidelines for Resilience Systems Analysis*, OECD, Paris, 2014.

H. van Houtum, O. Kramsch & W. Zierhofer, "Prologue b/ordering space", *In* H. van Houtum, O. Kramsch & W. Zierhofer (eds.), *B/ordering Space*. Aldershot, Ashgate, 2005, pp. 1–13.

G. Wandji, "Rethinking the time and space of resilience beyond the West: An example of the post-colonial border", *Resilience: International Policies, Practices and Discourses*, vol. 7, 2019, pp. 288–303.

C. Wille & R. Kanesu (eds.), "Bordering in pandemic times. Insights into the COVID-19 lockdown", *UniGR-CBS Thematic Issue*, Bd. 4, 2020.

Part 5

Hungarian-Romanian Cross-Border Cooperation: A Model for Central Europe?

Border Control on the Romanian-Hungarian Border in the Socialist Era

János Sallai

Hungary's borders and border control have changed several times during the 20th century. The establishment and the first precise demarcation of the Romanian-Hungarian border dates back to the second half of the 19th century. Onwards, border guarding became the centre of attention, particularly after the new borders were defined by the Treaty of Trianon, after which the institutional system of small border traffic was prevalent along the borders between the two countries. Even though during the Socialist era Hungary and Romania were allies, the permeability of the common state border was tightened, mainly because of the Romanian side. This study presents the border control practices of the Romanian-Hungarian border during the Socialist era.

The Antecedents of the Demarcation of the Romanian-Hungarian Border

The internal borders of the Monarchy[1] could be freely crossed by its citizens. At the same time, the common border between the Kingdom of Hungary and Romania was an external border for the Monarchy, as we understand it today. Therefore, border skirmishes became frequent due to the lack of clear demarcation. Interestingly, the common "external border" was negotiated and brought to an agreement with Romania by the Hungarian side, despite the joint foreign policy they agreed upon previously. This negotiation took place in 1887, followed by the Romanian-Hungarian border demarcation in 1888. Friedrich Ratzel wrote in his

1 The author refers to the Habsburg Empire (Austrian Empire, which became in 1867 the Dual Monarchy).

Political Geography that "the nature of the scientifically established border is expressed in the agreement ratified 16 April, 1888 by Austria–Hungary and Romania".[2] The finance guard inspectors were no longer able to ensure guarding the newly marked borders properly, therefore Article 1 of Chapter 1 of the common ministerial order No. 50341 of 15 July 1891, provided for the establishment of the Border Regions' Gendarmerie. "The tasks of the Border Regions' Gendarmerie included the supervision of the border line and its equipment, the prevention of unauthorised intervention and the arrest of those who committed this misdemeanour. They also had to prevent border violations and apprehend the border violators".[3] As a result of this situation, a criminal law (Act XLI of 1891) was passed on the protection of the border and land measuring markers. The first border service order was also issued with the purpose of the enforcement of the provisions of the Romanian-Hungarian border agreement. In legal history, it was after the Hungarian-Romanian border agreement of 1888, and after the Hungarian-Bukovina border agreement of 1897, that the border marks' protection through criminal law was first implemented and the institutional system of small border traffic was established. Subsequently, the common border was revisited every 5 years, with the financial expenses shared between the two countries.

The 1888 Romanian-Hungarian border was adjusted after the signing of the Treaty of Bucharest in 1918. The short-lived treaty meant an increase in territories for Hungary. The territorial gains of Hungary based on the report made for the Treaty of Trianon were as follows[4]: between the Danube and Cerna: 249,51 km^2; between upper-Cerna and the Parâng Mountains: 530,10 km^2; between the Parâng Mountains and Buceş: 1670,24 km^2; between Buceş and Nagybászkapatak: 448,65 km^2; between Lukové and the Oituz-Pass: 232,24 km^2; between the Oituz-Pass and Gyimes Pass: 304,25 km^2; between Tărhăuşi and the Békás Pass: 327,50 km^2; between Nagygerinces and Bucovina: 1347,25 km^2; and between Siret and Prut rivers: 517,62 km^2 (by which, Hungary reached its biggest size since the Treaty of Karlowitz in 1699).

---

2 Cited by J. Sallai, "Magyarország határai a trianoni döntés előtt és után", *Határrendészeti tanulmányok*, vol. 18, n°1, 2021.

3 Á. Suslik, "A honvéség szervezése a mai Kárpátalja területén 1868–1914 között", KMMI-Füzetek XVII., Káspátalja Magyar Művelődési Intézet, Ungvár, 2012.

4 The geographical names are indicated with the official names of the relevant period.

The border demarcation was carried out by four border commissions based on the order of the Armeeoberkommando ("Army Higher Command") of the Imperial and Royal Army Headquarters issued on the 15th of May 1918.[5] The border commissions were delegated in the following towns: Lugoj, with a jurisdiction between Caraş-Severin County and Hunyad County; Sibiu, with a jurisdiction between Szeben County, Fogaras County and Brassó County; Fogaras with a jurisdiction between Háromszék (today Covasna County) and Comitatul Ciuc (Ghimeş-Palanca Pass) and Háromszék with a jurisdiction between the Ghimeş-Palanca Pass and the border connection of Bucovina.

To join the commissions, the Military Assessment Command assigned professionally trained non-commissioned officers, ensigns, geodesists, who spoke well both Hungarian and German. The Minister of Finance, furthermore, assigned surveyors. In addition to the above, the Minister of Defence of the Kingdom of Hungary assigned to each committee the followings: a non-commissioned sergeant officer, a non-commissioned paymaster officer, two non-commissioned liaison officers, a non-commissioned officer, five military assistants, and a woodworker detachment. The woodworker detachment employed a non-commissioned officer, twenty men, three officer's servants, two cooks, a shoemaker and a tailor. According to the order on border demarcation: "The permanent or temporary participation of the political or administrative officials of the competent counties, as well as the delegation of gendarmerie and border police to the various border commissions, shall be decided by the Government of the Kingdom of Hungary. The Minister of Defence of the Kingdom of Hungary will participate in the work of the committees through special military representatives, as required."[6] The established border commissions were coordinated by the First General Headquarters (of Brassó) for military service. In the performance of other professional duties, they were treated in the same way as the detachments on military reconnaissance and recruiting duty. The main task of the border commissions was to urgently (due to time constraint) conduct the administration of the newly established border sections together with the delegated Romanian commissions.

---

5 "Documents of the Treaty of Bucharest", Kriegsarchiv B IX c 675-3 1, Vienna.

6 *Ibid.*

They were assigned with the following tasks: assessment of the border; prior demarcation of the border; making records of the border sections and fixed points. These tasks had to be carried out as defined in Article XIV of Law 1888, and their implementation took place county by county. And although the Bucharest Treaty was signed in May 1918, the Romanian political leadership seemed to have forgotten about it by 1919. It is known that in 1919, during the preparation of the Trianon borders, the issue of the Hungarian-Romanian border was discussed with the Romanian side, represented by Ion C. Brătianu, while the Hungarian side was not consulted at all. The validity of a secret agreement from 1916 was argued by many. "Although Clemenceau publicly distanced himself from the secret treaty, in private, he did his utmost to ensure the fulfilment of the promises the French made to Romania".[7] At the end, Romania was supported by the Italian delegation as well. Italy's interests led to the prioritisation of the Romanian claims regarding the delimitation of the Hungarian-Romanian border as per the 1916 secret agreement, practically disregarding the Treaty of Bucharest.

After the end of the First World War, the Paris Peace Conference defined the new Hungarian-Romanian border, which was demarcated between 1921 and 1923. The Council of Foreign Ministers communicated the defined borders to the Romanian side on the 11th of June 1919. The Hungarian government received Clemenceau's note regarding the borders, which was then accepted on the 16th of June. Romania, however, not only rejected the Conference's proposals on the Hungarian-Romanian border, but counter-attacked with its armed forces after the unsuccessful offensive of the Hungarian Red Army. By the 3rd of August they were already in the centre of Budapest, and they were moving further into the country. Even the Entente Powers disagreed with their actions, the Romanian leadership, however, did not follow the orders urging withdrawal. They only left Budapest later, on the 14th of November, and they pulled out of the line of the Tisza on the 25th of November 1919. They fully retreated from the Trans-Tisza region on the 31st of March 1920. This was followed by the demarcation of the Hungarian-Romanian border as proposed in the Peace Treaty. On the 21st of August 1921, the Hungarian-Romanian sub-committee was formed from the members of the Entente Commission who arrived to Budapest. They implemented

---

7 M. Ádám, "Nagyhatalmi elképzelések a trianoni határokról", *História*, vol. 1, 2002, p. 17.

the demarcation of the State borders along the 431 km and 531 m long section, based on border description and border maps. The Hungarian-Romanian border was divided into 10 border sections, each of them named after Latin capital letters[8]. The Hungarian-Romanian border demarcation process was closed with agreement, which was signed on the 27th of June 1925.

This new situation required a new solution, which led to the consolidation of the institutional system of the Romanian-Hungarian small border traffic. The conventions and agreements made with Romania on the 16th of April 1924 came into effect with the government decree No. 7.934 M.E. of 1924. The related "Resolutions on the facilitation of small border traffic", annexed to the above treaty, include the following: the territorial scope of small border traffic was limited to the municipalities within a 10 km territorial band of the border area, which could be extended "where local conditions require such extension". The agreement excluded cities from small border traffic. The border pass was issued for 3 months, after that, its validity could be extended. The agreement on the settlements and administrative units belonging to the Hungarian border districts was later declared by the Ministry of Finance in 1927.[9]

In 1940, as a result of the Second Vienna Arbitration, Northern Transylvania was reconnected to Hungary. However, the newly established Hungarian-Romanian border turned out to be short-lived, as in the autumn of 1944 the Soviet army together with Romanian and Bulgarian troops pushed the Hungarian and German forces out of Transylvania.

After the Second World War, the Paris Peace Treaty, signed in 1947, restored the common Romanian-Hungarian state border to the original status of the 1st of January 1938.

---

8 The border sections were defined as follows: Section L: 57 km and 402 m; Section K: 36 km and 680 m; Section H: 52 km and 358 m, Section G: 32 km and 0.91 m; Section F: 45 km and 327 m; Section E: 32 km and 772 m, Section D: 62 km and 423 m, Section C: 35 km and 0.25 m: Section B: 41 km and 567 m and Section A: 35 km and 886 m.

9 See 144.127/1927 Ministry of Finance provision.

Border Control of Romanian-Hungarian Borders after the Second World War

As a result of the Soviet Union's expansion, Hungary and Romania became part of the Socialist block. Despite belonging to the same economic and military alliance, both countries tightened security of their common borders and the border control was rather hostile initially.

After the Second World War, during the Socialist era, the following agreements were made between Romania and Hungary on border control, border management and border traffic control.

During the cross-border negotiations, the Hungarian party was authorised by the Hungarian Ministry of the Interior,[10] and agreed with the Romanian party (in order to ensure the performance of economic tasks) to extend the validity of the border pass of landowners – who had fields on both sides – by two years. The previous agreements were used as a basis for the customs and other procedural rules to incentivise small border traffic. The issuing of small border traffic border passes came to a halt in 1947 and then it was reintroduced in the 1960s.[11]

The guarding of the Romanian-Hungarian border was strongly affected by the different development of the political leaderships of Romania and Hungary and the Romanian political leadership's efforts to assimilate the Hungarian minority into Romania. Nothing describes better the Hungarian-Romanian political relations than the fact that the two neighbouring party leaders have barely met during their decades in power. Even one of these occasions, the meeting in Debrecen and Oradea in 1977, hardly brought about any results. Probably, this prompted the amendment of the Hungarian-Romanian small border traffic agreement. With the modification, the territorial scope of small border traffic agreement was fixed at 20 km from the state border of both States, while the temporal limit was set at 6 days of stay for 12 occasions. Although, an earlier Ministry of the Interior provision excluded Satu Mare and Oradea from the list of cities that could be visited. A separate agreement between the Romanian and Hungarian parties in 1962 regulated the"joint" border

---

10 See Ministry of the Interior in provision No.154.690/1945 IV-13.

11 See provision No 43 of 1969 of Hungarian People's Republic.

traffic control at the border crossing points of Borş[12] and Curtici.[13] As there are several rivers along the border, it was a common interest to facilitate the mobility of the water management officials across borders, which was regulated in a joint agreement in 1963.

The legislation on the Hungarian-Romanian border regime dates back to 1964 and its renewal to 1983.[14] The agreement includes the following main areas: demarcation of the state border, maintenance of the boundary line and boundary marks; joint preservation and control of boundary marks; border commissioner system; transfer of people, pets and other objects; as well as emergency crossing of state border.

The annex to the agreement outlined the cases of border violation, which are now considered obsolete. The contract in force at the time appointed on the order of the Hungarian-Romanian state border and cooperation in border matters, for the border commissioner of border section I: Colonel János Pál, his deputies: Lieutenant Colonel György Bálint and major Károly Dávid; for the border commissioner of border section II: Lieutenant Colonel József Seres, and his deputies: Lieutenant Colonel Ferenc Dóka, and Major László Szabó.

In the 1980s, the Hungarian-Romanian border relations were considered "friendly", which meant that the two countries would return illegal border crossers to each other as per a secret agreement.[15] The friendly relation was also characterised by restrictions on the use of weapons by the Hungarian border guards.

At this time, the Romanian-Hungarian state border on Hungarian territory was guarded from the border guard district headquarters in Nyírbátor and Orosháza. According to border guarding principles, within a friendly relation, small patrol units guarded large border sections. Due to the friendly relation, the border guarding security was mainly

[12] Agreement between the Governments of the Hungarian People's Republic and the Romanian People's Republic on the Establishment of a Joint Road Traffic Control Point at Borş.

[13] Agreement between the Hungarian People's Republic and the Romanian People's Republic on the Organisation of Border Services at the Joint Station of Curtici.

[14] Convention between the Government of the Hungarian People's Republic and the Government of the Romanian People's Republic on the regime of the Hungarian-Romanian state border, cooperation in border matters and mutual assistance.

[15] J. Sallai, "A változások kezdete, az erdélyi magyarok kiengedése a nyugati határon 1985-től", *Nemzetbiztonsági szemle*, vol. 7, n°3, 2019, pp. 70–80.

single-level. Only in certain main directions of border crossers – during their expected arrival time – was introduced a two/three-level security. Border patrols were usually led by two officers/guards, and patrol duty was carried out by an enlisted personnel. One patrol unit covered about 16–30-kilometre-long border sections. Every patrol unit was equipped with a truck (IFA) or an off-road vehicle (UAZ), and two-six Simson motorcycles or bicycles.

In addition to patrol equipment, the main tool of the border guards was the H-63 device with horses and dogs. It is fair to say that the Romanian-Hungarian border section was relatively poorly equipped, while the smugglers coming from the direction of Romania already used Chinese car radios with a range of 8–10 km as well as hand-held radios. Wired communications were only available on the patrol lines, while radio distribution was handled by Soviet Sokol radios.

Even in the early 1990s, the patrol officers mostly served on foot, and in many cases alone, often reinforced by a bicycle or a dog. The use of horses was gradually phased out for the patrol units, starting from the beginning of the 1990s. The fuel was delivered to the stations in barrels and the cars were refilled on site. The fuel budget, apart from visiting the district once or twice a month, and visiting the village once or twice a week, was not enough for inspections, even less for patrol control. At the stations, they had to farm, create vegetable gardens, keep pigs, which they did off-duty, in addition to the daily workload of the service.[16]

The Deterioration of the Situation on the Romanian-Hungarian Border

The impacts of domestic political and economic difficulties in Romania – such as the lack of fuel and restrictive measures – were already noticeable on the Romanian-Hungarian border at the end of the 1970s. These circumstances sometimes even led to congestion at border crossing points.[17]

---

[16] L. C. Papp, "A magyar-román államhatáron átnyúló rendészeti együttműködés korszerűsítési lehetőségeinek kutatása", Ph.D thesis at Nemzeti Közszolgálati Egyetem, Budapest, 2012.

[17] J. Sallai, "A változások kezdete, az erdélyi magyarok kiengedése a nyugati határon 1985-től", *op. cit.*

By the end of the 1980s, the deteriorated relationships between Romania and Hungary became a constant topic of discussion in the Hungarian party press and the political leadership of both countries, with some elements of the issue reflecting on the border.[18] Hungarian travellers and tourists were denied entry at the Romanian-Hungarian border by the Romanian guards without offering any explanation. In the late 1980s, even having a certain Hungarian newspaper, Népszabadság, could be a reason for refusal, if Romanian customs or border guards found it in someone's belongings.

On 27 May 1987, in the area of the Romanian-Hungarian-Yugoslav triborder, a Romanian border guard crossing into Hungarian territory chased an illegal border crosser and used his weapon on him, as a result of which the Romanian border crosser died on Hungarian territory. This served as a warning for the upcoming period of the Romanian-Hungarian border.

In 1988, before and immediately after the Romanian revolution, people caught illegally crossing the border – usually Hungarians living in Romania – were returned on the same day after being questioned where they suffered severe reprisal and physical abuse immediately after their return. The receiving officer beat them severely with a wooden stick or a rubber truncheon before putting them into the car.[19] Later, Romanian Hungarians instead of being returned were taken in. Consequently, they were no longer in hiding, instead, they went to the patrol stations on Hungarian territory and asked the local population for help.[20] Subsequently, in many cases, the Romanian intelligence services sent across their own people, who were very difficult to expose, even with Hungarian intelligence services being present during their interrogations as well.

Around November-December of 1989, the Romanian-Hungarian border hit the global headlines when Nadia Comăneci, a multiple Olympic champion Romanian gymnast, Romanian emblematic figure and favourite of the Romanian dictator, escaped from Romania to

---

[18] J. Sallai & P. Horváth, "The 'evaporation' of Romanian citizens towards Western Europe with the secret help of Hungary, between 1985 and 1989", *Studia Universitatis Babes-Bolyai Historia*, vol. 65, n°2, 2021, pp. 157–170.

[19] J. Sallai, "Magyarország határai a trianoni döntés előtt és után", *op. cit.*

[20] J. Sallai & P. Horváth, "The 'evaporation' of Romanian citizens towards Western Europe with the secret help of Hungary", *op. cit.*

Hungary at the green border around Kiszombor, accompanied by several other Romanian citizens. The story of her escape is also of imprints of the era, and her memories illustrates the situation of the Romanian-Hungarian border in the late 1980s, according to which: "The celebrated gymnast, without a single penny in her pocket and wearing a single dress, made her escape at nightfall with six of her friends. Comăneci said that her escape to the West was planned by her friend, Constantin Panait, a 34-year-old building contractor from California".[21] Constantin met Nadia Comaneci earlier, and arrived in Romania in 1989 to help with her escape and accompany her on the way to Vienna. They arrived to the Hungarian border in a crowded car. Nadia Comăneci recalls the night as follows: "We departed at midnight, trudging through the mud across the open terrain. Sometimes we stumbled, and we had to crawl and creep in icy puddles. I wore my usual outfit: trainers, canvas jacket[22] and jeans. I made the trip in the outfit you see me wearing right now.[23] I could not bring anything else with me, not even my precious gold medals or a photograph of my family. During the first three hours, we had a guide, a local man, who was a friend of Constantin and knew the area inside and out. He led us through frozen rivers and endless roads. We could carry no torches or any light otherwise the border guards could see us." Later they were caught around Kiszombor, and they were brought to the border petrol station. Then, following the usual schemed out interrogation, Nadia was accommodated at the Royal Hotel in Szeged, from where she escaped with the help of Constantin in the early morning hours of the next day. They headed for the western border, where, contrary to the location wrongly indicated in Nadia Comăneci's memories, she left Hungary at the border checkpoint in Rábafüzes and left with her "accomplice" by car for Vienna and then they travelled further to the USA.[24]

By the end of 1988, most border violations were committed from Romania towards Hungary, with more than 6700 illegal Romanian border crossers registered by the 23rd of December.[25] At this point, most of

---

21 P. Horváth & J. Sallai, "Nadia Comaneci útja a magyar határokon 1989-ben", Manuscript, 2022.

22 Author's note: probably, she wore a denim jacket.

23 Author's note: She said this in a press conference in New York, where she was wearing a denim jacket.

24 K. Halász, "Comaneci hiteles töténete?", *Határőr Magazin*, vol. 45, n°9, 1990.

25 J. Tarsoly, "Tízezer határsértő Magyarországon", *Zalai, Hírlap*, vol. 47, n°143, 1991.

them were not returned by the Hungarian government anymore. The issue of using guns against people attempting to cross the border on Romanian territory, already surfaced at the time. The Hungarian border guarding officers, however, had not detected anyone crossing with a gunshot wound.[26] Later, there were several rumours saying that people attempting to cross the border from Romania into Hungary were shot into the river. Although, the investigation conducted in Hungarian territory did not confirm these rumours.[27]

In August 1989, around Gyulavári the Romanian border guards pursued someone even across the state border into Hungarian territory and arrested him there, which was a serious border incident at the time. In September 1989, nearly thirty Polish tourists were assaulted at a Romanian-Hungarian border crossing point, and many of them had to be taken to hospital in Budapest.[28]

A significant proportion of mainly Hungarian and German[29] citizens living in Romania considered migration to their country of origin, resettlement or further travel to the West as the primarily solution. The Hungarians living in Transylvania felt oppressed by the discrimination against them and Ceausescu's village demolishing policies.

The severity of the situation is well indicated by the fact that the situation of the Hungarian minority in Romania and Transylvania formed part of the agenda even of the supreme political body of the North Atlantic Treaty Organization (NATO) (then in opposite alliance). According to which, "Hungarians mostly resent the overrepresentation of Romanian culture in education, not only because it is feared that it will corrupt the chances of Hungarians being employed in positions which require high qualifications, but also because it gradually jeopardises the existence of the Hungarian culture in Transylvania. President Ceauşescu's minority policy is a result of Romanian nationalism, which is the essence of both his foreign policy and his domestic policy endeavours of creating a strong and centralised nation state".[30]

---

26 K. Halász, "Nem létező sírok", *Határőr Magazin*, vol. 44, n°4, 1989.

27 *Ibid.*

28 *Ibid.*

29 The Transylvanian Germans were transferred to West Germany by the Romanians in exchange for a bounty.

30 B. Révész, "A Magyar-román viszony problématörténetéből az 1980-as években", *Acta Universitatis Szegediensis*, vol. 72, n°1–22, 2009, pp. 463–522.

Hungarian-Romanian border relations were already tarnished by the fact that an extradition treaty was in force between the two countries, according to which Romanian citizens who illegally crossed to Hungary, had to be returned to the Romanian border patrol. These Romanian citizens were then liable to imprisonment or corporal punishment.

During this period, legal border traffic soared: compared to the 60s and 70s, the number of people crossing the border legally at the Romanian-Hungarian border increased by several millions in the 1980s.

The number of Hungarians living in Romania crossing the (green) border also increased dynamically: while in the early 1980s it was around 2–300, in 1987 it was close to 1000, and in 1988 it exceeded 7000, rising to 30,000 by the end of 1989. The latter entered Hungary legally, but they tried to travel through Austria illegally towards the West.

Simultaneously with the above events, a transit route for Romanian citizens in Hungary was formed, induced by the domestic political situation in Romania and as a result of which: "Hungarian tourists were under a much severe control. The increasing severity was already evident in 1984, when between April and the 31st of December, 2790 Hungarian citizens were denied entry at the border because they were carrying unauthorised press products, books, etc. During the first five months of 1985, this number surged up to 1500".[31] Furthermore, the number of people resettling to Hungary continuously kept increasing.

A good example of the above is the circular issued by the National Commander of Border Guarding of the Ministry of the Interior, sent out in the spirit of change. It was qualified as "Top secret, especially important" and outlined a special procedure for Romanian citizens starting from the 15th of July 1985.[32]

Unlike the rules in force for foreigners, the circular issued states that those "certain Romanian" (as per verbal clarification: Hungarian) citizens who arrived to Hungary with a valid passport and intended to continue their journey through the Ferihegy border checkpoint or the West-South border checkpoints, but did not have a valid exit permit or visa for Yugoslavia, Austria or another Western State, were checked according

---

[31] G. Földes, *Magyarország, Románia és a nemzeti kérdés 1956–1989*, Napvilág Kiadó, Budapest, 2007.

[32] National Commander of Border Guarding of the Ministry of the Interior No. 0001/1985.

to the protocol in force. After recording their data, they had to be exited by the passport control officer.[33] In order to manage the process and avoid disruptions, the national leadership, in addition to the top leaders of the affected districts, involved the heads of the border checkpoints and the reconnaissance leaders. The passport control officers were only in possession of the fundamental information necessary for enforcement.

The personal data from their passport and the passport number of the exited Romanian citizens were registered. After recording the data, the passport control officer put the exit stamp on the sheet containing the recorded data and then forwarded it to the border checkpoint officer in charge. These data sheets were then collected by an officer on duty, and later forwarded in a sealed envelope to the head of the district's intelligence unit. The data was transferred from the districts to the Head of Border Guard Surveillance Department of the Ministry of the Interior, who collected and stored them in a separate folder.

Technological Changes in Border Control on the Two Sides of the Romanian-Hungarian Border

In the 1980s, the direction of the legal or illegal immigrants were clearly moving from Romania towards the Hungarian border, including the green border and the designated border crossing points. At the time, border guarding system on Romanian territory were mainly limited to the main unsealed roads leading to the state border. They mostly patrolled at night, with 1–2 patrol officers a day. Usually there was one high observation post maximum 4–500 m from these main roads in the area of each guard, which was used by a patrol officer during the day.[34]

In order to put a stop to this process, Romanian border guarding authorities tightened up border control at the lowlands and on the roads leading to the border. They strictly cracked down on illegal border crossers, often even physically abusing them. The Romanian border guarding authorities blocked access to the border with barriers, wire fencing, water

---

[33] Notably, in many cases, those who had been exited were not let in by the Austrian authorities and were sent back. They were then allegedly directed towards the lowlands by the Hungarian authorities, from where they usually attempted to cross the green border to the West, for example, North of Hegyeshalom or South along the border sections, which were then guarded by an electronic signalling system.

[34] Interview with Border Guard Lieutenant Colonel Dr. Csaba Pap, 6 May 2021.

and movement barriers, and even by deepening the beds of border rivers. Moreover, they also used high-powered binoculars and night vision equipment on the green border. The people doing agricultural work along the border were controlled by patrol officers with dogs, in order to prevent them from escaping across the border.

Some desperate Romanian citizens – even people of Romanian nationality in an increasing number – tried everything to get across the border. They came by cars, horse-drawn carriages, tractors, inflatable boats, trucks, on foot... in snow, in frost, crawling across icy rivers.[35]

The former Director of the Border Guard Directorate in Orosháza (Brigadier-General Károly Dávid) recalls this period as follows: "Previously, here, in the East, border guarding was performed according to a certain "friendly relation", which meant the lack of force and technological background. Our job was basically to guard the borders with a small staff, collaborating with the neighbouring countries. At least that was the official version. Each border patrol station had a staff of up-to 20–25 people, so essentially this was only symbolic border guarding. Roughly 60–70 % of the force and technological equipment was focused on the Western side, which was a result of the system. And then suddenly, the focus changed, as the number of refugees increased tenfold (from 330 to 3000). Neither did we have the necessary force, nor the technical equipment on this border section. Before, the number of people moving across the border was around a few hundred annually. We also had to acknowledge the fact, that these people are not criminals, only immigrants, therefore, we had to change our previous approach as well".[36]

Conclusion

Due to the complicated geographical and geopolitical situation, the Romanian-Hungarian border has always been the focus of political attention over the last two hundred years. The occasional political changeovers, world wars and the border removals had a great impact on the citizens of the two countries, especially in the 20th century, when hundreds of thousands of Hungarians were forced to live under the Romanian rule.

---

35 *Ibid.*

36 K. Halász, "Comaneci hiteles töténete?", *op. cit.*

Intriguingly, the two neighbouring countries were in the same political and military alliance for decades (and during the 2000s both became members of NATO and the EU again), but their citizens expressed their opinion about their political leaders by migrating. The situation was particularly complicated for the vulnerable Hungarian minority living in Romania during the period of the Romanian Socialist dictator, when they felt liberated after crossing the Romanian-Hungarian border.

What happens at the state border is actually a reflection of the public and political sentiment of the country: with the rise of the Romanian dictatorship, the number of people of both Romanian and Hungarian nationality crossing the Romanian-Hungarian border and leaving the country increased immediately.

Bibliography

M. Ádám, "Nagyhatalmi elképzelések a trianoni határokról", *História*, vol. 1, 2002.

"Documents of the Treaty of Bucharest", Kriegsarchiv B IX c 675-3 1, Vienna.

G. Földes, *Magyarország, Románia és a nemzeti kérdés 1956–1989*, Napvilág Kiadó, Budapest, 2007.

K. Halász, "Nem létező sírok", *Határőr Magazin*, vol. 44, n°4, 1989.

K. Halász, "Comaneci hiteles töténete?", *Határőr Magazin*, vol. 45, n°9, 1990.

P. Horváth & J. Sallai, "Nadia Comaneci útja a magyar határokon 1989-ben", Manuscript, 2022.

L. C. Papp, "A magyar-román államhatáron átnyúló rendészeti együttműködés korszerűsítési lehetőségeinek kutatása", Ph.D thesis at Nemzeti Közszolgálati Egyetem, Budapest, 2012.

B. Révész, "A Magyar-román viszony problématörténetéből az 1980-as években", *Acta Universitatis Szegediensis*, vol. 72, n°1-22, 2009, pp. 463–522.

J. Sallai, "A változások kezdete, az erdélyi magyarok kiengedése a nyugati határon 1985-től", *Nemzetbiztonsági szemle*, vol. 7, n°3, 2019, pp. 70–80.

J. Sallai, "Magyarország határai a trianoni döntés előtt és után", *Határrendészeti tanulmányok*, vol. 18, n°1, 2021.

J. Sallai & P. Horváth, "The 'evaporation' of Romanian citizens towards Western Europe with the secret help of Hungary, between 1985 and

1989", *Studia Universitatis Babes-Bolyai Historia*, vol. 65, n°2, 2021, pp. 157–170.

Á. Suslik, "A honvéség szervezése a mai Kárpátalja területén 1868–1914 között", KMMI-Füzetek XVII., Káspátalja Magyar Művelődési Intézet, Ungvár, 2012.

J. Tarsoly, "Tízezer határsértő Magyarországon", *Zalai, Hírlap*, vol. 47, n°143, 1991.

Changing Borders, Shifting Paradigms. The Romanian-Hungarian Border

Nicolae Păun & Enya-Andrea Tămaș

European integration has had a major impact on borders and border regions in Europe. Once regarded as the "scars of history", the borders of an integrating Europe are borders that disappear,[1] while border regions have become a fertile ground for territorial and institutional cooperation.[2] From people-to-people projects, through infrastructure investments to supporting institutional cooperation initiatives, the European Union (EU) has made a genuine difference in border regions and has contributed to their transformation. Over the last 70 years, cross-border regions have flourished across Europe, as the EU's integration policies have participated to peaceful borderlands and fuelled economic integration across all Member States.

Since the creation of the European Communities, several radical changes and paradigm shifts have taken place regarding the nature, content and forms of cross-border relations in Europe.[3] By adopting a timeline approach and a descriptive method, the present article has the aim of uncovering these paradigm shifts both in the wider, European context, and a more localised one, at the Romanian-Hungarian border. Although, for a long time, this border section has not been a part of the European Union, the changes in the EU narrative regarding borders and border

1 C. Yndigegn, "Between debordering and rebordering Europe: Cross-border cooperation in the Öresund region of the Danish-Swedish Border region", *Eurasia Border Review*, vol. 2, n°1, 2011, p. 50.

2 L. de Sousa, "Understanding European cross-border cooperation: A framework for analysis", *Journal of European Integration*, vol. 35, n°6, 2012, p. 1.

3 B. Baranyi, "Adalékok a határ mentiség újraértelmezéséhez Magyarországon", *Deturope: Central European Journal of Tourism and Regional Development*, vol. 6, n°2, 2014.

regions and the development of border specific policies and instruments have had a major impact on this particular region too. It is somewhat fortunate that the intensification of the EU's focus on border regions in the late 1980s and early 1990s coincided with the political, economic and social changes that took place in the Central and Eastern European countries, circumstances which enabled these countries to "hop on the train". Nonetheless, border regions from Western Europe enjoy a longstanding history of more or less informal cooperation, which undoubtedly constitutes a significant base for the current dynamics of cooperation and integration.

In the following, the paper adopts a chronological perspective, going through the various shifts in narrative and changes of paradigm regarding border and border regions, first exploring them at the European level, then continuing from the point of view of the Romanian-Hungarian border. The authors of the paper consider that these contributions are a welcome contribution to the academic literature, having in mind the lack of (recent) studies focusing on this specific border region.

From "Barriers" to "Bridges"

Borders and border regions have always been regarded as central to the development of the European Communities, but for a long time they have been viewed only in terms of political, military and historical dimensions.[4] Although the first informal cooperation initiatives started in the late 1950s in Western Europe, cross-border cooperation was largely marginal and unregulated until the 1980s.[5] In the early decades, it was the Council of Europe rather than the European Community that started to and helped promote an alternative image of borders as bridges and border regions as illustrative laboratories for the problems and potentialities of European integration.[6] Due in great part to the efforts of the Council

---

4 L. O'Dowd, "The changing significance of European borders", *Regional & Federal Studies*, vol. 12, n°4, 2002, p. 17.

5 E. Evrard & A. Engl, "Taking stock of the European Grouping of Territorial Cooperation (EGTC): From policy formulation to policy implementation", *In* E. Medeiros (ed.), *European Territorial Cooperation. Theoretical and Empirical Approaches to the Process and Impacts of Cross-Border and Transnational Cooperation in Europe,* Springer, Dordrecht, 2018, p. 211.

6 Liam O'Dowd, "The changing significance of European Borders", *op.cit.*, p. 17.

and local and regional advocates, the debate surrounding the importance of cross-border cooperation, and the political support thereof, intensified in the 1980s and by the end of the decade succeeded in convincing the European Commission of the necessity of defining an explicit European border regions' policy with commensurate financial resources.[7]

From Borders as "Barriers to Political Union" to "Barriers to Economic Integration"

The shift was also supported by the events brought about by the new decade, namely the proliferation of borders with the break-up of the Soviet bloc and the completion of the Single Market. Thus, borders became important both from the point of view of enlargement and the completion of the internal market, and their transformation into bridges became more central to EU policy.[8]

From the setting up of the Single Market, the EU thought of borders not so much as barriers to political union, but as barriers to the completion of an integrated economic space and market.[9] This provided the opportunity for the European Commission to develop new initiatives for border regions. With the launch of the Interreg programmes in 1990, now European Territorial Cooperation, a specific focus on border regions was established at the level of the European Union.

The creation of the Schengen area also represents a historic milestone in the creation of a "borderless" Europe. By eliminating controls at internal borders, the Schengen system managed to tear down barriers and bring previously separated populations closer, thus strengthening European economies by helping drive competitiveness and generate jobs and growth.[10]

7 A. L. Amilhat, V. Kolossov, I. Liikanen, D. Newman, P. Joenniemi, N. Yuval-Davis, S. Rosière & J. Scott, *EUBORDERSCAPES: State of the Debate Report 1*, 2012, p. 45. Retrieved from: <http://www.euborderscapes.eu/index.php?id=project_reports>. Accessed 23 May 2023.

8 Liam O'Dowd, "The changing significance of European Borders", *op.cit.*, p. 19.

9 *Ibid.*, p. 21.

10 European Commission, "State of Schengen Report 2022", Communication from the Commission to the European Parliament, the European Council, the Council, the European Economic and Social Committee and the Committee of the Regions,

From a Project Based Perspective to an Institutionalised One

Since 1990, Interreg funding has supported cross-border cooperation programmes along EU border regions and has financed thousands of projects and initiatives that have helped improve European integration.[11] The fact that Interreg was elevated into one of the major goals of EU Cohesion Policy in 2007 is in itself a supporting proof that cross-border cooperation is central to the very existence of the EU, and 32 years later it is still one of the EU's key instruments supporting cooperation across borders. Evaluation studies[12] repeatedly confirmed the positive results of Interreg interventions in furthering territorial development and deepening territorial co-operation.

However, the original paradigm of cross-border cooperation and integration through programmes and projects continues to operate today in various adapted forms, the EU offering an enriched toolbox for cross-border interaction, ranging from financial, to regulatory, to legislative instruments. In 2006, the EU created a new instrument that marked a shift in its approach to cross-border cooperation: the European Groupings of Territorial Cooperation (EGTC). This new instrument allows public entities of different Member States to come together under a new entity with full legal personality, thus being the first European cooperation structure with a legal personality defined by the European Law.[13] According to Gyula Ocskay, they were originally envisaged as a

---

COM (2022) 301 final/2, 2022, p. 1. Retrieved from: <https://ec.europa.eu/home-affairs/schengen-package_en>. Accessed 21 May 2023.

11 European Commission, "Boosting growth and cohesion in EU border regions", Communication from the Commission to the Council and the European Parliament, COM(2017)534. 2017, p. 2. Retrieved from: <https://ec.europa.eu/regional_policy/en/information/publications/communications/2017/boosting-growth-and-cohesion-in-eu-border-regions>. Accessed 21 May 2023.

12 European Union, "INTERREG III Community Initiative (2000–2006) Ex-Post Evaluation", 2010. Retrieved from: <https://files.nettsteder.regjeringen.no/wpuploads01/blogs.dir/473/files/2015/08/2010-06-23_final_report1.pdf>. Accessed 21 May 2023; European Commission, "Ex post evaluation of Cohesion Policy programmes 2007–2013, focusing on the European Regional Development Fund (ERDF) and the Cohesion Fund (CF)", 2016, Retrieved from: <https://op.europa.eu/en/publication-detail/-/publication/e9182ca7-7a40-11e6-b076-01aa75ed71a1/language-en>. Accessed 2 May 2024.

13 See the description of EGTCs on the portal of the Committee of the Regions: <https://portal.cor.europa.eu/egtc/Pages/welcome.aspx>. Accessed 21 May 2023.

tool to manage cross-border programmes and projects. 16 years later, however, it is obvious that the tool did not meet this original expectation, rather, it is generally used for the implementation of bottom-up, territorial projects.[14] Nonetheless, the creation of the EGTC instrument is being regarded as one of the most important milestones in terms of the European Territorial Cooperation, which has dramatically improved the general climate and legal environment for cross-border cooperation.

The popularity of the EGTC instrument varies across the European Union and so does the organisations' success. Since the creation of the instrument, a number of groupings have been set up, with 79 registered groupings in May 2022, according to data collected in the Committee of the Regions' register.[15] Many authors emphasise the uniqueness and suitability of the EGTC as a tool capable of tackling the legal obstacles encountered during cross-border cooperation arising from differing national laws and procedures. At present, the EGTC is regarded to be the basic legal solution for cross-border cooperation.[16]

The EU's Cohesion Policy and its deepening integration efforts have prompted several major changes – paradigm shifts – in the perception, nature and functioning of borders. European borders have undergone major transformations, from being markers of demarcation to becoming bridges, interfaces, with the adjacent regions becoming the *locus* of opportunities for the population on both sides of the border. However, as Anna Moraczewska highlights, the process of the functional transition of borders is not linear but circular, past functions of a border can be restored.[17] Until recently, it seemed that the era of closed borders would come to an end and the obstacles hindering integration would be eliminated. However, the COVID-19 pandemic has demonstrated how fragile European borders can be, and how quickly decades of good neighbourly

---

[14] G. Ocskay, "Changing interpretation of the EGTC tool", *In* G. Ocskay (ed.), *15 Years of the EGTCs. Lessons Learnt and Future Perspectives*, Central European Service for Cross-border Initiatives (CESCI), Budapest, 2020, pp. 37–54.

[15] European Committee of the Regions, "List of European Groupings of Territorial Cooperation", 2022. Retrieved from: <https://portal.cor.europa.eu/egtc/CoRActivities/Documents/Official_List_of_the_EGTCs.pdf?Web=0>. Accessed 21 May 2023.

[16] M. Krzymuski, "The EGTC as a legal solution of institutionalisation of cross-border cooperation", *In* G. Ocskay (ed.), *15 Years of the EGTCs, op.cit.*, p. 8.

[17] A. Moraczewska, "The Changing Interpretation of Border Functions in International Relations", *Revista Română de Geografie Politică*, vol. 12, n°2, 2010, p. 332.

relations and constructive cross-border cooperation can be disrupted by sudden and uncoordinated closure of the borders.

The negative impact of these measures quickly became visible in many border regions.[18] In a public consultation carried out by the Commission in 2020, 65 % of respondents stated that border closures increased their perception of the border as an obstacle.[19] Based on the lessons learnt, the need for more and deeper institutional cross-border cooperation is further evidenced.

Paradigm Shifts at the Romanian-Hungarian Border

Just as in the wider Europe, over the course of the last decades, the perception, the form and the functions of the Romanian-Hungarian border have shifted. The 448-kilometre-long Hungarian-Romanian border region has been for a long time separating the cross-border region, but with the end of socialism, the rigidity of the border eased and since it has become the site of extensive cross-border cooperation.

The Hungarian-Romanian border is an internal border in the context of the EU. In this relation, the traditional role of the state border is partially eliminated, but the region is still not considered to be part of the "Europe without borders", as Romania does not belong to the Schengen Area. This transitional border regime confers a special status to the border region. In the following synopsis we continue to apply a timeline approach to understand how the border, and the narrative about it, has been shifting in the last 30 years in this specific area as a reaction to certain (geo)political, economic and social changes.

Separation

Most European borders are unnatural, political constructions, the majority of them being drawn as a consequence of the two world wars in

---

18 J. Peyrony, J. Rubio & R. Viaggi, The effects of COVID-19 induced border closures on cross-border regions: An empirical report covering the period March to 2020, European Union Publications Office, Luxemburg, 2021.

19 A. Pinto, R. Ferreira & N. Verschelde, *Public consultation on overcoming cross-border obstacles – 2020*, Summary Report, Publications Office of the European Union, Luxemburg, 2021.

the 20th century.[20] The Romanian-Hungarian border is no exception, it was changed several times in the last two centuries, particularly in 1888, 1920, 1940 and 1947.[21] Its present form was established by the Peace Treaty of Trianon signed on the 4th of June 1920, at the meeting point of the historical area of Hungary and Transylvania.[22]

In spite that Hungary and Romania belonged to the same political power block, the "friendship and brotherhood" declared in the socialist era only existed in the slogans, in reality the relationship between the two countries was rather tense.[23] During the socialist era the Hungarian – Romanian border became a closed border, its separating function being dominant. The crossing of the borders and the establishment of any cross-border relations was completely blocked. The constant hostility between the two countries and the closure of borders made the establishment of any partnerships almost impossible.[24] Still, the socialist regimes allowed and supported twin settlement relations, mainly to strengthen the role of the socialist ideology. However, these relations were highly centralised, controlled and limited in activities as they could never go beyond the cultural and protocol levels.[25]

With regard to cross-border mobility, during this era the barrier function of the border reduced cross-border traffic to an absolute minimum. Local border traffic represented the only exception, which, after the easing of the political climate in the 1970s, enabled border citizens living in

---

20 C. Yndigegn, "Between debordering and rebordering Europe", *op.cit.*, p. 48.

21 J. Sallai, "A változások kezdete, az erdélyi magyarok kiengedése a nyugati határon 1985-től", *Nemzetbiztonsági Szemle*, vol. 7, n°3, 2019, p. 70.

22 B. Baranyi, "A magyar-román interregionális kapcsolatok sajátosságai", *Pro Minoritate.* 2007, p. 99. Retrieved from: <https://prominoritate.hu/wp-content/uploads/2019/05/ProMino07-2-06-Baranyi-Magyar-Roman.pdf>. Accessed 21 May 2023.

23 M. Benczi & G. Ocskay, "The evolution of cross-border cooperation in Hungary", *Észak-Magyarországi Stratégiai Füzetek*, vol. 18, n°2, 2021, p. 55; I. Balcsók, L. Dancs, & G. Koncz, *The Development of Hungarian-Romanian Cross-border Connections After the Change of Regime*, European Regional Science Association, ERSA conference papers, 2005, pp. 2–4. Retrieved from: <https://ideas.repec.org/p/wiw/wiwrsa/ersa05p301.html>. Accessed 21 May 2023.

24 Z. Bacsi & E. Kovács, *Határrégiók fejlődésének sajátosságai*, Hévíz Többcélú Kistérségi Társulás & Nyugat-Balatoni Társadalomtudományi Kutatóműhely, Keszthely, 2007, p. 199; B. Baranyi, "A magyar-román interregionális kapcsolatok sajátosságai", *op. cit.*

25 K. Czimre, "Recovery or discovery? Models and motives of cross-border co-operation along the Eastern Border of Hungary after 1989–1990", *Eurolimes*, vol. 26, n°26, 2018, p. 104.

a distance of 20 km from the borderline to cross the border with a special traveller's certificate valid for half a year with restricted justification of travelling.[26] This system of local border traffic remained significant even after the regime change, until the accession of the two countries to the EU.[27]

Transition and Consolidation

The political transformations of Hungary and Romania at the end of the 1980s restored a more favourable framework for bilateral relations. Following a transition period characterised by rather tense relations between the two countries, the second half of the 1990s brought about a period of historical reconciliation.[28] In terms of the border, an immediate change started to take place as new border crossing stations were established (from 7 in 1990 to 22 in 2021),[29] the visa regime gradually ceased to exist, and in fact, bilateral border crossing became possible.[30] Soon, different bottom-up informal forms of cross-border cooperation have started mushrooming along the border.[31] However, the cooperation that quickly started at lower levels was seriously hindered by the reviving nationalism on both parts, which brought to the surface previously suppressed problems.[32]

Given both countries European and Euro-Atlantic aspirations, and their recognition of the necessity of establishing bilateral relations and

---

26 M. Benczi & G. Ocskay, "The evolution of cross-border cooperation in Hungary", *op.cit.*, p. 55.

27 J. Sallai, "A változások kezdete, az erdélyi magyarok kiengedése a nyugati határon 1985-től", *op.cit.*, p. 71.

28 M. R. Vădean, *Relaţiile româno-ungare în contextul integrării în structurile europene şi euroatlantice*, Editura Ştiinţifică Lumen, Iaşi, 2011, pp. 218–225.

29 M. Benczi & G. Ocskay, "The evolution of cross-border cooperation in Hungary", *op.cit.*, p. 56.

30 I. Balcsók, L. Dancs & G. Koncz, *The Development of Hungarian-Romanian Cross-border Connections After the Change of Regime, op.cit.*, p. 18.

31 M. Benczi & G. Ocskay, "The evolution of cross-border cooperation in Hungary", *op.cit.*, p. 54.

32 I. Balcsók, L. Dancs, G. & Koncz, *The Development of Hungarian-Romanian Cross-border Connections After the Change of Regime, op.cit.*, p. 18; M. R. Vădean, *Relaţiile româno-ungare în contextul integrării în structurile europene şi euroatlantice, op.cit.*, pp. 227–229.

deepening cooperation, the untrusting and often tense relations gradually eased.[33] Melinda Benczi and Gyula Ocskay have identified three waves in the development and intensification of cross-border cooperation in Hungary,[34] which are echoed also in its relation to Romania. To this, the authors of the present paper add one more at the end.

The first wave of cross-border cooperation was dominated by the state level when the two governments signed bilateral treaties on good neighbourliness and cooperation with each other. The signing of the Treaty between the Republic of Hungary and Romania on Understanding, Cooperation and Good Neighbourliness addressed some of the core causes of decades of mistrust and suspicions between both governments, through the mutual recognition of the present border (Art. 4) and the rights of ethnic minorities in both States (Art. 15)[35]. The Treaty placed emphasis on the joint orientation of the two countries to the goal of European integration (Art. 7), and identified several areas of mutual cooperation without setting out to achieve concrete objectives, thus remaining a framework document at best.[36] Still, it paved the way for the creation of a modern, solid and binding framework for the Romanian-Hungarian relationship, which generated positive developments both for the cooperation between Romania and Hungary, and for regional stability, and 26 years later it remains a fundamental benchmark for the development of relations between the two countries.

---

[33] I. Balcsók, L. Dancs, G. & Koncz, *The Development of Hungarian-Romanian Cross-border Connections After the Change of Regime, op.cit.*, p. 2; M. R. Vădean, *Relațiile româno-ungare în contextul integrării în structurile europene și euroatlantice, op.cit.*, pp. 230–235.

[34] M. Benczi & G. Ocskay, "The evolution of cross-border cooperation in Hungary", *op.cit.*, pp. 59–60.

[35] I. Balcsók, L. Dancs, G. & Koncz, *The Development of Hungarian-Romanian Cross-border Connections After the Change of Regime, op.cit.*, p. 4. Hungary and Romania. *Treaty of understanding, cooperation and good neighbourliness*, Signed at Timisoara on the 16th of September 1996. Retrieved from: <https://treaties.un.org/doc/Publication/UNTS/Volume%201966/volume-1966-I-33604-English.pdf?fbclid=IwAR1OSb-Udn1ioQWcD6Soj7YWaAc2sc9_rMaNVYn02JimLjhkaSCCrGOYyq4>. Accessed 21 May 2023.

[36] A. Corpădean, "The Romanian-Hungarian Strategic Partnership 2002–2016", *In* N. Păun & S. Schirmann (eds.), *Borders, Identities, Communities. The Road to Reconciliation and Partnership in Central and Eastern Europe*, Nomos, Baden-Baden, 2016, p. 163.

The next major step in the normalisation of ties between Romania and Hungary took place in 2002 with the signing of the Declaration on the Strategic Partnership between the Governments of Romania and the Republic of Hungary, for 21st century Europe. According to Adrian Corpădean, geographical proximity, past rivalry and social arguments constitute the three main reasons justifying the need of the strategic partnership, and one cannot help but observe the similarity between the context of signing this and the previous document. What was unique to this partnership is that it both gave rise to the tool of common government meetings and set in place 8 Councils dealing with separate matters. During its 20 years of existence, there have been only four intergovernmental meetings between the two governments, which took place in 2005 in Bucharest, in 2006 in Budapest, in 2007 in Sibiu and in 2008 in Szeged. Although it seems that since 2008 this practice has been formally suspended, the meetings had some modest results in terms of infrastructure development, agriculture, culture, tourism and education.[37]

Seeing the clear improvement of inter-state relations and having in mind that subnational cross-border relations are strongly influenced by the relationship between the neighbouring countries, it is not surprising that the second wave witnessed the proliferation of town twinnings. Established generally between neighbouring localities along the border, these were determined by a shared geographical and/or economic position[38] and even by the existence of previous socialist-era twin relations.[39] Undoubtedly, the formation of these relations was also heavily motivated by the common ethnic element. Today, approximately 144 twinnings exist within the cross-border region, most of them focusing on cultural and sports activities and exchanges.[40]

37 *Ibid.*, pp. 157–172; Ministerul Afacerilor Externe, *Aniversarea a 20 ani de la semnarea Tratatului de Înțelegere, Cooperare și Bună Vecinătate între România și Ungaria*, Press Release, 16 September 2016. Retrieved from: <https://www.mae.ro/node/38399>. Accessed 21 May 2023.

38 I. Süli-Zakar & T. Kecskés, "The historical borders and the cross-border connections' effect on Debrecen's spirituality and cultural economy", *Eurolimes*, vol. 19, 2015, p. 110.

39 K. Czimre, "Recovery or discovery?", *op.cit.*, p. 105.

40 Interreg VI-A Romania-Hungary Programme, 2021, p. 13. Retrieved from: <https://interreg-rohu.eu/wp-content/uploads/2022/03/Interreg-VI-A-Romania-Hungary-Programme_March_2022_EN_f.pdf>. Accessed 21 May 2023.

Besides twinning relations, various actors in the border regions also started to establish more formal and structured forms of cross-border cooperation. Thus, in the third wave, euroregional organisations emerged along the Romanian-Hungarian border, following Western European patterns. Romania and Hungary are involved in several euroregions, the established three cross-border cooperation initiatives showing distinct particularities.[41]

The Carpathian Euroregion was established in 1993 with the participation of Hungary, Slovakia, Poland, Ukraine and Romania, being the first of its kind to involve only former socialist countries.[42] Its creation, along the geographical backbone of the Carpathian mountain range,[43] can be seen as the materialisation of the previously existing bottom-up local and regional bilateral relations in the region.[44] The desire for cooperation was also fuelled by a number of common historical features, the similarities in economic development and an aspiration for progress and European integration.[45] By building and fostering relations between its members, the Euroregion was meant to be preventing regional ethnic conflicts,[46] as well as creating a structure that would facilitate the implementation of pre-accession EU programs.[47]

Initially, the Euroregion was created by only four countries, the admission of the Romanian counties happened under contradictory circumstances a few years later. Initially, the application of the counties of Satu Mare and Maramureş was accepted in 1993, but the Romanian

41 R. Săgeată, *Euroregiunile de cooperare transfrontalieră din Bazinul inferior al Dunării. Studiu geografic*, Editura Academiei Române, Bucharest, 2015, p. 130.

42 I. Süli-Zakar, *Közös Múltunk-Közös Jövőnk, Együtt Európáért: 20 Éves a Kárpátok Eurorégió. Emlékek, Sikerek, Megújulás*, DIDAKT Kft, Debrecen, 2014, p. 365.

43 M. Smętkowski & J. Rok, "The Carpathian Euroregion from the perspective of economic cooperation in peripheral regions", MPRA Paper, n°82763, 2020, p. 6. Retrieved from: <https://mpra.ub.uni-muenchen.de/82763/1/MPRA_paper_82763.pdf>. Accessed 21 May 2023.

44 H. Tanaka, "Carpathian Euroregion and Cross-border governance", *The Journal of Comparative Economic Studies*, vol. 2, 2006, p. 64.

45 R. Săgeată, *Euroregiunile de cooperare transfrontalieră din Bazinul inferior al Dunării. Studiu geographic, op.cit.*, p. 117.

46 H. Tanaka, "Carpathian Euroregion and Cross-border governance", *op.cit.*, p. 64.

47 M. Perkmann, "Cross-border regions in Europe significance and drivers of regional cross-border co-operation", *European Urban and Regional Studies*, vol. 10, n°2, 2003, p. 166.

government declared this step annulled, forcing these counties to return to an observer status until 1997.[48]

Almost 30 years later, it has become clear what the benefits of its creation and what the obstacles to its functioning have been. In the first years of its existence, the Euroregion seemed to be successful in decreasing the distrust between nations and ethnic groups in the area,[49] thus contributing to the stabilisation of the region. Since then, the activity of the Carpathian Euroregion has become diverse and its cross-border social-cultural-economic role was strengthened, achieving significant results in terms of softening its internal borders through the opening of new border stations and the modernisation of the existing ones, through the evolvement of the economic relations within, but also through intensifying educational and cultural cooperations.[50]

However, the cooperation did not fulfil its expected mission despite the above listed results. The primary reason for this is caused by the financial problems the Euroregion has faced since its inception,[51] further worsened by its lack of clearly defined institutional structures.[52] The excessive size of the Carpathian Euroregion is being regarded as one of the main reasons for its failure,[53] the efficiency of the cooperation being further hindered by the varying levels of development and economic conditions of the participating regions.[54] The situation of the Euroregion is further complicated by the different status of the member countries with regard to their EU membership.

The Danube-Kriş-Mureş-Tisa (DKMT) Euroregion was established in 1997 on the Hungarian-Romanian-Serbian triplex border by the signing of the Regional Cooperation Protocol Danube-Mures-Tisa. The

48 I. Süli-Zakar, *Közös Múltunk-Közös Jövőnk, Együtt Európáért, op.cit.*, p. 372.

49 Z. Ludvig & I. Süli-Zakar, *A Kárpátok Eurorégió Együttmüködés Mérlege*, Oktatási Minisztérium, Budapest, 2002, p. 109.

50 I. Süli-Zakar, *Közös Múltunk-Közös Jövőnk, Együtt Európáért, op.cit.*, pp. 365–366.

51 *Ibid.*, p. 367.

52 M. Smętkowski & J. Rok, "The Carpathian Euroregion from the perspective of economic cooperation in peripheral regions", *op.cit.*, p. 6.

53 A. Tamaskó, *A DKMT Eurorégió Tevékénységének Bemutatása Ismertségének Tükrében*, Budapest Gazdasági Főiskola, Külkereskedelmi Kar, 2009, p. 20; R. Săgeată, *Euroregiunile de cooperare transfrontalieră din Bazinul inferior al Dunării. Studiu geographic, op.cit.*, p. 34.

54 I. Süli-Zakar, *Közös Múltunk-Közös Jövőnk, Együtt Európáért, op.cit.*, p. 374.

cooperation has its origins in the intensification of the bilateral relations between Csongrád and Timiș counties.[55] Throughout its existence, the membership of DKMT Euroregion has changed from a maximum extent in 1997 to its current shape, with 2 Hungarian counties, three Romanian counties, and 1 Serbian province, constituting today the organisation. The structure later came to be known as the Danube-Kriș-Mureș-Tisa Regional Cooperation.[56]

Behind the region's intention to cooperate lies its common past, which historically was a homogeneous unit, both from a geographical and an administrative point of view,[57] with the objective of broadening relationships among local communities and helping the region to join the process of European integration.[58]

In the first ten years of its existence, the DKMT was regarded more as an umbrella organisation with rather limited results, mainly being in charge of facilitating bilateral collaborations and symbolic manifestations but with few concrete results.[59] As a result, in 2003 changes were made to both its organisational structure and operational methods. Following the reforms, DKMT Euroregion was constructed as a two-part structure, one as an open consultative political forum, and the other as an operative working association with a registered legal-economic status.[60]

The DKMT Euroregion was involved in many cross-border projects, using multiple EU funding sources, which covered the main objectives of the Euroregion: promoting tourism and environmental protection, infrastructure development, increasing the social cohesion of the region.[61]

---

55 D. Abrudan, G. Pușcas & C. Duda-Daianu, "Cross-borders cooperation inside the euroregions. Case of DKMT Romanian Euroregion", *Nierówności społeczne a wzrost gospodarczy*, vol. 27, 2012, p. 84.

56 I. Grigorescu, M. Dumitrașcu, M. Sima & D. Micu, "Case study: The Euroregion Danube-Kris-Mures-Tisa", *In* P. Balogh et al., *Crossing the Borders. Studies on Cross-border Cooperation within the Danube Region*, CESCI, Budapest, 2016, p. 6.

57 A. Tamaskó, *A DKMT Eurorégió Tevékénységének Bemutatása Ismertségének Tükrében*, *op.cit.*, p. 18.

58 See for instance the website DKTM: <http://www.dkmt.net/en/index.php?bov=61361200392410>. Accessed 21 May 2023.

59 R. Crețan, "The evolution of cultural and economic activities in the DKMT Euroregion", *Review of Historical Geography and Toponomastics*, vol. 1, n°1, 2006, p. 63.

60 I. Grigorescu, M. Dumitrașcu, M. Sima & D. Micu, "Case study: The Euroregion Danube-Kris-Mures-Tisa", *op.cit.*, p. 16.

61 *Ibid.*, p. 31

Among others, the increase in the number of social and cultural activities, the establishment and strengthening of cross-border business relations, and the development of tourist routes in the region can be regarded as the most successful achievements of the cooperation.[62]

In spite that the cooperation was for a long time (and in different matters still is) overshadowed by the disadvantageous status of Vojvodina,[63] the DKMT is being regarded by as a functional cooperation and one of the most active areas of Eastern Europe.[64]

The Bihor-Hajdú-Bihar Euroregion was established in 2002 by the two participating County Councils of Bihor and Hajdu-Bihar, the cooperation being based on the long-standing Oradea–Debrecen cooperation axis. Although the Euroregion looks back on a history of 20 years and constituted the subject of several academic articles in the beginning, one cannot help but observe the lack of literature focusing on the Euroregion's analysis so far. Contrary to the other two Euroregions which have, at least an outdated, website, the Bihor-Hajdú-Bihar Euroregion or anything related to its activity cannot be found anywhere. Thus, making even a short analysis of the Euroregion's activity would require a more in-depth research on the matter.

The most recent literature focusing on the Euroregion dates back to 2014, according to which the main characteristics of the Bihor-Hajdú-Bihar Euroregion are those generated by its privileged position and by the consequences of a long-standing cooperation (physical-geographical complementarity; average size as area and number of inhabitants; balanced distribution of polarisation centres; compatibility of the administrative-territorial organisation system; balanced ethnic structures; well-developed system of communication channels), which are able to confer it a high degree of functionality.[65] However, seeing the lack of proof of the activity and even existence of the Euroregion makes it hard to believe that these assumptions yielded any results.

---

62 *A DKMT Eurorégió Tevékénységének Bemutatása Ismertségének Tükrében*, *op.cit.*, p 18.

63 R. Creţan, "The evolution of cultural and economic activities In The DKMT Euroregion", *op.cit.*, p. 63.

64 R. Săgeată, *Euroregiunile de cooperare transfrontalieră din Bazinul inferior al Dunării. Studiu geographic, op.cit.*, p. 142.

65 *Ibid.*, pp. 140–157.

Although the euroregions have had a significant role in helping the peripheral areas to adapt to the changes of transition and contributed to their development, solidified social, cultural and economic ties between the local actors, most scholars argue that they have not been able to fulfil their objectives, for several reasons (financial, structural, size-related, external, etc.)

The legitimacy of these euroregions is indisputable, they have contributed to the softening of the borders both physically, through the fluidisation of cross-border traffic by opening new border crossing points and by the modernisation of transport infrastructure, and spiritually by establishing and strengthening networks between actors across borders, the evolvement of economic relations and intensification of educational and cultural cooperation. Still, these forms of cooperation did not fulfil their expected missions and objectives despite the above listed achievements. It has to be kept in mind that the emergence of euroregions in this area began much later than in Western Europe, where, among others, these organisations enjoy a much greater degree of autonomy. These euroregions have not yet become such an integral part of the region as their Western counterparts did,[66] but they have become a stimulus for the establishment of other cooperation forms.

The development of cross-border cooperation was also considerably influenced by the availability of financial resources. Inarguably, during this period the Hungarian-Romanian Phare Cross-border Cooperation (CBC) programme had a significant role in intensification of connections among people living on both sides of the border.[67] For 8 years, the programme has been the prime vehicle for financial support to this border region, in the period of 1996–2003, 34 million Euro being allocated for cross-border cooperation projects to be implemented on the Hungarian side and 28 million Euro for the Romanian side.[68] The use of these funds has been centred around large scale physical infrastructure development (modernisation of border-crossing stations, roads, business infrastructure development) and environment and nature protection (water resource

---

66 A. Tamaskó, *A DKMT Eurorégió Tevékénységének Bemutatása Ismertségének Tükrében, op.cit.*, p. 38.

67 I. Balcsók, L. Dancs & G. Koncz, *The Development of Hungarian-Romanian Cross-border Connections After the Change of Regime, op.cit.*, p. 7.

68 *Ibid.*, p. 8. See the website Hungary-Romania Cross-Border Cooperation Programme 2027–2013: <http://www.huro-cbc.eu/en/overview/>. Accessed 23 May 2023.

management),[69] and smaller projects that included people-to people actions.[70] This "experimental" programme proved to be successful both in terms of establishing some key facilities in the border area[71] and in laying the foundations and improving the primary conditions for long-term collaboration.[72]

Integration

2004–2007

The year 2004, with Romania becoming a Norhtb Atlantic Treaty Organization (NATO) member and Hungary joining the EU, opened a new chapter in the two countries bilateral relations. This context conferred a special status to the Romanian-Hungarian border area which for two years became the exterior border of the EU.

For the period of 2004–2006, the European Commission has decided to launch a trilateral cross-border cooperation programme between Hungary, Romania and Serbia in order to contribute to "strengthening the spatial, physical and infrastructural integrity of the cross-border area" and to the "promotion of cooperation initiatives in order to facilitate the integration of markets and enhance coherence between local societies".[73] The program had a budget of nearly 32 million euros for Hungary (through Interreg) and nearly 20 million for Romania (through Phare CBC). During this period approximately 140 projects were financed,[74]

---

69 *Ibid.*

70 F. Feier & A. Bădulescu, "The HU-RO Cross-Border Cooperation Programme 2007–2013: Insights related to the effectiveness of using the European Funds", *Oradea Journal of Business and Economics*, vol. 1, n°1, 2016, p. 50.

71 Hungary-Romania Cross-Border Cooperation Programme 2027–2013: <http://www.huro-cbc.eu/en/overview/>. Accessed 23 May 2023.

72 F. Feier & A. Bădulescu, "The HU-RO Cross-Border Cooperation Programme 2007–2013", *op.cit.*, p. 50.

73 See G.-E. Csoka, "The Cross Border Cooperation between Romania and Hungary", University of Oradea, Romania, 2018. Retrieved from: <https://mpra.ub.uni-muenchen.de/92067/1/MPRA_paper_92067.pdf>. Accessed 23 May 2023.

74 Breco, "PROGRAMUL PHARE CBC RO-HU 2004–2006", Retrieved from: <https://brecoradea.ro/index.php/programe/phare-cbc>. Accessed 23 May 2023.

but due to this differentiated nature of the programme serious difficulties and delays have been met during their implementation.[75]

This period also came to know a stage with resurfacing tensions, centred around Hungary's intention to grant Hungarian citizenship to ethnic Hungarians in neighbouring states,[76] but the organisation of joint government meetings since 2005 has opened yet another stage for a return to normality.

2007-present

The cross-border co-operation between Hungary and Romania has arrived to a very important milestone with the accession of Romania to the European Union and Hungary becoming a part of the Schengen Area. Thus, 2007 marked a radical turning point in the nature, content and form of cross-border cooperation.

The Hungary-Romania Cross-Border Cooperation Program 2007–2013 continued the cross-border cooperation programs implemented in the past in the region, now being implemented within a common institutional structure with the usage of common funds.[77] Between 2007 and 2013, more than 211,006,439 Euro were awarded to the 455 selected projects which had several significant achievements, such as the construction of 48 km and modernisation of 79 km of road infrastructure; the construction and/or modernisation of 160 km of bicycle paths; the development of 130 tourist attractions; the establishment of 24 new research and/or technical centres; or the development of 34 health care institutions.[78]

According to Feier and Bădulescu, based on the analysis of the indicators' achievement and seeing the high absorption rate of the funds, the Hungarian-Rumanian CBC Programme 2007–2013 was a successful

---

75 F. Feier & A. Bădulescu, "The HU-RO Cross-Border Cooperation Programme 2007–2013", *op.cit.*, pp. 50–55.

76 M. R. Vădean, *Relațiile româno-ungare în contextul integrării în structurile europene și euroatlantice*, *op.cit.*, p. 230.

77 F. Feier & A. Bădulescu, "The HU-RO Cross-Border Cooperation Programme 2007–2013", *op.cit.*, p. 50.

78 HURO CBC Programme, "Co-operation Portraits: Hungary-Romania Cross-Border Co-operation Programme (RO)", YouTube video, 16 December 2014. Retrieved from: <https://www.youtube.com/watch?v=dGeUAYFQNsg>. Accessed 23 May 2023.

programme which managed to achieve its main objectives, namely bringing together actors from both sides of the border and strengthening the economic cohesion of the border region.[79] Still, the 33 different project categories resulted in a high fragmentation of the programme, meaning that the interventions have not reached critical mass in certain areas and impacts have been felt mostly at local rather than regional level.[80]

As a continuation of the above objectives, the Interreg V-A Romania-Hungary Programme was approved by the European Commission on 9th of December 2015 for the period of 2014–2020. The 109[81] projects contracted in the 2014–2020 period have a value of 151,334,312 Euro, representing 103.9 % of the entire programme allocation.[82] After a late start of implementation and delayed launching of the call for proposals, the programme implementation showed good progress in 2018 and 2019. Although the Covid-19 pandemic added significant risks to the achievement of the programme targets, having in mind that for almost three months a total lockdown was declared in the two countries and mobility restrictions were put in place, the programme succeeded to commit the entire funding.[83]

A main conclusion from the 2014–2020 programming period is that cross-border cooperation remains relevant to the needs of the participating counties as they have succeeded to use the funding opportunity and resolve the development needs they encountered, however, there are significant differences regarding the prioritisation of investments (health infrastructure development for Romanians, cultural heritage and economic development for Hungarians). The limited matching of the priorities of the two sides reflects the low degree of cross-border integration, which in turn limits the contribution of the programme to the

79 F. Feier & A. Bădulescu, "The HU-RO Cross-Border Cooperation Programme 2007–2013", *op.cit.*, pp. 51–55.

80 European Commission, "Ex post evaluation of Cohesion Policy programmes 2007–2013", *op.cit.*, p. 27.

81 For details of the projects, see the websites Keep and Interreg Romania: <https://keep.eu/ and https://interreg-rohu.eu/ro/home/>. Accessed 23 May 2023.

82 Lideea Development Actions & Archidata srl., "Final Evaluation Report. Services for evaluating the implementation of The Interreg V-A Romania – Hungary Programme", 2020, p. 14. Retrieved from: <https://interreg-rohu.eu/wp-content/uploads/2020/12/ROHU-Final-Evaluation-Report.pdf>. Accessed 23 May 2023.

83 *Ibid.*, p. 61.

cohesion of the cross-border area, a process also challenged by the fact that Romania did not enter the Schengen area.[84]

In March 2022, the Programming Committee approved the Interreg VI-A Romania-Hungary Programme. Upon the European Commission's approval, the Programme will have a total budget of 175,940,025 Euro[85] and has the objective of creating a greener, more resilient and cohesive cross-border region, with a better understanding of cooperation opportunities, increased trust and reduced barriers to a more sustainable cooperation.[86]

EGTCs represent a growing cooperation reality not just in Western Europe, but also in the Romanian-Hungarian border region. Recognising the problem encountered by the euroregional organisations, the local actors started establishing smaller and, in theory, more effective institutional organisations. Since 2009, four EGTCs have been established along the Romanian-Hungarian border, namely the Banat-Triplex Confinium EGTC, Gate to Europe EGTC, European Border Cities EGTC, European Common Future Building EGTC.

There is a stark difference between the two countries in terms of EGTCs. In Hungary, the territorial coverage of EGTCs is particularly high, out of the 24 groupings in which Hungarian entities are involved, 21 are based there. In contrast, the current number of EGTCs in Romania is 6, none of them having their seat in the country (in the case of Romanian-Hungarian EGTCS this is mainly due to a contradiction in the national legislation of the two countries). In our understanding, this might be attributed to the clear-cut distinction between the attitude of the two governments –supportive vs. passive. Based on the findings of an unpublished empirical research conducted in 2020, the experience of the Romanian-Hungarian EGTCs is that it is very difficult to obtain the approval of the Romanian government, which in the beginning viewed the grouping with suspicion. On the other hand, the Hungarian government has paid special attention to EGTCs from the very beginning,

---

84 *Ibid.*, pp. 8–10.

85 Interreg Romania-Hungray. Retrieved from: <https://interreg-rohu.eu/en/approval-of-the-interreg-vi-a-romania-hungary-programme-2021-2027/>. Accessed 23 May 203.

86 Program Interreg VI-A Romania-Hungary. Retrieved from: <https://interreg-rohu.eu/en/home-en/> Accessed 23 May 2013.

supporting their development and contributing financially to their operation.

The four EGTCs show great diversity and are characterised by different levels and directions of development. It can be unequivocally stated that only half of the EGTCS along the Romanian-Hungarian border are successful in their activities and have an undeniable role in the development in the region, while the other two groupings have so far failed to deliver any meaningful results. In terms of their activity, among the four groupings the Banat-Triplex-Confinium and the Gate to Europe EGTCs clearly stand out. Their projects so far have generally focused on 'soft' areas, areas such as culture, education and capacity building, tourism and trade, with a few exceptions of larger scale infrastructure investments. The groupings contribute to the development of the area not only through their own initiatives and projects, but also through their project writing and management activities with which they assist their members in attracting funds.

Despite the above mentioned achievements, the 10 years' experience was not enough for the EGTCs to establish themselves as a major actor in their regions' development. Their activities so far have focused on increasing networking in the region, boosting economic and social relations and encouraging cooperation in all areas. Overall, the Romanian-Hungarian EGTCs show only limited results in the general stimulation of the integration of the Romanian-Hungarian cross-border region, also having only a negligible influence on the relevant Interreg Cross-border cooperation programmes.

These findings are in line with the situation analysis conducted for the Programming document of the Interreg VI-A Romania-Hungary, namely that the financial and human resources capacities of the EGTCs are severely differentiated and most of them highly depend on top-down financing.[87] Based on our findings, the operation of all four EGTCs is highly dependent on the financial support provided by the Hungarian Ministry of Foreign Affairs and Trade and it is not expected that any of them will become more independent in the future.

---

[87] *Ibid.*, p. 13.

Conclusions

Since the creation of the European Communities, multiple paradigm shifts regarding the perception, the form and the functions of borders, border regions and cross-border cooperation can be distinguished. In the greater European context we can distinguish the following paradigm shifts: from seeing borders as barriers to seeing them as bridges; from seeing borders as barriers to political union to seeing them as barriers to economic integration; and from seeing cross-border cooperation as a project based perspective to favouring an institutionalised one.

Similarly, in the Romanian-Hungarian relation, over the course of the last decades, the perception, the form and the functions of the border have shifted. The 448-kilometre-long Hungarian-Romanian border region has been for a long time separating the cross-border region, but with the end of socialism the rigidity of the border eased and since it has become the site of extensive cross-border cooperation. Thus, we can identify a transformation of the border, shifting from its barrier function to a connecting bridge one, a shift which, undoubtedly, (re)ignited the relations between the two sides.

In spite of the fact that during the socialist era Hungary and Romania belonged to the same political power block, the relationship between the two countries was rather tense. Even after the changes of regime, the "shadow of Trianon" and the reviving nationalism on both parts brought to the surface previously suppressed problems. Still, given both countries European and Euro-Atlantic aspirations, and their recognition of the necessity of deepening cooperation to further these aspirations, the untrusting and often tense relations gradually eased, the second half of the 1990s bringing about a period of historical reconciliation, followed by the gradual intensification and proliferation in nature and in scope of the cooperation. Therefore, the second paradigm shift saw the transformation of tense (first of all inter-state) relations into harmonious, cooperative ones. The clear improvement of the general context influenced to a great extent the lower-level, subnational cross-border relations. This gave way to the third shift, from informal cooperation to structural institutional cooperation, starting from local twinning initiatives to the establishment of Euroregions, and from the failure of Euroregions to the creation of EGTCs.

Following the timeline approach adopted in the present paper, it is clear that several stages can be identified in the Romanian-Hungarian relations, going from a great degree of separation, through transition and consolidation towards integration. The area is characterised by a long history of cooperation, with a gradual transition from informal forms to more institutionalised ones, in line with European practices. There have been many good examples of positive cooperation along the way, both at higher and lower levels.

At the national level, the Treaty and the Declaration on a Strategic Partnership have created a binding framework for the Romanian-Hungarian bilateral relations, but joint meetings have been informally suspended. It is unfortunate that despite the existence of these instruments there is no will at the political level to use them, especially in a period when recent international events proved how much good neighbourly cooperation would mean. On the local level, twinnings have remained centred on mostly 'soft' areas of cooperation, such as cultural, educational and sports activities and exchanges. Still, local ties have been strengthened by the joint implementation of cross-border cooperation projects through Interreg, especially at the administrative levels. At the intermediate level, the Euroregions present little concrete results, however, their symbolic value is very important. In the first years, Euroregions had a significant role in helping the peripheral areas to adapt to the changes of transition and contributed to their development, solidifying social, cultural and economic ties between the local actors. Still, it seems that they have not been able to fulfil their objectives set out at their creation, they did not manage to establish themselves as a major actor in their regions' development. Nor did the EGTCs. Although they have been around for more than a decade now, their activities so far have focused on increasing networking in the region, boosting economic and social relations and encouraging cooperation in all areas. Overall, the EGTCs so far have generally focused on 'soft' areas of cooperation and show only limited results in the general stimulation of the integration of the region. It is evident that the Romanian-Hungarian cross-border region is characterised by a high degree of interconnectedness, but still a low level of integration and significantly divergent interests and priorities of the two sides.

The next paradigm shift, namely Romania's accession to the Schengen Area, could inarguably serve the further spiritualisation of the border, while also contributing to the physical realisation of a "Europe without borders". Romania's accession to the Schengen Area would be a historic achievement, affecting the daily lives of millions of people. This next step would unify the region by establishing an area without frontiers, both in terms of the elimination of border controls, thus reducing its remaining barrier effect, and through the symbolic meaning of establishing a sense of unity, of togetherness. Currently, Romania actively applies the Schengen acquis and awaits the Decision of the Council, in a context where the Commission has confirmed (since 2011!) that Romania fulfils the necessary conditions to join and called upon the Council to take the necessary steps[88] for Romania to take its place in the borderless Europe.

Bibliography

D. Abrudan, G. Pușcas & C. Duda-Daianu, "Cross-borders cooperation inside the euroregions. Case of DKMT Romanian Euroregion", *Nierówności społeczne a wzrost gospodarczy*, vol. 27, 2012, pp. 80–87.

A. L. Amilhat, V. Kolossov, I. Liikanen, D. Newman, P. Joenniemi, N. Yuval-Davis, S. Rosière & J. Scott, *EUBORDERSCAPES: State of the Debate Report 1*, 2012.

Z. Bacsi & E. Kovács, *Határrégiók fejlődésének sajátosságai*, Hévíz Többcélú Kistérségi Társulás & Nyugat-Balatoni Társadalomtudományi Kutatóműhely, Keszthely, 2007.

I. Balcsók, L. Dancs & G. Koncz, *The Development of Hungarian-Romanian Cross-border Connections After the Change of Regime*, European Regional Science Association, ERSA conference papers, 2005, pp. 2–4.

B. Baranyi, "A magyar-román interregionális kapcsolatok sajátosságai", *Pro Minoritate*. 2007, pp. 98–138.

B. Baranyi, "Adalékok a határ mentiség újraértelmezéséhez Magyarországon", *Deturope: Central European Journal of Tourism and Regional Development*, vol. 6, n°2, 2014, pp. 26–45.

M. Benczi & G. Ocskay, "The evolution of cross-border cooperation in Hungary", *Észak-Magyarországi Stratégiai Füzetek*, vol.18, n°2, 2021, pp. 54–67.

A. Corpădean, "The Romanian-Hungarian Strategic Partnership 2002–2016", *In* N. Păun & S. Schirmann (eds.), *Borders, Identities, Communities. The Road to Reconciliation and Partnership in Central and Eastern Europe*, Nomos, Baden-Baden, 2016.

[88] European Commission, "State of Schengen Report 2022", *op.cit.*, p. 11

R. Creţan, "The evolution of cultural and economic activities in the DKMT Euroregion", *Review of Historical Geography and Toponomastics,* vol. 1, n°1, 2006, pp. 61–78.

G.-E. Csoka, "*The Cross Border Cooperation between Romania and Hungary*", University of Oradea, Romania, 2018.

K. Czimre, "Recovery or discovery? Models and motives of cross-border co-operation along the Eastern Border of Hungary after 1989–1990", *Eurolimes*, vol. 26, n°26, 2018, pp. 97–112.

L. de Sousa, "Understanding European cross-border cooperation: A framework for analysis", *Journal of European Integration*, vol. 35, n°6, 2012, pp. 1–19.

E. Evrard & A. Engl, "Taking stock of the European Grouping of Territorial Cooperation (EGTC): From policy formulation to policy implementation", *In* E. Medeiros (ed.), *Theoretical and Empirical Approaches to the Process and Impacts of Cross-Border and Transnational Cooperation in Europe*, Springer, Dordrecht, 2018, pp. 209–227.

F. Feier & A. Bădulescu, "The HU-RO Cross-Border Cooperation Programme 2007–2013: Insights related to the effectiveness of using the European Funds", *Oradea Journal of Business and Economics*, vol. 1, n°1, 2016, pp. 47–57.

I. Grigorescu, M. Dumitraşcu, M. Sima & D. Micu, "Case study: The Euroregion Danube-Kris-Mures-Tisa", *In* P. Balogh et al., *Crossing the Borders. Studies on Cross-border Cooperation within the Danube Region*, CESCI, Budapest, 2016.

M. Krzymuski, "The EGTC as a legal solution of institutionalisation of cross-border cooperation", *In* G. Ocskay (ed.), *15 Years of the EGTCs. Lessons Learnt and Future Perspectives*, Central European Service for Cross-border Initiatives (CESCI), Budapest, 2020, pp. 7–35.

Z. Ludvig & I. Süli-Zakar, *A Kárpátok Eurorégió Együttmüködés Mérlege*, Oktatási Minisztérium, Budapest, 2002.

A. Moraczewska, "The changing interpretation of border functions in international relations", *Revista Română de Geografie Politică*, vol. 12, n°2, 2010, pp. 329–340.

L. O'Dowd, "The changing significance of European borders", *Regional & Federal Studies*, vol. 12, n°4, 2002, pp. 13–36.

G. Ocskay, "Changing interpretation of the EGTC tool", *In* G. Ocskay (ed.), *15 Years of the EGTCs. Lessons Learnt and Future Perspectives*, Central

European Service for Cross-border Initiatives (CESCI), Budapest, 2020, pp. 37–54.

M. Perkmann, "Cross-border regions in Europe significance and drivers of regional cross-border co-operation", *European Urban and Regional Studies*, vol. 10, n°2, 2003, pp. 153–171.

J. Peyrony, J. Rubio & R. Viaggi, *The effects of COVID-19 induced border closures on cross-border regions: An empirical report covering the period March to 2020*, European Union Publications Office, Luxemburg, 2021.

A. Pinto, R. Ferreira & N. Verschelde, *Public consultation on overcoming cross-border obstacles – 2020*, Summary Report, Publications Office of the European Union, Luxemburg, 2021.

R. Săgeată, *Euroregiunile de cooperare transfrontalieră din Bazinul inferior al Dunării. Studiu geografic*, Editura Academiei Române, Bucharest, 2015.

J. Sallai, "A változások kezdete, az erdélyi magyarok kiengedése a nyugati határon 1985-től", *Nemzetbiztonsági Szemle*, vol. 7, n°3, 2019, pp. 70–80.

M. Smętkowski & J. Rok, "The Carpathian Euroregion from the perspective of economic cooperation in peripheral regions", MPRA Paper, n°82763, 2020.

I. Süli-Zakar, *Közös Múltunk-Közös Jövőnk, Együtt Európáért: 20 Éves a Kárpátok Eurorégió. Emlékek, Sikerek, Megújulás*, DIDAKT Kft, Debrecen, 2014.

I. Süli-Zakar & T. Kecskés, "The historical borders and the cross-border connections' effect on Debrecen's spirituality and cultural economy", *Eurolimes*, vol. 19, 2015, pp. 99–115.

H. Tanaka, "Carpathian Euroregion and cross-border governance", *The Journal of Comparative Economic Studies*, vol. 2, 2006, pp. 59–80.

A. Tamaskó, *A DKMT Eurorégió Tevékénységének Bemutatása Ismertségének Tükrében*, Budapest Gazdasági Főiskola, Külkereskedelmi Kar, 2009.

M. R. Vădean, *Relațiile româno-ungare în contextul integrării în structurile europene și euroatlantice*, Editura Științifică Lumen, Iași, 2011, pp. 218–225.

C. Yndigegn, "Between debordering and rebordering Europe: Cross-border cooperation in the Öresund region of the Danish-Swedish border region", *Eurasia Border Review*, vol. 2, n°1, 2011, pp. 47–59.

General Trends and Evolutions from the Past Century Regarding the Perception of the Romanian-Hungarian Border in the Context of the European Territorial Cooperation

The Case of Bihor and Hajdú-Bihar Counties

István Polgár & Mircea Brie

During the 17th and 18th centuries the Carpathian Basin was known as one of the most troubled regions of Europe concerning ethnic and religious conflicts, in so much that the ideological and social systems that ruled the region during the last century were not able to remove the existing tension between the different ethnic groups.[1]

The end of the World War I resulted in a new situation in international politics. The victorious European powers and their overseas allies were prepared to redraw on the political map of the world without any substantial restrictions. The disappearance of multinational empires, such as Russia and Austria-Hungary, at the end of World War I led to a Central and Eastern Europe in the construction of national states on the basis of ethnic and demographic criteria, first and also according to the economic and strategic interests of the powers of the moment.

History has proven that the way in which the managing of a new European architecture was envisaged during the interwar period was doomed from the start, as the myriad of ethnic enclaves marring the national unity of the successor states were as many reasons to trigger revisionist actions, which in reality were only the mask that concealed

1 K. Károly & K. H. Eszter, *Ethnic Geography of the Hungarian Minorities in the Carpathian Basin*, Geographical Research Institute Hungarian Area, Budapest, 1998, pp. 13–14.

economic interests. This situation was due to the fact that between citizens of different ethnic groups in the same country internal frontiers were raised, separating groups from each other, limiting their areas of availability, values and ideas.[2]

A large part of the researchers focusing on this topic considers that these tensions were aggravated by the new rigid borders, which limited the successor states after World War I. Also, these new borders discontinued the old historical and cultural ties and traditional economic contacts pre-existing in the former empires in such a way that no Communist internationalism proved to be able to solve the problem. The political behaviour from the last centuries, regarding the territory and the border of the European states, often generated tensions and conflicts. These disagreements have led to the creation of barriers in the border areas.[3]

The main conflicts of the 20th century occurred mainly because of political and ideological reasons. The borders of states became real insurmountable barriers not only for enemies, but several times also for their own citizens. This character of the borders created rather the role of elements that prevented the emergence of common activities and values. Although from a strategic point of view, strictly from the perspective of border regions, they were directly interested in developing a strategy based on cooperation with neighbouring regions. Nevertheless, the specificity of Europe is given not only by the diversity of cultural and historical realities, but also by the existence of numerous borders between States.[4] Most often because of strategic and/or military reasons border areas, or peripheries were strongly connected to the center. Moreover, in most cases migration flows from the periphery to the centre could be observed.[5] People in the border regions, who were mostly affected by the consequences of historical conflicts, have gained mutual fears

---

[2] A. Marga, *Filosofia unificării europene [The Philosophy of the European Unification Process]*, Biblioteca Apostrof, Cluj-Napoca, 1997, p.140.

[3] I. Polgár, "The impact of territorial governance structures from Bihor County on cross-border cooperation during 2014–2020", *Journal of Romanian Literary Studies*, n°25, 2021, pp. 392–405.

[4] C. Ricq, *Manual de cooperare transfrontalieră pentru uzul colectivităților locale și regionale în Europa*, Ediția a 3-a, Bucharest, 2000, pp. 8–15.

[5] A. Ilieș, *România între milenii. Frontiere, areale frontaliere și cooperare transfrontalieră*, Editura Universității din Oradea, Oradea, 2003, pp. 92–97.

and animosity over time. The conditions that created these relationships undermined the availability of cooperation and contacts.[6]

The borders gone through allot of changes in terms of understanding the space of a culture, race, religion, city or nation. Growing from physical walls, intense militarised areas to a formal understanding of regions with no physical boundaries or restrictions of any kind, the frontier or the border, has been an area of the utmost importance to the different branches of science. The border region is the collision point of cultures, religions, languages, administrations, legislations and economic influences. All these factors have substantial contribution to the development of the phenomenon called cross-border cooperation. The hard border, which in the view of the European institutions would ensure the protection of its citizens, has thus become primarily the responsibility of the new members of the European Union (EU).[7]

The concept of "Europe without borders" can provide the solution to the ethnic problem on the continent. Some politicians and political scientists consider that the model of the "nation state" has not been able to provide sustainable finality through the citizenship institution. Limited in its defining pattern, the classic citizenship offered only a kind of equality principle, abstract, to residents within the borders of a national state, but the same borders did split many ethno-cultural communities, dividing the population into two categories, the majority and the inhabiting minorities.[8]

As the EU takes on new members and its external boundaries gradually shift, socio-economic and political transformations are taking place at the borders that not only adumbrate new regional development opportunities but also many potential problems and tensions. In an enlarged Europe long-term commitment are necessary to support local and regional initiatives of cross-border cooperation.[9] During the process of European construction, respectively the integration of European states in

---

6 I. G. Bărbulescu, M. Brie & N. Toderaş, *Cooperarea transfrontalieră între România şi Ucraina, respectiv între România şi Republica Moldova. Oportunităţi şi provocări în perioada 2014–2020*, Institutul European din Romania, Bucharest, 2016, pp. 22–26.

7 I. Polgár, *The impact of territorial governance structures from Bihor County on cross-border cooperation during 2014–2020, op.cit.*

8 R. Weber, *Un concept românesc privind viitorul Uniunii Europene*, Editura Polirom, Iaşi, 2001, p. 53.

9 A. Athanasiu, "Foreword", *In* A. C. Popoviciu & D. Cigan (eds.), *The Frontier Worker – New Perspectives on the Labor Market in the Border Regions*, C.H. Beck, Bucharest, 2013, pp. 7–8.

the community structures, a change of perception can be noticed which focus on the borders of the national states. These changes firstly appeared in the process of opening up, economically in the space of the single market. Secondly the changes appeared at political and social level, in the space of the EU.

Nowadays almost a third of EU citizens live and work in Europe's border regions. The management and perception of these borders have both a direct and an indirect impact on their lives. People living in border regions often face specific challenges, either in finding a job or having access to healthcare and other public services, or in terms of daily commuting and overcoming administrative problems.[10]

At present, cross-border cooperation has proven to be the most effective tool for overcoming the barrier effect and the role of border separation and for strengthening the territorial cohesion of border regions.

The Romanian-Hungarian Border: Facts and Numbers

Border area supposes good neighbourhood. As a doctrine, good neighbourhood is defined as "a state of facts", as "unavoidable" (neighbours are given by history and geography) existing ever since peoples and states have been established and generating two aspects at borders: preserving security by managing conflicts and preserving cooperation between neighboring states.[11] Good neighbourhood is tightly connected with the concept of "border", yet the role of border has altered in time. It has changed from barrier to meeting point between different realities discovering the opportunity of complementarity beyond rigid conceptions on national sovereignty.

The Romanian-Hungarian border, the one drawn at the end of the peace process that followed the First World War, was from the beginning, and continues to this day, a controversial topic in the historical research of the two

---

10 European Committee of the Regions; "Cross-Border Public Services in Europe", COR-2020-02615-00-00-AC-TRA (EN), Bruxelles, 2021. Retrieved from: <https://edz.bib.uni-mannheim.de/edz/doku/adr/2020/cdr-2020-2615-en.pdf>. Accessed 23 December 2023.

11 M. Brie, *The European Neighborhood Policy, Mass-media and Cross-border Cooperation*, *In* Analele Universităţii din Oradea, Seria Relaţii Internaţionale şi Studii Europene, Oradea, 2009, p. 83.

neighbouring countries. The intensity with which each of the two dialogue partners has defended over time, more or less argued, the theses launched regarding the border, has also fluctuated, being influenced by a complex of domestic and international political, ideological or economic factors.

From the beginning, the attention of the reader, passionate about the subject is attracted by the massive production of works of this kind coming from Hungarian researchers, a volume that far surpasses the Romanian one both quantitatively and, in some cases, qualitatively. Somewhere it is natural to be so, as long as Hungary was one of the main losers of the world conflict, and Romania was in the camp of the winners. This reality, psychologically left its mark on the thinking of the two antagonistic categories of historians and implicitly on the content and the message conveyed by their creations. Over the historical tensions, there is an internationally recognised border between the two countries, based on the following structure (See Figure 1).

Figure 1: The Romanian-Hungarian border (map view).[12]

[12] Source: map prepared by the author.

The border between Romania and Hungary has a total length of 448 km, of which 415.9 km lend and 32.1 km river (the Mures, Cris, Someş rivers).[13] On the Romanian side there are four counties, Satu Mare, Bihor, Arad and Timiş. On the Hungarian side there are Szabolcs-Szatmár Bereg, Hajdú-Bihar, Békés and Csongrád. All the 8 counties are classified as NUTS-3 and are integrated in four regions of level NUTS-2.[14]

The Romanian Hungarian Border covers the South-Eastern and Eastern part of Hungary and the North-Western and Western part of Romania. The eight counties have a total surface of 50,454 km^2, of which 43.7 % Hungarian and 56.3 % Romanian area. The Hungarian territory is 23.7 % of the total surface of Hungary, and the Romanian part is 11.9 % of Romania. The total population in 2011 was more than 4 million, of which slightly less than half lives in Hungary, and slightly more than half lives in Romania.[15]

Socio-Political Features of the Romanian-Hungarian Border 1918–1945

At the turn of the century, both countries were attracted and then dragged into armed clashes between the military blocs of the time, whether they were called the Entente and the Central Powers, or later the Allied Powers and the Berlin Rome-Tokyo Axis. After these armed clashes the perception and the role of the borders has become increasingly negative. The end of the First World War and the breakup of the

---

13 National Institute of Statistics, "Statistical yearbook 2011", p. 11. Retrieved from: <https://insse.ro/cms/en/content/statistical-yearbooks-romania>. Accessed 23 December 2023.

14 The counties Szabolcs-Szatmár-Bereg and Hajdú-Bihar belong to Észak-Alföld Region; the counties Békés and Csongrád are part of Dél-Alföld Region; counties Arad and Timiş belong to West Region; counties Satu Mare and Bihor are part of the North-West Region. For further details see Eurostat, "Regions in the EU, Nomenclature of territorial units for statistics, NUTS 2006 /EU-27". Retrieved from: <https://ec.europa.eu/eurostat/web/products-manuals-and-guidelines/-/ks-ra-07-020>. Accessed 20 February 2023.

15 Hungary-Romania Cross-Border Cooperation Programme. Retrieved from: <http://www.huro-cbc.eu/en/programme_area>. Accessed 20 February 2023.

multinational empires in Europe created the conditions for drawing new borders.

June 4th marks the 105th anniversary of the signing of the Treaty of Trianon, the document that established the borders after World War I and through which Hungary lost three quarters of the territory it had administered. Transylvania then became, from a legal point of view, part of Romania, after the *de facto* union took place on the 1st of December, 1918. In this period, mainly as a post-war consequence we can observe the volatile nature, character of the borders, which is specific to the inter-war period. The main causes of the volatile character of the borders are mainly coming from the several and radical political and administrative changes implemented in this period.

Another phenomena representative for this period, which contributed to the volatility of the borders is the migration. In this period migration of more than 200.000 people was registered. The above-mentioned effects and several other period specific issues were supported the must by the population from the border are. Even though a new border was drawn, even though several changes were implemented by the new Romanian administration, the objective was the same, to keep the relations between the communities, to keep the flow of the goods between the two sides of the border. This was especially important since the demographic composition from the cross-border area did not changed significantly after the Trianon decisions and the establishment of new borders. This way razing social tension were avoided in an already confused and instable context and territory.

Another important objective was to assure a peaceful transition in the new Romanian territories, with a special regard on the border area. Before the Trianon Treaty there were several proposals regarding the new border line between Romania and Hungary. The map below represents an inventory of the proposal (See Figure 2).

After World War I, as the Austro-Hungarian monarchy was disassembled and Transylvania passed under Romanian administration, the cross-border area between the two countries and its features passed through radical changes. The historical and the geopolitical context presented above and the need of future cooperation demand the necessity of further research and investigation on the topic, especially on the way, how the border between the two countries was drawn.

Figure 2: Variants for tracing the Romanian-Hungarian border proposed at Trianon (1920).[16]

Socio-Political Features of the Romanian-Hungarian Border 1945–1989

Communist Europe was constituted in its turn in a separate structure. Unlike Western Europe, the binder of this superstructure was the common ideology, not interpersonal relations, maintained when needed including through armed interventions.[17] Such events demonstrate the inability of communist ideology to build a social order based on justice and freedom. The ethnic intolerance that marked the history of Europe over the centuries did not find its cure even under the dome of proletarian internationalism.[18] The communist system did nothing but

---

16 Source: R. Sageata, "The evolution of the ethnic and political Romanian-Hungarian border as reflected in sources", *Transylvanian Review, XXVIII* (4), 2022, pp. 3–22. **Legend**: 1. Requested by Romania in 1916; 2. Proposed by British experts; 3. Proposed by French experts; 4. Proposed by Acad. Gyula Varga in 1912; 5. Proposed by American experts; 6. Proposed by Italian experts; 7. Proposed by Serbian experts; 8. Proposed by Count Stephen Bethlen in 1912; 9. Present border, after Trianon (1920).

17 Ş. Delureanu, *Geneza Europei comunitare: mesajul democraţiei de inspiraţie creştină*, Editura Paideia, Bucharest, 1999, p. 368.

18 A. Marino, *Revenirea în Europa: idei şi controverse româneşti: 1990–1995*, Editura Aius, Bucharest, 1996, p. 398.

freeze and preserve the seeds of inter-ethnic adversity. The results of this pseudo-policy were seen as soon as the system collapsed.[19]

After 1945, the Romanian-Hungarian relations were officially presented as normal, ordinary connections between two socialist, friendly states. But beyond official relations of good neighbourliness, in certain historical situations several points of divergence existed (See Figure 3). These were generated by the numerous differences inherited from the specific history of the two states and nations. One of the determinant factors in bilateral relations was the issue of minorities, especially the situation of the Hungarian minority from Romania. During the communist years, to cross the state border was extremely rare, reserved for only a few people, under certain conditions, and those in communist states approved after many checks and generally for organised trips with the group. Permanent possession of a passport was not possible, so the freedom of movement of people beyond the borders of the country was seriously restricted.[20]

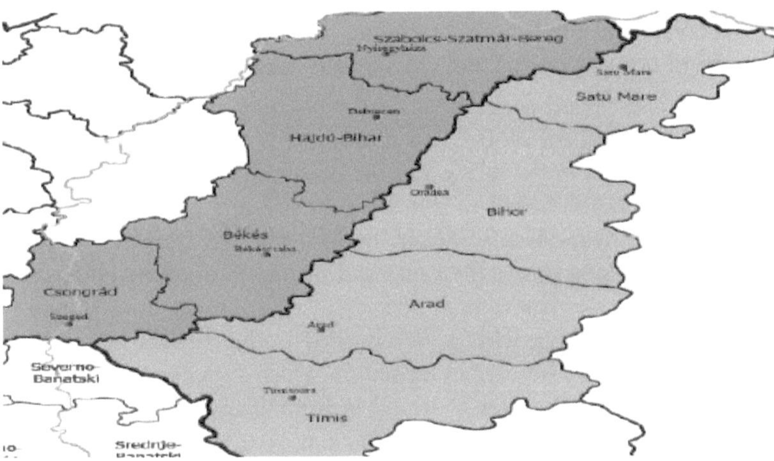

Figure 3: The Romanian-Hungarian border between 1945–1989.[21]

[19] Ş. Delureanu, *Geneza Europei comunitare, op.cit.*, pp. 349–353.

[20] A. Marga, *Filosofia unificării europene, op.cit.*, p. 92.

[21] Source: Retrieved from: <https://interreg-rohu.eu/en/eligible-area/> (black&white edition). Accessed 23 December 2023.

In terms of domestic affairs but also in terms of foreign affairs, neighbourly relations and border control we divide the so-called communist period in 3 main "chapters" or periods which affect the perception and the rigidity or flexibility of the border.

The first period is focusing on the perception of the border in the spirit of the Marxist-Leninist ideology 1948–1960. In the first half of the 1950s similar political and social processes took place in both countries: consolidation of the Stalinist dictatorship, socio-economic transformations and so on.[22] The 1956 Revolution represented a turning point in the relations between the two States.[23] Over the diplomatic negotiations and constraints, the Romanian side considered that revolutionary events represent a revival of nationalism and Hungarian revisionism.

On the Hungarian side the Kadar regime, ideologically was based on an internationalist, anti-nationalist discourse. At the beginning of the period we can speak about a certain Kadar-Dej relation and political help (regarding the deportation of Imre Nagy, Prime Minister of Hungary during the revolution from 1956). This political relation, attitude and support made the question of the Hungarian minority ignored, especially in terms of border crossing and cross-border relations.

The second period is dedicated to the national unity in the national spirit of communism 1960–1985. During this period several changes were implemented in the relation of the two countries. The political system from Romania started to put a higher pressure on the Hungarian minority from Transylvania.[24] Even though the two countries were in the same ideological block, the issues and the question of the Hungarian minority from Transylvania made bilateral relations almost non-existent.[25] Borders became frozen with a very limited traffic. In this period new laws appeared in the Romanian legislation which stipulated the visit and accommodation of Hungarian citizens in Romania.

The third period was characterised by a deep deterioration of the relations and strengthening of borders 1985–1989. Travelling, border crossing

---

22 L. Boia, *Istorie şi mit în conştiinţa românească*, Humanitas, Bucharest, 1997, p. 67.

23 N. Ionescu-Gură, *Stalinizarea României. Republica Populară Română 1948–1950: transformări instituţionale*, Editura All, Bucharest, 2005, p. 329.

24 T. Solomovici, R. Theodorescu & S. R. Stănescu, *Securitatea şi evreii: despre călăi şi despre victim*, Editura Ziua, Bucharest, 2004, p. 81.

25 L. Boia, *Istorie şi mit în conştiinţa românească*, *op.cit.*, p. 110.

was very difficult. During an official visit to Budapest of a Romanian delegation, led by Emil Bobu (vice PM of Romania) it was mentioned that travelling is so difficult, that it was easier in the Horthy-Antonescu period, during the II World War. A permanent issue which created tensions between the two administrations, making cross-border cooperation impossible was the Hungarian community from Transylvania. Another issue was focusing on the problem of the political refugees.[26] After 1986 there were some attempts for de-tensioning the situation and to make border crossing easier, at least for the population from the cross-border area. This action was focusing on the permission of the so called, small border traffic. This action was introduced first on the border with Yugoslavia, but starting from 1986 it was practiced also on the Romanian-Hungarian border.

The perception of the border during the communist times changed in accordance with the so called societal development process, from the "proletarian internationalism" of the 1948–1960, when Romanian-Hungarian brotherhood and understanding in work and daily life were imposed, to the "national communism" of the 1960–1989, when the Romanian-Hungarian political conflicts during the first half of the 20th century were shown and explained closer to the reality of the time.

Socio-Political Features of the Romanian-Hungarian Border 1989-Present

The events, which took place in 1989–1990, caused a new situation on both sides of the Romanian-Hungarian border. While the border between the two countries has been transformed into a formal border, their cooperation has become even closer. We witnessed the change in the functional role of borders, by changing their perception from barriers that divide to borders that connect.[27]

Related to this phenomenon of border transformation, we can say that the new political and economic situation promoted at the European level aims to standardise the policies regarding the role of the border

---

26 L. Nastasă, A. Varga, A. Andreescu & A. Zub, *Minorități etnoculturale, mărturii documentare: maghiarii din România (1956–1968)*, Centrul de Resurse pentru Diversitate Etnoculturală, Bucharest, 2003, p. 61.

27 I. G. Bărbulescu, *Noua Europă. Identitate și model European*, Editura Polirom, Iași, 2015, pp. 305–310.

between countries and transform them into areas of convergence, of major economic and cultural interest. In parallel, the military function of the border lost its supremacy in favour of a purely administrative function. The Romanian-Hungarian border has changed from a hard, closed, forbidden border to a soft, open and welcoming border (See Figure 4). In the three decades since the events of 1989–1990 we have witnessed not only a shaping of the geographical border in terms of stimulating cross-border traffic flow or cross-border cooperation, but also a mutation from ethno-national to cosmopolitan mindsets.[28]

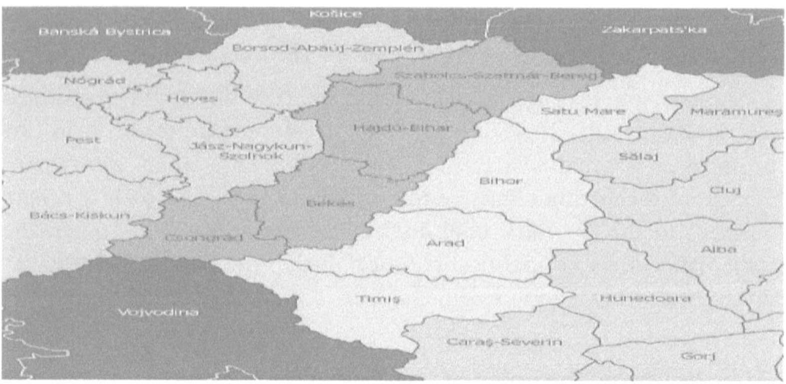

Figure 4: The Romanian-Hungarian border between 1990–present.[29]

In the process of EU-integration the Hungarian-Romanian state border is becoming even more permeable as well. Despite the fact, that Romania is not member of the Schengen Agreement, crossing the borderline between the two neighbouring countries is much easier than a few years earlier. The growing permeability of the European borders contributes to the unification of the economic potentials of the neighbouring areas supplying benefits on regional and local level.[30]

---

28 C. S. Săraru, "Considérations sur les accords de coopération transfrontalières entre les unités administratives-territoriales limitrophes des zones frontalières de la Roumanie et les structures similaires dans les pays voisins", *Curentul Juridic*, vol. 2, n°45, 2011, p. 88.

29 Source: Retrieved from: <https://interreg-rohu.eu/en/eligible-area/>. Accessed 23 December 2023.

30 I. Polgár, "Romanian-Hungarian cross border cooperation and the impact on transnational institution building and EU fund absorption" *In Analele Universităţii din*

Even if Hungary joined the EU on the 1st of May 2004 and Romania on the 1st of January 2007, by the common European destiny of the two states, they are obliged to achieve a strengthened partnership focused on increasing their welfare and harmonious economic development.[31] After the change of regime, from the beginning of 1990s the citizens of Central and Eastern Europe could move freely. At the beginning of the 1990s new border stations were opened one after the other and the frontier traffic increased considerably. Mainly the shopping tourism's growth was behind the increase of the trips' number and especially the border regions' inhabitants were concerned in it (See Figure 5).

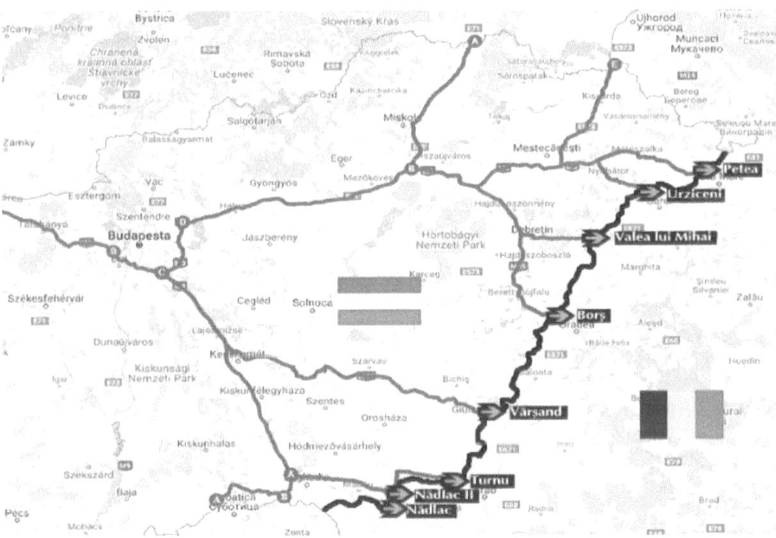

Figure 5: Border crossing points on the Romanian-Hungarian borderline between 1990–2004.[32]

Oradea, Seria Relații Internaționale și Studii Europene, TOM XIV, Oradea, 2022, pp. 23–34.

[31] A. C. Popoviciu, "The frontier worker. Romania-Hungary Study Case", *In* A. C. Popoviciu & D. Cigan (eds.), *The Frontier Worker – New Perspectives on the Labor Market in the Border Regions*, C.H. Beck, Bucharest, 2013, pp. 25–27.

[32] Source: Retrieved from: <https://www.politiadefrontiera.ro/en/main/i--map-with-the-itinerary-of-the-open-border-crossing-points-at-the-romanianhungarian-border-7592.html>. Accessed 23 May 2023.

Different relationships built on the cultural and financial similarities and interests were formed successively among those settlements, which lie on both sides of the border. Naturally the fact that there are communities on both sides of the border, which belong to the same ethnic groups, was a motivational factor in the deepening of the border regions' relationships. It had a positive effect on the concerned regions relationships and it deepened the cooperation among them.[33]

Based on the common elements mentioned above, based on the needs of the local community, based on the European context and the demand from the economy the development and the opening of new border points continued. The central authorities together with the local authorities from both countries established an intensive program for crossing the Romanian-Hungarian border on ten new border points with cross-border roads, all built with European funding, allocated within the Hungarian-Romanian cross-border cooperation program 2007–2013. The aim of these actions was to ensure the use of these cross-border roads as widely as possible within the limits allowed by the applicable legal framework, including the Agreement between the Government of Romania and the Government of Hungary regarding road links across the Romanian-Hungarian state border, until Romania's accession to the Schengen area.[34] Based on the consultation between the two states authorities, these border points were supposed to have a permanent regime. These border points were: Nădlac – Csanadpalota, Peleş – Zajta, Bercu – Garbolc, Horea – Omboly, Variaşu Mic – Dombegyhaza, Grăniceri – Elek, Cheresig – Körösnagyharsány, Iarmata Neagră–Dénesmajor, Roşiori – Pocsaj, Voivozi – Bagamer.[35]

---

33 I. Polgár, *The impact of territorial governance structures from Bihor County on cross-border cooperation during 2014–2020, op.cit.*

34 Ministry of Internal Affairs of Romania, "Transit of operational border points on the border with Hungary under permanent regime". Retrieved from: <https://www.mai.gov.ro/en/transit-of-the-operational-border-points-at-the-border-with-hungary-in-permanent-regime-the-time-intervals-that-generated-the-queues-at-the-border-with-romania-were-eliminated/>. Accessed 22 February 2023.

35 Border crossing points on the Romanian-Hungarian borderline: Data retrieved from the Romanian Border Police. <https://www.politiadefrontiera.ro/ro/traficonline/>. Accessed 22 February 2023.

The Border and Cross-Border Cooperation: Theoretical and Conceptual Analysis

Although the existence of borders is one of the basic elements of human existence, today borders represent a political-administrative demarcation line between countries. The formation process of national states in Europe was accompanied by multiple and significant changes at the level of borders.

Regarding the definition of the areas close to the state border, according to Hansen we can affirm that it is the part of the natural space, where the economic and social life is directly affected by the presence of an international border.[36] The rigidity or openness of a border is determined by the economic and social characteristics of the border area. The border area must be treated and examined differently, considering that it has other characteristics and behaves differently than the central or peripheral internal areas. These differences also apply to the analysis of border areas, the examination of each border area requires a unique approach.[37]

From a typological point of view, we can highlight four categories of borders. The first refers to the barrier type border, which divides the physical space. The second category refers to the filter type border. The third category is represented by the periphery type border or buffer zone and the fourth category is the contact zone type border, contact element.[38] Among the four categories, only the first represents a rigid, dividing line, the others representing a different role and character of the border in the same space. It is important to mention that the first category, the barrier-type border, has the ability to play the role of a filter or buffer zone or even a contact zone.

During the examination of border types, it is necessary to analyse the so-called mental borders. A mental frontier represents a so-called border that takes shape in the minds of those who live in the border area, in terms of the position and location of their own country vis-à-vis neighbouring

---

36 N. Hansen, "Border regions: A critique of spatial theory and a European case study", *Annal of Regional Science*, vol. 15, 1977, pp. 255–270.

37 A. Ilieş, *România între milenii. Frontiere, areale frontaliere şi cooperare, op. cit.*, pp. 92–97.

38 J. Nemes-Nagy, *A tér a társadalomkutatásban*, ETR, Budapest, 1998, p. 141.

regions or countries.[39] By the mental border we mean an imaginary line that defines the space of a person's territorial identity. Within this space, the population is classified into two categories, the group called "mine" and the group of those who are different from an ethnic point of view, who live outside the first group.[40]

During the typological analysis we emphasised the above mentioned five border types, since, through all of these types, the Romanian-Hungarian border can be easily characterised. In order to delimitate the border areas, we use the indicators established by the NUTS system used in the EU. The volume and the nature of cross-border cooperation can be correctly examined by demarcating the region or the cross-border space, respectively taking into account its socio-economic specifics. Cross-border cooperation can be defined in relation to direct collaboration between neighbouring regions, found along a border, regardless of the level at which this relationship is achieved.[41] Collaboration can be achieved in all fields, between all national, regional and local authorities, involving all actors. The efficiency, reliability, authenticity and legitimacy of the European construction, of European integration are also linked to the success of cross-border cooperation.[42] In this context, we use the definition of Charles Ricq:

> "...The basic principle of cross-border cooperation is the creation of contractual spaces in the border areas in order to allow finding common solutions to similar problems, where state entities are not ignoring their peripheral communities and the particularity and specificity of the neighbourhood problems they face".[43]

The efficiency of cross-border cooperation depends on the coordination of all actors involved in this process, but also on cooperation at all levels,

---

39 H. Tamas, *Az államhatár mint térbeli jelenség, In* H. Tamas et al., *Határok és városok a Kárpát-medencében*, MTA RKK, Győr-Pécs, 2009, p. 20.

40 I. Polgár, "The Phenomen of Migration and Integration of Minorities in the European Union", *In* I. Polgár, I. Horga & M. Brie (eds.), *Migration and European Integration of Minorities*, Lambert Academic Publishing, Saarbrucken, 2017, pp. 9–11.

41 L. Pantea & M. Brie, "The role of International Cooperation in Bihor County's Sustainable Development strategy", *In Analele Universității din Oradea*, Seria *Relații Internaționale și Studii Europene*, Oradea, 2020, p. 47.

42 I. G. Bărbulescu, M. Brie & N. Toderaș, *Cooperarea transfrontalieră între România și Ucraina, op. cit.*, pp. 21–27.

43 C. Ricq, *Manual de cooperare transfrontalieră pentru uzul colectivităților locale și regionale în Europa op. cit.*, p. 10.

doubled by the existence of an adequate legal framework. Dysfunctionality at one of the levels can generate its failure.[44] Therefore, cross-border cooperation is no longer the strict prerogative of states, as international actors.[45]

An important role in the process of cross-border cooperation is played by the local and regional administrative structures from the border areas, the development associations created by these, but also the transnational corporations interested in promoting an economic and political environment favourable to business development[46]. Bihor County and Hajdú-Bihar County are such administrative-territorial units, and in order to be successful in the border area it is a must to cooperate with cross-border partners on an equal basis and respect the interests of each involved party.

The EU encourages and stimulates the development of partnerships in cross-border cooperation. The EU's interest is not only to assure a harmonious development of the border regions, but also the strengthening of the European spirit and identity among its citizens, beyond their ethno-national identities.[47] Following the political and economic changes of the last 20 years in Central and Eastern Europe, the basic conditions for this type of cooperation are now ensured in most border regions. The intention and need for cooperation are indicated by the growing number of Euroregion's and other forms of cross-border cooperation. The EU Member States from Central and Eastern Europe have developed different strategies and legal frames regarding cross-border cooperation. These small differences make the process more efficient and functional in certain States than in others.[48]

During the last 15 years, Romania and Hungary have built a structure and a similar institutional system for the functionality of the cross-border

---

44 L. Pantea & M. Brie, "The role of International Cooperation in Bihor County's Sustainable Development strategy", *op.cit.*, p. 47.

45 M. Brie, "Rethinking the European Neighbourhood Policy. Eastern Partnership", *In Analele Universității din Oradea, Seria Relații Internaționale și Studii Europene*, Oradea, 2017, pp. 55–71.

46 C. S. Săraru, "Considérations sur les accords de coopération transfrontalières entre les unités administratives-territoriales limitrophes des zones frontalières de la Roumanie et les structures similaires dans les pays voisin", *op.cit.*

47 M. Brie & D. Blaga, *Identity rematch in the European space*, *In* A. Stoica, I. Horga, M. Brie, *Cultural Diplomacy at the Eastern and Western Borders of the European Union*, Editura Universității din Oradea, Oradea, 2015, pp. 255–273.

48 C. Ricq, *Manual de cooperare transfrontalieră pentru uzul colectivităților locale și regionale în Europa*, *op.cit.*, p. 10.

cooperation process. This system is based on the European legislative framework, on the best practices from other European border regions, as well as on the interpretation of the historical past, respectively the ethno-linguistic affinities of the communities from the border area.[49]

It should also be noted that cross-border cooperation between administrative-territorial units from Romania's and Hungary's border areas depended and depends on the degree of openness allowed by the political regime, the standard of living and the cultural, historical and ethnic affinities of the neighbouring environment.

Cross-Border Cooperation: Drivers and Examples from Bihor and Hajdú-Bihar Counties

In the Romanian-Hungarian border region the frontier had a permanent hindering effect over the social-economic development of the area. Since it was drawn, it permanently had a deep dividing character.

Nowadays, we can observe that the dividing character of the border is more subtle, but the periphery position and the low development level of the area, which mainly was generated by the immediate vicinity of the border are difficult to change overnight. However Western European examples demonstrate that the borderland position does not mean disadvantage for the social-economic development in itself, moreover it can be definitely advantageous too. Bihor and Hajdú-Bihar counties and the two urban centres, Oradea and Debrecen have common historical and cultural roots. These centuries-old ties after Trianon and during the communist period became weaker. Still, the two cities established closer links during the last two decades. Currently, this type of multilateral cooperation, which includes all areas of life, characterises the relationship between the two cities. The cooperation between the two cities, which lie about 60 kms from each other, is promoted by the similar size, the similar number of inhabitants, the similar geographical (borderland) position and the problems, which come from it (periphery position within the countries).

European cross-border cooperation aims to approach challenges, which are jointly identified and solved in border regions. Also, it proposes to exploit untapped growth potential of border areas, through the

49 I. G. Bărbulescu, M. Brie & N. Toderaş, *Cooperarea transfrontalieră între România şi Ucraina*, *op. cit.* pp. 21–27.

cooperation process, for a harmonious development of the whole EU.[50] Cross-border cooperation agreements and partnerships have recently become increasingly important in the context of the EU's cooperation policies for Member States or for the European Neighbourhood Policy. Cross-border cooperation can be defined in relation with direct collaboration between neighbouring regions, found along a border, regardless of the level at which this partnership is achieved. Regional and cross-border cooperation operates with the principle of subsidiarity, of "reallocation of authority"[51] from the state level to the administrative-territorial units situated in the border areas having the main goal to balance regional imbalances. Thus, cross-border cooperation agreements are a way to promote good neighbourliness, stimulate balanced economic development and social stability by capitalising on local and regional resources in joint projects.

Romania's accession to the EU in 2007 changed the perspective on cross-border cooperation with Hungary. If before there was cooperation at the external border of the EU, after 2007 cooperation between two Member States is being considered. European policies, programs, instruments and mechanisms of cross-border cooperation are able to amplify cross-border cooperation and the development of public services in the cross-border space.

The debate regarding regional development as a concept, but also as a direction or guideline for public policies, is a very complex one. Several components are needed to create an advantageous environment in which a Euroregion can be established and developed.[52] Economic stability, common cultural elements, common historical heritage, respectively partnerships between administrative units belonging to EU member states. Territorial administrative units, public law bodies and associations

---

50 M. Brie, European instruments of cross-border cooperation – Case study: The Romanian-Ukrainian Border. *In* F. Maron & G. Pozarlik (eds.) *Identities, citoyennetes et democratie, 20 ans apres*, Bruylant, Brussels, 2010, pp. 265–280.

51 B. Dicharry & J. Hamm, "*L'Union européenne est-elle toujours au service du défi de la convergence?*", *Bulletin de l'Observatoire des politiques économiques en Europe*, n° 39, 2018, pp. 3–8.

52 Council of Europe, "European Outline Convention on Transfrontier Co-operation between Territorial Communities or Authorities, Madrid, Council of Europe, Strasbourg, 1980. Retrieved from: <http://conventions.coe.int/Treaty/en/Treaties/Html/106.htm>. Accessed 28 February 2023.

that are part of Euroregion's have at their disposal several instruments of territorial cooperation, regulated by Community law.[53]

In order to achieve the proposed European goals and using European best practice, with the creation of common, cross-border bodies the Bihor – Hajdú-Bihar Euroregion was founded by the two counties (See Figure 6). The Bihor – Hajdú-Bihar Euroregion is a structure with legal personality established in 2002, at the initiative of the Bihor County Council and the Hajdú-Bihar Local Self-Government to organise and stimulate new ways of cooperation between local communities. Its establishment is based on numerous antecedents of cooperation between the two counties.

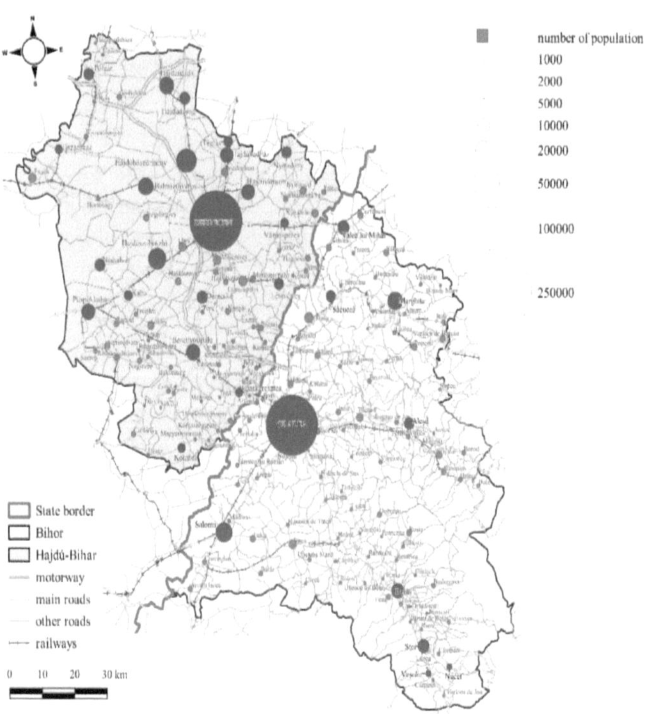

Figure 6: Bihor-Hajdú-Bihar Euroregion containing the network of urban settlements on both sides of the border.[54]

---

[53] I. Polgár, "The impact of territorial governance structures from Bihor County on cross-border cooperation during 2014–2020", *op.cit.*

[54] Source: I. Polgár, "Romanian-Hungarian cross border cooperation and the impact on transnational institution building and EU fund absorption", *op.cit.*, p. 30.

The Euroregion is made up of two counties at NUTS 3 level that fall into 2 NUTS2 development regions.[55] The decisions and activities undertaken within the Euroregion are validated and coordinated by the Steering Committee, respectively by the Council of the Euroregion.[56] The Council of the Euroregion decides the acceptance of new members. In case if an existent member of Euroregion expresses its will to exit the Euroregion, the rights and obligations of the member which leaves the Euroregion will be deleted only after the approval of the Council of the Euroregion. Applications for joining the Euroregion and the declaration of acceptance, as well as the application regarding the intention to leave the Euroregion, must be submitted to the Steering Committee, which, after expressing an opinion on these applications, forwards them to the Council. Admission to the Euroregion takes effect from the moment the application is approved by the Council. The newly admitted member will pay the same fee as all other members, but proportional to the time period remaining in the current year. The size of this amount will also be established in the Council's decision to accept the accession of new members.[57]

Most of the objectives of the European territorial cooperation programs focus on solving cross-border problems and jointly developing the potential of various territories. The cooperation activities undertaken are financially supported by the European Regional Development Fund (ERDF) through three key components. These components are cross-border cooperation, transnational cooperation and interregional cooperation. In the case of the Bihor – Hajdú-Bihar Euroregion, the Interreg Romania-Hungary Program is not only the promoter of the activities but provides the necessary financial support for the implementation of the initiatives. During two development periods, 2007–2013 and 2014–2020, respectively, the program supported the implementation of integrated interventions with a strong cross-border and strategic focus, with a view to exploiting common potential and managing common challenges.

---

55 Ministerul Afacerilor Externe, *Tratat de Înțelegere, Cooperare și Bună Vecinătate între România și Republica Ungară*. Retrieved from: <http://www.mae.ro/node/56691>. Accessed 26 February 2023.

56 "Hotărârea Consiliul Județean Bihor privind funcționare Euroregiunii Bihor- Hajdú-Bihar", 2013, p. 38. Retrieved from: <https://www.cjbihor.ro/informatii-publice/hotarari-ale-consiliului-judetean/hotarari-ale-consiliului-judetean-anul-2013/>. Accessed 26 February 2023.

57 *Ibid.*

For the future development period 2021–2027, the Interreg Romania-Hungary Program aim is to continue financing joint Romanian-Hungarian projects, which address needs identified on both sides of the border and which require a common approach, as well as innovative solutions, thus contributing to the sustainable development of the eligible area.[58] In the future, the main task of the Euroregion consists in the joint exploitation of the advantages offered by the border area. The aim of this approach is to develop the economy, tourism, environment and human resources in the border areas. An essential condition for achieving these objectives is improving the interoperability of borders and the accessibility of the border area. In this regard the realisation of new border crossing points is necessary (See Figure 7). Over the infrastructural development, the final decision over the accession of Romania in the Schengen Area is vital.

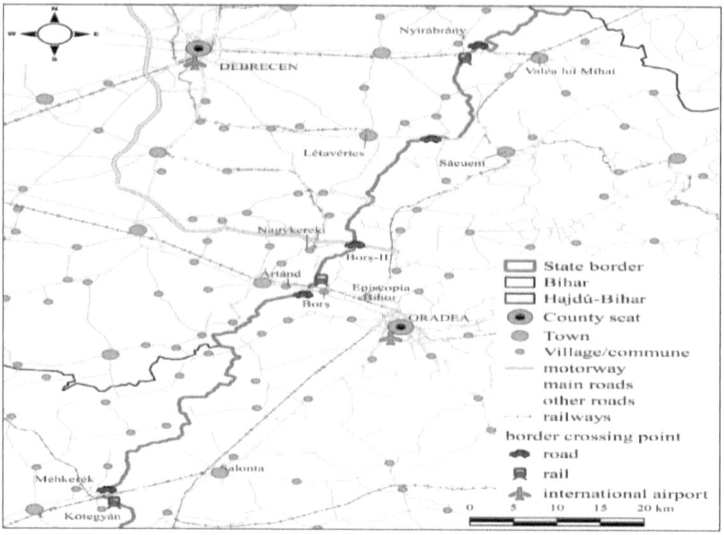

Figure 7: Guide for the Development of International Cooperation of Bihor County.[59]

---

[58] Interreg Romania-Hungary Program overview. Retrieved from: <https://interreg-rohu.eu/en/programme-overview/>. Accessed 26 February 2023.

[59] Source: M. Brie, M. Marcut & I. Polgar, "Ghid pentru Dezvoltarea Cooperarilor Internationale al Judetului Bihor [Guide for the Development of International Cooperation of Bihor County]", Oradea, 2022, p. 31.

Beyond the institutions which facilitate Bihor and Hajdú-Bihar County's access to regional development programs and cross-border cooperation, the two counties have the opportunity to participate in international organisations, but also to make cooperation agreements with other similar entities. The Bihor – Hajdú-Bihar Euroregion, it is an entity created to develop cross-border cooperation and with the stated purpose of stimulating new ways of cooperation between the communities of the two countries, starting from the principles of good neighbourliness. The Euroregion proposed the development of joint projects and to support the two members in the process of European integration.[60]

Although there is an institutional framework at Community level, which is present in the Member States through different implemented EU policies and programs, the level and the forms of cooperation between territorial governance structures differ from state to state.[61] There are Member States that built and developed relations for a stable cross-border cooperation at institutional level. On the other side, there are states, which are still guided by a general scepticism regarding cross-border cooperation between territorial governance structures. Another factor, which makes cooperation difficult, is the lack of communication between the actors.[62] The cross-border cooperation process will generate significant and effective results only if the actors involved set common goals and allocate the necessary resources.

In the following, we propose a synthesis of all contracted projects by public institutions from Romania, Bihor County, from the eligibility area of the Interreg Romania-Hungary Program in the period 2014-2020 (See Figure 8). The source of the presented data is the web archive of the Romania-Hungary Interreg cross-border cooperation program and a research article published by the author in the Journal of Romanian Literary Studies n°25/2021.Based on the information obtained from the

60 Consiliul Județean Bihor, "Statutul Județului Bihor", 2013, pp. 37–39. Retrieved from: <https://www.cjbihor.ro/wp-content/uploads/2020/07/Statutul-Judetului-Bihor.pdf>. Accessed 26 February 2023.

61 D. Bădulescu, A. Bădulescu & C. A. Bucur, "Considerations on the effectiveness of cross-border cooperation in public order and civil protection services. The case of the Romanian-Hungarian border area in LEX LOCALIS", *Journal of Local Self-Government,* vol. 13, n°3, 2015, pp. 559–578.

62 I. Polgár, "The impact of territorial governance structures from Bihor County on cross-border cooperation during 2014–2020", *op.cit.*

Figure 8: List of contracted projects by Romanian territorial governance structures from Bihor County as Lead Partner, First Open Call - December 2019[63]

Nr. crt.	eMS code	Project title	Acronym	Lead beneficiary	Partners	Total budget euro	Total ERDF euro
1.	ROHU68	Creating a joint tourist destination in the crossborder area of Lugaşu de Jos and Komádi, through cross-border protection of natural heritage values along the Crişul Repede Valley	PRO-CRV	Comuna Lugaşu de Jos	Komádi Városi Önkormányzat Asociaţia "Pro Valea Crişului Repede – Pro Sebes Koros Volgy"	904,905. 47	769,169. 64
2.	ROHU126	Nature preservation, protection and promotion on both sides of the Romainian-Hungarian border	PRONA-TURE	Bihor County Council	Primaria Vadu Crisului Asociaţia Centrul pentru arii protejate şi dezvoltare durabilă Bihor Körösszegapáti Nagyközségi Önkormányzat	2,500,000. 00	2,125,000. 00
3.	ROHU48	Development of an integrated response capacity to cross-border emergency situations	DIRCCES	Town of Alesd	Sinteu Comune Szarvas Town Self Government	485,400. 00	399,532. 74
4.	ROHU14	Conservation, protection and promotion of the natural values from the Salonta-Békéscsaba crossborder area	The nature corner	Salonta Municipality	Körösök Völgye Natúrpark Egyesület Asociaţia Milvus Transilvania Vest Békés Város Önkormányzata	2,997,387. 86	2,547,779. 68
5.	ROHU53	Conservation and protection of the cross border natural heritage of Bihor - Hajdú-Bihar counties	Na-Tu-Re	Comuna Sinmartin	Asociatia Pescarilor Sportivi Aqua Crisius Fundatia pentru cultura si educatie ecologista Ecotop Zold Kor -a Fold Baratai Magyarorszag Tagja	2,835,216. 60	2,409,934. 10
6.	ROHU115	From One Bridge to another - De la un pod la altul – Hídtól hídig	FOBTA	Comuna Sălacea	Hortobágy Község Önkormányzata	1,611,547. 26	1,369,815. 16

[63] Source: Romanian-Hungarian Interreg. Retrieved from: https://interreg-rohu.eu/wp-content/uploads/2019/12/Contracted-projects-12.12.2019.pdf. Accessed 26 February 2023.

I Interreg Romania-Hungary Program data base, after the first open call in December 2019 a total number of 21 projects were contracted. From the 21 contracted projects, there are 6 projects contracted as Lead Partner, by Romanian territorial governance structures from Bihor County (See Figure 9).

Figure 9: List of contracted projects by Romanian territorial governance structures from Bihor County as Lead Partner, Second Open Call – March 2021[64]

Nr. crt.	eMS code	Project title	Acronym	Lead beneficiary	Partners	Total budget euro	Total ERDF euro
1.	ROHU179	Administrative bridge between towns in the Romania - Hungary cross border region	ABBTROHU	Town of Alesd	Comuna Sinteu Orasul Marghita Szarvas Varos Onkormanyzata Szarvas Town Self-Government Berettyoujfalu Varos Onkormanyzata/ Berettyoujfalu Town Self-Government	362,500.00	308,125.00
2.	ROHU183	SAFE BORDER, SAFER LIFE SAFEBOR	SAFEBOR	The Prefect's Institution - Bihor county	Hajdú-Bihar Megyei Rendőrfőkapitányság/ Hajdú-Bihar County police Inspectoratul Teritorial al Politiei de Frontiera Oradea Inspectoratul de Politie Judetean Bihor	287,904.85	244,719.12
3.	ROHU200	Crossborder events for crossborder citizens	CCC	Toboliu Commune	Körösnagyharsány Local Self Government Girisu de Cris Commune Nojorid Commune	77,000.00	65,450.00
4.	ROHU265	LET'S CELEBRATE OUR TRADITIONS TOGETHER TOGETHER TOGETHER	TOGETHER	COMUNA CETARIU	KABA Város Önkormányzat - KABA City Local Government	70,400.00	59,840.00

(continued)

[64] Source: Romanian-Hungarian Interreg. Retrieved from: https://interreg-rohu.eu/wp-content/uploads/2021/03/Signed-contracts-2nd-open-call-website-19.03.2021.pdf. Accessed 26 February 2023.

Figure 9: Continued

Nr. crt.	eMS code	Project title	Acronym	Lead beneficiary	Partners	Total budget euro	Total ERDF euro
5.	ROHU271	The Minority Question in the ROHU crossborder area: Slovak Culture, Tradition and History without Borders	SCTH	Comuna Sinteu	Csabai Szlovákok Szervezete/Organization of Slovaks from Csaba	71,500. 00	60,775. 00
6.	ROHU280	Cultural Cooperation among Citizens from the Salonta-Gyula Cross-border Area Co-cultured	Co-cultured	The Municipality of Salonta	Gyula town	80,000. 00	68,000. 00
7.	ROHU283	Sustainable cross-border cooperation for citizens between Marghita and Berettyóújfalu municipalities	CCC_MB	Municipiul Marghita	Berettyó Cultural Center	80,000. 00	68,000. 00
8.	ROHU289	Connecting the traditions of the Crisul Repede Valley	CCRV	Comuna Vadu Crişului	Körösszegapáti Nagyközségi Önkormányzat - Körösszegapáti Local Government	79,999. 00	67,999. 15
9.	ROHU198	Open Doors for Culture and Tradition in Europe	ODCTEU	Town of Alesd	Kaba Town Self-Government	74,055. 00	62,946. 75
10.	ROHU319	Joint program for cross - border cooperation between young people	4 YOUTH	Paleu Commune	Hosszúpályi Commune Cetaiu Commun Kaba town	200,600. 00	170,510. 00

In the second open call a total of 37 projects were contracted. There was a number of 10 projects contracted as Lead Partner, by Romanian territorial governance structures from Bihor County (see Figure 10).

Figure 10: List of contracted projects by Romanian territorial governance structures from Bihor County as Lead Partner, Third Open Call – June 2021[65]

Nr. crt.	eMS code	Project title	Acronym	Lead beneficiary	Partners	Total budget euro	Total ERDF euro
1.	ROHU425	Joint Development of Human Resources in CrossBorder AreaJoint Development of Human Resources in Cross-Border Area	HRArea UAT	UAT Comuna Paleu	Derecske Város Önkormányzata Asociatia Pro Cariere Asociatia de Dezvoltare Intercomunitara Zona Metropolitana Oradea	1,337,577.8	1,136,941.13
2.	ROHU-421	Improve employment in Bihor County, Oradea and Hajduboszormeny through the development of the local potential infrastructures	CBC-EMPLOY-MENT	Municipality of Oradea	Development Agency of Bihor County Local Government of Hajduboszormeny City County Agency for Employment Bihor	7,507,873.22	2,549,531.04
3.	ROHU-426	Support to the local economy by encouraging alternative use of agricultural products	ALTERAGRI	Directorate for Bihor County Agriculture	Asociatia „TRANSILVANIA TECHNOLOGY TRANSFER" Oficiul de Studii Pedologice si Agrochimice Bihor Science for Environment Public Foundation	1,679,320.00	1,427,422.00

(*continued*)

[65] Source: Romanian-Hungarian Interreg. Retrieved from: https://interreg-rohu.eu/wp-content/uploads/2021/06/Signed-contracts-Open-call-3-23.06.2021.pdf. Accessed 26 February 2023.

Figure 10: Continued

Nr. crt.	eMS code	Project title	Acronym	Lead beneficiary	Partners	Total budget euro	Total ERDF euro
4.	ROHU-359	Promoting an integrated approach regarding the improving of the employment rate and business environment at the level of Marghita and Beretyoujfalu microregions	CROSSL-OCALDEV	Marghita Municipality	Municipality of Berettyóújfalu	2,615,165.66	2,222,890.81

In the third open call a total of 31 projects were contracted. There was a number of 4 projects contracted as Lead Partner, by Romanian territorial governance structures from Bihor County. Based on the data presented in the three charts, we can affirm that even though the number of territorial government structures present in the list of beneficiary/lead partners is fluctuating, still these are playing a very important role in the implementation of the cross-border programs.P.580

Based on the information obtained from the I Interreg Romania-Hungary Program data base, after the first open call in December 2019 a total number of 21 projects were contracted. From the 21 contracted projects, there are 6 projects contracted as Lead Partner, by Romanian territorial governance structures from Bihor County (See Figure 9).

In the second open call a total of 37 projects were contracted. There was a number of 10 projects contracted as Lead Partner, by Romanian territorial governance structures from Bihor County (See Figure 10).

In the third open call a total of 31 projects were contracted. There was a number of 4 projects contracted as Lead Partner, by Romanian territorial governance structures from Bihor County.

Based on the data presented in the three charts, we can affirm that even though the number of territorial government structures present in the list of beneficiary/lead partners is fluctuating, still these are playing a very important role in the implementation of the cross-border programs.

Conclusions

The border between Romania and Hungary has gone through allot of changes in terms of understanding the space of a culture, race, religion, city or nation. Growing from physical walls, intense militarised areas to a formal understanding of regions with no physical boundaries or restrictions of any kind, the frontier or the border, has been an area of the utmost importance to the different branches of science.

The border region is the collision point of cultures, religions, languages, administrations, legislations and economic influences. All these factors have substantial contribution to the development of the phenomenon called cross-border cooperation. Border regions have a peripheral geographical position and often an unfavourable economic situation. Due to these characteristics, they start with a significant disadvantage in the development and integration process. The issue of cross-border cooperation in general goes beyond geographical, economic and legal approaches. If we intend to treat the phenomenon from a specific point of view, we will find ambiguities and uncertainties regarding the role and functioning of some institutions in the process of cross-border cooperation and partnerships.

Cross-border cooperation is an important element of the European integration process. Based on the research carried out in the border regions, we can say that the development of joint programs oriented towards innovation and institutional cooperation can represent the solution for the success of the territorial development of the border areas. On the other hand, there are specific problems in each border region. The role and level of involvement of the various local or regional actors in the construction and realisation of partnerships and strategies regarding cross-border cooperation differ. In this context, we tried to investigate the situation of cross-border cooperation in terms of institutional cooperation, focusing on the specifics of the Bihor – Hajdú-Bihar Euroregion.

The Hungarian-Romanian border, from the fall of communism and from the Trianon Treaty, has changed its status of a hard, close, exclusive border to that of a soft, open, inclusive frontier. In the process of EU-integration the Hungarian-Romanian state border is becoming even more permeable as well.

Bibliography

A. Athanasiu, "Foreword", *In* A. C. Popoviciu & D. Cigan (eds.), *The Frontier Worker – New Perspectives on the Labor Market in the Border Regions*, C.H. Beck, Bucharest, 2013, pp. 7–8.

D. Bădulescu, A. Bădulescu & C. A. Bucur, "Considerations on the effectiveness of cross-border cooperation in public order and civil protection services. The case of the Romanian-Hungarian border area in LEX LOCALIS", *Journal of Local Self-Government,* vol. 13, n°3, 2015, pp. 559–578.

I. G. Bărbulescu, *Noua Europă. Identitate şi model European*, Editura Polirom, Iaşi, 2015.

I. G. Bărbulescu, M. Brie & N. Toderaş, *Cooperarea transfrontalieră între România şi Ucraina, respectiv între România şi Republica Moldova. Oportunităţi şi provocări în perioada 2014–2020*, Institutul European din Romania, Bucharest, 2016.

L. Boia, *Istorie şi mit în conştiinţa românească*, Humanitas, Bucharest, 1997.

M. Brie, *The European Neighborhood Policy, Mass-media and Cross-border Cooperation*, *In* Analele Universităţii din Oradea, Seria Relaţii Internaţionale şi Studii Europene, Oradea, 2009, pp. 81–86.

M. Brie, European instruments of cross-border cooperation – Case study: The Romanian-Ukrainian border, *In* F. Maron & G. Pozarlik (eds.), *Identities, citoyennetes et democratie, 20 ans apres*, Bruylant, Brussels, 2010, pp. 265–280.

M. Brie, "Rethinking the European Neighbourhood Policy. Eastern Partnership", *In Analele Universităţii din Oradea, Seria Relaţii Internaţionale şi Studii Europene*, Oradea, 2017, pp. 55–71.

M. Brie & D. Blaga, *Identity rematch in the European space*, *In* A. Stoica, I. Horga & M. Brie (eds.), *Cultural Diplomacy at the Eastern and Western Borders of the European Union*, Editura Universităţii din Oradea, Oradea, 2015, pp. 255–273.

M. Brie, M. Marcut & I. Polgar, "Ghid pentru Dezvoltarea Cooperarilor Internationale al Judetului Bihor [Guide for the Development of International Cooperation of Bihor County]", Oradea, 2022.

Ş. Delureanu, *Geneza Europei comunitare: mesajul democraţiei de inspiraţie creştină*, Editura Paideia, Bucharest, 1999.

B. Dicharry & J. Hamm, "*L'Union européenne est-elle toujours au service du défi de la convergence?*", *Bulletin de l'Observatoire des politiques économiques en Europe*, n° 39, 2018, pp. 3–8.

N. Hansen, "Border regions: A critique of spatial theory and a European case study", *Annal of Regional Science*, vol. 15, 1977, pp. 255–270.

A. Ilieş, *România între milenii. Frontiere, areale frontaliere şi cooperare transfrontalieră*, Editura Universităţii din Oradea, Oradea, 2003.

N. Ionescu-Gură, *Stalinizarea României. Republica Populară Română 1948–1950: transformări instituţionale*, Editura All, Bucharest, 2005.

K. Károly & K. H. Eszter, *Ethnic Geography of the Hungarian Minorities in the Carpathian Basin*, Geographical Research Institute Hungarian Area, Budapest, 1998.

A. Marga, *Filosofia unificării europene [The Philosophy of the European Unification Process]*, Biblioteca Apostrof, Cluj-Napoca, 1997.

A. Marino, *Revenirea în Europa: idei şi controverse româneşti: 1990–1995*, Editura Aius, Bucharest, 1996.

L. Nastasă, A. Varga, A. Andreescu & A. Zub, *Minorităţi etnoculturale, mărturii documentare: maghiarii din România (1956–1968)*, Centrul de Resurse pentru Diversitate Etnoculturală, Bucharest, 2003.

J. Nemes-Nagy, *A tér a társadalomkutatásban*, ETR, Budapest, 1998.

L. Pantea & M. Brie, "The role of International Cooperation in Bihor County's Sustainable Development strategy", *In Analele Universităţii din Oradea*, Seria *Relaţii Internaţionale şi Studii Europene*, Oradea, 2020.

I. Polgár, "The Phenomenon of migration and integration of minorities in the European Union", *In* I. Polgár, I. Horga & M. Brie (eds.), *Migration and European Integration of Minorities*, Lambert Academic Publishing, Saarbrucken, 2017, pp. 9–11.

I. Polgár, "The impact of territorial governance structures from Bihor County on cross-border cooperation during 2014–2020", *Journal of Romanian Literary Studies*, n°25, 2021, pp. 392–405.

I. Polgár, "Romanian-Hungarian cross border cooperation and the impact on transnational institution building and EU fund absorption" *In Analele Universităţii din Oradea, Seria Relaţii Internaţionale şi Studii Europene,* TOM XIV, Oradea, 2022, pp. 23–34.

A. C. Popoviciu, "The frontier worker. Romania-Hungary study case", *In* A. C. Popoviciu & D. Cigan (eds.), *The Frontier Worker – New Perspectives on the Labor Market in the Border Regions*, C.H. Beck, Bucharest, 2013, pp.25–27.

C. Ricq, *Manual de cooperare transfrontalieră pentru uzul colectivităților locale și regionale în Europa*, Ediția a 3-a, Bucharest, 2000.

R. Sageata, "The evolution of the ethnic and political Romanian-Hungarian Border as reflected in sources", *Transylvanian Review* vol. *XXVIII*, n°4, 2022, pp. 3–22.

C. S. Săraru, "Considérations sur les accords de coopération transfrontalières entre les unités administratives-territoriales limitrophes des zones frontalières de la Roumanie et les structures similaires dans les pays voisins", *Curentul Juridic*, vol. 2, n°45, 2011, pp. 87–95.

T. Solomovici, R. Theodorescu & S. R. Stănescu, *Securitatea şi evreii: despre călăi şi despre victim*, Editura Ziua, Bucharest, 2004.

H. Tamas, *Az államhatár mint térbeli jelenség*, *In* H. Tamas et al., *Határok és városok a Kárpát-medencében*, MTA RKK, Győr-Pécs, 2009, p. 20.

R. Weber, *Un concept românesc privind viitorul Uniunii Europene*, Editura Polirom, Iași, 2001.

Outlining the Main Perception Strands on the Hungarian-Romanian Border[1]

MELINDA BENCZI

The analysis of borders between countries, societies and ethnic groups had been long in the forefront of academic scrutiny and public discourse. The theoretical narrative, however, did not remain unchanged throughout the decades. Whereas first it focused more on describing the location and history of the borders, later shifted towards the understanding of their functional content. Even though as early as in the 1960s scholars already pointed out that what mattered was the function and process of borders rather than their form and location and called for analysing the perceptions of the borders,[2] this direction only gained real traction in the 1990s. It is the study volume "The Geography of Border Landscapes"[3] that is considered to mark of a new approach that places the emphasis on the border landscape (instead of the border line) understood as the product of the totality of cultural, economic and political interactions.[4]

---

1 The present article relies on some of the research activities carried out within the framework of Jean Monnet network's project called *Frontières en Mouvement* (Frontem). The cartographic project of mental maps was implemented by Dr. Pauline Pupier from the Artois University in France, while some of the preparatory work regarding the literature review was carried out by Océane Menu, intern at Central European Service for Cross-Border Initiatives (CESCI), Budapest.

2 J. Minghi, "Boundary Studies in Political Geography", *Annals of the Association of American Geographers*, 1963, pp; 407–428.

3 D. Rumley & J. Minghi, *The Geography of Border*, Routledge, London, 1991.

4 G. Popescu, *Bordering and Ordering the Twenty-first Century. Understanding Borders*, Rowman & Littlefield Publishers, INC. UK, 2012.

One strand of this school of thought[5] is concentrating exclusively on the perception of the borders. Several studies theorised the topic,[6] but less so about the actual manifestation of the theory on the ground. This is despite of the fact that due to the process of globalisation and the EU's integration process in particular "has created demand for such studies of its internal borders and the changing views about them".[7] Accordingly, in compliance with the change in the separating and connecting role of Romanian borders as a result of the country's expected joining the Schengen area and emerging socio-economic cross-border cooperation, "adequate knowledge of the border areas is becoming more and more relevant nowadays".[8] Consequently, analysing the perception of border people on the Hungarian-Romanian border is a compelling exercise that attempts to understand how the border and the people living in the neighbouring country are perceived and what factors this perception depends on.

When analysing the citizens' perception of the Hungarian-Romanian border geographical, historical, linguistic, cultural, economic and even psychologic elements are to be taken into consideration all in the refraction of the personal and inter-generational observations and experiences. Literature distinguishes between two main types of tools which are used to evaluate and analyse these complexities. On the one hand, there are tools designed to visualise the perception of a border mainly through mental and cognitive mapping. "A mental map is a unique, personal and selective representation of reality. (…) Mental map is used as a reference for orientation and movement throughout a territory (routing), but also for associative processes and judgment valuation. Mental map is based upon

5 This is partly inspired by H. Lefebvre, *The Production of Space*, Blackwell, Oxford, 1991.

6 For example, see H. van Houtum, "The Development of Cross-Border Economic Relations", Thela Thesis Publishers, Amsterdam, 1998; T. Hardi, Z. Hajdú & I. Mezi, "Határok és városok a Kárpát-medencében", *MTA Regionális Kutatások Központja*, Győr, Pécs, 2009; A. Faludi, *The Poverty of Territorialisms, A Neo-Medeival View of Europe and European Planning*, Edward Elgar Publishing, Cheltenham, 2018.

7 H. van Houtum, "III European perspectives on borderlands", *Journal of Borderland Studies*, vol. 15, n°1, 2000, pp. 57–83.

8 B. Balázs & A. Farsang, "A szegedi középiskolások országhatár-képzete Magyarország délkeleti határáról", *Földrajzi Közlemények*, vol. 140, n°3, 2016, p. 258.

personal experience with an area, but by lack of this upon indirect information from mass media or a certain reputation".[9] However, this method is not without criticism with regards to its measurement since "drawings, pictures or sketches cannot be read objectively".[10] Nevertheless, more and more scholars decide to use it.[11] On the other hand, the perception of a border can also be analysed and understood through polls, surveys, workshops or focus groups. These tools and events give the opportunity to shed light on how the members of the society perceive the border on a more qualitative and detailed way. It gives a more in-depth overview on the subject especially by giving the opportunity to the target group to discuss and exchange their experiences and feelings. Even though this technique is somewhat impressionistic, it is still one of the best ways to tap into the unique experiences and perceptions that people have about the borders separating two countries.[12] Consequently, the present paper follows these scholarly traditions in assessing the perception of the border and the neighbours as well as the factors that these perceptions that are generated by it.

Methodology

The present paper is aligned with constructivist philosophy stating that there is no single reality or truth regarding how the border is or can be perceived but rather it is created by individuals in groups.

---

9 W. A. Sulsters, "Mental mapping, viewing the urban landscapes of the mind. Conference proceedings", Workshop 2, session 2.3, Mapping home, 2005, p. 1.

10 A. Székely & B. Kotosz, "From fence to wall? Changes of mental space of border zones in Eastern Europe", *Regional Science Policy & Practice*, vol. 10, issue 4, 2018, oo. 269–282.

11 See for example J. Le Rider, "Mitteleuropa, Zentraleuropa, Mittelosteuropa: A mental map of Central Europe", *European Journal of Social Theory*, vol. 11, n°2, 2008, pp. 155–169; L.A. Broadhead, "Canadian sovereignty versus Northern Security: The case of updating our mental map of the Arctic", *International Journal*, vol. 65, n°4, 2010, pp. 913–930; M. Rédei, Á. Kincses & Á. Jakobi, "The world seen by Hungarian students: A mental map analysis", *Hungarian Geographical Bulletin*, vol. 60, n°2, 2011, pp. 91–104.

12 M. Bruter, "On what citizens mean by feeling 'European': Percpetions of news, symbols and borderless-ness", *Journal of Ethnic and Migration Studies*, vol. 30, 2004, pp. 21–39.

Constructivism recognises "that reality is a product of human intelligence interacting with experience in the real world",[13] consequently, in order to be able to discover the meaning of events and activities, the reality needs to be interpreted. For the purposes of the present study, three different methods were used: literature review, mental mapping and a focus group discussion.

First of all, a preliminary literature review was carried out in order to gain sufficient understanding on previous relevant research results. Literature review targeted works which were either focusing on the Romanian-Hungarian border area or made universally valid observations that were not territorially bound. The author placed the biggest emphasis on the works written in Hungarian and Romanian about the subject in order to tap into the border perceptions as directly as possible. The findings of the literature review were used throughout the paper to compare and contrast with the findings of the primary data collection methods.

Secondly, the paper relies on the focus group discussion and the mental mapping exercise carried out within the framework of the Jean Monnet network project "Borders in motion" (FRONTEM) launched in 2019. Under the leadership of Sciences Po Strasbourg, the network had seven partners: the University of Southern Denmark, the Euro-Institute of Kehl, the Centre for Cross-Border Studies in Armagh, the Babeş-Bolyai University from Cluj-Napoca, the Catholic University of Louvain and the University of Victoria in Canada. The project aims to benchmark of and to provide a tool-kit on different border management systems and the evolving perception of borders by the citizens.

As part of the project, a focus group discussion was held on the 19th May 2022 at Cluj-Napoca, Romania with the active contribution of 10 participants from Hungary and Romania coming from various municipalities and backgrounds (academia, public policy, border services, civil society, etc.). The focus group was dedicated to assessing citizens' border

---

[13] D. Elkind, "Response to objectivism and education", *The Educational Forum*, 2005, p. 334.

perception. The names of the participants are not disclosed, the direct quotes are anonymised.

The cartographic project of mental map itself was implemented by Dr. Pauline Pupier from the Artois University in France between April and May 2022, however, its interpretation is the result of the author's own work. The mental map is created based on 22 online answers where primarily but not exclusively students were asked to mark those territories on the map displaying Romania and Hungary, which they consider border regions. For this purpose, the map was built up of grids representing areas of 20x20 km. The number and location of cells they could identify was not limited.

Whereas this research design allows to gain some valuable understanding of the Romanian-Hungarian border perception which fills a gap in the literature, the results are not representative neither territorially, nor temporally and cannot be extrapolated to other border territories in Europe or globally. Nevertheless, the results can serve as a memento of what was possible to be captured at this moment which later on can serve as a starting point for further inquiries as well as a basis for comparison in recording possible changes in the perception of the border (for instance after Romania's joining the Schengen area when similar research with the same methods can be repeated).

Citizens' Border Perception

Geographical Perception

One of the biggest international mental mapping projects was realised between 2009 and 2012 designed by the European Commission in order to explore various representations of Europe. In the project called "European Union & the world seen from abroad" (in short: EuroBroadMap) a quantitative survey involving 9,300 undergraduate students in 43 cities and 18 countries was carried out based on which an in-depth analysis of mental maps of the world inside and outside the European Union was prepared. Even though at least 40 scientific publications were drafted based on the research project, none of them focused specifically on the Hungarian-Romanian border, which at the time was an internal European Union, but external Schengen border.

At the same time, Kincses and Jakobi (2012)[14] explore the most important Hungary-related findings of the project in the light of the survey based on which important information can be gained in relation to the Hungarian-Romanian geographical border perception. There are two, concurring, mentions of Romania in the study. The first one is where the surveyed Hungarian students rate the top ten countries where they would not like to live. The list is led by Romania (112 answers) and followed by Iraq (79), China (69), Ukraine (68), Russia (63), Slovakia (58), Iran (50), Afghanistan (45), the USA (45) and Serbia (36). The second mention is where the authors group the countries based on the answers and place Romania in the "high awareness, negative asymmetry" group, since this country was mentioned by "many students and usually in a negative sense. [...] The countries here are divided into two groups: on the one hand, the countries neighbouring Hungary, with which Hungary had conflicts, and on the other hand, the countries far which are the source of global conflicts".[15] Even though this survey is not exclusively focusing on the border itself, it clearly demonstrates the high level of mistrust among Hungarian children towards Romania which is a basic factor when speaking about cross-border relations.

Furthermore, 93 responses of a mental map survey carried out among high school students in Szeged, by the South-Eastern border of Hungary (i.e. the border with Serbia and Romania), demonstrated that "the dissolution of the rigid role of the state border does not appear in the students' border concepts".[16] It seems like the national border poses an impediment in the high school students' mental maps, which is underlined by the fact that two-thirds of the students marked settlements only on the Hungarian side and no-one in the neighbouring countries despite that there was nothing encouraging them to do this. Furthermore, the survey also showed that the students felt the settlements on the other side of the border more distant, except in the case of the more well-known municipalities.[17]

---

[14] Á. Kincses & Á. Jakobi, "A világ a diákok szemével", *Területi Statiszika*, vol. 3, 2012, pp. 264–267.

[15] *Ibid.*, p. 271.

[16] B. Balázs & A. Farsang, A szegedi középiskolások országhatár-képzete Magyarország délkeleti határáról, *op. cit.*, p. 267.

[17] *Ibid.*

In comparison, the mental mapping exercise of the Frontem network to some extent concurs with the above observations (Figure 1). First of all, the participants who filled-in the survey exposed a wide range of individual views on what the border area actually consists of. The smallest region selected was only 1 cell (400 km^2), while the largest selected area covered 138 cells (55,200 km^2). While the former one might be only a mistake (as it is hard to imagine that somebody reduces the border region of a border that is 443 km long to a single 20 x 20 km square), the later one demonstrates an extremely generous approach as the respondent selected a 6.5 times bigger region then the territory where the border actually lies. The average area selected covered 46.4 cells (18,545 km^2) which more-or-less indicates a scale of a territory that mathematically would be needed to pad the border with a 40 km distance on both directions.

However, the intriguing issue in the mental mapping is that the respondents did not necessarily mark the cells in a symmetrical or mathematically precise way, from which information on their perceptions can be gained. Based on the frequency of choices, it is noticeable that the highest frequency is quite low (73 %), meaning that there is no area along the border section which would be identified by every participant as a border section. The highest frequency selections are located towards Oradea (Romania), Nagykereki (Hungary) and Pocsaj (Hungary), which indicates that there is no consensus on cross-border centrality. Respondents selected a section of the border only, therefore, the aggregate mental map of the Romanian-Hungarian cross-border region corresponds rather with a diffuse cross-border strip of about 50–150 km wide recognised by 2/3 of the respondents only.

Secondly, there are 86 cells marked on the Hungarian and 95 marked on the Romanian side, indicating that for some reason the geographical perception of the border is more extensive on the Romanian side. Finally, it is quite intriguing to see which are the furthest points on each side still considered as parts of the border region. On the Romanian side, territories east from Alba Iulia are still marked, which, depending on the selected route, lie in a distance of 320–350 km from the closest border crossing point. On the Hungarian side, the furthest marked point is only half of the above distance, as Baja's region is of 120–150 km from the closest border crossing point depending on which route one choses.

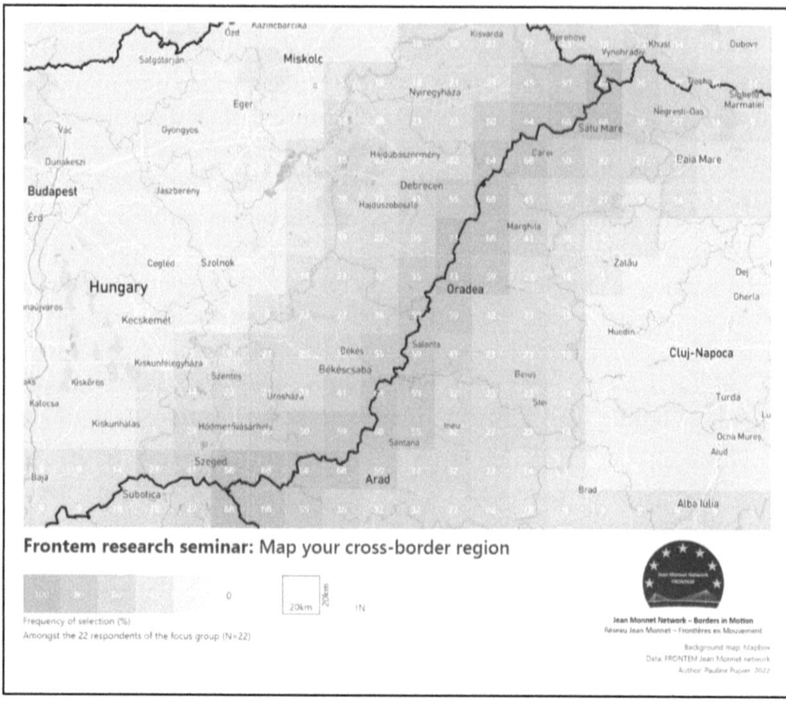

Figure 1: Results of the mental mapping exercise on the Romanian-Hungarian border.[18]

Borders as Barriers Versus Borders as Bridges

Borders are traditionally perceived as the physical manifestation of the separation of two entities, usually countries. However, this separation often brings along negative connotations such as alienation, obstructed economic development and poor social cohesion.[19] To put it shortly,

---

[18] Source: University of Strasbourg/Euro-Institute, *Toolkit on Models of Border Management and Perception in the EU. 5 Case Studies*, Strasbourg, 2023, p. 154. Retrieved from; <https://centre-jean-monnet.unistra.fr/2023/09/05/toolkit-modeles-de-gestion-de-la-frontiere-et-de-la-perception-sur-lunion-europeenne/>. Accessed 18 December 2023.

[19] B. Spierings & M. van der Velde, "Cross-border differences and unfamiliarity: Shopping mobility in the Dutch-German Rhine-Waal Euroregion", *European Planning Studies*, vol. 21, n°1, 2013, pp. 5–23.

the "main effect of borders is that they discourage spatial interaction".[20] Based on the traditional geographic border discourses and the sociological categorisation of binary distinctions between two different entities, the "border is emphasised as a line dividing 'us' from 'them' ".[21] Interestingly enough, the "perceptions which emphasize notions of difference, mutual fear and threat are, more often than not, socially constructed from the centre with walls and fences constituting the mechanism through which difference is perpetuated".[22] In other words, from the literature it seems that the perception of the borders focusing on separation and discouragement of interactions is a phenomenon generated by the central or national level.

This observation is in line with one of the conclusions of the focus group meeting where some of the participants emphasised that the border between Hungary and Romania indeed can act as a barrier, an obstacle standing in the way of cross-border cooperation. Each time when border citizens are impeded to live their lives in a 360-degree radius, i.e. to use urban functions such as healthcare services, education facilities, business opportunities, entertainment possibilities etc. all around their micro-environment, their quality of life is deprived and their perception of the border might become more negative. A participant of the focus group mentioned that "legal and economic regulations affect everything, for example patient care, and they hinder our lives as border citizens, in many ways".

Furthermore, the general consensus of the focus group was that – validating the findings in the literature – these obstacles that turn the border into a barrier are engrained in the differences in the administrative systems of the two countries and the lack of satisfactory level of harmonisation between them despite of the initiatives such as Association for European Border Regions (AEBR)'s b-solutions project[23] or Central European

---

20 P. Rietveld, "Barrier effects of borders: Implications for border-crossing infrastructures", *European Journal of Transport and Infrastructure Research*, vol. 12, n°2, 2012, p. 151.

21 A. Moraczewska, "The changing interpretation of border functions in international relations", *Revista Română de Geografie Politică*, vol. 12, n°2, 2010, p. 334.

22 D. Newman, "On Borders and Power", *Journal of Borderland Studies*, vol. 18, n°1, 2003, pp. 13–25.

23 B-solutions is a pilot initiative to tackle legal and administrative border obstacles along EU internal land borders. It is promoted by the European Commission's Directorate-General for Regional and Urban Policy (DG REGIO) and managed by

Service for Cross-Border Initiatives (CESCI)'s Legal Accessibility initiative[24]. In fact, b-solutions include an obstacle from the Hungarian-Romanian border section where the small-scale farmers of the border region were impeded in selling their products at the local markets on the other side of the border as the border constituted a barrier due to the absence of harmonised rules and regulations.

During the focus group discussion, the participants reflected on the fact that before the political system has changed, the border was perceived to be dangerous, an area that was better to be kept away from since it was patrolled by armed guards and dogs. These experiences also added to the fact that people still might see the border as a barrier. At the same time one of the participants who lived close to the green border in the 1980–1990s said that since the border zone was regarded as a forbidden territory "you had to stay away, but it was a meeting place for people divided by the border, and a place where people could exchange goods and interact with each other". This statement demonstrates the complexity of border perception as for people living at the same place and time the border can manifest both a barrier and a bridge.

"The cross-border areas are perceived as progressive limits by the Hungarians and Romanians. They constitute a wide intermediary area, a meeting point for two (or more) communities of citizens exchanging exogenous and endogenous influences".[25] Perceiving the borders as bridges is not a newfound metaphor, it is being used for several decades now. Bridges are images of boundaries that both separate and connect,[26] or in other words "bridges are perceived as phenomena of connections".[27]

---

the Association of European Border Regions (AEBR). See more: <https://www.b-solutionsproject.com/>. Accessed 23 May 2023.

24 Legal accessibility was launched in 2016 as a project of the Central European Service for Cross-Border Initiatives. The project supported by the Hungarian Ministry of Justice aimed at identifying the pertaining legal and administrative obstacles along the Hungarian borders and drafting recommendations on the potential solutions based on the experiences of best practices. See more: <https://legalaccess.cesci-net.eu/en/about-the-initiatives/>. Accessed 23 May 2023.

25 E. Bioteau, *Everyday Life near State Borders: The Social Effects of the Changing Status of Borders in Central Europe (Romania-Hungary)*, University of Angers, Angers, 2007.

26 G. Simmel, "Bridge and Door", *Theory, Culture & Society*, vol. 11, 1:6, 1994, pp. 5–10.

27 H. van Houtum & A. Srüver, "Borders, strangers, doors and brudges", *Space & Policy*, vol. 6, n°2, 2012, pp. 141–146.

Some authors also take a step further and claim that when borders are perceived as bridges then they signify a connection or a contact "through which the politically constructed division can be overcame".[28]

This is an extremely important point in the observation and understanding of the perception of borders because it is at this stance where the two spheres (i.e. the country-level political discourse and the everyday experiences of the border citizens) meet which is important because geopolitical discourses on cross-border cooperation are relatively separated from the everyday lives of local people.[29] The diverse perceptions of borders depend on the different actors where the non-state actors (such as international organisations, integrating groupings etc.) are "promoters of perceiving borders as bridges than barriers".[30]

This standpoint was strengthened by some of the participants of the focus group discussion. Similarly to the above quoted participant who simultaneously perceived the borders as barriers and bridges, another participant claimed that he sees the borders not as a dividing line, but as a meeting line where people can exchange experiences and good understanding. They also implied that people who live in the border regions are often bilingual who have knowledge on the culture of both Hungary and Romania, who also often travel between the two countries and have an extended, cross-border network. Consequently, these border people (and thus in a way the border) can act as real bridges between the two countries and the people who live further away from the border and might not have first-hand experience with it. The bridge role can be manifested through the establishment of joint businesses, the launch of cross-border public services or the creation of regular cultural, sport and people-to-people activities among others.

---

28 Z. Simon, "'Ezdélyiség' napjainkban, a kulturális antropológiai érdeklődés vonzatában", *Uj Hegyvidék*, vol. 1, n°2, 2007, p. 132.

29 A. Paasi, "Europe as a process and social discourse: considerations of place, boundaries and identity", *European Urban and Regional Studies*, vol. 8, 2001, pp. 7–28.

30 A. Moraczewska, "The changing interpretation of border functions in international relations", *op. cit.*, p. 335.

Citizens' Perception of their Neighbours

Arguably in the recent times one of the most prominent factors affecting local citizens' perception of their neighbours at the Hungarian-Romanian border is a paradox resulting from the EU's discourse concerning border controls. Namely that "on the one hand, the EU promotes good neighbourly relations, while on the other hand it emphasises the need to strictly implement the Schengen acquis on border controls and visa regimes".[31] It is true that the open-borders policy "has affected thousands of ordinary citizens on both sides of the border and has significantly contributed to efforts to overcome the historical legacy of prejudice, stereotypes and resentment. Among the numerous examples one may single out the difficult and blood-stained relationships between (…) Hungary and Romania".[32] Consequently, the fact that Romania is still not granted a Schengen membership despite the fulfilment of the mandatory formal criterii causes frustration and the potential deceleration of trust building and proliferation of good neighbourly activities.

At the focus group discussion, these sentiments clearly came forward and several participants claimed that the still persisting Schengen border has a separating effect between the people living on either side of the border when posing obstacles during the design and implementation of joint initiatives. However, it is important to point out that the issue around the Schengen area is not directly influencing the perception of the neighbours, as the Hungarian side is being known to openly support Romania's accession to the Schengen area. Furthermore, one of the focus group members also mentioned that the border control services on the Hungarian-Romanian border are thought to be working together well, thus it seems that the delayed Schengen entrance is not actively driving the parties away, it just simply slows down the further development of good neighbourly relations.

During the focus group discussion as well as in previous research, two main strands of border people's perceptions regarding the neighbours are observed. One of the prevailing perceptions of the neighbours was that of cooperative partner. One of the focus group participants mentioned that

---

31 J. Apap & A. Tchorbadjiyska, "What about the Neighbours? The Impact of Schengen along the EU's External Borders", *Centre for European Policy Studies*, n°210, 2004, p. 1.

32 *Ibid.*

"at the level of the citizen, things are much simpler. There have always been interactions, the historical framework has led to various approaches to this exchange between the two sides. This has determined the emergence of a cooperation at all levels and areas: administrative, cultural, educational, health." It was also – albeit somewhat probably too naively – mentioned that "there is no competition, only very good collaboration between neighbours". Another participant offered practical examples to support this view on the neighbour as a partner with whom cooperation is possible when he mentioned joint projects (involving the development of cross-border bike paths and regular bus lines) in order to facilitate cross-border labour mobility.

Even though an objective need for cooperation both at the institutional and personal level were found to be recognised at the Hungarian-Romanian border,[33] it was also shown that there is a certain level of only "conformist attitudes and a formal sincerity"[34] when probed for the possibility of accelerating cross-border cooperation. This observation based on the high percentage of positive but unsubstantiated answers might be in line with the results of the focus group since the second prevailing perception regarding the neighbours was that of a competitor. A participant put this dichotomy succinctly when he said that "The relations are characterised by both collaboration and competition". To support this view, he offered the example of the airports in Oradea and Debrecen, two relatively similarly sized airports in close proximity on the two sides of the border. Indeed, it was already showed that instead of cooperating through destination management and fare packages, they rather behave like competitors targeting the same audience with similar offers.[35]

Regarding the perception of the neighbour as a competitor, another participant extended this observation from the private company-dominated business sphere to the level of municipalities and public urban planning when saying that "the 8 big cities of the border area are in strict competition with each other, and they will be as long as they feel

---

[33] D. Badesculu, A. Badesculu & C. Bucur, "Considerations on the effectiveness of cross-border cooperation in public order and civil protection services. The case of the Romanian-Hungarian border area", *Lex Localis*, vol. 13, 2015, p. 575.

[34] *Ibid.*

[35] L. Virág, "Regionális repülőterek müszaki paraméteeinek összehasonlítása a debreceni repülőtér versenyképességének megítéléséhez", *Agrártudományi Közlemények*, vol. 46, 2012, p. 103.

growth potential exists." Interestingly, this person also added that "their catchment area expects the same from them". This claim might show that people living in border cities would also rather regard their neighbours and the neighbouring settlements at the other side of the border as competitors and push this view on their elected officials to represent what they consider as their best interest.

It is an often-repeated claim that "in the European Union peoples tend to consider each other as neighbours – close from the point of view cultural and political as well as geographical – and not as strangers".[36] However, border studies have shown that, even if in recent decades certain national or political borders have disappeared or weakened, other borders have been newly demarcated or frozen, which are processes not independent at all from the perceptions of those citizens who live in the border area, the role they assume to themselves and the borders and the views they assign to their neighbours. These views can be translated into the agenda of the institutions, organisations and other entities that act on behalf of the border citizens, and thus it is of vital importance whether the prevailing perception is reflecting cooperation or competition.

Factors Influencing Border Perception

Shedding light on citizens' perceptions on borders and their neighbours living on the other side of the border would be somewhat incomplete without also examining the factors that influence these perceptions. Although the topic would deserve its own separate article, here only those three factors are explored that the participants of the focus group brought up, namely history, permeability as well as political discourse and its institutionalisation.

For a better understanding of the local citizens' relationships with the border, it "appears necessary to take into consideration (…) the historical and cultural aspects",[37] especially in the case of a border, such as the one dividing Hungary and Romania, that has a long and turbulent history

---

[36] P. P. Viazzo & G. Fassio, "Borders et frontiers: definitions théoriques et experience subjective d'un concept à géométrie variable", *Migrations Société*, vol. 140, n°2, 2012, p. 263.

[37] E. Bioteau, "Everyday Life near State Borders", *op. cit.*, p. 15.

which inherently shapes the perception of the people living in the border area.

The current border is older than a century and was drawn in 1920 by the Treaty of Trianon and reaffirmed by the Paris Peace Conference in 1947. The frequent changes of the border during the 20th century created numerous conflicts. These conflicts and especially the aftermath of the First World War are deeply engrained in generations of people through the visual images, textual expressions, and rhetorical figures related to "the image of the wounded, mutilated, and maimed country which became the cultural legacy of the Treaty of Trianon in Hungary".[38] This narrative shapes collective identity, even unconsciously, in a traumatic way on both sides of the border as the changes in border affected those the most who – or whose families – were living there.

Since the Treaty, the border has always been considered 'a sensitive border', as according to some sources it was periodically contested by Hungary,[39] "either through political irredentism or through nostalgia in Hungarian cultural circles".[40] The situation was not helped by the pursuit of Romania's "one people one country" principle either[41] which is unequivocally formulated to this day in the first line of Romania's constitution stating that "Romania is a national state, sovereign and independent, unitary and indivisible",[42] without any mention or regard to the national minorities living in its territory.

After the system transformation in 1989 and the signing of the Treaty of Good Neighbourliness in 1996 between Romania and Hungary, it appears that the historical events were more favourable to cross-border

---

38 A. Menyhért, "The image of the 'Maimed Hungary' in 20th-century cultural memory and the 21st century consequences of an unresolved collective trauma", *Environment, Space, Place*, vol. 8, n°2, 2016, p. 69.

39 Ş Ş. Manciulea, *Frontiera politică şi etnică româno-maghiară*, L'imprimerie du Journal Universul, Bucharest, 1938; P. Deică, "Preocupari geopolitice actuale ale geografilor maghiari", *Revista Română de Geografie Politică*, vol. 1, n°1, 1999, pp. 36–43; R. Sageată, "Graniţa de nord-vest a României şi argumentele trasăriio sale", *Geographica Timisiensis*, vol. 10, 2001, pp. 119–127.

40 N. Popa, "Frontières et regions transfrontalières en Roumanie: Entre territoires, cultures et fonctions", *Geographica Timisiensis*, vol. 13, n°2, 2004, p. 87.

41 D. J. Andersen, M. Klatt & Sandberg, *The Border Multiple. The Practicing of Borders between Publiv Policy and Everyday Life in a Re-scaling Europe*, Routledge, London, 2016.

42 Constituţia României, n°429/200.

cooperation. Both countries joined the North Atlantic Treaty Organization (NATO) (Hungary in 1999, Romania in 2004) and later the European Union (EU) (Hungary in 2004, Romania in 2007) as well as Euroregions (for example: the Biharia Euroregion or the Danube-Criş-Mureş-Tisa Euroregion) and European Groupings of Territorial Cooperations (such as Gate to Europe European Grouping of Territorial Cooperation (EGTC), Banat-Triplex Confinium EGTC) were established.

Whereas these recent events are usually regarded as factors influencing border perception in a more positive way, some authors go as far to claim that Hungary "has been extremely active in this field due to its concern for the preservation of its cultural identity and the promotion of the economic and political interests of the Hungarian minorities of the neighbouring countries".[43] It was observed that the high political stakes of these initiatives have given rise to some reluctance on the part of neighbouring countries such as Romania and have contributed to the formation of very large Euroregions, which have spread far to the east. However, it is worth underlining that these structures were not only about the Hungarian minority since there is also a Romanian minority living in Hungary, and also their main focus is rather on the development of cross-border cooperation benefitting both sides.

However, regardless how certain historical events are interpreted and presented by the different actors, what seems to be certain is that history itself does influence the perception on the borders both on the level of individuals and narratives.

The second factor concerns the permeability of borders. Contrary to the popular saying according to which good fences make good neighbours, more and more scholars are arguing that in fact permeable fences are those that make good neighbours.[44] From the 1990s scholarship various disciplinary fields "widely suggested that the permeability of borders has intensified to a degree that their role in politics and society has been

43 N. Popa, "Frontières et regions transfrontalières en Roumanie", *op. cit.*, p. 82.

44 See for example B. Feinstein & M. S. Dajani-Daoudi, "Permeable fences make good neighbors: Improving a seemingly intractable border conflict between Israelis and Palestinians", *American University International Law Review*, vol. 16, n°1, 2000, pp. 1–176.

substantially transformed".[45] The recent years' crises (such as migration crisis, Brexit, COVID-19, Ukrainian-Russian war), however, showed that the de-bordering processes are not quite as fast as they would have been envisaged and the Nation-State paradigm proves to be more resilient that it was predicted to be.

Nevertheless, permeability itself still influences border perception. "Taking into consideration the perception of state border according to particular paradigms one can see the gradation of its permeability from the line of division into the inner and outer environment, to a porous line, where the border is perceived as an element of the system with a set of numerous entrances and exits. Eventually the border becomes a virtual line, which does not mean its disappearance but a high degree of permeability".[46]

In other words, the very fact whether a border can be crossed or under what circumstances can be crossed have a serious impact on the way people see the border itself. If the border is a hard line as it was during the communist era between Hungary and Romania, then people are more likely to perceive the border to be something dangerous as it was recorded in many "complaint stories" in the narratives related to travel during that time. Usually, the subject of the complaint was the humiliation at the country's borders: unfair customs inspections, protracted verifications, and long waiting times, all relating experiencing the act of crossing the border. Even after the system transformation many people still tell stories about these experiences which show that the "permeability of national borders does not make the border irrelevant or border crossing an unreflective routine".[47] This is probably why the focus group discussion also pointed out that the measures linked to the permeability of the border (such as Romania entering the Schengen area) bears a special importance to the people living at the border region. Furthermore, permeability also has a physical aspect, which can be quantified through the relatively rare

---

45 O. Schmidtke, "Borders in public perception", *In* O. Schmidtke & S. Ozcurumez (eds.), *Of States, Rights, and Social Closure*, Palgrave MacMillan, New York, 2008, p. 92.

46 A. Moraczewska, "The changing interpretation of border functions in international relations", *op. cit.*

47 M. Feischmidt, "A határ és roman stigma", *In* N. Kovács, A. Osvát & L. Szarka (eds.), *Tér és terep. Tanulmányok az etnicitás és az identitás kérdésköréből III*, Akadémiai, Budapest, 2005, pp. 43–57.

crossing density. Even though there are new border crossings, but they not necessarily increase border traffic because of their restricted opening times.

The third factor relates to the role political discourse and institutions play in border perception. Albeit to a certain degree it is a question of chicken and egg as it is impossible to entirely dissect whether citizens based on their own views create the political discourse and institutions or the prevailing political discourse and the institutions shape citizens' perceptions, probably it is safe to say that the regional, national and international political discourse and institutions do play a decisive role in the way people perceive the border since they have the power and authority to regulate the mobility and even the identity of the citizens.

If a state is governed by a unilateralist politics, borders will mostly be closed and the relation with the neighbour will be less important than in the case of a multilateral approach. Examples for both approaches can be found throughout the history of the Romanian-Hungarian borderland, as mentioned above.

The leading political direction is represented at every level of the institutions. Both Romania and Hungary are quite hierarchical and centralised with institutions that are more reluctant to the idea of openness and cooperation. However, since "these institutions were positively forced by the need for change and modernisation, and, at regional level, by the specificity of the communities among they are operating (…) the results already obtained in a relatively short time by the Hungarian-Romanian cooperation actions under the Euroregional framework and under the cross-border cooperation specific programs are certainly encouraging".[48]

Finally, other institutions also influence border perception both in the border region among the border citizens as well as in a wider sense. During the focus group discussion, this aspect was evidenced when one of the participants emphasised the role of a given university that shaped the development of "new ways of knowledge and offering many opportunities for cooperation in all domains". This person claimed that the university is fostering cooperation not only in the academic sphere but also between people and countries. Thus, evidencing how institutions whose primarily raison d'être might not be tightly linked to cross-border

48 D. Badulescu et al., "What about the Neighbors?", *op. cit.*, p. 575.

cooperation still play a role in influencing the border perception if they carry out relevant activities.

Conclusions

Understanding the perceptions of borders is not only an intriguing intellectual challenge and an exciting methodological puzzle, but also a necessity in a world that constantly redraws and reinterprets the separating lines between countries. Regardless of the fact that the Hungarian-Romanian border is reconciled for more than a century, it still poses difficulties when it comes to theoretical analysis and practical, on-field realisation of cross-border cooperation. The present research widened the discourse by reviewing and juxtaposing the main findings of the literature regarding the border perception as well as interpreting the results of a focus group and a mental mapping exercise carried out on the Romanian-Hungarian border.

The result showed that the border perceptions appear on a heterogeneous scale, both in geographical and in symbolic way. Even people living close to the border seems to define differently what territories should be considered as border regions and also their perceptions differ on whether borders are more barriers separating people and countries and obstructing development, or, in contrast, they are bridges linking the actors together and opening new avenues for cooperation. Similarly, the citizens' perception of their neighbours also varies between perceiving the other as a competitor or as a potential cooperating partner.

Three main elements were found to influence border perception. Through socialisation, History imprints certain memories on the mindset of subsequent generations that might turn into even unconscious perceptions. Permeability of the borders also influence the perception as the border crossing is the most evident place where people come across the border and the experiences they gain there shape their overall perception of the border, the border region and the neighbours from the other side of the border. Similarly, political discourse and the institutions are factors affecting border perceptions as these have certain power and authority to shape the political decisions that in turn influence the perceptions of people.

In conclusion, the main perception strands on the Hungarian-Romanian border are characterised by heterogeneity among the individuals, which can only be typified up to a certain degree. Citizens living in this border sections are not hostile to each other, but cooperative spirit is sometimes replaced by competitiveness. The Hungarian-Romanian border itself often still appears as a barrier and not as a bridge which could be improved through the sensitive narration of history, the increase of the permeability and the prevalence of friendly politics.

Bibliography

D. J. Andersen, M. Klatt & Sandberg, *The Border Multiple. The Practicing of Borders between Publiv Policy and Everyday Life in a Re-scaling Europe*, Routledge, London, 2016.

J. Apap & A. Tchorbadjiyska, "What about the neighbours? The impact of Schengen along the EU's external borders", *Centre for European Policy Studies*, n°210, 2004.

D. Badesculu, A. Badesculu & C. Bucur, "Considerations on the effectiveness of Cross-border cooperation in public order and civil protection services. The case of the Romanian-Hungarian border area", *Lex Localis*, vol. 13, 2015, pp. 559–575.

B. Balázs & A. Farsang, "A szegedi középiskolások országhatár-képzete Magyarország délkeleti határáról", *Földrajzi Közlemények*, vol. 140, n°3, 2016, pp. 258–269.

E. Bioteau, *Everyday Life Near State Borders: The Social Effects of the Changing Status of Borders in Central Europe (Romania-Hungary)*, University of Angers, Angers, 2007.

L. A. Broadhead, "Canadian sovereignty versus northern security: The case of updating our mental map of the Arctic", *International Journal*, vol. 65, n°4, 2010, pp. 913–930.

M. Bruter, "On what citizens mean by feeling 'European': Perceptions of news, symbols and borderless-ness", *Journal of Ethnic and Migration Studies*, vol. 30, 2004, pp. 21–39.

P. Deică, "Preocupari geopolitice actuale ale geografilor maghiari", *Revista Română de Geografie Politică*, vol. 1, n°1, 1999, pp. 36–43.

D. Elkind, "Response to objectivism and education", *The Educational Forum*, 2005, pp. 328–334.

A. Faludi, *The Poverty of Territorialisms, A Neo-Medeival View of Europe and European Planning*, Edward Elgar Publishing, Cheltenham, 2018.

B. Feinstein & M. S. Dajani-Daoudi, "Permeable fences make good neighbors: Improving a seemingly intractable border conflict between Israelis and Palestinians", *American University International Law Review*, vol. 16, n°1, 2000, pp. 1–176.

M. Feischmidt, "A határ és roman stigma", *In* N. Kovács, A. Osvát & L. Szarka (eds.), *Tér és terep. Tanulmányok az etnicitás és az identitás kérdésköréből III*, Akadémiai, Budapest, 2005, pp. 43–57.

T. Hardi, Z. Hajdú & I. Mezi, "Határok és városok a Kárpát-medencében", *MTA Regionális Kutatások Központja*, Győr, Pécs, 2009.

Á. Kincses & Á. Jakobi, "A világ a diákok szemével", *Területi Statiszika*, vol. 3, 2012, pp. 264–267.

J. Le Rider, "Mitteleuropa, Zentraleuropa, Mittelosteuropa: A mental map of Central Europe", *European Journal of Social Theory*, vol. 11, n°2, 2008, pp. 155–169.

H. Lefebvre, *The Production of Space*, Blackwell, Oxford, 1991.

Ş. Manciulea, *Frontiera politică şi etnică româno-maghiară*, L'imprimerie du Journal Universul, Bucharest, 1938.

A. Menyhért, "The image of the 'Maimed Hungary' in 20th-century cultural memory and the 21st century consequences of an unresolved collective trauma", *Environment, Space, Place*, vol. 8, n°2, 2016, pp. 69–97.

J. Minghi, "Boundary studies in political geography", *Annals of the Association of American Geographers*, 1963, pp. 407–428.

A. Moraczewska, "The changing interpretation of border functions in international relations", *Revista Română de Geografie Politică*, vol. 12, n°2, 2010, pp. 329–340.

D. Newman, "On borders and power", *Journal of Borderland Studies*, vol. 18, n°1, 2003, pp. 13–25.

A. Paasi, "Europe as a process and social discourse: Considerations of place, boundaries and identity", *European Urban and Regional Studies*, vol. 8, 2001, pp. 7–28.

N. Popa, "Frontières et regions transfrontalières en Roumanie: Entre territoires, cultures et fonctions", *Geographica Timisiensis*, vol. 13, n°2, 2004, pp. 79–102.

G. Popescu, *Bordering and Ordering the Twenty-first Century. Understanding Borders*, Rowman & Littlefield Publishers, INC., UK, 2012.

M. Rédei, Á. Kincses & Á. Jakobi, "The world seen by Hungarian students: A mental map analysis", *Hungarian Geographical Bulletin*, vol. 60, n°2, 2011, pp. 91–104.

P. Rietveld, "Barrier effects of borders: Implications for border-crossing infrastructures", *European Journal of Transport and Infrastructure Research*, vol. 12, n°2, 2012, pp. 150–166.

D. Rumley & J. Minghi, *The Geography of Border*, Routledge, London, 1991.

R. Sageată, "Granița de nord-vest a Româniej și argumentele trasăriio sale", *Geographica Timisiensis*, vol. 10, 2001, pp. 119–127.

O. Schmidtke, "Borders in public perception", *In* O. Schmidtke & S. Ozcurumez (eds.), *Of States, Rights, and Social Closure*, Palgrave MacMillan, New York, 2008.

G. Simmel, "Bridge and door", *Theory, Culture & Society*, vol. 11, 1:6, 1994, pp. 5–10.

Z. Simon, "'Ezdélyiség' napjainkban, a kulturális antropológiai érdeklődés vonzatában", *Uj Hegyvidék*, vol. 1, n°2, 2007, p. 132ff.

B. Spierings & M. van der Velde, "Cross-border differences and unfamiliarity: Shopping mobility in the Dutch-German Rhine-Waal Euroregion", *European Planning Studies*, vol. 21, n°1, 2013, pp. 5–23.

W. A. Sulsters, "Mental mapping, viewing the urban landscapes of the mind. Conference proceedings", Workshop 2, session 2.3, Mapping home, 2005.

A. Székely & B. Kotosz, "From fence to wall? Changes of mental space of border zones in Eastern Europe", *Regional Science Policy & Practice*, vol. 10, issue 4, 2018, pp. 269–282.

University of Strasbourg/Euro-Institute, *Toolkit on Models of Border Management and Perception in the EU. 5 Case Studies*, Strasbourg, 2023.

H. van Houtum, "*The Development of Cross-Border Economic Relations*", Thela Thesis Publishers, Amsterdam, 1998.

H. van Houtum, "III European perspectives on borderlands", *Journal of Borderland Studies*, vol. 15, n°1, 2000, pp. 57–83.

H. van Houtum & A. Srüver, "Borders, strangers, doors and brudges", *Space & Policy*, vol. 6, n°2, 2012, pp. 141–146.

P. P. Viazzo & G. Fassio, "Borders et frontiers: definitions théoriques et experience subjective d'un concept à géométrie variable", *Migrations Société*, vol. 140, n°2, 2012, pp. 255–264.

L. Virág, "Regionális repülőterek müszaki paraméteeinek összehasonlítása a debreceni repülőtér versenyképességének megítéléséhez", *Agrártudományi Közlemények*, vol. 46, 2012, pp. 103–106.

Part 6

Cross-Border Cooperation from a Comparative Euro-Atlantic Perspective

Europe: Towards Cross-Border Regions and Multi-level Governance of Borders

Jean Peyrony

The present chapter derives from the "Analysis of the impact of border-related measures taken by Member States in the fight against Covid-19", carried out by Mission Opérationnelle Transfrontalière (MOT) for the European Commission in 2022,[1] as a follow-up of a first study.[2] This study assesses two years of border management and its impact on border regions and their inhabitants, then draws some recommendations for their management in the future. Its findings are based on an online survey addressed to MS and local cross-border cooperation structures, supplemented by bilateral interviews; an analysis of an extract file presenting anonymised questions submitted to the "Your Europe Advice service" by European citizens who tried to cross internal borders; and a literature review.

In the context of the European Union (EU), the paradigm initiated by Jacques Delors was a virtual erasure of internal borders. When in March 2020 they were brutally closed by States, without coordination between them, the inhabitants of border regions were affected in different dimensions of their daily life: as economic actors (cross-border workers or

---

1 J. Peyrony, T. Devillard, J. Rubio & R. Viaggi, "Analysis of the impact of border-related measures taken by Member States in the fight against COVID-19", EU commission impact analysis, July 2022. Retrieved from; <https://ec.europa.eu/regional_policy/en/information/publications/studies/2022/analysis-of-the-impact-of-border-related-measures-taken-by-member-states-in-the-fight-against-covid-19>. Accessed 30 May 2023.

2 J. Peyrony, J. Rubio & R. Viaggi, "The effects of COVID-19 induced border closures on cross-border regions", EU Commission report, 2021. Retrieved from: <https://op.europa.eu/en/publication-detail/-/publication/46250564-669a-11eb-aeb5-01aa75ed71a1/language-en>. Accessed 30 May 2023.

consumers) –cross-border mobility was interrupted or severely restricted; as users of cross-border public services (CPS), the first to be interrupted; as citizens, suddenly only nationals – restrictions were based on nationality, not health criteria, and European citizenship forgotten; as informed individuals – information was suddenly no longer available across the border; as individuals inspired by their culture or ethic, because trust was suddenly broken, and the resident on the other side of the border became a danger; as individuals connected within couples, families – because some were separated by the border.

Measures were quickly taken to safeguard cross-border mobility of goods and workers, which was considered essential for the economy. But many inhabitants of border regions have painfully experienced what it means to be reduced to their economic role, and stigmatised as foreigners in the neighbouring country. Suddenly, authorities forgot that each inhabitant might be at once a user of CPS, a citizen, an informed and inspired individual, a parent, and not only an "essential worker"; in short, a "person". These 6 dimensions of the person correspond to 6 spheres of collective action, 6 dimensions of cohesion, and symmetrically 6 categories of obstacles to cross-border integration: economic, functional, institutional, informational, cultural and interpersonal, that we use in this chapter as a common thread, inspired by Boltanski and Thévenot.[3]

The crisis has revealed the nature of cross-border interdependencies, of which public authorities had no clear view; their first reactions when the crisis occurred have mainly taken place within domestic frameworks through trade-offs between these 6 spheres, turning their backs on each border, which resulted in major difficulties of coordination across borders, and inappropriate, disproportionate, unfair, if not violent border management measures. Border regions were therefore more severely impacted than other territories. Moreover, borders' real nature and daily functioning have proved to be largely misunderstood, by public authorities and opinion.

Although most of the constraints linked with Covid-19 have been removed, several States have maintained border controls. Hence the situation of fully open borders prevailing before 2015 does not seem likely to come back. In the future, crises of different natures (sanitary, security,

---

[3] L. Boltanski & L. Thevenot, *On Justification. Economies of Worth*, Princeton University Press, Princeton, 2006.

migration) may recur, requiring more than ever to improve the European governance of borders, as the revision of the Schengen Borders Code intends to do.

In this chapter, we will first develop, in part one, what the Covid-19 crisis has revealed of the reality of persons' lives within cross-border living areas, and the role of different levels of government and key stakeholders for border management. In part two we will suggest how CB living areas can better integrate in the future, drawing lessons from the crisis, if States agree to exercise their sovereignty together, at the local level of cross-border regions; and with the support of a bottom-up multilevel governance.

Lessons of the Effects of Covid-19 Related Measures on Border Regions and Their Inhabitants

We will first adopt a matrix approach, considering different spatial patterns: national territories in the context of restricted opening of the border line, border zone when the border line is open, and with a European intervention; for each of the 6 dimensions of cross-border integration. Then we will adopt a systemic approach.

Lessons for Each of the 6 Dimensions of Cross-Border Integration

The return of border controls due to the public health crisis has created a counterfactual of cross-border integration. Let us first draw some lessons for each of the 6 dimensions of persons' lives within cross-border living areas.

Economic Sphere (Market and Public Economy)

As regards the economic sphere, each national territory encloses a domestic market, regulated by the State, and subject to taxation allowing to fund public policies. An open border is a resource for the market. Domestic productions are in competition with each other across the border; demographic, price and tax differentials create gradients generating cross-border flows of people, goods, capital and services, contributing to cross-border socio-economic integration. Productive individuals (such as cross-border workers, or more broadly, people who are 'multi-situated' between several national territories) are inserted into cross-border territorial systems, not as citizens, but as adjustment variables of border flexibility.[4] Moreover, if this socio-economic integration is not complemented by institutional integration, the absence of common management of the border by the two States concerned creates spill-over effects for local authorities, companies and inhabitants remaining outside the cross-border game.[5]

During the crisis, free movement of goods has proved to be better acknowledged – through measures such as "green lanes" – than free movement of people, which States severely interrupted in March 2020. Only later did States realise that there were "essential workers ", not only in a domestic context, but also across borders, for whom it was necessary to keep the borders open.

The crisis has drawn attention to the way cross-border work is subject to taxation and social security costs. Everywhere, teleworking, when possible, has contributed to maintaining economic activity; it has also been the case for cross-border workers, which has led States to adopt provisional bilateral agreements maintaining usual taxation rules, even if some thresholds were no longer respected. Many workers and firms now wish to maintain teleworking at a higher level than before the crisis. It can be a win-win game for territories, notably by reducing congestion linked with the use of cars. But pre-existing bilateral agreements, or their absence, are challenged.[6] The crisis has revealed inconsistencies resulting from non-harmonised State rules on taxation – a field where competence lies at national and not European level, whereas social security is already harmonised through EU regulation.

Another issue is linked with recovery plans. As the crisis has particularly hit border regions, and revealed the importance of developing and maintaining CPS, many voices have asked for cross-border projects to be

---

4 A.-L. Amilhat Szary, "Géopolitique des frontières, Découper la terre, imposer une vision du monde," *Le Cavalier Bleu*, Paris, 2020.

5 K.-H. Lambertz, "Fair distribution of taxes in transfrontier areas," Congress of local and regional authorities report of the 37th session, 29 October 2019. Retrieved from: <https://rm.coe.int/fair-distribution-of-taxes-in-transfrontier-areas-potential-conflicts-/168097f09d>. Accessed 30 May 2023. Only a few bilateral agreements have so far defined rules for sharing levied taxes between two States. Moreover, CB workers, when they pay their income tax at the place of work – which is recommended by OECD framework (to avoid double taxation), are denied their CB citizenship, as they do not vote for elections in the country they work in.

6 MOT, "Impacts du télétravail frontalier. Le télétravail frontalier: de marginal à indispensable," Paris, 2022. Retrieved from: <http://www.espaces-transfrontaliers.org/la-mot/publications/etudes-mot/impacts-du-teletravail-frontalier-de-marginal-a-indispensable/>. Accessed 30 May 2023.

prioritised within recovery plans. Unfortunately, our survey has revealed only little concretisation of such wishes.

In short, the crisis has revealed gaps in the achievement of cross-border integration in the economic sphere. Schengen is one of the "jewels" of the EU, but the crisis has shown the extent to which the single market is incomplete. Moreover, border regions' inhabitants are acknowledged as cross-border economic agents, but not as cross-border citizens. In the future, a further stage of cross-border cooperation should consist of a more complete, win-win integration, based on co-development: not only a mature cross-border market (border fully open, 4 freedoms) but also a cross-border common approach for economic policies serving the cross-border regions and their inhabitants.

Functional Sphere (Public Services)

In the functional sphere, each national territory is a bundle of domestic public services. Border lines, where public services meet with little or no interoperability, act as more or less porous filters. EU policies have been developed in sectoral fields (functional approach), depending on the States' will to delegate these policies to the EU (early in a field like transport, later in the field of health for instance).

In cross-border regions, the integration of CPS should tend to be as complete as in a national territory. Greater Geneva with the Leman Express public transport system is a good example. But too often, public services (for mobility, energy, health, environment, risk prevention, employment, education, innovation, etc.) stop at the border.

This has been confirmed by the Covid-19 crisis, revealing CPS' weaknesses: they were the first to be interrupted. Some have been difficult to reopen. In some cases, this has even led to their definitive closure.

Gaps concern also investment and daily operation. While CPS integration should be seen as favouring economies of scale and of agglomeration, this potential remains largely untapped. Moreover, States do not seem to be respecting their commitments made at the height of the crisis, as to implement some "missing links" rail projects.[7] In a context of

[7] MOT, "Retour sur la conférence 'liaisons ferroviaires transfrontalières' des 10 et 11 juin", Paris, 2021. Retrieved from: <http://www.espaces-transfrontaliers.org/actualites/news/news/show/retour-sur-la-conference-liaisons-ferroviaires-transfrontalieres-des-10-et-11-juin/>. Accessed 30 May 2023.

renewed attention on administrative burden, cross-border integration is often seen as an overburden, not an asset.[8] In the future, cross-border regions should define a common approach to transition,[9] through cross-border public investments and shared services, jointly managed locally with the support of the two or more States concerned, forming a fully integrated system. CPS should be co-produced by inhabitants, simultaneously cross-border economic agents, users and citizens.

Institutional Sphere (Public Policies Under Democratic Control)

The institutional sphere is naturally embedded within the framework of the nation State, founded on national democracy – limited to national citizens, where the State finds its legitimacy through ensuring security,[10] solidarity, and providing public goods and services via public policies.[11]

In the European integration context, Member States have chosen to build together a single market (economic sphere) and develop sectoral policies (functional sphere), under the control of an emerging democratic system (institutional sphere), where national citizenship also provides European citizenship – one of its concrete proofs being free movement of people.

However, by drastically limiting mobility across borders, the crisis has revealed the lack of effective recognition of such a European citizenship – not to mention the weakness of cross-border democracy. The border suddenly appeared as a double skin separating two bodies, being again the place of possible conflicts. The border line forces individuals to position

---

8 CESCI, *A koronavírus hatásai a határrezsimre. Értékelő elemzés a magyarországi határrezsim változásáról 2020 máciusa és 2021 augusztusa közt* [Impacts of the coronavirus pandemic on the border regime. Assessment of the changes of the Hungarian border regime between March 2020 and August 2021. – in Hungarian], CESCI, Budapest, 2021.

9 See the Territorial Agenda Pilot Action around "Luxembourg in Transition": <https://territorialagenda.eu/pilot-actions/cross-border-spatial-planning/>. Accessed 30 May 2023.

10 Classical analyses of Hobbes (Leviathan) and Weber (monopoly of legitimate violence).

11 J. Peyrony, *Frontières et cohésion territoriale, Actes de la conférence Construire des ponts à travers les frontières, vers une cohésion territoriale en Europe*, Steiner Verlag, Stuttgart, 2014.

themselves either inside or outside.[12] It is the "undemocratic condition of democracy"?[13] In the future, cross-border governance bodies should become more like cross-border governments, under cross-border democratic control. Full mandates, with clear objectives and responsibilities to develop cross-border integration should be given to them, and to local and State actors of cross-border cooperation (such as prefects in France). The border zone should become a place for transnational democracy, involving cross-border, sometimes bi-national, European citizens.

When dealing with European and cross-border integration, the economic, functional and institutional "hard" spheres, addressing individuals as economic agents, users of public policies, or citizens – holders of rights – are generally put forward, because EU institutions have their main mandate in these fields. But three other spheres have appeared relevant during the crisis: those relative to information, culture and interpersonal relations.

Information Sphere

Within a national territory, the sphere of information is regulated by domestic institutions. States – particularly where they are monolingual – exercise a "monopoly of evidence",[14] of the construction of reality, via the statistical apparatus, the production of social sciences, education and media. Henceforth, each border is a place of contradiction between states.[15]

Cross-border and European integration are supposed to overcome this national prism.[16] The crisis has been a crash test in this regard. In March 2020, information was suddenly missing everywhere about what

---

12 A.-L. Amilhat Szary, "Géopolitique des frontières", *op. cit.*

13 E. Balibar, "Europe as borderland," *Environment and Planning D: Society and Space*, vol. 27, n°2, 2009, pp. 190–215.

14 J. Peyrony, "Integration in cross border regions. Obstacles and solutions" *In* B. Wassenberg (ed.), *Castle Talks on Cross border Cooperation. Fear or Integration? The Pertinence of the Border*, Franz Steiner Verlag, Stuttgart, 2018.

15 "The truth on this side of the Pyrenees, error on the other", as Pascal said, to express the arbitrariness of the border between France and Spain, just designed by the Pyrénées Treaty in 1659.

16 Through "linguistic hospitality" (Ricoeur); translation allowing to correct the deceptive nature of any language (Balibar), and becoming the language of Europe (Umberto Eco).

was happening, risks, possible solutions, the right to proceed or not with the actions of daily life. A lot of information circulated through domestic and global media and social networks, but it was often inappropriate, and fostered chaos rather than coherence. Everybody felt as if they were an expert, while institutions' legitimacy was severely challenged.

Gradually, everybody obtained information concerning their rights and obligations within the national context, and about situations in other countries – including on the other side of the world, but always through a national prism[17] considering countries as monads, territorial boxes that can be either opened or closed. And yet, it was suddenly impossible not only to cross the border, but also to understand what was happening on the other side; the sanitary situation, the way decisions were taken, the possibility or not to cross the border were made opaque.

The lack of available information about borders (both data and appropriate concepts); the limited evidence, and the fact that this evidence is far from being shared by public opinion clearly require learning to think the border: cross-border territorial monitoring, scientific research and training. The way information is used in the public regulation decision process should also be considered. The best effect of the European Commission's guidelines on border related measures has been "informative", as a way to softly influence national policies.

A positive externality of the crisis has been the increased visibility of borders, and awareness raising about what happens there. While borders are generally apprehended through many biases and preconceptions, a number of unanswered questions became suddenly burning, and now require a better, common understanding.

What is an open/closed border? During the crisis, the media have propagated a simplistic vision, generating confusion between different sorts of borders. However, the closure of an intercontinental border, e.g. between a European country and China, correlated with the suspension of international flights, has little to do with European terrestrial internal borders, which have been subject to a wide range of measures from complete closure of certain border points to fully open borders. Moreover, borders are always subject to a dual regime from country A to B and B to A. This situationcan furthermore be asymmetric and complicate the lives

[17] Coined by Faludi as "territorialism"; A. Faludi, *The Poverty of Territorialism: A Neo-medieval View of Europe and European Planning*, Edward Elgar, Cheltenham, 2018.

of border inhabitants, who travel from and return to their homes across the border on a daily basis.[18]

Are controls on the border line the best way to filter secondary movements of migrants, to manage security or pandemics? Or should cooperation between neighbouring states and at EU level be preferred?[19] What are the cross-border interactions, flows, interdependencies across the border, and how should they be regulated?

They are generally not well known, as the issue of mobility shows: border inhabitants' movements – just as for anyone – are generally not limited to work-related travels and can include a stop for buying fuel, food, or dropping a child off at school, sometimes on the other side of the border.

In what cross-border functional areas do people live? Is it possible to objectively define "cross-border regions", local cross-border living areas? In some cases – monocentric cross-border functional areas like Greater Geneva – it seems easy to define such areas, more or less corresponding to the cross-border public transport system (in that case, the "Léman Express"), but in a majority of cases, cross-border mobility takes place within a strip along the border. How to identify it?

During the crisis, states have sometimes prohibited, in domestic as well as cross-border contexts, "non-essential" trip purposes. But on what grounds? On which criteria should such norms be based? A specific question concerns the "legitimacy" of cross-border movements in the case of a crisis, as the examples found in the Your Europe Advice entries show: which ones should be acknowledged as requiring protection? The definition of a "cross-border worker" is based on a criterion of minimum weekly travel. Should a worker commuting each weekend between two European, possibly remote, cities be protected? Is it possible and desirable to define a local cross-border mobility, which would be considered legitimate and protected, because it is more sustainable/cohesive than more long-distance mobility? What does "local" mean in that case? Does it make sense, given that there may be a continuum of cases

---

18 B. Wassenberg, "Return to mental borders: A diary of COVID 19 closures between Kehl, Germany, and Strasbourg, France", *BIG Review*, vol. 2, 2020, pp. 114–120.

19 C. Leuprecht, E. Brunet-Jailly, T. Hataley & T. Legrand, "Patterns in nascent, ascendant and mature border security: Regional comparisons in transgovernmental coordination, cooperation and collaboration", *Commonwealth and Comparative Politics*, Vol. 5, Issue 4, 2021, pp. 349–375.

between "proximity" and Europe-wide range, especially in the context of the extension of teleworking? Or should the EU single market prevail, and a fair EU ecotax regulate mobility?

Should a functional approach (based on distance, time) be preferred? Or a more institutional one, based on the status of a cross-border worker or resident, living in the perimeter of an existing cross-border governance structure, such as Eurodistricts or European Groupings of Territorial Cooperation (EGTC)? A conceptual change is needed: a border should be considered as a zone, not only a line; the future will require progress on cross-border data on flows, interdependencies, Functional Urban Areas, but also common understanding of the border geography (concepts,[20] not only data); defining "cross-border regions";[21] identifying various scales:[22] local, regional, macro regional; new common norms and visions about borders, shared by their inhabitants[23] and by local, national and EU institutions, allowing the building of sound public policies for cross-border regions.

Cultural Sphere

Each national territory hosts a community united by a national narrative and vision, equipped with a common heritage and history,[24] and generally a common language. Borders are places where different cultures and languages clash, or on the contrary where a common culture and language pre-exist across the border. Cooperation projects have been developed for decades with the support of the EU (free movement, Interreg) or even before, thanks to regional cultures. All this has contributed to develop a culture of cross-border respect, mutual trust and cohesion.

---

[20] As functional CB regions, defined by D. Rougemont, D., *L'avenir est notre affaire*, Stock, Paris, 1977.

[21] As proposed by the draft Schengen Code regulation, Proposal for a regulation of the European Parliament and of the Council amending Regulation (EU) 2016/399 on a Union Code on the rules governing the movement of persons across borders {SEC(2021) 440 final} – {SWD(2021) 462 final} – {SWD(2021) 463 final} 2021.

[22] J. Peyrony et al., "French border cities and their cross-border strategies", *In* E. Medeiros, *Border Cities and Territorial Development*, Routledge, New York, 2021, pp. 119–148.

[23] B. Latour, *Down to Earth: Politics in the New Climatic Regime*, Polity Press, Cambridge (MA), 2018.

[24] A.-M. Thiesse, *La Création des identités nationales, Europe, XVIIIe-XXe siècle*, Seuil, Paris, 1999.

In each country, Covid has been a crash test of national cohesion (distrust *vis-à-vis* governmental measures, such as vaccination), but common references provided resources in the context of uncertainty – trust and information reinforcing each other. The Covid crisis and resulting measures have reactivated mental borders and challenged trust, between states (the "capitals"), bilaterally and at EU level, and between local populations across each border. Not only different understandings (see previous point about information), but also different narratives and visions have clashed at the border[25] – as it has been the case on the Germany-France border (see below). In the future, mutual trust should be (re) built from the bottom up, by developing a better knowledge of the past, and a vision of the common future, allowing the emergence of a cross-border society, which are preconditions for cross-border integration and its maintenance in times of crisis.

Interpersonal Sphere (Interlinked Persons)

As already explained, public policies in the field of cross-border or European integration generally consider individuals: economic agents (cross-border workers), holders of rights, national citizens (with their identity card).[26] But persons are not only individuals, they are also interlinked within couples or families, whose life is generally regulated within domestic framework (civil status…).

Last but not least, the interpersonal sphere has been put upfront by the crisis. Restrictions of mobility, at least in the first phase of the pandemic, have been focused on individual rights (attached to national identity, or status linked to work). They have severely impacted couples or families living across the border, whose reality has been denied by uncoordinated domestic measures.[27]

The ultimate goal of public action should be to guarantee the rights of persons (beyond their economic or national citizen status), also in

---

[25] Impact on discourse on borders, see CESCI, *Impacts of the coronavirus pandemic on the border regime, op. cit.op. cit.*

[26] A. Lamassoure, "Le citoyen et l'application du droit communautaire", Rapport au Président de la République, 8 June 2008 Retrieved from: <https://www.vie-publique.fr/rapport/29870-le-citoyen-et-lapplication-du-droit-communautaire>. Accessed 30 May 2023.

[27] B. Wassenberg, "Return to mental borders", *op. cit.*

the border context. This has consequences for the daily management of borders, which should have more consideration for couples and families.

Another aspect revealed by the crisis concerns public action in the cross-border context. The *Instituto di Sociologia Internazionale* (ISIG) classification of obstacles to cross-border integration[28] has identified, beyond obvious obstacles related to economy, institutions or culture, a residual obstacle they have named "lack of propensity to cooperate". Conversely, in March 2020, after the initial border closures unilaterally decided by states, persons acting beyond their remit, and not institutions, were the first to react beyond borders, and relaunched cooperation between institutions.

Several reports have underlined the role of informal, interpersonal links across the border, going beyond professional obligations (resulting from existing domestic or CB routines), as resources for public action. Such informal networks of persons, already playing an important role in current domestic systems, also exist in cross-border systems, but require strengthening.[29] In the future, public policies should not consider only individuals, but also persons living in couples/families, sometimes separated by the border. Family rights should be better acknowledged in a European and cross-border perspective. And interpersonal cross-border links between politicians and civil servants should also be considered as key for the management of borders.

Lessons in a Systemic Approach

Beyond these lessons for the six dimensions of persons' lives within cross-border living areas, let us now draw lessons, about the systems these six dimensions form in each country, the way they have clashed on each border, and how cross-border actors and the EU have reacted.

Border restrictions that took place between the first Covid-19 wave, and subsequent waves, have not been applied consistently across borders. In the first wave, they were much more drastic. Federal states closed

---

[28] Instituto di Sociologia Internazionale, *Manual On Removing Obstacles To Cross-border Cooperation, manual for the Council of Europe*, 2013. Retrieved from: <https://rm.coe.int/16806f93c9>. Accessed 30 May 2023.

[29] J. Beck, *Grenzüberschreitende Zusammenarbeit nach der Pandemie/La coopération transfrontalière après la pandémie*, Peter Lang, Brussels, 2021.

some of their borders under pressure from their regions. Some states closed their borders unilaterally and pursued a policy of limiting border crossings without consultation with neighbouring states. Smaller states, such as Luxembourg, highly dependent on border crossings, have hardly implemented any restrictive measures at their borders. Some states (such as the Baltic States) cooperating at the regional level have coordinated to ease border restrictions or to reopen their borders at the same time; most have not. Flexibility measures have been introduced by certain states, addressing "*bassins de vie transfrontaliers*" (cross-border living areas) in France, or "border communities" in the Nordic countries.

Despite the existence of European rules (e.g. the Schengen Code – even if several states had notified the Commission of movement restrictions, and extended them over time, already before 2020), border management was suddenly renationalised; "methodological nationalism" prevailed. It is only later that engagement of persons and local/regional/cross-border actors allowed borders to reopen and cooperation to restart.

At the beginning of the crisis, the "national gravitational pull" played out in full, with all its dimensions: material (such as economic, functional, institutional), but also immaterial (informed, cultural, interpersonal) – in each country with its specific cognitive framework, a combination of these dimensions.

Crises at the borders have unveiled contradictions between cross-border reality and inadequate domestic institutions. All regions suffered from the public health crisis, but border regions experienced a double penalty, with border closures in addition to other containment measures. Most states do not have a holistic, inter-ministerial view of the measures put in place on their borders, and of their effects on border regions. The crisis revealed how public policies, suddenly reduced to their domestic, vertical and bureaucratic routines, have struggled to take into account the cross-border lives of persons and their complex chains of movement. Borders are and will remain lines separating the territories of sovereign states (each with its own specific legislation, political and administrative culture, social protection systems, etc.).

But when only apprehended as lines between two states, they deprive actors of the necessary scope for their daily activity, and divide a dense milieu of economic and social relations. They are also functional areas, in which the people living there cross the border line regularly. The administrative regime of cross-border cooperation should consist of erasing the harshness of the border in this transition zone, e.g. to establish CPS through bilateral agreements.[30]

This is supposed to be facilitated by European integration: freedom of movement[31] and territorial cohesion recognising "cross-border regions",[32] with Interreg as a catalyst. The crisis, and the previous ones since 2015, whether migratory or security-related, have brought to light obstacles already well known to cross-border actors.

The six spheres we have just explored are mainly organised within each nation State, where they form a system of checks and balances, forging national cohesion.[33] Promoting cross-border or European integration, which means building a new system at another level, at first requires understanding and questioning national cohesion.

The crisis has been a learning machine, in each country, on each border and at the European level. In each country, the public health crisis management (health policy; constraints on mobility, including management of borders) has required the same typology of trade-offs between the six spheres; but these trade-offs have not been the same in each country.

- 1st trade off: Human cost, security (in terms of functional sanitary criteria, i.e. minimizing death numbers) vs. economy, freedom of movement (also considering human cost in the longer term)
- 2nd trade off: Efficiency – in functional or economic (utilitarian) terms – vs equality or equity (in institutional terms – protection of rights – and inter-personal terms – consideration of family links). Should hard, uniform measures or adaptation to personal cases, prevail? This has been particularly tangible in the case of border management. With a limited staff able to carry out border controls,

30 As de La Pradelle had already demonstrated in his 1928 thesis, see B. Perrier, "La frontière selon Paul de La Pradelle," *Borders in Globalization Review*, vol. 1, n°1, pp. 127–132, 2019. He promotes an international border regime in which sovereign borders are transformed into functional borders.

31 Provided with the Maastricht Treaty, 1992.

32 Lisbon Treaty, 2007; Article 174 Treaty on the Functioning of the European Union (TFEU).

33 The cohesion of national societies is so strong that it can be a cause of violence. The problem is not that we do not like our neighbours on the other side of the border, but that our attention is focused on our fellow national citizens. P. Dumouchel, *The Barren Sacrifice: An Essay on Political Violence*, Michigan State University Press, Chicago, 2015.

should a utilitarian approach prevail (concentrating staff on a limited number of crossing points, which means closing others, and hindering the daily life of a limited number of persons in small CB living areas), or an approach respecting such areas?
- 3rd trade off: Sophistication of measures (functional sphere) vs their comprehension by the population (information sphere)

Moreover, at what territorial scale should trade-offs be appreciated, e.g. functional adaptation to different situations combined with equality of treatment? At the national or the regional scale? And what for a border region concerned by two states?

The feeling that rules were sometimes inequitable has been a domestic problem (cf. sophisticated measures for progressive easing of lockdown taken by France: movement was permitted within a radius of 1 km, 10 km, then 30 km from home; but how should those rules apply to persons living at the border between two regions?). This feeling was multiplied by comparisons made with the other side of the border, or with other borders (e.g. Germany's western borders remaining open with Netherlands and Belgium, but not with France).

The European typology of states is diverse: small or big; centralised or not; involved in bilateral or regional cooperation (like Benelux, Nordic Council…) or not. Diverse political, socio-economic, cultural backgrounds, plus path-dependency on different Covid infection trajectories, led to various domestic trade-offs and policy responses, contradicting – and unavoidably clashing – on borders, all the more that borders are themselves diverse (strong or weak cross-border flows and integration, geographical nature of the border…).

Let us take some examples. Germany is a large federal country, characterised by a culture of respect of rules, search of compromise and consensus, subsidiarity, multi-level responsibility of persons and institutions such as the German *Länder* – the latter having played an important role in managing the crisis. The independent Robert Koch Institute had a major role in monitoring the public health crisis: the combination of the data it produces, and rules (such as thresholds accepted by social consensus), have automatically[34] generated some decisions (lockdown,

[34] Like a central bank for money and interest rates – in an approach coined as "ordo-liberalism", or – with a critical tune – as in A. Supiot: *Governance by Numbers: The Making of a Legal Model of Allegiance*, Bloomsbury Publishing Oxford/Portland, 2017.

restrictions on mobility…). German approaches towards borders have been differentiated at each border, taking into account the rural border with Denmark, and the urban border with the Netherlands; and have stuck to objective evolutions throughout the crisis.

France is a large unitary country (even if decentralised), characterised by political/functional deals mainly at national level. Political decision (even if informed by facts that are objective and public) is preponderant. Throughout the crisis, the national government, facing strong social contestation (e.g. about vaccines), has tried to define rules, for instance successive limitations of mobility, that would combine efficiency and social acceptance, and apply to all regions. As for border regions, decisions have also been national, like with the decree acknowledging "*bassins de vie transfrontaliers*" (see below).

Countries of Benelux, Northern or Central Europe,[35] whether centralised or not, are relatively small countries where the crisis management has been centralised. When the country's whole territory is close to a border, there is no need to decentralise crisis border management – even if in a federal country like Austria, national decisions have been taken in concertation with the *Länder*.

Such different approaches fatally clashed on borders. The Nordic community[36] usually promotes its model of regional integration based on trust between states and populations ("the Nordic gold"). But in a context of uncertainty, fear, and contrasting approaches taken by states, due to different traditions (e.g. liberal approach in Sweden, contrasting to its neighbours' approaches), the model has been damaged by the crisis, and nationalism has reappeared. Moreover, governments seemed to be "seeing borders for the first time".

Let us develop the case of the French-German border. At the beginning of the sanitary crisis in March 2020, France was more impacted than Germany, for contingent reasons; but this created distrust from the German side. Moreover, the explicit, organised (even if difficult) dialogue between Bund and Länder, on the German side, contrasted with

---

35 CESCI, *Impacts of the coronavirus pandemic on the border regime, op. cit.*

36 A. Giacometti & M. Wøien Meijer, "Closed borders and divided communities: status report and lessons from Covid-19 in cross-border areas", *Nordregio Report*, vol. 6, Nordregio, Stockholm, 2021; A. Giacometti& M. Wøien Meijer, "Nordic border communities in the time of Covid-19", *Policy Brief*, vol. 3, Nordregio, Stockholm, 2021.

unclear roles and internal competition between State and regions, on the French side.

On another note, it was difficult for French authorities to understand the contradictory approaches (at times) taken on the French border by the Bund and the 3 *Länder*, with their different political coalitions leading to contrasted approaches of the crisis.[37]

The crisis has been characterised by a general lack of understanding of what was happening, in the domestic context, and all the more in the cross-border context, for cumulative reasons: novelty, intensity and infectiousness of the virus; global context of dissemination; ever-changing nature of the pandemic – all these factors creating much uncertainty.[38] As regards border management in the EU context of regional integration, this uncertainty was multiplied by the diversity of national approaches that should have been, but were not, coordinated on each border.

Rather than new obstacles, it is their exacerbation, their much larger visibility, which has characterised the crisis: it has been a crash-test, raising awareness of already existing problems.

Our study looked into the role of all levels of government and of key stakeholders such as EGTCs or euro-regions in facilitating cross-border life in the Covid-19 context. Intergovernmental forums such as the Benelux Union, the Nordic Council or the Baltic Council of Ministers allowed for further coordination and for a proper exchange of information at a multi-national level. In order to tackle border issues, new 'crisis management task forces' at national and regional level and informal exchanges between local authorities and other key stakeholders at the local level were the most common coordination forms, even though not all of them were set up from the very beginning of the crisis. Although cross-border coordination has not systematically improved in every border area in the consecutive waves of the pandemic, we could generally witness a real learning process in crisis management along most of the EU's borders. Existing local or regional cross-border committees were activated in most cases: while their impact on softening national level

---

[37] L. Coatleven, F. Hublet, T. Rospars, *Subsidiary crisis management in the COVID-19 pandemic. Germany's federalist experiment in transborder perspective*, Groupe d'études géopolitiques report, Paris, 2021.

[38] H. Böhm, "Challenges of Pandemic-Related Border Closures for Everyday Lives of Poles and Czechs in the Divided Town of Cieszyn/Český Těšín: Integrated Functional Space or Reemergence of Animosities?" *Nationalities Papers*, 2022, pp. 1–15.

restrictions was rather low, they were useful in the implementation of common information tools, or in carrying out advocacy activities for raising awareness at the national level of the issues identified locally related to border restrictions. Many stakeholders also mentioned their ability to keep the cooperation spirit alive in a difficult time, while re-building cross-border trust between institutions and citizens.

The European Commission has reacted appropriately, within the scope of its too limited competences. Our assessment of the effectiveness of the European Commission's guidelines on free movement of cross-border workers and on healthcare assistance shows that their effect on border closures has not been systematic. For most of the responding states, the guidelines came after certain measures had been put in place to facilitate the passage of frontier workers, carers or emergency services; and for others they had no impact on the national measures taken by their State. But through its guidelines, and by showcasing good practices already in place on certain borders, the European Commission has encouraged states not to systematically close their borders and, when they had to re-establish border controls, to take measures mitigating the effects of such controls on border regions: it has provided a framework that it should develop in the future.

Different lessons can be drawn for border management. Persons living in cross-border areas have been impacted, when they have been struck with prohibitions or fines in the neighbouring country, while they felt they were acting reasonably (e.g. buying fuel on their way back to work). when border management was processed by civil servants unaware of regional border contexts – e.g. Swiss federal police, not speaking French-local language in Greater Geneva; or when measures did not seem to be "proportionate" and clearly motivated, for instance borders closed, not taking account of small local cross-border communities (Spain-France border in the Pyrenees).

Even if most civil servants in charge have probably done their best, when enforcing rules, to make accommodations for particular situations, some border inhabitants have experienced a sort of ambivalence, violence, injustice of the border.

In the future, there is a need to respect persons in their cross-border living areas (right to live at 360°); to build a feeling of justice: people can accept constraints, provided that they believe them to be necessary, and humanly applied; to manage decisions on borders, as regions hosting local communities, subject to equity, and not as parameters of utilitarian management of flows;[39] to develop a culture of accountability vis à vis border inhabitants.

Coordination has lacked within domestic systems, both vertically between local and national authorities, and horizontally inside State systems themselves. While proceeding with our research, and trying to find appropriate contacts in national administrations, we observed that several ministries were generally concerned, dealing respectively with national sovereignty (border control), considering the border as a line; regional development (in charge of monitoring the situation of border regions during the crisis, but not considering its cross-border dimension); with little information and coordination between them. In the future, border management will require in each State, as for any region, an integrated, holistic approach to cross-border regions.

And of course, coordination problems have occurred between neighbouring states. Cumulative effects of distrust, favouring the development of rumours, accelerated by pre-existing prejudice *vis-à-vis* other nations; by uncertainty and lack of information (general and across the border; about facts, different and evolving rules).

In the future, there is a need to build a simultaneously vertical (in each country), horizontal (across the border and towards cross-border citizens), and also diagonal coordination, taking into account the asymmetry of territorial governance across the border. There also is a need to develop a systemic vision of communication, including proper connections between institutions, media, social networks, non-formal and interindividual networks ("address book"); all this with increased assistance from EU.

Let us consider a table (Figure 1) with 4 columns (types of spaces: national territory, border line, border zone, Europe), and 6 rows (the 6 dimensions of cohesion: economic, functional, institutional, informational, cultural, interpersonal; and obstacles related to each of them). These dimensions spread out at first in each national territory (1st column); then they meet on the "Westphalian" border line separating two national territories (2nd column); they deploy in the border zone, where two States agree to go beyond their sovereignty in a cooperation aimed at cross-border integration (3rd column); within a European, multi-level perspective (4th column).

39 This raises the issue of the nature of spaces subject to "justice" (in the perspective of John Rawls' Theory of justice): national territories, or wider spaces: CB regions, Europe (reference to a European citizenship), or even the world (refugees' rights)?

Types of space/ dimensions of cohesion	**<u>National territories</u> in a context of limited opening of borders**	**<u>Border line</u> not fully open, as during the sanitary crisis**	**<u>Border zone</u> as it can develop when the border is open**	**<u>Multilevel/European scale</u> as a frame condition for open borders**
<u>Economic</u> (Market; public economy: tax and social security contributions) - Market-regulating institutions, addressing **individual workers - consumers** - **Economic obstacles**	States are only concerned with the preservation of essential flows within national markets. National citizen = taxpayer, financing public policies, only in domestic context	CB gradients -> economic flows CB integration only economic; if border closure: only CB flows of "essential" workers and goods considered No common CB management -> spill over effect impacting non-CB actors CB workers generally paying income tax abroad, but no public investment funding across the border	Consideration of the cross-border living area (not only workers, consumers; also citizens, functional dimension) CB market, and CB economic policies **Co-development:** Bilateral agreements on CB spatial planning and tax retransfer (CoE); "Financial compensation" in Greater Geneva CB recovery plans	European approach to cross-border work, "green lanes" for goods, … European framework for taxation of CB work, like for social security (already harmonised through EU regulation)
<u>Functional</u> - Administrations, policies, public service managers addressing **users** - **Technical and administrative obstacles**	National territory: bundle of domestic public services maintained by State	Public services end at the border Limited interoperability Untapped potential for economies of scale and agglomeration Borders as filter In case of crisis, reopening on a case-by-case basis	Cross-border integration **Cross-border public services** guaranteed in case of crisis: cross-border living area card; managed by EGTCs with competences Health criteria coordinated on the border	CPS European framework (Branda report), beyond sectoral regulations EU Green pass integrating the needs of CB living areas European health criteria

<u>Institutional</u> - States, local authorities and CB groupings, addressing **citizens** - Institutional obstacles	Institutional sphere embedded within nation State, founded on national democracy-limited to national citizens State provides security, solidarity, public goods and services via public policies Crisis management within the national (and/or sub-national: federal States...) democratic framework State monopoly of legitimate violence	Existing cooperation structures, but not mandated for crisis management Theoretical European citizenship- but limitation of free movement in case of crises No or little coordination on the border	CB governments or structures, under democratic control, with full mandates and competences Territorial structures with crisis management mandates Transnational democracy: CB citizen debate forums; towards CB election Towards Cross-border citizenship Dual citizenship recognised and encouraged	Multi-level governance of borders, regulated at European level (Schengen Code+) EU tool to solve obstacles (ECBM +) Network of European and national border coordination points Concrete European citizenship in particular in the CB context European competence in crisis management, public health
<u>Information</u> - Media aimed at **informed individuals** **- Obstacles related to lack of in-formation**	Statistics, production of social sciences, education and media organised at national (or sub-national) level State monopoly of evidence	Border, place of contradiction between States Lack of cross-border information, redoubled in case of crisis Misunderstanding	CB Media Cross-border information organised on the basis of cross-border and multilevel structures	Multi-level European information European media

Types of space/ dimensions of cohesion	**<u>National territories</u> in a context of limited opening of borders**	**<u>Border line</u> not fully open, as during the sanitary crisis**	**<u>Border zone</u> as it can develop when the border is open**	**<u>Multilevel/European scale</u> as a frame condition for open borders**
<u>Culture</u> - Civil society institutions (civic, cultural, etc.) bringing together **inspired individuals** - **Obstacles related to culture**, language	Civil society structured within national framework Each national territory is a community united by a “national narrative”, heritage, common language Covid crash test of national cohesion; common references provided resources for trust	Borders: places where different cultures, visions, languages, clash Boundary between 2 national universes turning their backs on each other Covid crisis has reactivated mental borders, distrust	Emergence of a **cross-border civil society** People to people, educational, cultural projects Learning the neighbour's language Common history teaching Knowledge of the past, vision of common future, preconditions for CB integration and its maintenance in case of crises	Emergence of a European civil society European Observatory of history teaching (CoE) Towards an EU competence on education

<u>**Inter-personal**</u> - Institutions protecting the rights of families, couples, …bringing together individuals - Interlinked persons: not only individuals as holders of rights, or agents - **Obstacle: lack of interpersonal links; lack of willingness to cooperate**	Life of couples and families regulated within domestic framework (civil status) Recognition of national citizens only Networks in a predominantly national framework	Only economic agents (CB workers) or national citizens considered Non-recognition of non-nationals (including in case of CB couples or families) Distrust Networks back-to-back at the border (boundary) with some exceptions	Guarantee the rights of persons, couples, families, living in CB living areas (beyond economic or national citizen status) CB living areas cards Informal, interpersonal CB links, going beyond routines: resources for CB public action, to be reinforced Multi-level structured cross-border networks of persons	Family rights acknowledged in a European and CB perspective European framework "27e scheme" (Alain Lamassoure) tested in the CB living areas Extended networks at European scale

Figure 1. Lessons of the effects of Covid-19 related measures for the six dimensions of cross-border integration.

Recommendations for the Future of Cross-Border Living Areas and Their Multi-Level Governance

Many cross-border stakeholders share the feeling, reinforced by the crisis that cross-border cooperation 1.0, generally defined as voluntary, has reached its limits, and that we should move to cross-border cooperation 2.0, where it would be an obligation, not just an option. Some authors characterise the evolution of border policies as moving from coordination to cooperation, and then to collaboration.[40]

What should be the tools to fulfil such an evolution? Let us consider a flagship CPS, the hospital of Cerdanya.[41] It has been funded with the support of Interreg. Its governance is supported by an EGTC. To ensure the daily operation of the hospital, a number of domestic sectoral policies (where the EU has limited competence, such as health, civil status, rescue, etc.) have to be coordinated, through an adapted regulatory framework. Such CPS are obstacle-solving machines. Identification of needs; elaboration, signature and ratification of necessary bilateral agreements all take much time (more than 10 years for the Cerdanya hospital). This shows that new procedural tools (bilateral or European) are required to solve obstacles and develop sectoral cross-border regulatory frameworks. Hence such a CPS calls for three different tools: governance tools (such as EGTCs); funding tools (such as Interreg); and sectoral frameworks.

Let us now present recommendations for these three types of tools at the local level of cross-border regions; and within a multi-level approach (EU framework, implemented within multi-level governance), deriving from lessons learnt in part 1. Contrary to part one where we adopted a six dimension analyse, here we adopt a cross cutting, systemic approach.

Recommendations at the Local Level of the Cross-Border Region

After the initial shock of March 2020, persons were the first to react. Then local institutions, in particular cross-border governance bodies

[40] C. Leuprecht, E. Brunet-Jailly, T. Hataley & T. Legrand, "Patterns in nascent, ascendant and mature border security", *op. cit.*

[41] J. Peyrony & X. Faure,"Border vs Cross-border integration: The case of the Cerdanya hospital" *In* B. Wassenberg (ed.), "Cross-border circulation and networks in the European space", *Fare cahiers* n° 18, L'Harmattan, Paris, 2020, pp. 125–149.

(Eurodistricts, Euroregions, EGTCs), prompted the resumption of cooperation, showing, if proof was needed, that "if nothing is possible without men, nothing lasts without institutions", in the words of Jean Monnet. Their role should be reinforced.

Governance

As for the pattern that local cross-border governance should follow, diverse options are on the table. Should EGTCs (or equivalent organisations) be entrusted with competences, and not only missions[42] – as the EGTC regulation prescribes in its present wording? If they were to be given competences, under what democratic control? In other terms, should we move from cross-border governance to government? Our study does not allow a clear-cut conclusion – it should depend on local contexts, and different understandings across the border – as the Germany-France border shows.

But in any case, cross-border local governance (more or less formalised), should innovate, both continually and in view of future crises, and be given a more prominent role to manage borders, with increased capacities. This means contributing to building a cross-border administrative space in line with the cross-border socio-economic space.[43] Secondly, it means be a facilitator, a place of coordination, also in the case of a crisis (joint crisis strategies). Thirdly, develop CPS, e.g. in the field of health, and ensure their continuity in the case of a crisis. Finally, to create cross-border spaces of interactions between politicians and citizens, while tending towards permanent democratic cross-border councils.

Strengthening mutual trust requires acting also on immaterial aspects, and developing local cross-border observatories delivering harmonised data at the level of cross-border living areas – with the support of national and EU levels but also cross-border information about projects (ESPOO convention[44]) then cross-border information centres for cross-border

---

[42] J. Peyrony, "Should EGTCs have competences and not only tasks?" *In* CESCI, *15 Years of the EGTCs. Lessons Learnt and Future Perspectives, op.cit.*, pp. 219–243.

[43] See J. Beck's analyses about numbers of people working in cross-border vs domestic administrations.

[44] The Convention on Environmental Impact Assessment in a Transboundary Context (informally called the Espoo Convention) is a United Nations Economic Commission for Europe (UNECE) convention signed in Espoo, Finland, in 1991.

workers, consumers, businesses, maintained in times of crisis, thirdly intercultural training,[45] as to say training on language, knowledge of the system of the neighbouring country. Furthermore, this means extending Erasmus for civil servants and politicians, but also promoting cross-border media, cross-border education and funding of people-to-people projects to (re)-build trust, as well as personal networks. All this should also be completed by regular crisis exercises involving cross-border politicians and civil servants at managerial level.

A common teaching of history, a common – pre-existing or emerging – cross-border culture, would contribute to overcoming diverging narratives, and build a vision of a common future of the cross-border region, and Europe.

Cross-border integration can be supported by tools addressing concrete, daily needs of cross-border residents, both in ordinary situations and in times of crisis. A promising concept consists of a "mobility card" or "cross-border living area card", as it has been developed on the Spanish-Portuguese border, linking for instance CPS access with belonging to a cross-border living area. Ideally, such cards would have the same format for all inhabitants, regardless of their nationality. They could be the medium for a variety of CPS, such as a common transport ticket for the entire cross-border living area. In the event of a crisis and border restrictions, such cards would replace or be combined with certification systems, which would facilitate checks. They could be issued by Infobests, EGTCs or equivalent bodies and should avoid being presented in terms of "status" in order to avoid recreating border effects.

Another perspective would be to link such a card with an EU green card, like the EU Covid certificate, developed during the pandemic.[46] The advantage would be to connect a concrete measure for inhabitants of cross-border regions with the EU, demonstrating synergy between national and EU citizenship, through a sort of functional cross-border citizenship.[47]

---

45 As Euro institute develops it on the Germany-France-Switzerland border.

46 The idea of an EU regime of mobility (28th regime), that would be experimented in cross-border regions, has been developed by A. Lamassoure, "Le citoyen et l'application du droit communautaire", *op. cit.*

47 Approach close to Denis de Rougemont's vision of cross-border functional regions (1977).

New Approaches to Resolve Sectoral Obstacles to Cross-Border Integration

Many cross-border stakeholders, while developing cross-border projects, need to adapt to divergent (economic, institutional…) parameters in the two countries. They claim for any measure, going in the direction of adaptations, derogations of legal or administrative domestic rules; or mutual recognition between them. Such differentiated measures seem to be of common sense, but may appear contradictory to equality in domestic contexts.

Some authors propose to secure such cross-border local solutions: derogations concentrated on a limited cross-border area, thus having a limited impact, via a "*de minimis* rule".[48] Derogations granted by States to solve cross-border obstacles have always existed, and are possible, at least through bilateral treaties. An example of these is given by article 13 of the Aachen treaty between France and Germany.

Funding: Moving Towards Cross-Border Co-Development

Interreg should not be considered as the only source of funding for cross-border projects. Even if Interreg has contributed to funding CPS, and will do in the future, required investments often go beyond Interreg envelopes, and other EU or domestic funding is needed.

Moreover, the economic situation and funding capacity can be asymmetric –as in case of borders separating an employment centre and an area with more residential characteristics. At least in strongly integrated cross-border areas; common strategy and funding are required, to implement cross-border "co-development". There comes the importance of tools such as the "Agglomeration projects" in Trinational Eurodistrict Basel or Greater Geneva, mixing domestic, local and federal funding (Swiss "policy for agglomerations"), while moving towards cross-border investment funds.

The Organization for Economic Cooperation and Development (OECD) should monitor the application of its "model bilateral convention" on the non-double taxation of persons, and encourage it to

[48] J. Beck, *La coopération transfrontalière après la pandémie, op. cit.*

re-examine its model in view of border situations, that are sometimes very unbalanced, in the light of the emergence of border telework since 2020.

Recommendations for Multi-Level Governance within the EU Framework

During the crisis, most problems have come from lack of State coordination across the border. In a crisis context, sovereign States have legitimacy to limit mobility, but should coordinate, vertically with local authorities and the EU, and horizontally across borders. The EU's limited mandate should be increased on missions such as checking that State action is fair and proportionate, providing information and coordination.

As we have done for the local level, let us now explore, in a multi-level perspective, governance, sectoral policies and funding.

Towards a Bottom-Up, Multi-Level Governance of Borders and Cross-Border Regions

The EU is clearly expected to play a more important role in the future, and monitor a multi-level governance of borders, in a holistic perspective. The European Commission has already made proposals, through two draft regulations, within two different areas: management of borders (revision of the Schengen Code); and solution of obstacles to cross-border integration (European Cross-border Mechanism (ECBM)).

Revision of the Schengen Code

On 14 December 2021, the European Commission published its proposals for new rules to strengthen the governance of the Schengen Area. The draft regulation, which sets forth a "Union Code on the rules governing the movement of persons across borders", provides the obligation for Member States to designate "cross-border regions", that would be subject to derogations from measures restricting free movement at internal borders in the event of crises. Some key lessons on the impact of the pandemic have been taken into account. MOT's first study is quoted by the Commission's impact assessment report. It is not for the present chapter to comment on the regulation or its negotiation, but to provide preliminary ideas and propositions from cross-border regions' point of view, on the basis of lessons presented earlier.

A better knowledge of cross-border flows, and interdependencies they reveal in the daily life of cross-border territories, is an absolute priority, as a basis to develop local public policies, and to raise awareness of national and EU authorities, so that they take into account cross-border territories in their own policies. An evidence-based description of functional regions is required, but far from being available right now. How should the "cross-border regions" (as mentioned by the draft Schengen Borders Code revision) be defined?

Functional criteria based on accessibility and time[49] ("24 hours" rule;[50] or 3 hours maximum per day for travel to work and back) seem better than criteria based on distance. The important idea here is that only persons who return daily to their home should be considered, in order to distinguish such mobility from weekly mobility for instance. Nevertheless, activities in cross-border regions cannot be circumscribed to cross-border work.

Should cross-border regions be defined as the perimeters of existing EGTCs (or equivalent structures)? This would create side effects, and be difficult to manage. Is then the NUTS3 level border region the best scale? Even if it raises issues, such as their heterogeneity, this is already a common compromise.

Given the diversity of borders, there are doubts that unequivocal design of such "cross-border regions" by Member States or EU would ever be possible. In any case, it should be up to Member States to define cross-border regions bilaterally. "Small border traffic zones", which have been designed for non-Schengen borders (Hungary-Romania, Poland-Ukraine, Poland-Russia, Poland-Belarus, etc.), can be considered as a possible solution for facilitating the lives of internal borders' inhabitants in times of border closures.

Acknowledgement of cross-border regions and "status of frontier workers" are intertwined questions, as the example of teleworking, explained

---

49 The Joint Reasearch Center (JRC) study on driving time to border crossings (30 mins, 60 mins, 90 mins) is a good reference. Retrieved from: <https://publications.jrc.ec.europa.eu/repository/handle/JRC116859>. Accessed 30 May 2023.

50 See for instance French decree n° 2021-99, 30 January 2021.

From 31 January 2021, anyone entering France, including from other European countries, needed to be in possession of a negative Covid test obtained within the previous 72 hours. However, this obligation did not apply to the inhabitants of cross-border living areas (for journeys lasting less than 24 hours in a perimeter defined by a radius of 30 kilometres around place of residence) or to cross-border workers.

earlier, shows. The aim is now to find new rules allowing teleworking on a higher level than before the crisis. According to the results of a study carried out by the MOT,[51] some recommendations can be made. In the context of renegotiations of bilateral conventions, simplify as far as possible the definition of the "tax status" of frontier workers, and bring it closer to the definition of 'status of frontier workers' in Regulation 883/2004[52] on social security rules. It should also encourage the European Commission to initiate an amendment of this regulation, in order to take into account this new way of working, a fortiori cross-border, and rethink the notion of "substantial activity".

In the daily management of borders, public action should guarantee the rights of persons (including couples, families...). Hence, the Schengen Borders Code revision should provide procedural guarantees of information, explicit decisions, accountability and fairness. The burden of proof has to be reversed. Systematic border controls have been suppressed by the Schengen legislation. National authorities are entitled to re-establish controls in certain conditions. In that case citizens have to prove their right to cross the border, but authorities also have to prove that they respect these conditions.

European Cross-Border Mechanism (ECBM)

Procedural tools (bilateral or European) are required to solve obstacles and develop sectoral cross-border regulatory frameworks. As said earlier, local authorities and their cross-border groupings can be given new possibilities to act, in times of crisis and in normal times, through decentralisation, deconcentration, differentiation, simplification. Nevertheless, many obstacles require changes in national or even EU legislation, and/or bilateral treaties. The experience has shown that such legislation or treaties often require too much time to be elaborated, negotiated and ratified.

This is where the Franco-German Aachen Treaty can be considered as a good practice, with its chapter dealing with cross-border cooperation. Its article 14 establishes a channel to solve cross-border obstacles when

51 MOT, "Impacts du télétravail frontalier. Le télétravail frontalier: de marginal à indispensable", *op. cit.*

52 Article 1 (f) "frontier worker" means any person pursuing an activity as an employed or self-employed person in a Member State and who resides in another Member State to which he/she returns as a rule daily or at least once a week.

local solutions cannot be found, under a multi-level democratic control, through a cross-border cooperation committee involving local/regional authorities on both sides of the border, Eurodistricts (local cross-border governance bodies), national governments, and Members of Parliament (MP).[53]

Such multi-level bilateral mechanisms exist on other borders: the Franco-Italian border (Quirinal Treaty), the Franco-Spanish border (Barcelona Treaty) and the Spanish-Portugese border. Organisational asymmetries across borders require "diagonal"[54] coordination, within a multi-level governance framework.

The study has shown the difficulty to identify those national administrations having a holistic view of borders. It seems clear that most Member States should engage in better coordination of border management at national level.

This finds an echo with the draft regulation on ECBM – even if blocked so far by a working group in the Council – which includes the obligation for Member States to create national border coordination points. Beyond their role for solving obstacles, such coordination points should coordinate with: sectoral administrations at national level (as head of a network of contacts in each ministerial department); national Parliaments (to deal with cross-border impacts of domestic legislation); domestic regional authorities; neighbouring States through a bilateral/multi-lateral approach (e.g. to coordinate transposition of EU directives).

The ECBM would be a generalisation of the Aachen Treaty's principles. The EU border coordination point created by the European Commission in DG REGIO, playing such a role while interacting with sectoral DGs within the European Commission, with EU institutions, and cross-border cooperation stakeholders, shows the way.

---

53 J. Peyrony, "Le Traité d'Aix la Chapelle: potentiels et défis du chapitre sur la coopération transfrontalière", *In* J. Beck, *La coopération transfrontalière après la pandémie*, *op. cit.*, pp. 359–385.

54 T. Alberts, V. Köbele-Ennaji, J. Ross, & V. Wolfart,"Quand les mécanismes de coopération sont mis à l'épreuve. L'épidémie du Covid-19 et la fermeture de la frontière franco- allemande au printemps 2020 comme stress test pour la coopération transfrontalière", *Les papiers de la recherche de l'ENA*, Paris, 2021.

Greater EU Role for Sectoral Policies

Taking stock of problems that occurred and solutions that were developed during the crisis, the EU should also be entrusted with new roles or competences. This directly results from lessons learnt during the crisis, in fields such as the harmonisation of crisis management from the EU level, including cross-border emergency plans that could be implemented locally, health policy, including cross-border health integration,[55] the definition of common health criteria (thresholds, quarantine duration…) and cross-border data (comparable data across borders).

The crisis has however shown the need to progress on immaterial dimensions favouring cross-border trust, such as information and culture. More EU intervention would be welcome also in the field of education – a wish also expressed through the Conference on the Future of Europe that would contribute to the emergence of an EU society, also experimented in cross-border regions.[56]

Funding Tools: The Role of Interreg

Even if insufficient for significant cross-border investments, Interreg remains of course a reference. Interreg programmes now provide an incentive to develop synergies between programmes, integrated territorial approaches (objective 5 : A Europe closer to citizens); strengthen cross-border governance, cross-border data, and mutual trust (Objective ISO 1 : Better cooperation governance).

Such an approach for Interreg is at the core of Cohesion policy, intersecting bottom-up approaches,because each border is specific, requires a customised approach, and knowledge is on the ground; and top-down approaches,because States still regulating borders and controlling the largest fiscal resources, now have to exert their sovereignty in common, with the EU as framework to build the single market and common goods, and face global transitions.

---

55 CESCI, *Impacts of the coronavirus pandemic on the border regime*, *op. cit.*

56 On themes such as history; see Council of Europe's history observatory <https://www.coe.int/en/web/education/-/observatory-on-history-teaching-in-eu-rope-starts-its-work>. Accessed 30 May 2023.

A New Role for the EU

The present study has shown the extreme diversity, both of borders themselves, and of domestic contexts for border management (from large countries, federal or not, where borders can be far away from the capital) to small countries where borders are a day-to-day topic.

Arrangements should be specific to each country and border, making it difficult to design a "one size fits all" EU solution. It means that EU normative, top down solutions should not prevail, but rather approaches where the EU has a softer role – as it is already the case for the two draft regulations mentioned above.

The EU should instead design European frameworks, in fields like CPS and their maintenance in times of crisis.[57] The EU is also awaited to play a general coordination role, through the action of DG HOME in the field of border management, and DG REGIO in the field of territorial cohesion. To create processes of convergence (e.g. for taxation of cross-border work, as it is already the case for social security) and ensure that Member States respect EU requirements on open borders (Schengen Code) and support to cross-border integration (ECBM) are two other possibilities for the EU. Also, it should facilitate a network of national border contact points, then promote experimentation and peer reviews, favour collecting and disseminating good practices, benchmarks, in particular through Interreg programmes. Also, the EU should promote and support the development of Border Information Points, to give appropriate reports to the competent institutions about the obstacles users are faced with, and to improve communication to inhabitants about cross-border policies implemented. The EU should also develop and communicate evidence about borders and cross-border regions, through its own action (statistic work: Eurostat, REGIO, JRC), applied research and network programmes (ESPON). Finally, it should provide information, EU standards, check lists, harmonise data; develop cross-border Territorial Impact Assessment (TIA) of EU legislation and contribute to promoting on each border TIA of legislation and measures taken.[58]

---

57 See Branda report. Retrieved from: <https://cor.europa.eu/en/news/Pages/strengthening-cross-border-public-services.aspx>. Accessed 30 May 2023.

58 See ITEM's cross border impact assessment tool; "CB check". Retrieved from: <https://www.maastrichtuniversity.nl/research/item/research/item-cross-border-impact-assessment>. Accessed 30 May 2023.

Member States should be free to choose the way they implement EU objectives, coordinate on each border: bilateral approaches like on the Franco-German border (Aachen Treaty); multilateral approaches (Benelux, Nordic Council). They should report to the EU, on the way they take account of cross-border regions in their policies and contribute to the resolution of cross-border obstacles.

Member States could be asked to provide annual reports, taking into account "cross-border regions" that would be assessed by the European Commission (via an Open Method of Coordination (OMC)), both in the field of the Schengen Code, and in the field of cross-border integration. The European Committee of the Regions could be given a role in such a process.

Conclusion

During the first wave of the pandemic, States restricted free movement of persons, in particular by closing or strictly controlling international borders. While this was justified, to some extent, for intercontinental travel, it was much less so for internal European borders. Borders came suddenly back, and split up cross-border living areas, without any coordination between States. Given the widespread uncertainty in which humanity found itself in the spring of 2020, States can be given the benefit of the doubt. This has however contributed to validating a mistaken view of the nature of borders, and has created negative expectations and deteriorated confidence.

Borders are ambivalent institutions, which reveals their sacred nature.[59] They are limits supposed to guarantee internal cohesion within national territories, but they reject violence outside. Like any institution, the border is useful and respectable as long as its cost is not too high. The challenge is not to abolish it, but to desacralize it: to unveil it by making its operation explicit.

Nation States, whatever their deficiencies, have been the dominant framework for collective action and financial solidarity during the public health crisis and in the resulting economic and social crisis. National

---

[59] J. Peyrony, "Frontières et sacré. Crises et nouveaux défis", *In* B. Wassenberg, *Frontières, acteurs et représentations d'Europe (Fare): Mélanges en l'honneur de Sylvain Schirmann*, Peter Lang, Brussels, 2022, pp. 105–132.

systems frame everyday life and the way people think, and this has played a positive role through the crisis.

But the crisis has revealed deep cross-border interdependencies. Public intervention modes have been called into question, in general and in the cross-border context. The problem is less the nation State as such, than the overemphasis on a solitary exercise of sovereignty. Borders should be overcome through cross-border and European integration. New visions of cross-border territories and of Europe, going beyond national visions are required. This involves the construction, beyond borders, of each of the six dimensions of cohesion: a common cross-border territory, a common European territory, as a collective project capable of including persons.

A paradigm shift is needed. Persons must be placed at the centre of reflection and action, in order to break the vertical and bureaucratic logics of institutions that have shown their limits. For border regions, the starting point should not be administrative procedures, but territorial ecosystems, "cross-border living areas", where persons' daily lives can be either disrupted by closing borders or facilitated by cooperation. On this condition, the old Westphalian linear border, undermined by the crisis, will be overcome.

Public authorities should be accountable to citizens for a proportionate management of the border, and this in a multi-level framework, where the different levels are jointly responsible. In the future, for economic, social and simply human reasons, the reflex of closing borders without justification will not be acceptable.

After two years of experiential learning, border management clearly requires reinventing, considering borders not only as lines, but also as zones, as cross-border regions where people live. Future policies should be based on a 360-degree vision; personal responsibility; vertical and horizontal subsidiarity; horizontal, vertical and diagonal coordination; moving from voluntary co-operation to compulsory collaboration, with cross-border mandates given to institutions.

Cross-border local governance should be given a greater role in developing cross-border markets, and managing CPS, under cross-border democratic control, building trust through cross-border information and common vision, shared by persons interconnected through the border. Innovative policies such as cross-border resident cards giving access to CPS, maintained in case of crises, would make EU citizenship more visible. Local actors should be given more legal capability through decentralisation, differentiation, derogation, experimentation, and increased funding resources beyond Interreg, through coordinated domestic and other European funds.

A bottom-up multi-level governance for the management of borders is required. Two European regulations should allow progress. The first one, on the basis of the ECBM draft regulation, should facilitate the solution of cross-border obstacles. The second one is the Schengen Code revision, which proposes the designation of cross-border regions by Memebr States that would be protected from arbitrary measures of border control. New EU competences should be developed in policy areas such as spatial planning, taxation, health, crisis management, and education.

The crisis has revealed the real, but limited relevance of national borders. Member States, instead of holding on to exclusive sovereignty within their borders, should exert their sovereignty in common on each border and at the EU level. Only on this condition will they achieve the single market and develop public goods, managing the transitions they have to face, while fully involving citizens.

Bibliography

T. Alberts, V. Köbele-Ennaji, J. Ross & V. Wolfart,"Quand les mécanismes de coopération sont mis à l'épreuve. L'épidémie du Covid-19 et la fermeture de la frontière franco – allemande au printemps 2020 comme stress test pour la coopération transfrontalière", *Les papiers de la recherche de l'ENA*, Paris, 2021.

A.-L. Amilhat Szary, "Géopolitique des frontières, Découper la terre, imposer une vision du monde," *Le Cavalier Bleu*, Paris, 2020.

E. Balibar, "Europe as borderland," *Environment and Planning D: Society and Space*, vol. 27, n°2, 2009, pp. 190–215.

J. Beck, *Grenzüberschreitende Zusammenarbeit nach der Pandemie/La coopération transfrontalière après la pandémie*, Peter Lang, Bruxelles, 2021.

H. Böhm, "Challenges of pandemic-related border closures for everyday lives of poles and Czechs in the divided town of Cieszyn/Český Těšín: Integrated functional space or reemergence of Animosities?" *Nationalities Papers*, 2022, pp. 1–15.

L. Boltanski & L. Thevenot, *On Justification. Economies of Worth*, Princeton University Press, Princeton, 2006.

CESCI, *A koronavírus hatásai a határrezsimre. Értékelő elemzés a magyarországi határrezsim változásáról 2020 máciusa és 2021 augusztusa közt.* [Impacts of the coronavirus pandemic on the border regime. Assessment of the changes of the Hungarian border regime between March 2020 and August 2021. – in Hungarian], CESCI, Budapest, 2021.

L. Coatleven, F. Hublet & T. Rospars, *Subsidiary Crisis Management in the COVID-19 Pandemic. Germany's Federalist Experiment in Transborder Perspective*,Groupe d'études géopolitiques report, Paris, 2021.

P. Dumouchel, *The Barren Sacrifice: An Essay on Political Violence*, Michigan State University Press, Chicago, 2015.

A. Faludi, *The Poverty of Territorialism: A Neo-medieval View of Europe and European Planning*, Edward Elgar, Cheltenham, 2018.

A. Giacometti & M. Wøien Meijer, "Closed borders and divided communities: Status report and lessons from Covid-19 in cross-border areas", *Nordregio Report*, vol. 6, Nordregio, Stockholm, 2021.

A. Giacometti & M. Wøien Meijer,"Nordic border communities in the time of Covid-19", *Policy Brief*, vol. 3, Nordregio, Stockholm, 2021.

Instituto di Sociologia Internazionale, *Manual on Removing Obstacles to Cross-border Cooperation, Manual for the Council of Europe*, 2013.

A. Lamassoure, "Le citoyen et l'application du droit communautaire", Rapport au Président de la République, 8 June 2008

K.-H. Lambertz, "Fair distribution of taxes in transfrontier areas," Congress of local and regional authorities report of the 37th session, 29 October 2019.

B. Latour, *Down to Earth: Politics in the New Climatic Regime*, Polity Press, Cambridge (MA), 2018.

C. Leuprecht, E. Brunet-Jailly, T. Hataley & T. Legrand, "Patterns in nascent, ascendant and mature border security: Regional comparisons in transgovernmental coordination, cooperation and collaboration", *Commonwealth and Comparative Politics*, vol. 5, Issue 4, 2021, pp. 349–375.

MOT, "Retour sur la conférence 'liaisons ferroviaires transfrontalières' des 10 et 11 juin", Paris, 2021.

MOT, "Impacts du télétravail frontalier. Le télétravail frontalier: de marginal à indispensable," Paris, 2022.

J. Peyrony, *Frontières et cohésion territoriale, Actes de la conférence Construire des ponts à travers les frontières, vers une cohésion territoriale en Europe*, Steiner Verlag, Stuttgart, 2014.

J. Peyrony,"Integration in cross border regions. Obstacles and solutions", *In* B. Wassenberg (ed.), *Castle Talks on Cross Border Cooperation. Fear or Integration? The Pertinence of the Border*, Franz Steiner Verlag, Stuttgart, 2018.

J. Peyrony, "Should EGTCs have competences and not only tasks?", *In* CESCI, *15 Years of the EGTCs. Lessons Learnt and Future Perspectives*, CESCI, Budapest, 2020, pp. 219–243.

J. Peyrony, "Frontières et sacré. Crises et nouveaux défis", *In* B. Wassenberg, *Frontières, acteurs et représentations d'Europe (Fare): Mélanges en l'honneur de Sylvain Schirmann*, Peter Lang, Brussels, 2022, pp. 105–132.

J. Peyrony, T. Devillard, J. Rubio & R. Viaggi, "Analysis of the impact of border-related measures taken by member states in the fight against COVID-19", EU commission impact analysis, July 2022.

J. Peyrony et al., "French border cities and their cross-border strategies", *In* E. Medeiros (ed.), *Border Cities and Territorial Development*, Routledge, New York, 2021, pp. 119–148.

J. Peyrony, J. Rubio & R. Viaggi, "The effects of COVID-19 induced border closures on cross-border regions", EU Commission report, 2021.

J. Peyrony & X. Faure,"Border vs cross-border integration: The case of the Cerdanya hospital", *In* B. Wassenberg (ed.), "*Cross-border Circulation and Networks in the European Space*", *Fare cahiers* n° 18, L'Harmattan, Paris, 2020, pp. 125–149.

B. Perrier, "La frontière selon Paul de La Pradelle," *Borders in Globalization Review*, vol. 1, n°1, pp. 127–132, 2019.

D. Rougemont, *L'avenir est notre affaire*, Stock, Paris, 1977.

A. Supiot, *Governance by Numbers: The Making of a Legal Model of Allegiance*, Bloomsbury Publishing Oxford/Portland, 2017.

A.-M. Thiesse, *La Création des identités nationales, Europe, XVIIIe-XXe siècle*, Seuil, Paris, 1999.

B. Wassenberg, "Return to mental borders: A diary of COVID 19 closures between Kehl, Germany, and Strasbourg, France", *BIG Review*, vol. 2, 2020, pp. 114–120.

(Mis)Perceptions of the Canada-US borderland

Tatsiana Shaban & Emmanuel Brunet-Jailly

Introduction

Today, typical issues about the Canada-United States (US) border regions may primarily be concerned with everyday logistics, cross-border traffics and trade fluidity. All are enhanced by issues of security as well as by apprehensions about how open the border should be to refugees and immigrants; all those greatly affect the daily lives of borderland communities. Indeed, 576 billion worth of goods cross annually, that is about 13.5 million trucks. Also, about 24.2 millions Canadians and 11.5 millions Americans make a single day return across the boundary line each year.[1] These numbers point to the particularly relevant effect of the border on Canadians, because 90 % of the population lives within 160 miles from the boundary line. It is also true of indigenous communities and of larger urban centres such as the greater Seattle region, the greater Detroit and Buffalo regions and the large suburban regions that are connected to two of Canada's urban regions: Vancouver and Montreal.[2] Hence, perceptions from those borderland regions and their governance is of importance for the general public and local authorities. Following the core research question raised by the *Frontières en mouvement* (Frontem) Jean Monnet network, this chapter addresses perceptions and misperception of those who deal with the Canada-US border on a daily basis since the 1990s;[3] it considers the context and looks at the impact of

1 *CBC News*, 15 May 2009. Retrieved from: <https://www.cbc.ca/news/canada/by-the-numbers-1.801937>. Accessed 15 September 2023.

2 E. Brunet-Jailly, "US-Canada border cities and territorial development trends", *In* Medeiros (ed.), *Border Cities and Territorial Development*, Routledge, New York, 2021.

3 In 1989 Canada-US Free Trade Agreement (CUS FTA) placed Canada and the US at the forefront of trade liberalisation. It was brought into force (then superseded by

key governments' policies on their lives. In many instances, these have resulted in changing perceptions of the border.

Concurrently, we ask the following questions: How do citizens perceive the border(s) with the neighbouring country(s)? Where does the border become obvious for citizens? How do citizens feel about the border(s)? What does it represent for citizens? Our findings suggest a profound transformation of the attitudes and perceptions[4] of the border, whereby those attitudes and perceptions further "distantiate" borderland communities, i.e. increases the sense of distance across the boundary line.

Since the 2008 financial crisis, Canada remains the United States' largest world trade partner. However, since the Free Trade and North American Free Trade Agreements, trade progression is stable but stagnant.[5] Borderland communities find it more difficult to bridge the boundary line while the US-Canada federal level relationship continues to deteriorate. Indeed, border politics continually interfere.

Because past and present perceptions of the border vary, these perceptions in borderland regions and their respective regional border governances, including the shifting borders of airports and preclearance, have gained more importance over time for both public authorities and the general public. Our analyses of perception of the Canada-US borderland focuses on the national (federal), state/provincial and local (municipal) levels. All in all, it seems that border policies further divide and undermine public trust.

First, our work briefly provides a historical and geographical contextual analysis of the border region and of the impact of those major crises on the life of the borderlanders. Then, we review developments in

---

NAFTA in 1992/1994, which included Mexico as a signatory state of the Agreement). Key elements of the agreement included the elimination of tariffs, the reduction of many non-tariff barriers, and it was among the first trade agreements to address trade in services. It also included a dispute settlement mechanism for the fair and expeditious resolution of trade disputes.

4 For the purpose of this study, we understand "attitude" as "a settled way of thinking … that is reflected in a person's behaviour" (Oxford Languages), and, we understand "perception" as "the ability to see, hear, or become aware of something through the senses" (Oxford Languages).

5 Global Affairs Canada, Office of the Chief Economist, "Canada's State of Trade, the early impacts of COVID-19 on trade", 2020. Retrieved from: <https://www.international.gc.ca/gac-amc/assets/pdfs/publications/State-of-Trade-2020_eng.pdf>. Accessed in fall 2023.

governance policies, including trade, at different levels of the Canadian and American systems. Finally, we underline the reasons behind people's differences and similarities in perceptions on both sides of the border by uncovering their reaction to those policies in both countries and by analysing interviews with stakeholders carried out in the summer 2023.

Harvard professor and famed political analyst Seymour Martin Lipset suggested that the differences between the two countries were minor and that political differences could be measured in a range between 5 % and 10 %[6]; however, this work points to growing differences. One major difference is for example, national identity, because it remains a challenge for Canadians.[7] In the United States, national identity is based on a vision of a single nation defined by democratic principles and united by a common public culture. In Canada, on the contrary, the two founding colonial settler communities of English and French origin colonised and oppressed indigenous communities, yet officially designated the country as 'multicultural'. Also, Canadians believe that they live in the shadows of their larger, more populous and powerful neighbour, whose cultural and economic influence is often seen as an infringement on Canada's autonomy and concurrent difficult collective identity: an identity that struggles with its colonial past.[8] The boundary lines of this colonial past, nearly 9,000 km long, extend from the Arctic, south, along the Pacific and across the continent to the Atlantic and have a mythic significance, at least in Canada.[9] In addition, the boundary lines are and continue to be perceived as the longest undefended borders in the world. Furthermore, because nearly all Canadians live within a few miles from the border, according to Gibbins, such proximity has made Canada a "borderland society". This contrasts with the US where the northern border plays a small role in the formation of national consciousness. For Gibbins the border has a defining significance for Canadians that "penetrates

---

6 S. M. Lipset, *Continental Divide: The Values and Institutions of the United States and Canada*, Routledge, New York, 1990.

7 G. Custred, "North American Borders: Why They Matter", Report, *Backgrounder*, Center for immigration Studies. Washington, DC, 2003.

8 B. Hoy, *A Line of Blood and Dirt*, Oxford University Press, Oxford, 2021.

9 R. Gibbins, "The meaning and significance of the American-Canadian border", *In* P. Ganster et al. (ed.), *Borders and Border Regions of Europe and North America*, San Diego State University Press, San Diego, 1997, pp. 315–333.

the Canadian consciousness, identity, economy, and polity to a degree unknown and unimaginable in the United States".[10]

In sum, the Canada-US border is considered a bridge between two essentially liberal democracies, which, although markedly different in their political make-up and economic clout, are generally perceived as more similar than different.[11] Thus, forms of cooperation have driven the relationship for nearly a century in policy areas such as environment for example on monitoring the movement of waste materials or maintaining the quality of waterways as illustrated by the Boundary Waters Treaty of 1909 or the Great Lakes Water Quality Agreement of 1978. In the field of security, there is the North American Defence Agreement (NORAD) of 1957 or more recently the Smart Border Declaration (SBD) of 2001, on health, the Covid-19 Travel Restrictions agreement of 2020, and on migration, the renewed Safe Third Country Agreement of 2023.

Free trade agendas have integrated both economies successfully also thanks to important collaborative efforts that produced both cooperative closeness and on occasion also conflicts. Such partnerships are demonstrated for instance, by the Canada United States Free Trade Agreement of 1987. It was then superseded by the North American Free Trade Agreement in 1994 and the more recent United States-Mexico-Canada Trade Agreement that entered into force in 2020.

What the Canada – US border regions are may be loosely described as "bounded territorial units composed of the territories of authorities participating in a cross-border cooperation initiative".[12] Indeed, adjacent Canadian Provinces and US States, as well as local communities, have signed numerous agreements in diverse policy areas. In the areas of defence and security alone, more than 2,500 agreements link Canada and the US.[13] Although these agreements illustrate a long and well-established tradition of close cooperation between the two federal administrations, agencies and other lower-level governments in various areas of public

---

[10] *Ibid.*, p. 317.

[11] D. Stirrup & J. Clarke, "Straddling boundaries: Culture and the Canada-US border", *Comparative American Studies. An International Journal*, 13 (1–2), 2015, pp. 1–15.

[12] M. Perkmann, "Cross-border regions in Europe: Significance and drivers of regional cross-border co-operation", *European Urban and Regional Studies*, 10 (2), 2003, pp. 153–171.

[13] E. Brunet-Jailly, "A new border? A Canadian perspective of the Canada-US border post-9/11", *International Journal*, 67 (4), 2012, pp. 963–974.

policy, they have never led to any firm and sustainable international institutional developments.[14] Rather, they have established functional linkages of cooperation that rely on shared values of expertise and efficiency. While there is a wide variety of organisations on both sides of the border, which "coordinate, sometimes cooperate, and rarely collaborate activities in a multitude of policy arenas"[15] cross-border cooperation, for example, in the Cascadia region lacks a structured institutional skeleton.[16]

Conversely, with European cross-border cooperation processes – mainly driven by public authorities – private and non-profit organizations' involvement is much more intense in the North American context.[17] This is the case of the Pacific Northwest Economic Region (PNWER): a non-profit organization, which, since 1991, has been a "platform for fluid, public-private networks that are often constructed for single purposes".[18] The discussion on border blocking effects discloses that the current US federal political climate and resultant immigration restrictions yield to instability heavily impacting cross-border economic interaction across the US-Canadian border.[19] This instability can work both as a barrier to cooperation but also as a driver for US companies to locate on the Canadian side of the border. For Konrad and Nicol, the borderland is a "zone of interaction" and a "prosperity space for Canada and the US"

---

14 E. Brunet-Jailly, "NAFTA, Economic Integration, and the Canadian-American Security Regime in the Post-September 11, 2001 Era: Multi-Level Governance and Transparent Border?", *op.cit.*; E. Brunet-Jailly, "In the Increasingly global economy, are borderland regions public management instruments?", *International Journal of Public Sector Management*, vol. 25, 6/7, 2012, pp. 483–491.

15 E. Brunet-Jailly, "Cascadia in comparative perspectives: Canada-US relations and the emergence of cross-border regions", *Canadian Political Science Review*, 2/1, 2008, p. 114; E. Brunet-Jailly, "Introduction" and "Conclusion", *In* Ch. Leuprecht & T. Hataley, *Security, Cooperation, Governance,* University of Michigan Press, Michigan, 2023.

16 B. Dupeyron, "Cascadia Revisited from European Case Studies: The Socio-Political Space of Cross-Border Networks", *Canadian Political Science Review*, 2 (2), 2008, pp. 84–103; F. Cappellano & T. Makkonen, "The Proximity Puzzle in Cross-Border Regions", *Planning Practice & Research*, 35 (3), 2020, pp. 283–301.

17 J. Blatter, "From 'spaces of place' to 'spaces of flows'? Territorial and functional governance in cross-border regions in Europe and North America", *International Journal of Urban and Regional Research* 28 (3), 2004, pp. 530–548.

18 K. Zimmerbauer, "Supranational identities in planning", *Regional Studies*, 52 (7), 2018, p. 917.

19 J. van den Broek, P. Benneworth, & R. Rutten, "Border blocking effects in collaborative firm innovation", *European Planning Studies*, 26 (7), 2018, pp. 1330–1346.

as perceived by Canadians.[20] According to them, tricky balance between closure (security) and openness (trade) requires "smart" borders that enable border management to be "faster and better".[21] In short, until the late part of the 20th century, old North American geopolitical borders seemed to have been replaced with a new geo-economics of global borders; signs of globalisation seem to be everywhere; Brunet-Jailly had argued then that the Canada-US border was "transparent to trade".[22] This was indeed the case with for instance 90 % of all Ontario trade with the world being done with Michigan (US) and 65 % vice versa; then 9/11 happened starting a new era for borders worldwide including in North America.[23] These two eras will be detailed in the following sections.

Border Perceptions before 9/11

Since the First World War, the United States and Canada have enjoyed a stable relationship. The two countries have shared the longest non-militarised international border in the world and progressively and steadily stronger security and economic relations. The North American Free Trade Agreement (NAFTA) signed in 1994 has been perceived as a major free-trade deal and; by 1998, trade between both nations had further increased, so much that by 2000, the Canada – US border seemed to be part of a-world-without-borders; or, at least a world without borders to goods, services and investment flows.[24]

---

20 V. Konrad, & H. N. Nicol, *Beyond Walls: Re-Inventing the Canada-United States Borderlands*, Ashgate, Burlington, 2008, pp. 12, 19, 32, 33.

21 *Ibid.*, p. 19; V. Konrad & G. Correa-Cabrera, "Borders in Transition: North American Borders in Comparative Perspective", *L'Espace Politique*, 42 (3), 2020, pp. 1–22.

22 E. Brunet-Jailly, "Globalization, Integration and Cross-Border Relations in the Metropolitan Area of Detroit (USA) and Windsor (Canada)", *International Journal of Economic Development*, 2 (3), 2000, pp. 379–401; E. Brunet-Jailly,"Toward a Model of Border Studies: What Do We Learn From the Study of The Canadian American Border," *Journal of Borderland Studies*, 19 (1), 2004, pp. 1–13; E. Brunet-Jailly, "Comparing Local Cross-Border Relations Under EU and NAFTA", *Canadian American Public Policy*, University of Maine, Maine, U.S.A. 2004, pp. 1–60.

23 T. Courchene & C. Telmer, *From Heartland to North American Region State: The Social, Fiscal and Federal Evolution of Ontario,* University of Toronto Press, Toronto, 1998; E. Brunet-Jailly, "NAFTA and Cross-Border relations in Niagara, Detroit and Vancouver", *Journal of Borderland Studies,* 21 (2), 2006, pp. 1–19.

24 E. Brunet-Jailly, "A new border? A Canadian perspective of the Canada-US border post-9/11", *International Journal* 67 (4), 2012, pp. 963–974; E. Brunet-Jailly, "In

Indeed, historically, the economic activities that set the very foundations of the development of the North American core industrial heartlands take place along the boundary line, with some of the largest and most prosperous cities and urban regions at the core of that industrial expansion: Chicago, Detroit, Toronto, Montreal and Quebec.[25] For nearly a century, thanks to industries leading to the development of the auto motor industries, Detroit has remained the limelight of the North American heartland. And, despite the lack of a harbour or linkage to the Atlantic coast, Detroit and the Great Lakes attracted immigrants and investors who moved thanks to water-and-railway infrastructures to Detroit and through Detroit into the center of the North American continent as it had developed in the 1800 and 1900.[26]

At inception, bilateral cooperation was initiated thanks to a few treaties and agreements: such as the Reciprocity Treaty of 1854, the 1902 and 1909 agreements on water issues, as well as the negotiated but failed free trade agreements of 1911 and 1948. Those, however, led to the very successful Autopact of 1965, which then expanded into the 1987 Free Trade Agreement and subsequently the 1994 North American Free Trade Agreement.

Also, after the second world war, the Canada-United States defence cooperation has been intense involving the creation of the Permanent Joint Board of Defence in 1940, which still exists today.[27] Provinces and States, and communities that straddle the border, have signed agreements in diverse policy areas as well. The Permanent Joint Board on Defence established in 1940 and the 1957 North American Aerospace Defense Agreement (NORAD), as well as the 1994 Defence White Paper,[28] have provided Canada with regular consultations on Ballistic Missile Defence

the Increasingly global economy, are borderland regions public management instruments?" *op.cit.*

25 E. Brunet-Jailly, "US-Canada Border Cities and Territorial Development Trends", *op. cit.*

26 E. Brunet-Jailly, "NAFTA, Economic Integration, and the Canadian-American Security Regime in the Post-September 11, 2001 Era: Multi-Level Governance and Transparent Border?", *Journal of Borderlands Studies*, 19 (1), 2004, pp. 18–37.

27 J. Noble, "Fortress America or Fortress North America", *Law and Business Review of the Americas*, 11, 2005, pp. 462–477.

28 *White Paper on Defence* Government of Canada publications, 1994. Retrieved from: <https://publications.gc.ca/collections/collection_2012/dn-nd/D3-6-1994-eng.pdf>. Accessed 10 August 2023.

(BMD) with the US and other allies. In the same vein, when, in the post-war era, America became an atomic superpower, it extended its nuclear shield to Canada, its northern neighbour. Canada's security has always been in the national interest of the US and has earned Canada a guarantee of automatic protection, a commitment that stands as the basis of Canadian defence policy.[29]

Since then, from the late 1990s, border management and joint institutional structure have been active across the border regions and, consequently, the Canada-US Integrated Border Enforcement Teams (IBETs*)* were created.[30] Interestingly, most of these initiatives bring local, regional, provincial/ states and central or international government levels together. As early as in the 1990s, Guiliano Zaccardelli, a former Commissioner of the Royal Canadian Mounted Police (RCMP), has promoted such strategy of multi-lateral cooperation. He argued that a progressive integration of law enforcement agencies at all levels of government would increase the effectiveness of strategies to combat crime. The IBETs concept was tested in 1996 in the Vancouver-Seattle border region and the first results were introduced at the Canada-US Cross-Border Crime Forum and then generalised.[31]

In sum, it seems clear today that the signature of many bilateral trade and security treaties has drawn both countries much closer. As suggested by Thomas Courchene,[32] more economic and geographic proximity also contributed to social and cultural proximity.[33] Indeed, US-Canadian cross-border economic regions are not only structured by the exchanges of goods and services trading networks, as well as their serving waterways, railways and highways, and concurrent urban sprawl. The evidence suggests also forms of functional and non-territorial interaction of networked organisation that represent and advocate for those market flows

---

29 *Ibid.*

30 Government of Canada, IBETs, "Royal Canadian Mounted Police", .Retrieved from: <https://www.rcmp-grc.gc.ca/en>. Accessed 10 November 2023.

31 E. Brunet-Jailly, "Security and border security policies: Perimeter or smart border. A comparison of the European Union and Canadian-American border security regimes", *Journal of Borderland Studies*, 21 (1), 2006, pp. 3–22.

32 T. Courchene, "Renegotiating Equalization: National Polity, Federal State", *International Economy C.D. Howe Commentary No. 113*, C.D. Howe Institute, Toronto, 1998.

33 E. Brunet-Jailly, "US-Canada Border Cities and Territorial Development Trends", *op. cit.*

across the Canada-US boundary line.[34] These are clusters of public private sector managers that continually interact to facilitate trade and other sectors of cooperation. This, however, changed in 2001 due to the international terrorist crisis that struck the US.

Post 9/11 Perception of the Border

The 11 September 2001 terrorist attacks on New York and Washington, D.C. changed the Canada-US border with great significance. Indeed, despite the very successful NORAD partnership since 1957 and its handling of the 9/11 aerospace security when 240 planes destined to US airport were allowed to land at one of 17 Canadian airports, for many Americans, the longest undefended border in the world started to look like a long portal for terrorists.[35] Different political cultures made the "harmonization" of necessary policies for a North American security zone extremely difficult.[36] Although no terrorists involved in the September attacks had entered the US via Canada, terrorists in the past had. The terrorist Ahmed Ressam for example attempted a crossing in Port Townsend, in Washington State, in 1999. He had taken advantage of loose border security on the Canadian side but was caught on the US side.[37] As a result, steep measures reinforced security on the US side also involved increased cooperation with Canadian authorities. The introduction of a Smart Border Action Declaration (SBD) was signed in 2001. Interestingly, rather than setting up an agency of cooperation in security matters, the SBD led Canada and the US to sign as many security-related agreements as necessary, involving federal, state/provincial and municipal security agencies. The SBD itself focused on greater cooperation in subareas of customs, immigration and security. While announcing the terms of the SBD, the Canadian Foreign Affairs Minister John Manley and

[34] E. Brunet-Jailly, "In the increasingly global economy, are borderland regions public management instruments?", *op.cit.*

[35] W. Ferguson, "The 49th Parallel: Our defining line", *Canadian Geographic*, 2020. Retrieved from: <https://canadiangeographic.ca/articles/the-49th-parallel-our-defining-line/>. Accessed November 2023.

[36] G. Custred, "North American Borders: Why They Matter", *op.cit.*

[37] H., Bernton, M., Carter, D. Heath & J. Neff, "The Terrorist Within: The Story Behind One Man's Holy War Against America. A Special Report", *The Seattle Times*, 2002. Retrieved from: <https://special.seattletimes.com/o/news/nation-world/terroristwithin/chapter9.html>. Accessed 15 August 2023.

the Director of US Homeland Security Governor Tom Ridge explained that secure flows of people and goods, secure infrastructure, coordination and information-sharing were necessary because of the high volume of trade between the two countries.[38] By 2011, more than 2500 agreements linked Canada and the US in the areas of defence and security alone.[39]

At federal levels, the SBD was followed by the Security and Prosperity Partnership (SPP) in 2005. Both agreements set major precedents in the security history of the two countries because they set out 4 major themes of cooperation and 32 clear objectives thus raising the question whether security trumped trade or vice versa. Indeed, the agreements were intended to secure flows of people, goods and infrastructures, as well as to increase coordination and information sharing. They also aimed at establishing clear policy objectives for public agencies to collaboratively implement a common security policy.

Hence, post 9/11 radical border security reforms on both sides of the 49th parallel were quickly implemented.[40] As recognised by Heyman and Ackleson in 2010, the SBD provided a framework for a post-9/11 risk paradigm shift whereby possible threats were identified prior to the physical border itself. They also included ongoing security procedures once a person had passed through a port of entry.[41] Such a new risk paradigm was also evident through an examination of the operational approaches taken by the border security agencies newly created in both Canada and the US. In 2003, the Government of Canada announced the creation of a new Ministry Public Safety and Emergency Preparedness Canada and of the Canada Border Services Agency (CBSA). They merged most security and intelligence facilities, as well as traditional customs and immigration bureaucracies to facilitate the flow of legitimate travellers and trade. Once established in Canada, the CBSA became the second largest federal law enforcement agency, after the Royal Canadian Mounted

---

38 L. Ch. Savage, "Security trumps trade at the US border", *Maclean's World*, 7 September 2011. Retrieved from: <https://macleans.ca/news/world/security-trumps-trade-at-the-border/>. Accessed 15 December 2023.

39 E. Brunet-Jailly, "A new border? A Canadian perspective of the Canada-US border post-9/11", *op.cit.*

40 J. A. Winterdyk & K. W. Sundberg, "Assessing public confidence in Canada's new approach to border security", *Journal of Borderlands Studies*, 25 (3–4), 2010, pp. 1–18,

41 J. McC. Heyman & J. Ackleson, "United States Border Security After September 11", *In* J. Winterdyck & K. Sundberg (eds.), *Border security in the Al-Qaeda era*, CRC Press, Boca Raton, 2009, pp. 37–74.

Police (RCMP).[42] It became one of six agencies under the umbrella of the new Ministry on Public Safety Canada (PSC).[43] According to the official statement, by consolidating these six agencies under the authority of a single Minister of Public Safety, the Government of Canada attempted to "create important synergies between national security and emergency management, corrections and crime prevention, justice information networks and law enforcement agencies"[44] in hopes of enhancing Canada's overall level of safety and security.[45] The CBSA, in conjunction with its domestic and international partners implemented new technologies and processes to identify, monitor and address possible threats to border security before a person or cargo actually reaches the border itself. Such a layered approach to border security became a cornerstone and remained a major element of the post 2001 SBD within the new border security paradigm of Canada as well as the US.[46]

In 2015, the Canadian Public Safety Minister Steven Blaney and the US Homeland Security Secretary Jeh Johnson signed an agreement designed to protect domestic security, to preserve international trade and to maintain strong diplomatic relations between Canada and the US in the wake of 9/11, by announcing pre-clearance custom procedures at the Canada-US border. Presented under the general banner of "Beyond the Border", an initiative established in February 2011, pre-clearance has been just one of a number of measures. Their aim was to "pursue a perimeter approach to security, working together within, at and away from the borders of our

---

[42] Government of Canada, "Public Service Commission 2009–2010 estimates, performance report for the period ending March 31", 2010. Retrieved from: <https://publi cations.gc.ca/site/eng/382876/publication.html>. Accessed November 2023.

[43] Government of Canada, Public Service Commission of Canada. Retrieved from: <https://www.canada.ca/en/public-service-commission.html>. Accessed 15 November 2023.

[44] A. McLellan, PC MP Solicitor General of Canada and Minister of Public Safety and Emergency Preparedness, "Public Safety and Emergency Preparedness Canada", Report on Plans and Priorities, 2005. Retrieved from: <https://publications.gc.ca/coll ections/collection_2015/sp-ps/PS1-11-2005-eng.pdf>. Accessed 23 December 2023.

[45] The other five agencies were the Royal Canadian Mountain Police, the Canadian Security Intelligence Service (CSIS), the Correctional Services of Canada (CSC), the National Parole Board (NPB) and the Canada Firearms Centre (CFC).

[46] J. McC. Heyman & J. Ackleson, "United States Border Security After September 11", *op.cit.;* US Department of State. "US – Canada Smart Border/30 Point Action Plan Update. Archive", 2002. Retrieved from: <https://2001-2009.state.gov/p/wha/ rls/fs/18128.htm>. Accessed 15 November 2023.

two countries to enhance our security and accelerate the legitimate flow of people, goods and services between our two countries".[47] To date, while pre-clearance has been in operation at a number of airports in the US and Canada, the same has not been true for those entering either country via land and seaports. Increasing delays at such locations following the major increases in security since 9/11 have contributed to a significant decline in cross-border traffic of up to 34 %. While factors such as currency fluctuations also come into play, waiting times and added bureaucracy at the border have clearly played their part.[48] In terms of car vehicle passengers alone, the US Department of Transportation accounted for a drop of over 30 million annually between 2000 and 2014, with 90,046,948 crossings in 2000, and 59,664,192 in 2014.[49] These figures point to a significant change between the two countries that share what was once the longest undefended border in the world. The major drop in the number of crossings illustrates the exceptional transformation of this border; whether literally or psychologically, this outstanding drop reflects the transformation of the Canada – United States boundary line as a barrier. Indeed, for many borderlanders, there is a sense of urgency for a return to the days of free flow, which pre-clearance could provide.[50]

Migration is another area where numbers and policies point to a transformation of the border. The migration management regime in both Canada and the United States did not cope well with the 2015 migration crisis. According to Young, migration management regimes produced border crises while simultaneously masking and revealing contested histories and geographies of migration.[51] Between January 2017 and December 2019, 54,000 people walked across the Canada-US border

---

47 Government of Canada, "Public Safety Canada. Beyond the Border Action Plan. Beyond the Border: A Shared Vision for Perimeter Security and Economic Competitiveness", 2016, Retrieved from: <https://www.publicsafety.gc.ca/cnt/brdr-strtgs/bynd-th-brdr/ctn-pln-en.aspx#part1>. Accessed 15 November, 2023.

48 D. Stirrup & J. Clarke, "Straddling boundaries: Culture and the Canada-US border", *op.cit.*

49 Bureau of transportation Statistics, US Department of transportation, "Transportation in the United States: Highlights from 2015 Transportation Statistics Annual Report", Retrieved from: <https://www.bts.gov/archive/publications/transportation_in_the_united_states_2015/index>. Accessed 15 September, 2023.

50 Stirrup & J. Clarke, "Straddling boundaries: Culture and the Canada-US border", *op.cit.*, p. 1.

51 J. E. Young, "Seeing like a border city: Refugee politics at the borders of city and nation-state", *Environment and Planning C: Politics and Space*, 37 (3), 2018, p. 407.

to make refugee claims.[52] Canadian officials, policymakers and media reports framed these movements as a crisis requiring a decisive response from the state. As Carastathis, Kouri-Towe, Mahrouse and Whitley suggest, the language of crisis shifted the focus from the "experience of displacement as a crisis for refugees, to the perception of their entry as a crisis for nation-states".[53] This shift from crisis as the cause of forced migration to the construction of crisis as an effect of human mobility had, according to them, "a number of important political effects, not least of which is that it enables accelerated border militarisation…and the closure of paths to safety".[54] In this sense, sovereign states used crisis as a strategic and deliberate approach to the governance of migration.[55]

The Safe Third Country Agreement (STCA), which came into effect in 2004 between the US and Canada, prevents people from using official border entry points to make a refugee claim. Each time they are turned away and denied entry. Canada acknowledged these crossings by erecting pop-up border stations like a well-publicized one at Roxham Road, which facilitates the movement of migrants. On the occasion of US President Joe Biden's visit to Canada on 24 March 2023, the two countries announced the expansion of the STCA across the entire land border, including internal waterways. Today, if anyone crosing the border to make an asylum (refugee) claim and does not meet one of the Agreement's exceptions, they are returned to the U.S. Canada-US under the terms of the Safe Third Country Agreement.[56]

However, according to Loyd and Mountz, policies like the STCA are short-sighted because stricter border policies, militarisation and surveillance do not stem migration. Instead, desperate for protection, people

---

52 Government of Canada, Immigration and Refugee Board of Canada (IRB), Retrieved from: <https://www.publicsafety.gc.ca/cnt/brdr-strtgs/bynd-th-brdr/ctn-pln-en.aspx#part1>. Accessed 15 November, 2023.

53 A. Carastathis, N. Kouri-Towe, G. Mahrouse & L. Whitley, "Introduction", *Refuge: Canada's Journal on Refugees*, 34 (1), 2018, p. 5.

54 *Ibid.*

55 A. Mountz & N. Hiemstra, "Chaos and crisis: Dissecting the spatiotemporal logics of contemporary migrations and state practices", *Annals of the Association of American Geographers*, 104 (2), 2014, pp. 382–390.

56 Government of Canada, "Canada-US Safe Third Country Agreement". Retrieved from: <https://www.canada.ca/en/immigration-refugees-citizenship/corporate/mandate/policies-operational-instructions-agreements/agreements/safe-third-country-agreement.html>. Accessed 15 September, 2023

simply take more dangerous routes, leading to loss of life, broken families and lifelong trauma. Such government strategies of deterrence and pre-emption only widen inequities – racialized, classed, gendered – with regards to mobility.[57] Indeed, Anna Pratt argues that states have turned jurisdiction and authority into a "mobile resource",[58] which speaks to the ways in which mobility has been integrated into the border itself in order to affect not only the present but the future. This makes crossing the border more dangerous and more inhuman.[59]

In November 2023, two days before the US Thanksgiving weekend, a car exploded in Niagara Fall, between New York State and Ontario: the Bentley was on the US side of the Rainbow Bridge. The explosion triggered a strong counterterrorist response that demonstrated the currency of the security partnership in both countries. Former Assistant Secretary Homeland Security, Alan Bersin, commented on the issue to suggest that "the deep historic friendship between Canada and the United States is not matched by modes of regular official communication and established mechanisms for bilateral planning in advance of crisis".[60]

In sum, post 9/11, despite NORAD and seamless aeronautical security coordination across the boundary line, Canada and the United States remain at odds. This lack of coordination also affects trade in unexpected manners, especially during and after the COVID-19 pandemic.

Cross-border Trade before and after Covid-19

Despite being highly economically integrated, the Canada – United States region does not form the most interdependent economy in the

57 J. M. Loyd & A. Mountz, *Boats, Borders, and Bases: Race, the Cold War, and the Rise of Migration Detention in the United States* University of California Press, Berkeley, 2018.

58 A. C. Pratt, *The Canada–US Shiprider Programme, Jurisdiction and the Crime–Security Nexus, In* R. K. Lippert et al. (eds.), *National Security, Surveillance and Terror,* Palgrave, Macmillan, New York, 2016, pp. 249–272.

59 J. Liew, P. Molnar & J. Youn, "The new US-Canada border deal is inhumane — and deadly Opinion", 2023. Retrieved from: <https://www.aljazeera.com/opinions/2023/4/19/the-new-us-canada-border-deal-is-inhumane-and>. Accessed December, 2023.

60 A. Bersin, "Somewhere Over the Rainbow Bridge: The Story That Wasn't and the Problem There Is", *Insights and Analysis,* Wilson Center, 2023. Retrieved from: <https://www.wilsoncenter.org/article/somewhere-over-rainbow-bridge-story-wasnt-and-problem-there>. Accessed December, 2023.

world. It comes third behind the EU and its Member States in Western Europe and the Greater Mekong Sub-region in South Asia. Thanks to free trade agreements and infrastructure developments, both the European and South Asian regional economies trade with each other's much more than with the rest-of-the-world: nearly 60 % in East Asia and Europe, rather than 50 % in North America.[61] Yet from a Canadian perspective, with about 90 % of the Canadian population living within a one-hour drive of the border, and about 3 million people crossing the countries' 119 shared border-crossings monthly, and, with about 800,000 Canadians living in the United States, the perception of economic interdependency is strong, has been strong and remains strong. Many statistics explains this perception. For instance, the Bureau of Transportation Statistics data for the period 2019–2022 reveals that millions of individuals cross the border each month in a year (See Figure 1). In all, in 2019, 522,954 individuals walked across the border, 76,525 buses and 6,733,089 personal vehicles also crossed the border. In addition, the number of container-trains and truck were 2,624,644 and 5,681,155 respectively.[62] Interestingly though, vehicle crossing the boundary line is concentrated on a relatively small number of border-crossings. Of all bilateral flows between Canada and the United States, the Detroit-Windsor trade corridor, sees one-in-four-crossings each day, specifically, in 2019 it saw 27 % of all trucks, 31 % of all trade, and 15 % of all personal vehicle traffic.

Indeed, crossings decreased due to COVID-19. For instance, for 20 months, Point Roberts, a US enclave in Canada, became a ghost town, right after and during the pandemic. When authorities started to gradually remove restrictions, traffic across the border came back. Thus, the total value of transborder freight drooped in 2020 but recovered by 2022. For instance, North American Transborder Freight from July 2021–22 was up by 19.1 %.[63]

---

[61] E. Brunet-Jailly, "Cross-border cooperation: A global overview", *Alternatives*, 47 (1), 3–17, 2022, pp. 3–17.

[62] US Department of Transportation. "Border Crossing/Entry Data: Time Series Analysis", 2019, 2020, 2021, 2022. Retrieved from: <https://data.bts.gov/stories/s/TransBorder-Freight-Data/myhq-rm6q/>. Accessed 15 September, 2023.

[63] *Ibid.*

	2019	2020	2021	2022
U.S.-Canada Border	94,661,870	29,507,038	24,617,052	55,389,412
U.S.-Mexico Border	275,538,145	171,024,678	199,542,060	254,884,072
Total	370,200,015	200,531,716	224,159,112	310,273,484

Figure 1: Number of U.S.-Canada/Mexico Border Crossings: 2019–2022[64]

Indeed, regarding trade, Canada and the United States may have the largest trading relationship in the world with CA$3.4 billion worth of goods and services crossing the border daily. In all, this is about CA$1.2 trillion worth of exchanges yearly, or about 2/3 of all of Canada's trade with the rest of the world.[65]

For context, in 1977, at the time of discussion regarding the Free Trade Agreement, trade was stable at 45.6 billion. During the first ten years of NAFTA, cross border trade in goods between both countries increased faster than GDP growth.[66] Before Covid-related events, trade was stable at about 718 billion.[67] Thanks for just-in-time-production in the great lakes regions, for instance, the auto-motor industries are integrated across the boundary line, and just-in-time production-chains straddle the boundary line. In the Windsor (Canada) and Detroit (United States) crossings or across the Niagara regions, the manufactures of cars and trucks have transformed cross-border infrastructures.[68] According to Brunet-Jailly, free trade primarily benefited to the larger urban centres

---

64 Source: Bureau of Transportation Statistics, Border Crossing/Entry Data. Retrieved from: <https://www.bts.gov/content/border-crossingentry-data>. This data only covers legal crossings. For data on illegal crossings, refer to Customs and Border Protection Data Portal: <https://www.cbp.gov/newsroom/stats/cbp-public-data-portal>.

65 Government of Canada, "Canada-United States relations – bilateral relations and partnershis and organisations", 2022. Retrieved from: <https://www.international.gc.ca/country-pays/us-eu/relations.aspx?lang=eng>. Accessed 15 November 2023.

66 E. Brunet-Jailly, "A new border? A Canadian perspective of the Canada-US border post-9/11", *op.cit.*

67 Government of Canada, "Statistics Canada", Retrieved from: <https://www150.statcan.gc.ca/n1/en/type/data?MM=1>. Accessed 15 September, 2023.

68 J. B. Sutcliffe & W. P. Anderson, "The Canada-US Border in the 21st Century: Trade, Immigration and Security in the Age of Trump", Routledge, 2019, pp. 1–280; J. Nelles & J. B. Sutcliffe, "On the Boundary: Local Authorities, Intergovernmental Relations and the Governance of Border Infrastructure in the Detroit-Windsor Region", *Regional and Federal Studies*, 23 (2), 2013, pp. 213–232.

where economic activities have transformed the borderlands and perception of interdependency. Historically, it was indeed economic activities, which set the very foundations of the development of the core industrial heartlands of North America, with some of its largest and most prosperous cities: Chicago, Detroit, Toronto and Montreal/Quebec.[69] Thus, the political geography of free trade since the signature of the NAFTA has left infrastructural and developmental marks on those large cross-border urban regions; Those are larger than just transportation networks.[70]

For example, at the core of the Great Lake's region, Ontario (Canada) and Michigan (United States) are today strongly economically integrated. This was already clear in the 1990s, when according to Courchene and Telmer (1992): both Ontario and Michigan had become economically interdependent.[71] Since Courchene's works became known, key cities such as in the greater Toronto urban region (in the Golden Horseshoe, the most populated region and industrialised region of Canada) – Buffalo, Niagara Falls, Hamilton, Mississauga and Toronto, all expanded already strong economic connections straddling the border. Thus, Ontario and Michigan are each other's first international trading partners, with 92 % of all of Ontario's trade with the-rest-of-the-world going to Michigan, and with two-thirds of Michigan's trade going to Ontario; Ontario and Michigan have the greatest trade relationship between any Provinces or States in North America. In 2023, total trade between them reached 88.6 billion, with Michigan importing 31.3 billion worth of goods from Ontario.

The Vancouver-Seattle region is at the centre of similar economic interdependency that started in the 1970s with Vancouver expanding as a diversified harbour and services industries center, and Seattle expanding from its own harbour to include aerospace and computer electronics. Today, both harbours continue to compete, but the computer and electronics industries use each side of the border for their relative competitive advantages. Vancouver remains attractive to highly educated international labour, while Seattle exploits its proximity to the large computer science hubs of Northern California.[72] In 2016, Microsoft bought 142,000

---

69 E. Brunet-Jailly, "US-Canada Border Cities and Territorial Development Trends", *op. cit.*

70 *Ibid.*

71 Th. Courchene & C. Telmer, *From Heartland to North American Region State: The Social, Fiscal and Federal Evolution of Ontario, op.cit.*

72 *Ibid.*

square feet office space downtown Vancouver and by 2022, the company had expanded with additional leases for 400,000 square feet downtown towers (six new buildings) now named "Bentall 6 office tower".[73]

In sum, trade contributes to a positive perception of the role of the border between both countries. After the signature of the Free Trade and the North American Free Trade Agreements, trade expanded across the boundary line in unexpected ways and led both national economies to become interdependent in specific sectors, but also urban regions and concurrent infrastructures expanded their reach across borderlands.

However, the Covid-19 crisis of 2020 modified those trends and perception of economic integration and of travels across the boundary line. The relatively porous border of 1990s progressively became a barrier to some-exchanges and to further crossings in particular of personal vehicles.

The Consequences of the Covid Pandemic on Border Crossings and Perception

Indeed, on 21 March 2020, for the first time since the War of 1812, a lack of coordination affected the border: in response to the Covid-19 pandemic, the United States closed its border to Canada. This measure was taken in consultation with the Canadian government but only a few professionals, mainly health professionals, were able to cross the border thereafter.[74] The Canadian government placed border restrictions on foreign nationals from 16 March 2020 onwards and then, in conjunction with the US, suspended non-essential travel across the border on 21 March 2020. The temporary restriction lasted for months and was finally only changed to accept travellers with masks on 1 October 2022.[75]

---

73 Michael, "A Tour of Microsoft's Sleek New Vancouver Office", *Officelovin'*, 2016. Retrieved from: <https://www.officelovin.com/2017/03/tour-microsoft-canada-excellence-centre/>. Accessed 15 November 2023; Th. Knowlton, *Techouver*, 2023. Retrieved from: <https://techcouver.com/2022/06/02/microsoft-additional-office-vancouver/>. Accessed 23 November 2023.

74 Prime Minister of Canada Justin Trudeau "Prime Minister announces temporary border agreement with the United States", 2020. Retrieved from: <https://www.pm.gc.ca/en/news/news-releases/2020/03/20/prime-minister-announces-temporary-border-agreement-united-states>. Accessed 15 November 2023.

75 P. Vieira & K. Mackrael "US Canada Border Restriction: What to know about crossing requirements – Canada to drop vaccination-entry rules, mask requirements", *The Wall Street Journal*, 26 September 2022..

The closure of the US-Canada border created a growing divide between the two countries and Covid related decisions greatly affected safety rules while enhancing mistrust among both Canadians and Americans.[76] One of the core reasons was the much lower number of Covid-related-deaths in Canada with 9000 for 38 million versus 120,000 a year in the US.

Halting leisure traffic and barring entry to non-citizens, the closure of borders also blocked consumers who wanted to cross the border to shop and take advantage of favourable exchange rates, lower gas prices or product selection. The closure also effectively suspended tourism and, with a few exceptions, it stopped the daily commuting for cross-border workers and property owners.[77]

Furthermore, Aylet Shachar observed that controlling the movements of people started to happen more often "elsewhere" and often not at the actual border[78]. In Canada, the idea of preclearance was extended by government through a logic impacting people bound to Canada all over the world. The country announced that, not only would passage across its borders be temporarily restricted to citizens and permanent residents, but also that anyone – including its own citizens – with Covid-19 symptoms would be barred from boarding a Canada-bound flight.[79] With this decision, Canada stretched its border outward both conceptually and legally. Canada was also perfecting its technique of interdiction abroad, relocating much of its border-control activities to overseas gateways, primarily in Europe and Asia. The result was that regulation of entry into Canada happened in foreign transit hubs that could be located thousands of miles away. The US also extended its state-outward-power to regulate mobility at a distance. That was especially evident on 11 March 2020 with the proclamation from the Oval Office in the US that the country would

---

76 D. Baldwin & A. Bitterman, "COVID-19 response highlights major differences between U.S. and Canada", *CBC News,* 3 June 2020. Retrieved from: <https://www.cbc.ca/news/canada/hamilton/american-canadian-covid-1.5596226>. Accessed 15 December 2023; "Covid-19 has strained Canada's relations with America", *The Economist,* 1 August 2020 Retrieved from: <https://www.economist.com/the-americas/2020/08/01/covid-19-has-strained-canadas-relations-with-america>. Accessed 15 November 2023.

77 D. Baldwin & A. Bitterman, "COVID-19 response highlights major differences between U.S. and Canada", *op.cit.*

78 A. Shachar, "Beyond open and closed borders: The grand transformation of citizenship", *Jurisprudence*, 11 (1), 2020, pp. 1–27.

79 *Ibid.*

suspend travel from 26 European countries for 30 days. While commercial travel was still broadly permitted, it had nonetheless been seriously obstructed by the Covid-19 crisis. Auto production, for example, stalled by reduced movement of parts between countries.

The Covid-19 crisis also revealed major differences between Canada and the United States: the high infection and death rates in the US was shocking to Canadians. In January 2021, Angus Reid, a not-for-profit-pooling institute, found that two out of three Canadians would "close the border to quell discretionary travel".[80] Following government policies, public opinion also shifted. On the one hand, Canadians felt more anxious about the future direction of the relationship between the two countries. On the other hand, one could also observe forms of cross-border solidarity.

To express grassroots support for Americans, private business owners and building managers in Canadian border cities like Niagara Falls and Windsor (Ontario) decorated for example the windows of tall office towers and hotels with lighted hearts as a symbol of hope. This message of encouragement was a symbol of the goodwill by Canadians for US citizens.[81] Additionally, diverging partisan opinions emerged in assessment of how the US was handling the pandemic. Canadians who identified with the New Democratic Party were more likely than supporters of the governing Liberal Party and the Conservative Party, to criticise the US pandemic response. Still, at least half of each party's supporters were critical of the American coronavirus response.[82]

Overall, statistics reveal a substantial drop in cross-border movement between Canada and the US thus highlighting the pandemic-induced effect on the economy and movement across borders. The Trump administration also contributed to an erosion of bilateral goodwill between the

---

80 Angus Reid Institute, "Covid 19: two-in-three Canadians say borders should be closed to keep sun-seekers close to home" 4 January 2021. Retrieved from: <https://angusreid.org/covid-international-travel/>. Accessed 15 November 2023.

81 D. Baldwin & A. Bitterman, "COVID-19 response highlights major differences between U.S. and Canada", *op.cit.*

82 Partisan divides emerge in assessments of how the US has handled the pandemic. Canadians who identify with the New Democratic Party are more likely than supporters of both the governing Liberal Party and the Conservative Party to criticise the US pandemic response. Still, half or more of each party's supporters are critical of the American coronavirus response. PEW Research Center Canada, "Global Attitudes Survey" *Politico*, 2021.

two nations. This required Canada to reconsider ways to promote its interests. According to Nimijean and Carment, the Covid crisis made it clear that Canada needed to examine their joint economic and geopolitical prospects as the "special relationship came to an end".[83]

In sum, while for nearly 50 years free trade enhanced the porosity of the border, 9/11, and in particular, the Covid-19 pandemic crisis changed the nature of the perception of Canada-US relations and perception of the border. In 2023, for Canadians, the border marks fundamental differences with the US.

Post-Pandemic Border Perception in Cascadia

The Jean Monnet network Frontem centers on the role of the border in European integration considering that there is no single model of perception and management of EU borders, but that borders depend on the specific context of each border area concerned. The focus is on perception: perception of what the border is and the impact it has for the broader European civil society. To address similar questions in North America we interviewed a selected group of representatives of civil society organisations, local and regional government and public and private officials, to ask them about the Canada/United States border. We asked: how do citizens perceive the border(s) with their neighbouring country(s)? How do they perceive their border neighbours on the other side? How important is mobility across the border, and how important are mobility restrictions? How do these factors affect the work of their organisation?

In the Cascadia region (at the crossroad of British Columbia, Canada, and Washington State, United States), the perception of proximity of the border is strong as well as the perception of living close to the United States. For both, Canadian or American citizens, the perception of the delineation is unambiguous because it is in many instances a physical space. It is delineated on land for instance by the Peace Arch or clear cuts of trees, 10 m wide and hundreds of kilometres long, or a clear delineation across waters, sea and air thanks to GPSs. Also, the border is

83 R. Nimijean & D. Carment, "Rethinking the Canada-US relationship after the pandemic", *Policy Opinions*, 7 May 2020. Retrieved from: <https://policyoptions.irpp.org/magazines/may-2020/rethinking-the-canada-us-relationship-after-the-pandemic/>, 2020. Accessed 10 September 2020.

very much present in people's minds for example regarding variations in the way English is spoken, different programs on television or in general different cultural references.

However, the perception of their neighbours does not vary enough to explain these cultural and territorial markers. For instance, trust is a function of similar socio-economic groups. They trust each other across the border. Hence, labour, business groups, educational networks play an important role in setting those high levels of trust. Generally, for many interviewees, border officials on the US side are particularly intimidating. On the US side, nearly 40 % of border regional residents do not have passports and are not interested in crossing the border ever. Consequently, many residents of borderlands think that border-crossing rules are unnecessary. Yet, some US citizens feel more secure because border guards are now also armed in Canada.

Still, in Cascadia, the border is less an impediment to travelling than water is for those living on an island. Indeed, 12 % of British Columbians cross the boundary line more than a few times a year, but of those, 55 % cross several times a years, and 12 % never cross the border. Some 40 % declare crossing less after the Covid pandemic, because of security measures, politics in the US and the lower Canadian. People cross for shopping, schooling, tourism. Cross-border mobility is important for people in the border region but for different reasons recreation/gas/grocery, major leagues – and there is more openness to mobility for recreation than for health, education or work, which are more controversial because of their unclear tax implication.

Each threat plays a role in restriction – 9/11, migration or the pandemic-, yet border rules and bureaucracies are not always supported. To the question "Are there any circumstances where you think restricting or preventing cross-border flows would be justifiable?" the consensus was "No, … maybe war?".

Conclusion

This article specifically focused on the gradual change of perceptions at the Canadian-American border and on the events that triggered such transformation: the trade agreements between Canada and US; 9/11, which put trade as a priority on governments agendas and enhanced security policy thereafter, and recently the Covid-19 pandemic, which strongly influenced decisions implemented by both governments and further "distanced" the attitudes of borderlanders.

Indeed, when crossing the border today, it is notable that 20 years of policy changes have contributed progressively to citizens' perceptions of uncertainty and insecurity. This is particularly well illustrated by our 2023 survey-results, which show that authorities do not make their citizens particularly confident by the various measures on the safety of people, which they implemented.

The border as an institution could be more accountable, professional and transparent in its operations, especially considering its new security and enforcement-focused mandate but also to explore a humble human rights approach protecting Canadian and American citizens. In their study of counter-terrorism in Australia Pickering, McCulloch and Wright-Neville suggest that one of the best ways to protect one's citizens from the threat of terrorism is for governments to pay credence to building bridges between communities and enhance the cultural literacy of those tasked with national security duties, rather than adopting more technologies of control.[84] Indeed, the results of our study questions the success of government agencies in charge of the border with regards to their ability to successfully and adequately educate and inform the public of their mandates, focus and activities.

Bibliography

A. Bersin, "Somewhere over the Rainbow Bridge: The Story That Wasn't and the Problem There Is", *Insights and Analysis*, Wilson Center, 2023.

J. Blatter, "From 'spaces of place' to 'spaces of flows'? Territorial and functional governance in cross-border regions in Europe and North America", *International Journal of Urban and Regional Research*, 28 (3), 2004, pp. 530–548.

E. Brunet-Jailly, "Globalization, integration and cross-border relations in the metropolitan area of detroit (USA) and Windsor (Canada)", *International Journal of Economic Development*, 2 (3), 2000, pp. 379–401.

---

84 S. J. Pickering, J. McCulloch & D. P. Wright-Neville, *Counter-Terrorism Policing: Community, Cohesion and Security*, Springer, Clayton, 2008.

E. Brunet-Jailly, "Toward a model of border studies: What do we learn from the study of the Canadian American border," *Journal of Borderland Studies*, 19 (1), 2004, pp. 1–13.

E. Brunet-Jailly, "Comparing Local Cross-Border Relations Under EU and NAFTA", *Canadian American Public Policy*, University of Maine, Maine, U.S.A., 2004, pp. 1–60.

E. Brunet-Jailly, "NAFTA, economic integration, and the Canadian-American security regime in the post-September 11, 2001 era: Multi-Level governance and transparent border?", *Journal of Borderlands Studies*, 19 (1), 2004, pp. 18–37.

E. Brunet-Jailly, "NAFTA and cross-border relations in Niagara, detroit and Vancouver", *Journal of Borderland Studies*, 21 (2), 2006, pp. 1–19.

E. Brunet-Jailly, "Security and border security policies: Perimeter or smart border. A comparison of the European Union and Canadian-American border security regimes", *Journal of Borderland Studies*, 21 (1), 2006, pp. 3–22.

E. Brunet-Jailly, "Cascadia in comparative perspectives: Canada-US relations and the emergence of cross-border regions", *Canadian Political Science Review*, 2 (1), 2008, pp.104–124.

E. Brunet-Jailly, "A new border? A Canadian perspective of the Canada-US border post-9/11", *International Journal*, 67 (4), 2012, pp. 963–974.

E. Brunet-Jailly, "In the increasingly global economy, are borderland regions public management instruments?", *International Journal of Public Sector Management*, 25 (6/7), 2012, pp. 483–491.

E. Brunet-Jailly, "A new border? A Canadian perspective of the Canada-US border post-9/11", *International Journal*, 67 (4), 2012, pp. 963–974.

E. Brunet-Jailly, "US-Canada border cities and territorial development trends", *In* E. Medeiros (ed.), *Border Cities and Territorial Development*, Routledge, New York, 2021, pp. 209–225.

E. Brunet-Jailly, "Cross-border cooperation: A global overview", *Alternatives*, 47 (1), 2022, pp. 3–17.

E. Brunet-Jailly, "Introduction" and "Conclusion", *In* Ch. Leuprecht & T. Hataley, *Security, Cooperation, Governance,* University of Michigan Press, Michigan, 2023.

F. Cappellano & T. Makkonen, "The proximity puzzle in cross-border regions", *Planning Practice & Research*, 35 (3), 2020, pp. 283–301.

A. Carastathis, N. Kouri-Towe, G. Mahrouse & L. Whitley, "Introduction", *Refuge: Canada's Journal on Refugees*, 34 (1), 2018, pp. 3–15.

T. Courchene, "Renegotiating Equalization: National Polity, Federal State", *International Economy C.D. Howe Commentary No. 113*, C.D. Howe Institute, Toronto, 1998.

T. Courchene & C. Telmer, *From Heartland to North American Region State: The Social, Fiscal and Federal Evolution of Ontario,* University of Toronto Press, Toronto, 1998.

G. Custred, "North American Borders: Why They Matter", Report, *Backgrounder,* Center for immigration Studies. Washington, D.C., 2003.

B. Dupeyron, "Cascadia revisited from European case studies: The socio-political space of cross-border networks", *Canadian Political Science Review*, 2 (2), 2008, pp. 84–103.

R. Gibbins, "The meaning and significance of the American-Canadian border", *In* P. Ganster et al. (ed.), *Borders and Border Regions of Europe and North America*, San Diego State University Press, San Diego, 1997, pp. 315–333.

B. Hoy, *A Line of Blood and Dirt*, Oxford University Press, Oxford, 2021.

V. Konrad & G. Correa-Cabrera, "Borders in transition: North American borders in comparative perspective", *L'Espace Politique*, 42 (3), 2020, pp. 1–22.

V. Konrad & H. N. Nicol, *Beyond Walls: Re-Inventing the Canada-United States Borderlands*, Ashgate, Burlington, 2008.

S. M. Lipset, *Continental Divide: The Values and Institutions of the United States and Canada,* Routledge, New York, 1990.

J. M. Loyd & A. Mountz, *Boats, Borders, and Bases: Race, the Cold War, and the Rise of Migration Detention in the United States*, University of California Press, Berkeley, 2018.

J. McC. Heyman & J. Ackleson, "United States border security after September 11", *In* J. Winterdyck & K. Sundberg (eds.), *Border Security in the Al-Qaeda Era*, CRC Press, Boca Raton, 2009, pp. 37–74.

A. Mountz & N. Hiemstra, "Chaos and crisis: Dissecting the spatiotemporal logics of contemporary migrations and state practices", *Annals of the Association of American Geographers*, 104 (2), 2014, pp. 382–390.

J. Nelles & J. B. Sutcliffe, "On the boundary: Local authorities, intergovernmental relations and the governance of border infrastructure in the

Detroit-Windsor region", *Regional and Federal Studies*, 23 (2), 2013, pp. 213–232.

J. Noble, "Fortress America or Fortress North America", *Law and Business Review of the Americas*, 11, 2005, pp. 462–477.

M. Perkmann, "Cross-border regions in Europe: Significance and drivers of regional cross-border co-operation", *European Urban and Regional Studies*, 10 (2), 2003, pp. 153–171.

S. J. Pickering, J. McCulloch & D. P. Wright-Neville, *Counter-Terrorism Policing: Community, Cohesion and Security*, Springer, Clayton, 2008.

A. C. Pratt, The Canada–US Shiprider Programme, jurisdiction and the Crime–Security Nexus, *In* R. K. Lippert et al. (eds.), *National Security, Surveillance and Terror,* Palgrave, Macmillan, New York, 2016, pp. 249–272.

A. Shachar, "Beyond open and closed borders: The grand transformation of citizenship", *Jurisprudence*, 11 (1), 2020, pp. 1–27.

D. Stirrup & J. Clarke, "Straddling boundaries: Culture and the Canada-US border", *Comparative American Studies. An International Journal*, 13 (1–2), 2015, pp. 1–15.

J. B. Sutcliffe & W. P. Anderson, "*The Canada-US Border in the 21st Century: Trade, Immigration and Security in the Age of Trump*", Routledge, 2019, pp.1–280.

J. van den Broek, P. Benneworth & R. Rutten, "Border blocking effects in collaborative firm innovation", *European Planning Studies*, 26 (7), 2018, pp. 1330–1346.

J. A. Winterdyk & K. W. Sundberg, "Assessing public confidence in Canada's new approach to border security", *Journal of Borderlands Studies*, 25 (3–4), 2010, pp. 1–18,

J. E. Young, "Seeing like a border city: Refugee politics at the borders of city and nation-state", *Environment and Planning C: Politics and Space*, 37 (3), 2018, pp. 407–423.

K. Zimmerbauer, "Supranational identities in planning", *Regional Studies*, 52 (7), 2018, pp. 911–921.

Conclusion

Frontières en mouvement (Frontem): No Models of Cross-Border Cooperation for the EU?

Birte Wassenberg

Dessine-moi une frontière (Draw me a border, please)[1] – this could have been the title of a chapter in Antoine de Saint Exupéry's book *Le petit Prince*, if he had written it in 2023. Indeed, when Ursula von der Leyen took office as President of the European Commission in 2019, she announced for the new programme as one of the priorities the intention to construct "solid borders" of the European Union (EU).[2] This discourse stands however in stark contrast to the one on a "Europe without borders" that has long been dominant during the European integration process and where border regions and cross-border cooperation occupied a central role.[3]

Indeed, since the end of the Second World War and parallel to the European integration process, cross-border cooperation has contributed to the construction of a Europe, where borders no longer exist as barriers, but where the borderland becomes a place of communication and exchange.[4] A borderless Europe was indeed an objective, to which

1 Instead of "draw me a sheep" (*Dessine-moi un mouton*).

2 U. Von der Leyen, "A Union that strives for more. My Agenda for Europe", European Commission, Political Guidelines for the next European Commission 2019–2024, Brussels, 2019. Retrieved from: <https://www.europarl.europa.eu/resources/library/media/20190716RES57231/20190716RES57231.pdf>. Accessed 23 December 2023.

3 B. Wassenberg, "Introduction – Frontiers in motion (Frontem): Comparative perspectives on European borders, cross-border cooperation, and integration", *Borders in Globalization Review*, vol. 4, Issue 2, 2020, pp. 8–13.

4 B. Wassenberg., "'Europa ohne Grenzen'. Mythus oder Realität? Eine historische Analyse"", *In* B. Wassenberg (ed.), *Frontières, acteurs et représentations d'Europe (Fare). Mélanges en honneur de Sylvain Schirmann*, Peter Lang, Brussels, 2022, pp. 415–439.

border regions contributed by "healing" European borders as "the scars of history", as Alfred Mozer, member of the Social Democratic Party of Germany (SPD) and Secretary to the Commissioner for Agriculture in the first European Commission, had already underlined in 1973.[5]

The implementation of a borderless Europe already started in 1957, when the Common Market of the European Economic Community (EEC) proceeded with the abolishment of customs duties within the Community until 1962. It was then pushed further with the project of the European Single Market, which not only aimed at eliminating economic tariffs on goods but also at ensuring the free mobility of people, services and capital.[6] This project was launched in 1985 by the new President of the European Commission, Jacques Delors, who put an emphasis on the elimination of economic borders to facilitate the creation of an internal market, where the border was no longer an obstacle to free movement.[7] Since the completion of the European Single Market in 1992, the effects of open borders have indeed been felt most directly in Europe's border regions and the latter have increasingly been identified, both in theory and in practice, as "models" of the process of European integration.[8]

The European citizens therefore increasingly felt the disappearance of the "Westphalian" functions of the border such as the systematic control of goods or people crossing the border.[9] The role of cross-border cooperation was to contribute to the tangible implementation of this "Europe

---

5 A. Mozer, "Entwicklungspolitik zu Hause", *In* Schöndube, C. *Entwicklungsregionen in der EWG. Ursache und Ausmaß der wirtschaftlichen Benachteiligung*, Bonn, Osang Verlag, 1973, pp. 14–25.

6 L. Warlouzet, "At the heart of the recovery in the 1980s, the internal market", *In The European Commission 1986–2000, History and Memories of an Institution*, Publication Office of the EU, Luxemburg, 2019, pp. 258–259.

7 C. Mestre, *Les incidences de l'achèvement du Marché unique sur les régions transfrontalières, Conseil de l'Europe, Strasbourg,* 1992.

8 K.-H. Lambertz (ed.), *Die Grenzregionen als Labor und Motor kontinentaler Entwicklungen in Europa. Berichte und Dokumente des Europarates sowie Reden zur grenzüberschreitenden Zusammenarbeit in Europa*, Dike/Nomos, Zürich, Baden-Baden, 2010.

9 M. Berezin & M. Schain (eds.), *Europe Without Borders: Remapping Territory, Citizenship, and Identity in a Transnational Age*, John Hopkins University press, Baltimore, 2003.

without borders".[10] The European Commission threfore associated the border regions to its regional policy, primarily through the *Interreg* programs, which, since 1991, have provided funds for cross-border projects, both at internal and external borders of the EU.[11] With the ongoing process of European integration, the national borders in the EU thus became "invisible" and borderlands open spaces of mobility and free movement where local and regional stakeholders could forge cross-border neighbourhood relations without obstacles, as the "Westphalian" border was no longer relevant.[12]

Today, almost 40 % of the EU's territory is covered by border regions containing approximately 30 % of the EU population. The progress of a "Europe without borders" can be lived here directly on the ground: horizontal mobility of goods, capital and people are very obvious, daily experiences in European border regions.[13] It is therefore not surprising, that, within the EU, the perception of borders seems to have evolved from a negative to a positive image, that borders as "scars of history" have healed and that borderlands are now largely compatible with the ideal of a "Europe without borders".

This positive perception of the borderland as an integrated space was mainly possible thanks to the suppression of border controls within the EU. Since 1995, the implementation of the Schengen Agreement and its subsequent integration into the EU Treaty of Amsterdam in 1997, allowed for an "open" border management, which has largely facilitated cross-border cooperation between local and regional actors situated on

---

10 B. Wassenberg., "Frontières, coopération transfrontalière et intégration européenne", *In* B. Wassenberg & M. Aballéa (eds.), "Frontières, géopolitique et relations internationales", *Cahier Fare* n°16, Harmattan, Paris, 2019, pp. 44–50.

11 T. Köhler, "Die Interreg-Programme und die Rolle der europäischen Kommission als wesentlicher Akteur der grenzüberschreitenden Zusammenarbeit", *In* B. Wassenberg & J. Beck (eds.), *Living and Researching Cross-border Cooperation (Volume 3): The European Dimension*, Steiner Verlag, Stuttgart, 2011, pp. 243–253.

12 J. Beck, "European Cross-Border Cooperation of the Future: Capacity-Building and the Principle of Horizontal Subsidiarity", *In* I. Pires (ed.), *Borders and Borderlands: Today's Challenges and Tomorrow's Prospects*, Proceedings of the ABS Lisbon Conference Centro de Estudos Geográficos, Lisbon, 2012, pp. 13–33.

13 J. Beck, "Introduction", *In* J. Beck & B. Wassenberg (eds.), *Vivre et penser la coopération territoriale en Europe (Volume 6) : vers une cohésion territoriale ?*, Steiner-Verlag, Stuttgart, 2014, pp. 37–43.

either side of the national border.[14] After the introduction of the Interreg-initiative by the European Commission in 1990, a multi-level governance system could thus be set into place between regional, national and European actors operating in the cross-border territories of the European Community.[15] The idea behind this system was the progressive establishment of a "Europe without borders", where border regions were associated to the process of European integration and where national borders as former places of control and separation would give way to an internal space of exchange and mobility.[16]

When compared to other areas in the world, this seems extraordinary, as international borders are in principle managed by national States and the competence for their control, opening or closure lies with the central governments. This is not *per se* different for European borders, but border management might be relegated to local and regional stakeholders in the border regions simply because within the EU, the project of a "Europe without borders" envisaged the disappearance of certain "Westphalian" functions of the border such as the systematic control of goods or people crossing the border.[17] With the process of European integration, border management in European border regions therefore became an issue of managing an open cross-border space of mobility and free movement which could be dealt with by local and regional stakeholders, as the "Westphalian" border was no longer relevant.[18]

Within the EU, one is therefore tempted to try identifying certain model-types of border perception and management that comply with the ideal of a "Europe without borders". The Jean Monnet Network *Fronières en mouvement* (Frontem) exchanged knowledge and practices on five different case studies of EU border management and perception: the border between France and Germany, the border between France and Belgium, the border between Ireland and Northern Ireland, the border

---

14 B. Wassenberg, "Frontières, coopération transfrontalière et intégration européenne", *In* Wassenberg, B., Aballéa, M. (eds.), "Frontières, géopolitique et relations internationales", *op.cit.*, p. 50.

15 J. Beck, "Introduction", *op.cit.*, p. 37.

16 B. Wassenberg., "Europa ohne Grenzen. Mythus oder Realität? Eine historische Analyse", *op.cit.*

17 M. Berezin, M. Schain (eds.), *Europe Without Borders, op.cit.*

18 J. Beck, "European Cross-Border Cooperation of the Future: Capacity-Building and the Principle of Horizontal Subsidiarity", *op.cit.*

between Rumania and Hungary and the border between Denmark and Germany.[19] A comparative analysis was also conducted with regard to the Canada/US border in order to examine whether the European management and perception of borders resembles that of other regions in the world. The partners of Frontem carried out an interdisciplinary cross-analysis of border perception and management in their border areas by organising each a scientific seminar with researchers and a focus group with local stakeholders.[20]

The starting point for the network was the assumption that the border is being perceived and managed differently according to the specific borderland concerned. In Western Europe, the Franco-German and the Franco-Belgian border regions could be labelled as models of open, integrated spaces where cross-border mobility is an evidence and where the "Westphalian" border as a line of separation between the sovereignty of two national States is no longer visible.[21] On the one hand, the Franco-German border also symbolises a model of a successful process of reconciliation after the end of the Second World War.[22] On the other hand, in Central Europe, the Hungarian-Rumanian border could represent a model for East-East reconciliation after the Cold War.[23] Cross-border cooperation in this area also helped to deal with the large Hungarian

---

19 See Frontem. Retrieved from: <https://centre-jean-monnet.unistra.fr/the-project/>. Accessed 23 December 2023.

20 The research seminars were organised: Hungary-Rumania online (21 Mai 2021); France-Germany in Kehl (6–7 October 2021); Denmark-Germany in Sonderborg (18 November 2021), Ireland-Northern Ireland in Belfast (14 March 2022), France-Belgium in Arras (16 November 2022), US-Canada in Victoria (6–7 July 2023). The focus-groups were organised for Denmark-Germany in Sonderborg (19 November 2021), Ireland-Northern Ireland in Argmagh/Dundalk (15 March 2022), France-Germany in Kehl (28 June 2022); Hungary-Rumania in Cluj (19 May 2022), France-Belgium in Mons (17–18 November 2022) and US-Canada in Victoria (6–7 July 2023).

21 T. Hölpel,"Der deutsch-französische Grenzraum als Modellregion der deutsch-französischen und europäischen Annäherung", *In* T. Höpel, *Der deutsch-französische Grenzraum: Grenzraum und Nationenbildung im 19. und 20. Jahrhundert,* Europäische Geschichte online, 2012. Retrieved <http://ieg-ego.eu/de/threads/crossroads/grenzregionen/thomas-hoepel-der-deutsch-franzoesische-grenzraum>. Accessed 23 August 2023.

22 N. Paŭn, & S. Schirmann (eds.), *Borders, Identities, communities. The Road to Reconciliation and Partnership in Central and Eastern Europe,* Nomos, Publications of the European Union Liaison Committee, Vol. 18, Baden-Baden, 2016.

23 *Ibid.*

minorities on the Rumanian side of the border and is generally regarded as a potential gateway to an enlarged Schengen space of free circulation

In a different perspective, the "invisible" border between Ireland and Northern Ireland could be identified as a model of appeasement and potential reconciliation. Being a marker of conflict reminding of the times of the "troubles", the border can serve to maintain peace in the conflictual area only when it becomes "invisible". Here, border management therefore explicitly aims at rendering the border invisible in order to build up peace following the Good Friday agreement in 2008.[24] The EU Peace programs, set up from the 1990s, enabled Northern Ireland and Ireland to participate in cross-border cooperation and establish a specific European borderland "without borders" where the peace function of the invisible border is essential.[25]

Finally, in Northern Europe, border management in the Danish-German border regions could be labelled as a model for successful management of border minorities. Cross-border cooperation between Denmark and Germany can be identified in this context as a tool for installing a "peaceful co-existence" between the respective Danish and German minorities,' which could be guaranteed through a kind of "parallel" border management.[26]

However, when the Frontem network was launched in 2019, it was already confronted with a questioning of the positive model functions of European borders. Indeed, since 2015, various crises in Europe have severely shaken the image of a borderless Europe.[27] In Europe, the Schengen crisis, spurred off by the unexpected migration flow in 2015, led to a re-questioning of European borders as best practice places of European integration.[28] Re-bordering policies were introduced by many

---

24 F. Mercier, "Making sense of the invisible: Cognitive mapping, affective realities and the Irish/Northern Irish Border", *Media, Culture & Society,* Vol. 45, Issue 1, pp. 126–14.

25 *Ibid.*

26 M. Klatt, "National minorities as a model for cross-border integration: Lessons from Schleswig", *In* J. Beck & B. Wassenberg (eds)., *Grenzüberschreitende Zusammenarbeit leben und erforschen: Vol.5 Integration und (trans-)regionale Identitäten*, Steiner Verlag, Stuttgart, 2013, pp. 301–320.

27 B. Wassenberg, "The Schengen Crisis and the End of the 'Myth' of Europe without Borders", *Borders in Globalization Review*, vol. 1, Issue 2, 2020, pp. 30–39.

28 S. Saurugger, "Crise de l'Union européenne ou crise de la démocratie", *Politique étrangère*, 1, 2017, pp. 23–33.

EU Member States in order to stop the massive inflow of refugees, not only creating problems for migrants to enter the EU, but also for the internal border flows of cross-border workers.[29] This process of (re)bordering was further reinforced by some Member States in order to counter the terrorist crisis, which had shaken Europe during the same period, especially in France, where a state of emergency was proclaimed, thus suspending the Schengen agreement for an unlimited time.[30]

The return of more or less systematic border controls at the internal and external borders of the EU had led to a prevailing logic of re-bordering, which endangered the principle of free movement within the internal market and the role of local and regional stakeholders in border regions as major actors involved in border management.[31] Indeed, the phenomenon of re-bordering led to a revival of the "Westphalian" functions of the border in forms of border control and security provisions, for which the competence was clearly allocated to the national States. This represented an obstacle to cross-border cooperation in Europe.[32] The "separation" function of the border was being reinforced and "the pertinence of the border" seemed to prevail.[33] One of the consequences of this crisis was also growing Euroscepticism and a new "fear of integration" which might even lead to the disintegration of the EU, as the Brexit referendum in 2016 largely confirmed.[34]

During the Covid-19 pandemic, this Westphalian re-bordering effect was even more brutal, as the closure of the border was more or less complete: in most Member States of the EU, borders were being closed

---

29 B. Wassenberg, "De-bordering (Re-)bordering in the EU during the 2015 migration crisis: The end of a 'Europe without **borders**'?", *In* M. J. Carpenter, M. Kelly & O. Schmidtke (eds.), *Borders and Migration in Canadian Comparative Perspective, University of Ottawa Press*, 2023, pp. 121–147.

30 E. Brunet-Jailly, "Borders and the EU: The consequences of the Schengen crisis", Communication at the 2nd ABS World Conference, Budapest/Wien, 10–14 July 2018.

31 B. Wassenberg, "The Schengen Crisis and the End of the 'Myth' of Europe without Borders", *op.cit.*, p. 30.

32 *Ibid.*

33 B. Wassenberg (ed.), *Castle-talks on Cross-Border Cooperation. Fear of Integration? The Pertinence of the Border*, Steiner Verlag, Stuttgart, 2017.

34 B. Wassenberg & N. Suzuki (eds.), *Origins and Consequences of European Crises. Global Views on Brexit*, Peter Lang, Bussels, 2020.

hermetically for the population during the first lockdown in spring 2020.[35] This meant that the freedom of movement of European citizens based on the European citizenship was suspended until the border was re-opened progressively from mid-2020 onwards.[36] In some border regions – for example alongside the Franco-German border – this also led to resentment against the neighbours putting into question the successful reconciliation process, for example alongside the Franco-German border.[37] The return of the border was now perceived as a proof of the failure of European integration and as a place where conflict and resentment against the neighbour reappeared.[38] The failed ideal of a "Europe without borders" stirred feelings of insecurity and fear of "the other" beyond the border, which was more specifically felt in border regions since it is there that close contacts are normally established with the other and otherness, which are disrupted in times of re-bordering.[39]

Therefore, the key question addressed during the research seminars of the Frontem network was not only to examine the assumed "model" functions of border perception and management in European border regions. In the light of re-bordering crises in Europe, one also had to assess the role of the border in the process of European integration when faced with the re-questioning of a borderless Europe. The objective was indeed to offer a critical reading of the ideal of a "Europe without borders".[40]

So what happens to the models of management and to the perception of borders in Europe in times of re-bordering, when faced with a return of the "Westphalian" border?

The research by Frontem reveals that the EU has developed a unilateral approach to borders, which essentially retains their economic

---

35 Mission Opérationnelle Transfrontalière (MOT), "The effects of COVID-19 induced border closures on cross-border regions. An empirical report covering the period March to June 2020", Paris, 2021, pp. 9–10.

36 C. Wille & R. Kanesu (eds.), *Bordering in Pandemic Times: Insights into the COVID-19 Lockdown*, Thematic Issue, Bd. 4. UniGR-Center for Border Studies, Luxemburg, 2020.

37 *Ibid.*

38 F. Berrod, M. Chovet & B. Wassenberg, "La frontière franco-allemande au temps du COVID-19: la fin d'un espace commun", *op;cit.*

39 J. Gabbe, "Sensibilität in europäischen Grenzregionen", *In* B. Wassenberg & J. Beck, *Vivre et penser la coopération transforntalière (Volume 4) : les régions frontalières sensibles*, Steiner Verlag, Stuttgart, 2014, pp. 93–105.

40 B. Wassenberg, "Introduction—Frontiers in Motion (Frontem)", *op.cit.*, p. 11.

dimension as a barrier, without sufficiently taking into account other aspects, symbolic and political, in particular. The hypothesis put forward that there exists a "Europe without (all) borders" cannot be validated, as for example political borders have never disappeared. Borders are complex and there is an ambivalence of EU borders both as places of contact and exchange and as means of protection and delimitation.[41] Therefore, the assumed model character of border management and perception in certain border regions cannot be confirmed. At the Franco-German border, during the Covid-19 pandemic, resentments against the neighbouring border population were apparently stronger than for example at the German-Polish border and it can thus hardly be argued that Franco-German cross-border cooperation can serve as a model of reconciliation for Central Europe.[42] The model of the "invisible" border between Northern Ireland and Ireland has been largely put into question by the Brexit process, where the Protocol on Northern Ireland had to shift the "border" into the Sea in order to avoid a (re)bordering process on land and the revival of conflict.[43] The German-Danish model of cross-border minority management is also questionable, as its implementation by means of "parallelism" or "peaceful coexistence" does not really correspond to the objective of cross-border integration or the objective of a "Europe without borders", if borders between the Danish and German minorities are carefully being maintained.[44] Its transposition to other border regions with minority issues remains doubtful as the Hungarian-Rumanian border region illustrates: cross-border cooperation is being supported here mainly on the Hungarian side and might be interpreted

[41] B. Wassenberg, "Le traité d'Elysée et la coopération transfrontalière du Rhin supérieur: quels liens?" *In* J. Beck (ed.), *Grenzüberschreitende Zusammenarbeit nach der Pandemie/La coopération transfrontalièrr après la pandémie,* Peter Lang, Bruxelles, 2021, pp. 65–85.

[42] E. Opilowska, "Resilienz grenzüberschreitender Zusammenarbeit im deutsch-polnischen Grenzgebiet in europäischer Perspektive", *In* D. Brodowski, J. Nesselhauf & F. Weber, *Pandemisches Virus – nationales Handeln. Covid-19 und die europäische Idee*, Springer, Wiesbaden, pp. 193–215.

[43] N. Wright, "Northern Ireland, the Backstop and No Deal", Policy Brief, UCL European Institute July 2019. Retrieved from: <https://www.ucl.ac.uk/european-institute/sites/european-institute/files/brexit_and_northern_ireland.pdf>. Accessed 23 December 2023.

[44] R. Tarvet & M. Klatt, "The impact of the Corona crisis on borderland living in the Danish-German border region with a special focus on the two national minorities", *National Identities*, 2021, pp. 1–18.

as a way to reincorporate a population lost by the borders traced by the Trianon Treaty of 1919. The model character of the Hungarian-Rumanian cross-border cooperation for Central Europe could therefore also be questioned, especially since its main motivation seems to be the prospect of membership in Schengen, which is felt as a potential solution for all cross-border problems, although there is no proof for this in terms of changing trends of mobility, perception of neighbours, etc.[45]

What remains is the EU model of cross-border cooperation with its multi-level governance approach, which stays unique when compared to other borderlands in the world. With a removal of customs borders and a (still) functioning regime of borderless crossings guaranteed by the Schengen Agreement, the US-Canada border definitely cannot compete.[46] If the effects of (re)bordering during the Covid-19 pandemic were comparable to Europe, the "Westphalian" border functions between the US and Canada have never been put into question in the same way as the European integration process has progressively done it for the EU Member States borders – even if there have been drawbacks during the (re)bordering crises since 2015.

Overall, the contributions to this collective work by the Frontem network validate two principles of borders: complexity on the one hand and specificity on the other. Thanks to the interdisciplinary comparative approach to management and perception of borders in European cross-border regions, the network has therefore produced a differential and multidimensional approach to borders in the five case-studies examined.[47] It takes in fact a new look at the role of the border in European integration, considering that there is not a single model of the perception and management of borders in the EU, but that both depend on the specific context of each border area concerned.

---

45 M. Benczi & G. Ocskay, "The evolution of cross-border cooperation in Hungary", *Észak-Magyarországi Stratégiai Füzetek,* vol. 18, n°2, 2021, pp. 54–67.

46 B. Dupeyron, "Cascadia revisited from European case studies: The socio-political space of cross-border networks", *Canadian Political Science Review*, 2 (2), 2008, pp. 84–103.

47 J. Beck & B. Wassenberg, "Prospects of cross-border cooperation in Europe – some building blocks for a transdisciplinary research agenda", *In* J. Beck (ed.), *Transdisciplinary Discourses on Cross-Border Cooperation in Europe*, Peter Lang, Brussels, pp. 527–539.

Bibliography

J. Beck, "European cross-border cooperation of the future: Capacity-building and the principle of horizontal subsidiarity", *In* I. Pires (ed.), *Borders and Borderlands: Today's Challenges and Tomorrow's Prospects*, Proceedings of the ABS Lisbon Conference Centro de Estudos Geográficos, Lisbon, 2012, pp. 13–33.

J. Beck, "Introduction", *In* J. Beck & B. Wassenberg (eds.), *Vivre et penser la coopération territoriale en Europe (Volume 6) : vers une cohésion territoriale ?*, Steiner-Verlag, Stuttgart, 2014, pp. 37–43.

J. Beck & B. Wassenberg, "Prospects of cross-border cooperation in Europe – Some building blocks for a transdisciplinary research agenda", In J. Beck (ed.), *Transdisciplinary Discourses on Cross-Border Cooperation in Europe*, Peter Lang, Brussels, pp.527–539.

M. Benczi & G. Ocskay, "The evolution of cross-border cooperation in Hungary", *Észak-Magyarországi Stratégiai Füzetek,* vol. 18, n°2, 2021, pp. 54–67.

M. Berezin & M. Schain (eds.), *Europe Without Borders: Remapping Territory, Citizenship, and Identity in a Transnational Age*, John Hopkins University Press, Baltimore, 2003.

E. Brunet-Jailly, "Borders and the EU: The consequences of the Schengen crisis", Communication at the 2nd ABS World Conference, Budapest/Wien, 10–14 July 2018.

B. Dupeyron, "Cascadia revisited from European case studies: The socio-political space of cross-border networks", *Canadian Political Science Review*, 2 (2), 2008, pp. 84–103.

J. Gabbe, "Sensibilität in europäischen Grenzregionen", *In* B. Wassenberg & J. Beck, *Vivre et penser la coopération transforntalière (Volume 4) : les régions frontalières sensibles*, Steiner Verlag, Stuttgart, 2014, pp. 93–105.

T. Hölpel,"Der deutsch-französische Grenzraum als Modellregion der deutsch-französischen und europäischen Annäherung", *In* T. Höpel, *Der deutsch-französische Grenzraum: Grenzraum und Nationenbildung im 19. und 20. Jahrhundert,* Europäische Geschichte online, 2012.

M. Klatt, "National minorities as a model for cross-border integration: Lessons from Schleswig", *In* J. Beck & B. Wassenberg (eds.), *Grenzüberschreitende Zusammenarbeit leben und erforschen: Vol. 5 Integration und (trans-)regionale Identitäten*, Steiner Verlag, Stuttgart, 2013, pp. 301–320.

T. Köhler, "Die Interreg-Programme und die Rolle der europäischen Kommission als wesentlicher Akteur der grenzüberschreitenden Zusammenarbeit", *In* B. Wassenberg & J. Beck (eds.), *Living and Researching Cross-border Cooperation (Volume 3): The European Dimension*, Steiner Verlag, Stuttgart, 2011, pp.243–253.

K.-H. Lambertz (ed.), *Die Grenzregionen als Labor und Motor kontinentaler Entwicklungen in Europa. Berichte und Dokumente des Europarates sowie Reden zur grenzüberschreitenden Zusammenarbeit in Europa*, Dike/Nomos, Zürich, Baden-Baden, 2010.

F. Mercier, "Making sense of the invisible: Cognitive mapping, affective realities and the Irish/Northern Irish Border", *Media, Culture & Society*, Vol. 45, Issue 1, pp. 126–141.

C. Mestre, *Les incidences de l'achèvement du Marché unique sur les régions transfrontalières, Conseil de l'Europe, Strasbourg*, 1992.

MOT, "The effects of COVID-19 induced border closures on cross-border regions. An empirical report covering the period March to June 2020", Paris, 2021, pp. 9–10.

A. Mozer.,"Entwicklungspolitik zu Hause", *In* Schöndube, C. *Entwicklungsregionen in der EWG. Ursache und Ausmaß der wirtschaftlichen Benachteiligung*, Bonn, Osang Verlag, 1973, pp. 14–25.

E. Opilowska, "Resilienz grenzüberschreitender Zusammenarbeit im deutsch-polnischen Grenzgebiet in europäischer Perspektive", *In* D. Brodowski, J. Nesselhauf & F. Weber, *Pandemisches Virus – nationales Handeln. Covid-19 und die europäische Idee*, Springer, Wiesbaden, pp. 193–215.

N. Paŭn & S. Schirmann (eds.), *Borders, Identities, Communities. The Road to Reconciliation and Partnership in Central and Eastern Europe*, Nomos, Publications of the European Union Liaison Committee, Vol. 18, Baden-Baden, 2016.

S. Saurugger, "Crise de l'Union européenne ou crise de la démocratie", *Politique étrangère*, 1, 2017, pp. 23–33.

R. Tarvet & M. Klatt, "The impact of the Corona crisis on borderland living in the Danish-German border region with a special focus on the two national minorities", *National Identities*, 2021, pp. 1–18.

U. Von der Leyen, "A union that strives for more. My agenda for Europe", European Commission, Political Guidelines for the next European Commission 2019–2024, Brussels, 2019.

L. Warlouzet, "At the heart of the recovery in the 1980s, the internal market", *In The European Commission 1986–2000, History and Memories of an Institution*, Publication Office of the EU, Luxemburg, 2019, pp.258–259.

B. Wassenberg (ed.), *Castle-talks on Cross-Border Cooperation. Fear of Integration? The Pertinence of the Border*, Steiner Verlag, Stuttgart, 2017.

B. Wassenberg, "Frontières, coopération transfrontalière et intégration européenne", *In* B. Wassenberg & M. Aballéa (eds.), "*Frontières, géopolitique et relations internationales*", *Cahier Fare* n°16, Harmattan, Paris, 2019, pp. 44–50.

B. Wassenberg, "Introduction – Frontiers in motion (Frontem): Comparative perspectives on European borders, cross-border cooperation, and integration", *Borders in Globalization Review*, vol. 4, Issue 2, 2020, pp. 8–13.

B. Wassenberg, "The Schengen crisis and the end of the 'Myth' of Europe without borders", *Borders in Globalization Review*, vol. 1, Issue 2, 2020, p.30–39.

B. Wassenberg, "Le traité d'Elysée et la coopération transfrontalière du Rhin supérieur: quels liens?" *In* J. Beck (ed.), *Grenzüberschreitende Zusammenarbeit nach der Pandemie/La coopération transfrontalièrr après la pandémie,* Peter Lang, Bruxelles, 2021, pp. 65–85.

B. Wassenberg, "Europa ohne Grenzen. Mythus oder Realität? Eine historische Analyse", *In* B. Wassenberg (ed.), *Frontières, acteurs et représentations d'Europe (Fare). Mélanges en honneur de Sylvain Schirmann*, Peter Lang, Brussels, 2022, pp. 415–439.

B. Wassenberg, "De-bordering (Re-)bordering in the EU during the 2015 migration crisis: The end of a 'Europe without borders'?", *In* M. J. Carpenter, M. Kelly & O. Schmidtke (eds.), *Borders and Migration in Canadian Comparative Perspective, University of Ottawa Press*, 2023, pp. 121–147.

B. Wassenberg & N. Suzuki (eds.), *Origins and Consequences of European Crises. Global Views on Brexit*, Peter Lang, Bussels, 2020.

C. Wille & R. Kanesu (eds.), *Bordering in Pandemic Times: Insights into the COVID-19 Lockdown*, Thematic Issue, Bd. 4. UniGR-Center for Border Studies, Luxemburg, 2020.

N. Wright, "Northern Ireland, the Backstop and No Deal", Policy Brief, UCL European Institute July 2019.

List of Authors

Dorte J. Andersen

Professor of European Ethnology and European Continental Philosophy at the University of Southern Denmark, Sonderborg, Department of Political Science and Public Management, Centre for Border Region Studies.

Melinda Benczi

International relations coordinator at the Central European Service for Cross-Border Cooperation, research fellow at the UPS-CESCI Border Studies Research Group and PhD student at the University of Sopron, Faculty of Economics.

Frédérique Berrod

Professor of Law at the University of Strasbourg, Sciences Po, vice-President of Finance, Jean Monnet Chair, member of the research unit *Centre d'études internationales et européennes* (CEIE) and the research federation *Europe en mutation.*

Mircea Brie

Professor of International Relations at the University of Oradea, Department of International Relations and European Studies, coordinator of the Jean Monnet Center of Excellence EUROINCLUS.

Emmanual Brunet-Jailly

Professor of Public Policy at the University of Victoria, School of Public Administration, Jean Monnet Chair, Director of the Borders in Globalization (BIG) Program.

Gaia Cardin

Graduate in BSc European Studies and Master's student in International Relations at Aalborg University.

Nora Crossey

Research Assistant in Human Geography at the Saarland University, Department European Social Research, research group European Studies.

Julia Dittel

Research Assistant in European Studies at Saarland University, Department of European Social Research, research group European Studies, member of the UniGR-Center for Border Studies.

Steen Bo Frandsen

Professor of Modern and Contemporary History at the University of Southern Denmark, Sonderborg, head of the Department of Political Science and Public Management, Centre for Border Region Studies.

Michael Frey

Professor of Public Law at the University of Applied Sciences Kehl, teaching European law, German constitutional law and German administrative law, leader of research projects on the legal and administrative issues of energy transition and Cross-Border Cooperation, director of the Institute for Applied Research of the University of Kehl.

Kathy Hall

Professor of Education in the School of Education, University College Cork.

Philippe Hamman

Professor of Sociology at the University of Strasbourg, Jean Monnet Chair, Faculty of Social Sciences, director of the Institute for Urbanism and Regional Development (*Institut d'Urbanisme et d'Aménagement Régional*

(IUAR)), member of the research unit *Sociétés, Acteurs, Gouvernement en Europe* (SAGE).

Katy Hayward

Professor of Socoiology at the Queen's University Belfast, School of Social Sciences, Education and Social Work, director of the Centre for International Borders Research.

Sarah Herrmann

Research Assistant in Communication Sciences at the Kehl University of Public Administration, Applied Research Institute (KIAF) on Cross-Border Cooperation.

Martin Klatt

Senior Researcher of minority issues in the Denmark-Germany border region at the European Centre for Minority Issues, Flensburg/ Associate Professor of Border Studies at University of Southern Denmark.

Milena Komarova

Lecturer in Human Geography at the Queen's University Belfast, School of Natural and Built Environment.

Catherine Ledig

Expert Consultant in Digital Innovation Strategy, Associate Professor of Digital Economy at the University of Strasbourg, Faculty of Law.

Fabienne Leloup

Professor of Political Sciences at the Catholic University of Louvain (UCLouvain FUCaM Mons), Institute of Political Sciences Louvain Europe (ISPOLE), member and co-founder of the Franco-Belgian Scientific Grouping *Institut Frontières et Discontinuités.*

Patrick Palmier

Head of the group mobility and territories at Cerema Hauts-de-France.

Nicolae Păun

Professor of Contemporary History at the Bogdan University, Cluj-Napoca, vice-Chancellor for Research and International Relations, Jean Monnet Chair.

Jean Peyrony

Director of the Mission Opérationnelle Transfrontalière (MOT), Paris.

István Polgár

Lecturer of International Relations at the University of Oradea, Department of International Relations and European Studies.

Bernard Reitel

Professor of political and urban Geography at the University of Artois, Department of History, Geography and Heritage, Jean Monnet Chair, Head of the research unit *Discontinuités,* member of the Franco-Belgian Scientific Grouping *Institut Frontières et Discontinuités.*

Cyprien Richer

Researcher in Geography and Urban Planning at Cerema-CY Cergy Paris Université, research unit MATRiS: *Mobilités, Aménagement, Transports, Risques et Société.*

János Sallai

Professor, Police Colonel at the University of Public Service, Budapest, Faculty of Law Enforcement, head of Department of Law Enforcement Theory and Law Enforcement History, Deputy Head of Doctoral School of Police Sciences and Law Enforcement at University of Public Service.

Tatsiana Shaban

Visiting Postdoctoral Scholar in International Relations at the University of Victoria, Centre for Global Studies, non-resident fellow, Borders in Globalization Program.

Anthony Soares

Director of the Centre for Cross Border Studies, Armagh.

Katarzyna Stokłosa

Professor of Contemporary History and Sociology at the University of Southern Denmark, Department of Political Science and Public Management, Centre for Border Region Studies.

Enya-Andrea Tămaş

PhD Student at the Babeş-Bolyai University Cluj – Napoca,"European Paradigm" Doctoral School.

Jennifer Todd

Professor of Politics (emeritus) at the University College Dublin, fellow at the Geary Institute for Public Policy, member of the Royal Irish Academy, member of the Steering Committee Analysing and Researching Ireland, North and South (ARINS).

Gunvor Vestergaard

Educational Advicer, Grænseforeningen, Copenahgen.

Reine Wakote

Senior Lecturer of Law at the University of Lorraine, Metz, Research Institute on the Development of the Nation and the State (Institut de Recherches sur l'Evolution de la Nation Et de l'Etat (IRENEE)).

Birte Wassenberg

Professor of Contemporary Hsitory at the University of Strasbourg, Sciences Po, director of the Franco-German Jean Monnet Excellence Center, Jean Monnet Chair, member of the research unit *Centre d'études internationales et européennes* (CEIE) and the research federation *Europe en mutation.*

Florian Weber

Juniorprofessor of European Studies at Saarland University, Department of European Social Research, research group European Studies, member of the UniGR-Center for Border Studies.

Ingo Winkler

Professor of International Business and Entrepreneurship at the University of Southern Denmark, Sonderborg, Department of Business & Management.

Borders and European Integration

Edited by

Joachim Beck (Professor of Administrative Studies, Hochschule Kehl)
Frédérique Berrod (Professor of Public Law, Université de Strasbourg)
Birte Wassenberg (Professor of History of International Relations, Université de Strasbourg)

The Series « Borders and European Integration » fills in a gap in Social Sciences, as it connects two so far independent research strands: European Studies and Border Studies. Mainly initiated by geographers and originally hosted in the United States, Border Studies primarily deal with the study of borders and borderlands, whereas European Studies analyse the process of European Integration, its actors, institutions and policy fields. Although the idea of a Europe without borders was part of the project of the European Economic Community, the multidimensional role of the border has not been sufficiently taken into account by researchers in European Studies. Inversely, Border Studies have only rarely examined the specificity of borders and borderlands in Europe in comparison to other regions in the world.

At the crossroads between Area Studies and International Relations, this Series therefore offers a pluri-disciplinary approach to borders and their role in the European construction. Taking into account the perspective of different disciplines in Social Sciences, the diversity of actors of European Integration and borderlands (local, regional, national), it allows a new multi-level and decentred view on conflicts and cooperation at European borders. The Series addresses researchers and university scholars of all disciplines in Social Sciences and wishes to tackle the challenging contemporary questions on borders in Europe.

Scientific board

Series titles

Vol. 1. Birte Wassenberg & Bernard Reitel, in collaboration with Mission opérationnelle transfrontalière (Jean PEYRONY & Jean RUBIO), *Critical Dictionary on Borders, Cross-Border Cooperation and European Integration*, 2020.

Vol. 2. Birte Wassenberg & Noriko Suzuki (eds.), *Origins and Consequences of European Crises: Global Views on Brexit*, 2020.

Vol. 3. Joachim Beck, *Grenzüberschreitende Zusammenarbeit nach der Pandemie / La coopération transfrontalière après la pandémie: Rechtlich-institutionelle Flexibilisierung im Kontext des Aachener Vertrags / Flexibilisation juridique et institutionnelle dans le cadre du Traité d'Aix-la-Chapelle*, 2022.

Vol. 4. Erhan Özdemir, *Socio-Economic Disparities in the Integration Process of Immigrants in Western Europe: A Comparative Study for Six EU Countries*, 2022.

Vol. 5. Birte Wassenberg (dir.), *Frontières, acteurs et représentations d'Europe (Fare) / Grenzen, Akteure und Repräsentationen Europas: Mélanges en honneur de Sylvain Schirmann / Festschrift für Sylvain Schirmann*, 2022.

Vol. 6. Birte Wassenberg (ed.), Frontières en mouvement *(Frontem): Which Models of Cross-Border Cooperation for the EU ? A comparative analysis from a Euro-Atlantic perspective*, 2024.

www.peterlang.com

www.ingramcontent.com/pod-product-compliance
Ingram Content Group UK Ltd.
Pitfield, Milton Keynes, MK11 3LW, UK
UKHW041834100926
8556IPUK00018B/958

9 782875 749260